T0398949

Nanotechnology in Water and Wastewater Treatment

Nanotechnology in Water and Wastewater Treatment
Theory and Applications

Edited by

Amimul Ahsan
Department of Civil Engineering, Uttara University, Dhaka, Bangladesh

&

Department of Civil and Construction Engineering,
Faculty of Science, Engineering and Technology, Swinburne
University of Technology, Melbourne, Australia

Ahmad Fauzi Ismail
Advanced Membrane Technology Research Centre (AMTEC),
School of Chemical and Energy Engineering, Faculty of Engineering,
Universiti Teknologi Malaysia (UTM), Johor, Malaysia

ELSEVIER

Elsevier
Radarweg 29, PO Box 211, 1000 AE Amsterdam, Netherlands
The Boulevard, Langford Lane, Kidlington, Oxford OX5 1GB, United Kingdom
50 Hampshire Street, 5th Floor, Cambridge, MA 02139, United States

Notices

Knowledge and best practice in this field are constantly changing. As new research and experience broaden our understanding, changes in research methods, professional practices, or medical treatment may become necessary.

Practitioners and researchers must always rely on their own experience and knowledge in evaluating and using any information, methods, compounds, or experiments described herein. In using such information or methods they should be mindful of their own safety and the safety of others, including parties for whom they have a professional responsibility.

To the fullest extent of the law, neither the Publisher nor the authors, contributors, or editors, assume any liability for any injury and/or damage to persons or property as a matter of products liability, negligence or otherwise, or from any use or operation of any methods, products, instructions, or ideas contained in the material herein.

British Library Cataloguing-in-Publication Data
A catalogue record for this book is available from the British Library

Library of Congress Cataloging-in-Publication Data
A catalog record for this book is available from the Library of Congress

ISBN: 978-0-12-813902-8

For Information on all Elsevier publications
visit our website at https://www.elsevier.com/books-and-journals

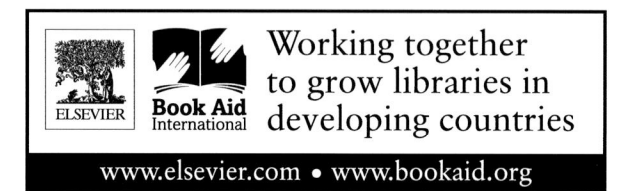

Publisher: Matthew Deans
Acquisition Editor: Simon Holt
Editorial Project Manager: John Leonard / Natasha Welford
Production Project Manager: Kamesh Ramajogi
Cover Designer: Miles Hitchen

Typeset by MPS Limited, Chennai, India

Contents

3   Nanoparticles as a Powerful Tool for Membrane Pore Size Determination and Mercury Removal   **63**

E. ARKHANGELSKY, V. INGLEZAKIS, V. GITIS AND A.V. KOROBEINYK

4   Overview of Potential Applications of Nano-Biotechnology in Wastewater and Effluent Treatment   **87**

RAMA RAO KARRI, SHAHRIAR SHAMS AND J.N. SAHU

List of Contributors

N. Adila Ab Aziz   Faculty of Civil and Environmental Engineering, Universiti Tun Hussein Onn, Malaysia, UTHM, Parit Raja, Batu Pahat, Johor, Malaysia

Norfadhilatuladha Abdullah   Advanced Membrane Technology Research Centre (AMTEC), School of Chemical and Energy Engineering, Faculty of Engineering, Universiti Teknologi Malaysia, Johor Bahru, Malaysia

O. Abdulrahman Adeleke   Division of Environmental Technology, School of Industrial Technology, University Sains Malaysia, Pulau Pinang, Malaysia

Sevinç Adiloğlu   Department of Soil and Plant Protection and Nutrition, Agricultural Faculty, Namik Kemal University, Tekirdağ, Turkey

Amimul Ahsan   Department of Civil Engineering, Uttara University, Dhaka, Bangladesh; Department of Civil and Construction Engineering, Swinburne University of Technology, Melbourne, VIC, Australia

Muhd Arif Aizat   Advanced Membrane Technology Research Centre (AMTEC), Universiti Teknologi Malaysia, Johor, Malaysia; Faculty of Chemical and Energy Engineering (FCEE), Universiti Teknologi Malaysia, Johor, Malaysia

Adel Al-Gheethi   Faculty of Civil and Environmental Engineering, Universiti Tun Hussein Onn, Malaysia, UTHM, Parit Raja, Batu Pahat, Johor, Malaysia

Najeeya Apandi   Division of Environmental Technology, School of Industrial Technology, University Sains Malaysia, Pulau Pinang, Malaysia

E. Arkhangelsky   Department of Civil and Environmental Engineering, Environmental Science & Technology Group (ESTg), Nazarbayev University, Astana, Republic of Kazakhstan

Farhana Aziz   Advanced Membrane Technology Research Centre (AMTEC), School of Chemical and Energy Engineering, Faculty of Engineering, Universiti Teknologi Malaysia, Johor Bahru, Malaysia

I. Azreen   Faculty of Engineering, Universiti Malaysia Sabah, Jalan UMS, Kota Kinabalu, Sabah, Malaysia

Arabinda Baruah   Indian Institute of Science Education and Research, Mohali, Punjab, India

Vandna Chaudhary   Center of Excellence for Energy and Environment Studies, D.C.R. University of Science & Technology, Sonepat, Haryana, India

Deniz İzlen Çifçi   Department of Environmental Engineering, Faculty of Çorlu Engineering, Namik Kemal University, Tekirdağ, Turkey

Zawawi Daud   Faculty of Civil and Environmental Engineering, Universiti Tun Hussein Onn, Malaysia, UTHM, Parit Raja, Batu Pahat, Johor, Malaysia

Ayeronfe Fadilat   Division of Environmental Technology, School of Industrial Technology, University Sains Malaysia, Pulau Pinang, Malaysia

J. Felijia   Faculty of Engineering, Universiti Malaysia Sabah, Jalan UMS, Kota Kinabalu, Sabah, Malaysia

V. Gitis   Unit of Energy Engineering, Ben-Gurion University of the Negev, Beer-Sheva, Israel

C. Gloriana   Faculty of Engineering, Universiti Malaysia Sabah, Jalan UMS, Kota Kinabalu, Sabah, Malaysia

P.S. Goh   Advanced Membrane Technology Research Centre (AMTEC), School of Chemical and Energy Engineering, Faculty of Engineering, Universiti Teknologi Malaysia, Johor Bahru, Malaysia

Hasrinah Hasbullah   Advanced Membrane Technology Research Centre (AMTEC), Universiti Teknologi Malaysia, Johor, Malaysia

H. Hasmilah   Faculty of Engineering, Universiti Malaysia Sabah, Jalan UMS, Kota Kinabalu, Sabah, Malaysia

Mahmood Hijab   Faculty of Civil and Environmental Engineering, Universiti Tun Hussein Onn, Malaysia, UTHM, Parit Raja, Batu Pahat, Johor, Malaysia

Monzur Imteaz   Department of Civil and Construction Engineering, Swinburne University of Technology, Melbourne, VIC, Australia

V. Inglezakis Department of Chemical Engineering, Environmental Science & Technology Group (ESTg), Nazarbayev University, Astana, Republic of Kazakhstan

Ahmad Fauzi Ismail Advanced Membrane Technology Research Centre (AMTEC), School of Chemical and Energy Engineering, Faculty of Engineering, Universiti Teknologi Malaysia, Johor Bahru, Malaysia

Norli Ismail Division of Environmental Technology, School of Industrial Technology, University Sains Malaysia, Pulau Pinang, Malaysia

S.S. Jie Faculty of Engineering, Universiti Malaysia Sabah, Jalan UMS, Kota Kinabalu, Sabah, Malaysia

Lau Woei Jye Advanced Membrane Technology Research Centre (AMTEC), Universiti Teknologi Malaysia, Johor, Malaysia

Rama Rao Karri Petroleum and Chemical Engineering, Faculty of Engineering, Universiti Teknologi Brunei, Gadong, Brunei

I. Khairunis Faculty of Engineering, Universiti Malaysia Sabah, Jalan UMS, Kota Kinabalu, Sabah, Malaysia

A.V. Korobeinyk Department of Chemical Engineering, Environmental Science & Technology Group (ESTg), Nazarbayev University, Astana, Republic of Kazakhstan; O.O. Chuiko Institute of Surface Chemistry, NAS of Ukraine, Kyiv, Ukraine

Vicky Kumar Faculty of Civil and Environmental Engineering, Universiti Tun Hussein Onn, Malaysia, UTHM, Parit Raja, Batu Pahat, Johor, Malaysia

Ab Aziz Abdul Latiff Faculty of Civil and Environmental Engineering, Universiti Tun Hussein Onn, Malaysia, UTHM, Parit Raja, Batu Pahat, Johor, Malaysia

Ritu Malik Synthesis and Real Structure Group, Technical Faculty, Institute of Materials Science, University of Kiel, Kiel, Germany

Süreyya Meriç Department of Environmental Engineering, Faculty of Çorlu Engineering, Namik Kemal University, Tekirdağ, Turkey

Omme Kulsum Nayna Department of Environmental Sciences, Jahangirnagar University, Dhaka, Bangladesh; Department of Environmental Science and Engineering, Ewha Womans University, Seoul, Republic of Korea

Mohammed Ndah   Faculty of Civil and Environmental Engineering, Universiti Tun Hussein Onn, Malaysia, UTHM, Parit Raja, Batu Pahat, Johor, Malaysia

B.C. Ng   Advanced Membrane Technology Research Centre (AMTEC), School of Chemical and Energy Engineering, Faculty of Engineering, Universiti Teknologi Malaysia, Johor Bahru, Malaysia

C.S. Ong   Water Desalination and Reuse Center, Division of Biological and Environmental Science and Engineering, King Abdullah University of Science and Technology, Thuwal, Saudi Arabia

Fahmida Parvin   Department of Environmental Sciences, Jahangirnagar University, Dhaka, Bangladesh

Sharmin Yousuf Rikta   Department of Environmental Science, Bangladesh University of Professionals, Dhaka, Bangladesh

Sarina Mat Rosid   Advanced Membrane Technology Research Centre (AMTEC), Universiti Teknologi Malaysia, Johor, Malaysia

Mohd Ikhram Roslan   Advanced Membrane Technology Research Centre (AMTEC), Universiti Teknologi Malaysia, Johor, Malaysia

M. Arif Rosli   Faculty of Civil and Environmental Engineering, Universiti Tun Hussein Onn, Malaysia, UTHM, Parit Raja, Batu Pahat, Johor, Malaysia

Nurul Shahira Mohd Sabri   Advanced Membrane Technology Research Centre (AMTEC), Universiti Teknologi Malaysia, Johor, Malaysia

J.N. Sahu   Faculty of Chemistry, Institute of Chemical Technology, University of Stuttgart, Stuttgart, Germany

Noresah Said   Advanced Membrane Technology Research Centre (AMTEC), Universiti Teknologi Malaysia, Johor, Malaysia

Mohammed Radin Saphira   Faculty of Civil and Environmental Engineering, Universiti Tun Hussein Onn, Malaysia, UTHM, Parit Raja, Batu Pahat, Johor, Malaysia

Shahriar Shams   Civil Engineering Programme Area, Faculty of Engineering, Universiti Teknologi Brunei, Gadong, Brunei

Yusuf Solmaz   Department of Soil and Plant Protection and Nutrition, Agricultural Faculty, Namik Kemal University, Tekirdağ, Turkey

Muhammad Hanis Tajuddin   Advanced Membrane Technology Research Centre (AMTEC), School of Chemical and Energy Engineering, Faculty of Engineering, Universiti Teknologi Malaysia, Johor Bahru, Malaysia

Shafi M. Tareq   Department of Environmental Sciences, Jahangirnagar University, Dhaka, Bangladesh

Sema Terzi   Department of Environmental Engineering, Faculty of Çorlu Engineering, Namik Kemal University, Tekirdağ, Turkey

Vijay K. Tomer   Berkeley Sensor & Actuator Center (BSAC), University of California, Berkeley, United States

C. Yoiying   Faculty of Engineering, Universiti Malaysia Sabah, Jalan UMS, Kota Kinabalu, Sabah, Malaysia

Norhaniza Yusof   Advanced Membrane Technology Research Centre (AMTEC), School of Chemical and Energy Engineering, Faculty of Engineering, Universiti Teknologi Malaysia, Johor Bahru, Malaysia

A.Y. Zahrim   Faculty of Engineering, Universiti Malaysia Sabah, Jalan UMS, Kota Kinabalu, Sabah, Malaysia

About the Editors

Associate Professor Amimul Ahsan has nearly 15 years research, teaching, and industry experience. He has published extensively in water and environmental engineering, including 9 books, 16 book chapters, and over 125 journal articles. He has received 14 international awards, including "Who's Who in the World 2015," "Leading Engineers of the World 2013," and the "Vice Chancellor Fellowship Award (Science and Technology)" from Sultan Selangor (Chancellor, UPM) in 2015. He is editor-in-chief of five journals in the United States, the United Kingdom, and Malaysia, and founder of the *Journal of Desalination and Water Purification and the Journal of Advanced Civil Engineering Practice and Research*. He is involved with several collaborative research projects globally and has a Scopus h-index of 22. He was a former faculty member of the Department of Civil Engineering and key researcher at the Institute of Advanced Technology (ITMA), University Putra Malaysia (UPM), Malaysia. Currently, he is an Associate Professor in the Department of Civil Engineering, Uttara University, Dhaka, Bangladesh and Adjunct Associate Professor in the Department of Civil and Construction Engineering, Faculty of Science, Engineering and Technology, Swinburne University of Technology, Melbourne, Australia.

Professor Ahmad Fauzi Ismail is the founding director of the Advanced Membrane Technology Research Center (AMTEC), School of Chemical and Energy Engineering, Universiti Teknologi Malaysia (UTM), Johor, Malaysia. He is the author and coauthor of over 500 refereed journals (cited over 16,000 times) with a Scopus h-index of 61. He has authored and edited 10 books, contributed 45 book chapters, and 11 patents granted with 19 patents filed. He has received more than 120 awards both nationally and globally, including "The Merdeka Award for Outstanding Scholastic Achievement" in 2014; "Malaysia Young Scientist" in 2000, and "ASEAN Young Scientist" in 2001. Professor Fauzi has been named as one of the most cited researchers in chemical engineering world by the Shanghai Jiao Tong University, Shanghai, China. Recently he received "The Malaysia Research Star," awarded by Clarivate Analytics for Frontier Materials in 2017 and for the international category in 2018. Professor Fauzi has previously served on numerous journal editorial boards and currently serves as editor of *Emergent Materials, associate editor of the Arabian Journal of Science and Engineering*, and *advisory editorial board member of Journal of Chemical Technology and Biotechnology*.

Acknowledgments

The editors would like to express appreciation to all who have helped to prepare this book. The editors express their gratefulness to John Leonard, Natasha Welford, Kamesh Ramajogi, Simon Holt, Elsevier Limited, United Kingdom, and MPS Limited. In addition, the editor appreciatively remembers the assistance of all authors and reviewers of this book.

Gratitude is expressed to Mrs. Ahsan, Mrs. Fadilah Abdullah, Dr. Goh Pei Sean, Ibrahim Bin Ahsan, Marium Binti Ahsan, Yahya Bin Ahsan, Yusuf Bin Ahsan, Mother, Father, Mother-in-Law, Father-in-Law, Brothers, and Sisters for their endless inspirations, mental supports, and also necessary help during times of difficulty.

The Editors

Amimul Ahsan
Department of Civil Engineering, Uttara University, Dhaka, Bangladesh

&

Department of Civil and Construction Engineering, Faculty of Science, Engineering and Technology, Swinburne University of Technology, Melbourne, Australia

Ahmad Fauzi Ismail
Advanced Membrane Technology Research Centre (AMTEC), School of Chemical and Energy Engineering, Faculty of Engineering, Universiti Teknologi Malaysia, Johor Bahru, Malaysia
10 September, 2018

1

Principles and Mechanism of Adsorption for the Effective Treatment of Palm Oil Mill Effluent for Water Reuse

O. Abdulrahman Adeleke[1], Ab Aziz Abdul Latiff[2],
Mohammed Radin Saphira[2], Zawawi Daud[2], Norli Ismail[1],
Amimul Ahsan[3,4], N. Adila Ab Aziz[2], Adel Al-Gheethi[2], Vicky Kumar[2],
Ayeronfe Fadilat[1], Najeeya Apandi[1]

[1]DIVISION OF ENVIRONMENTAL TECHNOLOGY, SCHOOL OF INDUSTRIAL TECHNOLOGY, UNIVERSITY SAINS MALAYSIA, PULAU PINANG, MALAYSIA [2]FACULTY OF CIVIL AND ENVIRONMENTAL ENGINEERING, UNIVERSITI TUN HUSSEIN ONN, MALAYSIA, UTHM, PARIT RAJA, BATU PAHAT, JOHOR, MALAYSIA [3]DEPARTMENT OF CIVIL AND CONSTRUCTION ENGINEERING, SWINBURNE UNIVERSITY OF TECHNOLOGY, MELBOURNE, VIC, AUSTRALIA [4]DEPARTMENT OF CIVIL ENGINEERING, UTTARA UNIVERSITY, DHAKA, BANGLADESH

1.1 Introduction

Wastewater is a general term used for the effluents generated from domestic, agricultural and industrial sources. The effluents contain organic and inorganic pollutants which are toxic to the ecosystem. Examples of organic pollutants are the volatile and chemical compounds with complex chain reactions. The discharge of volatile compounds and highly toxic chemical compounds have affected the quality of the water in some cases in Thailand, Vietnam and Columbia (Tran et al., 2015). The organic pollutants of palm oil-based waste can promote microbial growth affecting the flora and fauna of the water ways (Mukherjee and Sovacool, 2014). Example of the inorganic pollutants are the heavy metals deposit in water bodies due to the deposition of untreated waste from domestic, agricultural and industrial sectors into the water ways. Surface water degradation is caused by high toxicity of discharged waste pollutants into fresh water bodies (Essington, 2015). One of the most challenging problems of water pollution is the presence of organic pollutants and heavy metals in water bodies representing one of the greatest risks for the aquatic ecosystem (Galimberti et al., 2016). These pollutants are difficult to remove due to their low concentration in the effluent. When wastewater solutes are discharged into water ways, they are accumulated in sediments along the aquatic food chain (Mishra and Shukla, 2016). The toxic effects of the pollutants are not only limited to the water body but also on the environment without proper treatment before discharge (Guagliardi et al., 2013). The agro-based industries such as palm oil mill in Malaysia have become one of the largest contributors of water pollutants (Kamarudin et al., 2015). The palm oil mill effluent (POME) contains pollutant parameters such as chemical oxygen demand (COD), biological oxygen demand (BOD), oil and grease, suspended solids, ammonia-nitrogen and heavy metal concentration (Khemkhao et al., 2015).

Most countries in the world are developing regulations and strategies to ensure proper treatment of the water pollutants before discharge into the water bodies. Also, the level of awareness is on the increase to ensure that adequate protection and preservation of water is achieved. Wastewater management involves series of efforts that promotes effectiveness in the use of water (Cooper, 2016). There are three methods of wastewater treatment, which are the physical, chemical and biological methods. The physical methods include sedimentation, floatation, membrane filtration, and adsorption. Chemical methods include ozonation, advanced oxidation, electrolysis. Also, biological methods include the use of conventional activated sludge, anaerobic open ponding and anaerobic systems. However, some of the conventional techniques are costly and requires high maintenance to operate. For example the use of membrane have high potentials for the treatment of POME, but is expensive and have problems of membrane fouling (Azmi and Yunos, 2014). Some of the methods such as the anaerobic open ponding requires availability of large area for the treatment. Some other methods such as the anaerobic systems require routine maintenance of the reactors.

The method of adsorption is used for the removal of wastewater contaminant. Adsorption by solid reduces toxicity effect from industrial effluents (Haak et al., 2016). Adsorption has the advantage of low capital cost of adsorbents, easy to operate, minimum sludge generation and the ability of regeneration and reuse of spent adsorbents (Stawiński et al., 2017). Activated carbon adsorbent is applicable for wastewater treatment in the form of granular or powdered, it has been proven to be very effective for the removal of different types of contaminants in water ranging from industrial, municipal wastewater, landfill leachate and polluted groundwater. Activated carbon adsorption of pollutants of wastewater is recognized by USEPA (environmental protection agency) as one of the best methods of environmental control (Bautista-Toledo et al., 2014), this is due to large specific pore surface area which makes it a powerful adsorbent and has the ability to adsorb wide range of contaminants. The limitation of the use of commercial activated carbon as adsorbent is its high cost and problem of regeneration for reuse (Benhouria et al., 2015; Wei et al., 2012). However, the adsorptive capacity of activated carbon has necessitated low cost alternative adsorptive materials with similar composition as composite and the ability to have the potentials for regeneration. Activated carbon derived from cow bones have high potentials of both carbon and minerals composition, which is highly enriched in calcium and phosphorus, forming an insoluble precipitate known as hydroxyapatite (Medellin-Castillo et al., 2014). Adsorption on hydroxyapatite adsorbent material is effective for the treatment of both organic and inorganic pollutants (Patel et al., 2015). Activated carbon derived from cow bones have been processed for the treatment of POME (Adeleke et al., 2016). Waste materials from agriculture, domestic and industrial are synthesized to serve as replacement of commercial activated carbon for the treatment of wastewater. For example, coconut shell can be processed as activated carbon at elevated temperature and has been reported as a very good source of activated carbon because it contains cellulose, hemicellose and lignin. Materials containing lignocellulosic properties are a very good adsorbent material for the activation of carbonaceous materials. Adsorbent materials such as orange peel, mango waste have very good cellulose characteristics. However, the effectiveness of a prepared adsorbent material as activated carbon wholly depends on the ability of the adsorbent to reduce the pollutant of wastewater to a considerable low level.

1.2 Palm Oil Production and Processing

Oil palm is obtained from the fruit bunches which are sterilized by stripping and pressing at high pressure to separate the fruits from the bunch and to reduce the formation of free fatty acid. After the stripping of the bunch, the sterilized fruits are digested to loosen the mesocarp during pressing. The process is followed by the separation of the mesocarp and the nuts from the digester. The extracted crude oil contains both organic and dissolved matter in water. A centrifugation tank is used to separate the water from the crude oil. The crude oil refining process can be either physical or chemical as shown in Fig. 1-1 (www.google.com).

The process of degumming involves removing unwanted gums, such as phosphatide, which has the ability of affecting the stability of the processed crude oil. The crude palm oil is heated at between $90-110°C$ with phosphoric acid to decompose the phosphatides and making them to be easily removed by bleaching. The next stage is the bleaching which involves the treatment of the degummed oil which is usually achieved with bleaching earth at $100°C$ under continuous agitation with bleaching material until the contact time of 30 minutes is achieved. At this stage, the phosphatides are removed by the oxidative effect of the bleaching material (Rossi et al., 2003). The process of deodorization entails the removal of free fatty acid (FFA) in the form of palm fatty acid distillate (PFAD). In the chemical process, the phosphatides are removed by the addition of additives, the process is followed by naturalization with alkali for the removal of FFA. The naturalized oil is then bleached under vacuum using an agitator called bleacher at $90°C$. The oil obtained after bleaching is heated at $200°C$ with the evolution of volatile materials to obtain a deodorized oil blend.

The waste from oil palm industries consist of oil palm trunks (OPT), oil palm fronds (OPF), empty fruit bunches (EFB), palm processed fibers (PPF), palm kernel shells, fresh fruit bunch (FFB) and POME discharge (Rupani et al., 2010). The wastewater from the sterilization process of the FFB is known as the sterilizer condensate. The process of sterilization results to high amount of condensate which has effect between the fruit and the wastewater. It contains 35%−45% of POME (Liew et al., 2015).

1.3 Palm Oil Mill Effluents and the Treatment Methods

The major source of wastewater of POME is the clarification water, which is the discharge from the clarification process of the crude oil. The produced crude oil is a mixture of palm

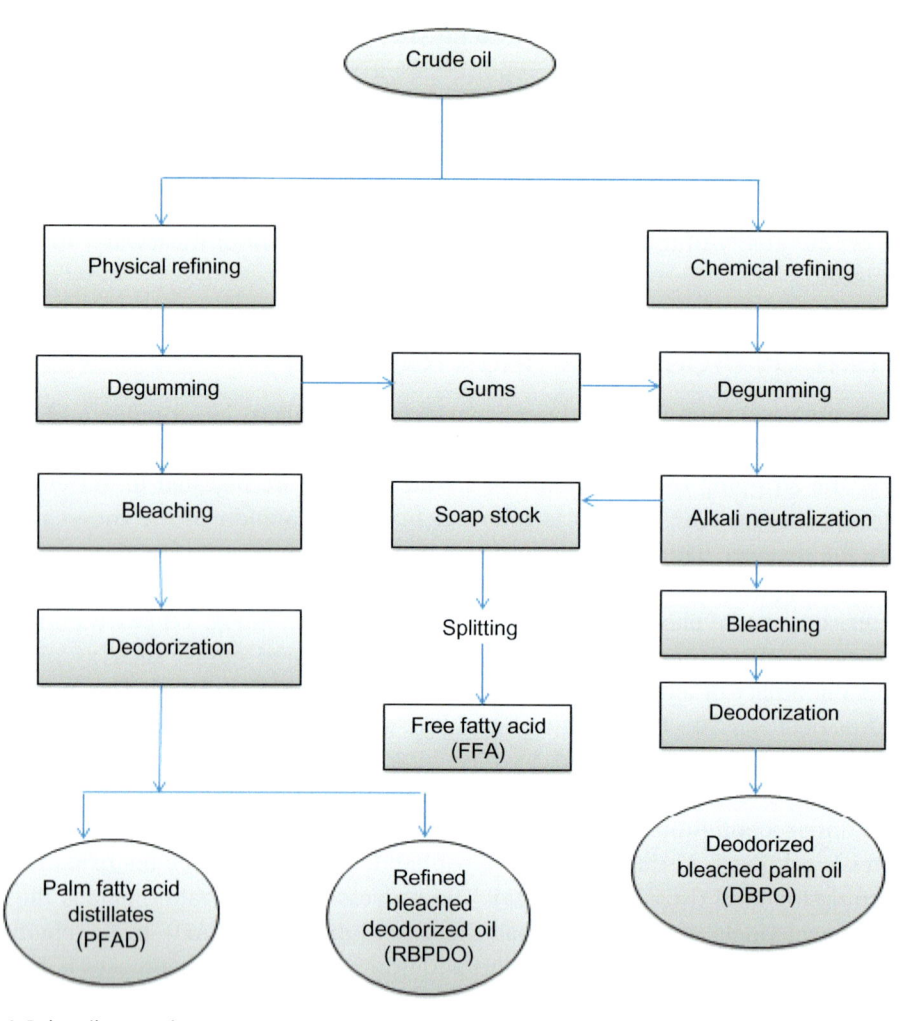

FIGURE 1-1 Palm oil processing.

oil, water and impurities comprising of organic matters (Ohimain and Izah, 2017) The wastewater from the clarification process contains about 60% of POME. The residual oil trapped in the press machine contains 4% of the POME and is known as the hydrocyclone wastewater (Liew et al., 2015). Palm oil mill effluent (POME) is a thick brownish colloidal slurry of water containing oil and fine cellulosic fruit residues. POME is usually generated from mill operation at a temperature ranging from 80 and 90 °C and is slightly acidic in nature. It has a pH of between 4 and 5 (Rupani et al., 2010). According to Ahmad (2016), POME is a very high strength industrial wastewater having 40,500 mg/L total solids, 4000 mg/L oil and grease, 50,000 mg/L COD and 25, 000 mg/L BOD. Detailed characteristics of the physicochemical parameters of POME are illustrated in Table 1-1.

Table 1-1   Characteristic of Palm Oil Mill Effluent

Parameters	Mean	Range
pH	4.2	3.5−5.2
Oil and grease	6000	150−18,000
Biochemical oxygen demand (BOD)	25,000	10,000−44,000
Chemical oxygen demand (COD)	50,000	16,000−100,000
Total solids (TS)	40,500	11,500−79,000
Suspended solids (SS)	18,000	5000−54,000
Total volatile solids (TVS)	34,000	9000−72,000
Ammonia nitrogen (AN)	35	4−80
Total nitrogen (TN)	750	80−1400
Phosphorus	180	
Magnesium	615	
Calcium	440	
Boron	7.6	
Iron	47	
Manganese	2.0	
Copper	0.9	
Zinc	2.3	

These physicochemical parameters must be reduced to appreciable low levels due to the danger they pose to human, animals and the ecosystem. The deposition of this effluent must comply with the prevailing effluent discharge standard for POME. The POME production per year increases sludge production and also high moisture content enriched with organic matter, which may constitute eutrophication effect. For this reason, it is considered one of the most harmful wastes for the environment if not properly treated. POME processing plant has increased over the years in Malaysia and Indonesia as a result of rapid industrialization, these has made both countries the two largest producers of oil palm. Consequently, POME produced in the oil palm processing mill has increased tremendously, thereby increasing the challenge of POME treatment before discharge.

Raw POME is colloidal in nature and contains 95%−96% water, 0.6%−0.7% oil, it also contains 4%−5% total solids and 2%−4% suspended solids having composition of solid matters from palm fruit mesocarp produced from sterilizer condensate, separator sludge and hydrocyclone wastewater (Ng and Cheng, 2017). POME is considered as the major contributor of water pollution by limiting the oxygen availability in water for aquatic respiration (Thangalazhy-Gopakumar et al., 2015), it contains elements like N, P, K, Mg, and Ca which are the major components of plant but the toxicity of this metals from effluents application in soil reduces the growth and developments of plants (Hossain et al., 2015). POME contains suspended solids and is acidic in nature with pH 4.5 (Khemkhao et al., 2015). According to Muhrizal et al. (2006), POME contains high concentration of Al and Pb at a concentration of >17.5 μg/g (Habib et al., 1997). Jameset al. (1996) stated that the presence of Pb in POME is as a result of pollution in pipes, tanks, and containers. The suspended particles and dissolved solids after the treatment of POME is known as palm oil mill sludge (POMS). The amount of POMS is dependent on the production of POME. It has high moisture content and pH of 8.4 and has elements such as nitrogen, phosphorus, and potassium in the concentration of 3.6, 0.9, and 2.1 mg/L respectively. The sludge, if not disposed, results in offensive odor and has the ability of polluting the surface and ground water. Therefore, it is very important for there to be effective technology at low cost for the disposal in order for the effluent to satisfy the prevailing discharge standard. According to (Chooi, 1984; James et al., 1996), POMS can be treated using open air drying and the dried product can be used as fertilizer because it has high nutrient value. However, the process is extremely difficult during the period of rainy season because of the slow drying rate and problem of over flow. Adsorption method is an effective method for the treatment of effluent; the method is applicable because it is less expensive compared to other treatment methods, in which the adsorbent materials are easily sourced and the treated effluent have no potential threat to the ecosystem.

1.3.1 Trends in POME Treatment Methods

The high amount of effluent generated in palm oil mill has increased research methods to minimize their effect of water pollution. Some of the methods that is applicable for the treatment of POME include the physical and chemical process such as filtration, floating sediment, coagulation, advanced oxidation precipitation and biological process which involves the use of either the conventional aerobic and anaerobic treatment systems as well as treatment using the anaerobic digestion. Some novel technologies are evolving for the treatment of raw POME such as biological digestion, membrane, Fenton technology and adsorption. The treatment efficiency of each method varies depending on the technology employed. The effectiveness of the treatment system is the ability of the treated effluent to satisfy the discharge standard. Some of the methods are classified into the physical-chemical methods and also the biological methods. The prevailing effluent discharge standard according to the Environmental Quality Act (EQA) Ministry of Environment (MOE) discharge limit Malaysia is illustrated in Table 1-1. It can be observed in Table 1-2, that the prevailing discharge of the physic-chemical parameters decreases over the years as the year increases except for pH and temperature from 1978 discharge standard to the present.

It was also observed that the prevailing discharge standard for the oil and grease deceases from 1978 and became constant from 1981 till present discharge standard. Similarly, the COD decreases as the year increases but as from 1982 onward, the concentration is not required in the effluent before discharge. This indicates that COD is a very important parameters that should be treated from the effluent before discharge. The concentration of BOD decreases from 1978 (5000 mg/L) to 100 (mg/L) since 1984. The reduction of the physic-chemical parameters especially the critical parameter is important in order to preserve the aquatic population and the water bodies.

Table 1-2   Environmental Quality Act 1974 for Palm Oil Mill Effluent Discharged

Parameters	DOE Discharged Limit (1986 Onwards)[a]	Environmental Quality Act[b]
BOD_3 (mg/L)	50	100
COD (mg/L)	1000	1000[c]
Suspended solids (mg/L)	1500	1500^3
Oil and grease (mg/L)	50	50
Ammoniacal nitrogen (mg/L)	100	150^3
Total nitrogen (mg/L)	200	200^3
pH	5.0	5.0–9.0
Temperature (°C)	45	45

[a]Malaysia Department of Environment (DOE).
[b]Parameters Limited of Environmental Quality (Prescribed premised) (crude Oil) Amendment.
[c]No new value stipulated since 1982.

1.3.2 Conventional Treatment Method

Biological treatment is the most common conventionally used method for the treatment of POME. The major principle of the biological method is to increase bacteria activities and other biodegradable organics to aid wastewater digestion. Biological treatment includes aerobic treatment making use of conventional activated sludge. It also includes the anaerobic open pond systems and the anaerobic treatment systems, known as the anaerobic digestion. The open ponding system is mostly used by palm oil mill operators around the world (Poh et al., 2015; Saswattecha et al., 2015). Some of the biological treatment methods is presented in Table 1-3.

The anaerobic pond is the most conventionally used POME treatment method. Open ponding system is the most popular treatment method of POME in Malaysia, utilized by more than 90% of the mills (Kuppusamy et al., 2017). It has a wide area of application because of its low cost of operation and capital. Anaerobic treatment has proven to be effective in reducing pollutants from high strength industrial wastewater. It involves biological reactions and sequences which enables one group of microorganism serving as substrate to another resulting in the conversion of organic matter to methane and CO_2. The treatment undergoes different stages of transformation of anaerobic microorganisms which produces CH_3 and CO_2. The anaerobic bacteria reduces the organic pollutants such as COD and BOD in the process of transformation. The open ponding system, also known as aerobic lagoons, are open shallow basin which treats wastewater with the use of bacteria and algae. It involves biological reactions and sequences which enables one group of microorganism serving as substrate to another resulting in the conversion of organic matter to methane and CO_2. A wide range of research efforts has been successfully conducted on reactor technology for industrial wastewater treatment such as food processing, textile industry, paper and pulp industry (Fiore et al., 2016).

Table 1-3 Treatment of Palm Oil Mill Effluent Using the Anaerobic System

Anaerobic Treatment System	Advantages	Disadvantages	Removal Efficiency (%)	Authors
Anaerobic pond	Higher emission of methane from tanks	Problem of land availability	97.8% removal of COD achieved before treatment in the facultative pond	Yacob et al. (2006)
Ultrasound membrane anaerobic system	High removal efficiency of COD at very short HRT	Expensive and requires routine maintenance	92.8%−98.3% COD removal achieved	Nour and Nour (2017)
Anaerobic palm oil mill effluent sludge	Pure culture bacteria for electricity generation	High rate of depletion of organic pollutant with increasing time, thereby reducing the generation of electricity in a batch system	High electricity generated and inversely proportional to the removal of COD	Nor et al. (2015)
Ponding system	High efficiency of the removal due to easy decomposition of microorganism	Longer HRT required	93.9% COD removal, 88.7% TSS removal achieved	Zahrim et al. (2014)
Sequencing batch reactor (SBR)	Very good settling ability and high biomass retention	POME samples used was from treated anaerobic pond	Total COD removal between 10−68% for total COD and 11%−94% soluble COD.	Fulazzaky et al. (2017)
Membrane bioreactor (MB)	Contains membrane fouling layer bio films which influences effluent quality by increasing bio-degradation	Prolonged retention time of POME in the MBR is needed for decolorization	30% COD removal rate achieved	Neoh et al. (2017)
Microbial fuel cell (MFC)-Adsorption hybrid system	The combustion catalytic oxidation can effectively analyze all organic compounds in the wastewater	Down flow mode results to clogging thereby minimizing organic removal	$90 \pm 0.3\%$ COD removal achieved	Tee et al. (2016)
Ultrasound cavitation-adsorption hybrid system	Higher efficiency of COD removal achieved at shorter operational time	Breakdown of the granular activated carbon by the energy released during cavitation time increases the concentration of the suspended solids	73.08% COD 98.33% TSS	Parthasarathy et al. (2016)
Upflow anaerobic sludge blanket (UASB)	High hydraulic loading rate. Low quantity of sludge produced	Lower organic loading rate If pH is not controlled, acidenogenic biomass buffers itself to a pH which depends on other environmental conditions	90% removal of COD achieved at initial loading of 1.1 $gL^{-1}d^{-1}$	Borja et al. (1996)

(Continued)

Table 1-3 (Continued)

Anaerobic Treatment System	Advantages	Disadvantages	Removal Efficiency (%)	Authors
Upflow anaerobic sludge fixed film (UASFF)	Hybrid reactor which combines the advantages of UASB	Lower OLR for the treatment of suspended solids	97% removal of COD at 11.58m^3/day OLR, 3 day HRT achieved	Poh and Chong (2009)
Upflow anaerobic sludge blanket —hollow centered packed bed (UASB-HCPB)	Shorter hydraulic retention time (HRT)	A shorter HRT causes problem of sludge wash out due to high upflow shear force.	88% COD, 90% BOD at an OLR 28.12 g COD L day	Poh et al. (2014)
Extended granular sludge blanket (EGSB)	Higher loading rates enhances the efficiency of the reactor	Requires the granulation of the anaerobic sludge. Also, the surface of the scum may affect pipes of the system	0.44 m^3 biogas/kg COD produced. 65%—70% CH_4, 25.36% CO_2 and 800—1500 ppm of H_2S.	Wang et al. (2015)
Integrated anaerobic—aerobic bioreactor (IAAB)	Higher loading rate and shorter HRT	Constraints of availability of sufficient land for pond and the length of HRT	>99% removal of COD, BOD and TSS achieved at 10.5 g COD/day with methane yield of 0.24 L CH_4/g	Chan et al. (2012)
Upflow anaerobic sludge fixed film (UASFF)	Combines the function of UASB reactor and the immobilized cell called fixed film	Instability at prolonged retention time and high influent concentration of COD. Sludge wash out occurrence due to the accumulation of TSS because of the inability of the fixed film bed to penetrate small size flocs	90% initial COD removal and at further increase, 82.4% COD removal achieved	Zinatizadeh et al. (2006)
Integrated baffled reactor inoculated with anaerobic pond sludge	High organic loading rate achieved	Increase in OLR results to the decrease in methane content	COD removal at 79% and 83% at HRT 4 and 6 days respectively	Malakahmad et al. (2014)
Upflow anaerobic sludge blanket reactor (UASBR)	It is very cheap and efficient at high OLR	The quality of sludge produced determines the stability of the treatment system	87% COD, 91% CH_3 achieved	Ahmad et al. (2005)
Continuous stirred tank reactor (CSTR)	Suitable for the treatment of substrates with high suspended solids	A deflector needs to be installed to promote the retention of suspended solids in the reactor	80% of COD removal achieved	Khemkhao et al. (2015)

Anaerobic systems are well acceptable technology because of its low construction and maintenance cost, low sludge production, small land requirements and biogas production as renewable energy. The treatment of POME undergoes different stages such as the sequences in the cooling pond, mixing pond, anaerobic pond and the facultative ponds. Yacob et al. (2006) treated POME using the anaerobic pond, in their findings, 97.8% removal of COD and between 35%−70% productions of CH_4/CH_2 was achieved. Also, Zahrim et al. (2014) achieved 93.9% removal of COD and 88.7% removal of TSS using the open ponding system. However, the limitation with the ponding system is the requirement of large land area for the treatment. Nowadays, 50% of mill operators treat POME using anaerobic digester systems (Ahmad, 2016). Nour and Nour (2017) treated POME using the ultra sound membrane anaerobic system. In their study, they achieved between 92.8%−98.3% removals of COD at short hydraulic retention time (HRT) between 8.2 and 500.8 days. High cost of setting up the membrane system makes the method not suitable for research.

Researches have demonstrated that anaerobic systems such as the sequencing batch reactor (SBR), membrane bioreactor (MB), up-flow anaerobic sludge blanket (UASB), up-flow anaerobic sludge fixed film (UASFF), up-flow anaerobic sludge blanket reactor (UASBR), continuous stirred tank reactor (CSTR) and extended granular sludge blanket (EGSB) can be used to treat high-strength industrial wastewater such as POME. Borja et al. (1996) were the foremost researchers who worked on UASB for the treatment of POME. In their findings, 90% COD removal was achieved at initial loading rate of $1.1\,g^{-1}d^{-1}$. This was obtained at high loading rate and low quantity of sludge was produced. The major challenge with the reactor was that if pH is not controlled, there is tendency for acidogenic biomass to buffer itself to a pH which is dependent on other environmental conditions. The UASBR was used by Ahmad et al. (2005) for the treatment of organic pollutants of POME. The reactor achieved 87% removal of COD and 91% production of CH_4. The method is cheap and very efficient at high OLR but the quality of sludge produced determines the stability of the reactor. The use of EGSB was reported by Wang et al. (2012) for the treatment of POME. It was observed that at high loading rate, $0.44\,m^3$ biogas/kg COD was produced. This contains 65%−70% CH_4, 25%−36% CO_2 and 800−1500 ppm of H_2S. Similarly, the CSTR was very suitable for the reduction of COD in the POME, even though there were high suspended solids, 80% removal of COD was achieved.

In some cases, reactors are combined and are used as hybrid system and have been reported in the literature for the treatment of POME. For example Poh et al. (2014) investigated the treatment of POME using the up-flow anaerobic blanket-hollow centered packed reactor (UASB-HCPB). The hybrid reactor achieved 88% COD removal and 90% BOD removal at an organic loading rate (OLR) of 28.12 g/COD. L.day. The result was achieved at a very short HRT. However, a reduced HRT may result to problem of sludge wash out due to high up-flow shear force. The use of integrated anaerobic-aerobic bioreactor (IAAB) achieved >99% removal of COD, BOD and TSS at 10.5 g/COD/day with methane yield of 0.24 L CH_4/ g. The constraint with the method is the problem of availability of sufficient land and the length of HRT. Also, Malakahmad et al. (2014) reported the use of integrated baffled reactor inoculated with anaerobic pond sludge. In their findings, it was observed that high organic loading rate favoured 79% removal of COD at HRT of 4 days and 83% removal at HRT of 6 days. The UASFF was used as hybrid reactor with the functions of the UASB (Poh and Chong, 2009). The result of the hybrid system achieved lower loading rate for the treatment of suspended solids but achieved 97% removal of COD at 11.58m^3/day OLR which was achieved at 3 day HRT. The performance evaluation of up-flow anaerobic sludge fixed film (UASFF) was compared at mesophilic temperature for the treatment of POME with UASB and AF (Ohimain and Izah, 2017). The performance of the reactor has high organic loading rate (OLR) better than UASB and AF, the reactor could produce 71.90% methane under OLR of 11.58 kg COD m^3/day. The anaerobic hybrid reactor was used to remove 64% of total COD higher than the removal rate in the UASB reactors.

The anaerobic hybrid reactor was used to remove 64% of total COD higher than the removal rate in the UASB reactors. The membrane anaerobic system (MAS) was utilized for the treatment of POME (Nasrullah et al., 2017). The ultra sound membrane anaerobic system was used in the study of (Nour and Nour, 2017) for the investigation of the adsorption of COD from POME. The removal of COD (92.8%−98.3%) demonstrated the effectiveness of the system. The membrane bioreactor in Neoh et al. (2017) resulted to 30% of COD removal which indicated less suitability of the treatment system. The treatment effort was less effective due to the deterioration of membrane flux rate as a result of membrane fouling, which have the possibility of affecting the treatment process. Furthermore, the periodical replacement of membrane as a result of fouling is very expensive and not sustainable for research. The membrane fouling can be reduced with faster cross flow feed velocities and regular membrane flushing.

The conventional activated sludge requires a lot of energy for the purpose of aeration. It also produces a large quantity of sludge that makes the cost of treatment and disposal very expensive. In the study of (Nor et al., 2015) anaerobic palm oil, mill effluent resulted in high electricity generation, which is inversely proportional to the removal of COD. Aerobic treatment method involves the presence of oxygen for the stabilization of organic matter in wastewater. In the investigation of Fulazzaky et al. (2017), POME samples obtained from the treated anaerobic pond was further treated using the sequencing batch reactor, very good settling ability of the reactor was observed with high biomass retention time. In the investigation, 94% COD removal was observed in the treatment process. Aerobic treatment can be combined with physical method of treatment, such as adsorption in a hybrid system, for the improvement of the treatment process. In the study conducted by Tee et al. (2016), the investigation of microbial fuel cell −adsorption hybrid system for the treatment of POME was conducted. It was observed that the removal of COD was effective due to the combustion catalytic oxidation of the organic matter in the POME. About $90.5 \pm 0.3\%$ of COD removal was observed. Parthasarathy et al. (2016) conducted an experiment for the removal of COD and TSS from POME using a hybrid system that combines ultrasound cavitation-adsorption system. It was observed that the breakdown of granular activated carbon by the energy release at increase cavitation time during cavity increases the concentration of suspended solids. From the result of the investigation, 73.08% and 98.33% of COD and TSS was removed respectively.

1.3.3 Physical and Chemical Processes of POME Treatment

Physical processes involved in the treatment of POME includes process screening, sedimentation, and oil removal before secondary treatment in biological treatment plants. Some of the other methods in Table 1-4 are the ultrafiltration, solvent extraction, reverse osmosis coagulation, electrocoagulation, coagulation-flocculation, floatation.

Acidification of pond and flocculation treatments are advanced pretreatment processes, also includes the use of membrane (Hojjat, 2009). The researchers demonstrated that the centrifugation and coagulation methods gave better pretreatment quality than filtration method. The separation of effluent from activated sludge can be done at a low-pressure using either the microfiltration (MF) or ultrafiltration (UF). Ultrafiltration has been reported in the literature as a useful technology for the treatment of POME. The choice of the selection of membrane depends on the effect it has on the target pollutants. The hydrophobic membranes have high retention capacity than hydrophilic cellulose membrane for the treatment of protein compounds in POME (Wu et al., 2007).

Hydrophobic membranes have more inclination to retain hydrophobic solutes on the surface of the membrane better than hydrophilic membranes. The result of the ultrafiltration using polysulphone membrane was investigated in Wu et al. (2007). The result obtained showed 97.7% removal efficiency of TSS, 88.8% reduction of turbidity, 6.5% TDS removal and 57% removal of COD. In addition, the effectiveness of the ultrafiltration membrane achieved 71.26% removal of SS (Azmi and Yunos, 2014). Mixed matrix membrane was also used in the study for the treatment of POME. According to Ho et al. (2017), the POME used for the treatment was secondary effluent after undergoing pretreatment processes.

The constraint with the use of membrane is the problem of fouling, which increases process down time due to damage. A variety of chemicals can be used for flocculation treatment purposes. Flocculation is the addition of chemicals (coagulants) to destabilize and aggregate colloidal particles in wastewater. A flocculants are usually organic chemicals added to wastewater to enhance flocculation, such chemicals are alum, aluminium chlorohydrate, aluminium sulphate etc. Natural materials such as chitosan can also be used for flocculation purposes as a replacement for the expensive chemicals. Since suspended solids in POME are related with organic matter composition, therefore coagulants can be used effectively for the removal of colloidal and suspended organic solids but may not be very effective in the removal of dissolved organic matter (Rupani et al., 2010). In the study conducted by Zinatizadeh et al. (2017), coagulation was used for the pretreatment of POME, the result obtained showed the removal efficiency 96.4 and 70.9% of TSS and COD respectively.

The combination of coagulation and flocculation methods can improve on the reduction of the pollutants of POME. The use of coagulation-flocculation achieved 87% recovery of sludge in the study conducted by Bhatia et al. (2007), although high dosage of coagulant and flocculants were required in their investigation. Also the effect of the combined coagulation and flocculation was studied by Shak and Wu (2015) for the treatment of POME. The result of their findings revealed that 81.58% removal of TSS and 48.22% removal of COD was achieved which indicated the effectiveness of the treatment process for the reduction of

Table 1-4 Physical-Chemical Method of Treatment of Palm Oil Mill Effluent

Method	Parameter Investigated	Removal efficiency	Limitation	Author
Ultrafiltration membrane	TSS, turbidity, TDS, COD	97.7%TSS, 88.5% turbidity, 6.5% TDS, and 57% COD	Raw POME was pretreated and the result of the pretreatment was removal of 97.3% TSS, 88.5% turbidity, 6.5% TDS, and 46.9% COD	Wu et al. (2007)
Ultrafiltration membrane	SS	71.26%	Raw POME was pretreated using adsorption before further treatment using membrane	Azmi and Yunos (2014)
Ultrafiltration membrane	Color	58.9%	Aerobically treated used for the investigation	Subramaniam et al. (2017)
Solvent extraction	Oil and grease	71.1%	Higher temperature needed to evaporate n-hexane from the residual oil which may result to thermal decomposition of carotene pigment and as a result lower carotene concentration	Ahmad et al. (2005)
Mixed matrix membrane	Color, TSS, turbidity, COD, and chlorine	75.46%−88.52% color removal, 98.59%−100% TSS, 79.10%−89.30% turbidity, 62.91%−75.5% COD, and 64%−76% chlorine	Diluted effluent from the aerobic pond used for the investigation and also problem of fouling of membrane fluxes	Ho et al. (2017)
Coagulation	TSS and COD	96.4% and 70.9% respectively	Pretreatment using polymer induced coagulant and physical treatment methods	Zinatizadeh et al. (2017)
Electrocoagulation	pH, COD	Satisfactory pH of discharge (7.6), 75.4% COD removal	Treated secondary effluent used for the investigation	Bashir et al. (2016)
Coagulation-flocculation	SS	87% recovery of sludge	High dosage of coagulant and flocculants required	Bhatia et al. (2007)
Coagulation-Flocculation	TSS and COD	81.58% and 48.22% respectively	Not very effective for the removal of COD	Shak and Wu (2015)
Sedimentation and centrifugation	Oil and grease	80% of oil recovered with 27.67 ± 0.10 FFA	High temperature needed for evaporation after centrifugation and sedimentation	Suwanno et al. (2017)
Flotation	COD	53.7% COD removal at 12.5 minutes contact time	Secondary treated POME was investigated and treated effluent not satisfactory for discharge	Poh et al. (2015)
Combined air floatation and membrane	COD	36.1, 26.8 and 26.6% removal achieved	Removal lower than micro bubble floatation used in the study of Poh et al., (2015)	Faisal et al. (2016)

(Continued)

Table 1-4 (Continued)

Method	Parameter Investigated	Removal efficiency	Limitation	Author
Photocatalysis	COD	Ag/ TiO$_2$ achieved better photocatalytic degradation of POME better than TiO$_2$	Adsorption −desorption equilibrium achieved before photo catalytic reaction	Cheng et al. (2016)
Photocatalysis	COD	55% removal of COD was achieved	POME from the discharge pond used for the treatment	Ng and Cheng (2017)
Photocatalysis (Tungsten oxide photocatalyst)	COD, pH and color	51.15% COD removal and 96.21% color removal	Adsorption process is needed for the organic pollutants before photocatalytic process occur.	Cheng et al. (2016)
Electro persulphate oxidation	COD, color, and SS	77.7% of COD, 97.96% of color and 99.72% SS.	Combined the effect of electro-oxidation, electro-coagulation. Also, secondary effluent was used for the treatment	Bashir et al. (2017)
Fenton oxidation	COD and color	85.15% COD removal, 92.1% color removal	Secondary effluent treated before discharge	Saeed et al. (2015)
Sono-Fenton oxidation Fenton process	COD	80% removal	Combined ultrasound and Fenton process	Taha and Ibrahim (2014a)
Aerated heterogeneous	COD	75% removal achieved	Secondary effluent after anaerobic treatment investigated	Taha and Ibrahim (2014b)
Ambient Fenton oxidation	COD, Color	75.2% COD and 92.4% color	Biologically treated POME was used for the treatment	Aris et al. (2008)

suspended solutes. The use of electrocoagulation was used in the investigation of Bashir et al. (2016) for pH and COD concentration in POME. However, the result of the investigation revealed that the pH after treatment satisfied the discharge standard while 75.4% removal was achieved for COD removal. Evaporation process can also be used for the treatment of POME when POME containing 3%−4% total solid as feed, about 85% water composition in POME can be recovered by distillation. The limitation with the process of evaporation is the energy requirement where the energy consumption rate is very high (kg of stream required 1 kg of evaporated water) (Hazlan, 2006).

The gravity type oil separator can be used for oil separation in POME with low suspended solids. The trapped oil is usually designed for maximum flow rate at a permissible surface loading rate of $2-6 \text{ m}^3/\text{m}^2\text{h}$. To ensure effectiveness of the oil separator, an automatic skimming device is usually installed to help in the recovery of good quality oil. The efficiency of oil separation by the gravity type POME wastewater stream is in the range of 60%−90% (Wang et al., 2013). Solvents extraction was reported by Hammed (2003) as very effective for the removal of residual oil from POME with the percentage removal to increase with increase of mixing time, mixing ratio and mixing rate for all solvents. Ahmad et al. (2005) conducted experiments for the removal of oil and grease using solvent extraction. In their findings, 71.1% removal efficiency was achieved after the treatment process of the POME. However, high temperature was needed to evaporate *n*-hexane from the residual oil, which may result to thermal decomposition of carotene pigment in the POME. Similarly, Suwanno et al. (2017) studied the removal of oil and grease using sedimentation-centrifugation process. The result obtained revealed that 80% of oil was recovered with $27.67 \pm 0.10\%$.

Chemical oxidation involves the use of oxidizing agent in the treatment of wastewater to oxidize the organic pollutants. The use of advanced oxidation is very common in the treatment of high strength wastewater such as POME. Some other advanced oxidation methods are photocatalysis, electro persulphate oxidation, Fenton oxidation, Sono-Fenton oxidation, Solar Fenton oxidation, aerated heterogeneous Fenton process, ambient Fenton oxidation.

The Fenton oxidation is a novel technology because of its simplicity and high removal efficiency of pollutant removal without any need for specialized equipment. The Fenton oxidation utilizes the principle of exchange of reactive hydroxyl radicals (. OH)

$$Fe^{2+} + H_2O_2 \rightarrow Fe^{3+} + OH^- + OH \quad (1.1)$$

$$Fe^{3+} + H_2O_2 \rightarrow Fe^{3+} + OOH + H^+ \quad (1.2)$$

A modified electron-Fenton process (EF) was developed to solve potential transportation risk ofH_2O_2, loss of reactivity and sludge production. The EF method effectively controls hydroxyl radical production and reduction of soluble Fe^{3+}cathodically toFe^{2+}. This principle is known as electrochemical catalysis (Barhoumi et al., 2016). The removal efficiency of EF is dependent on the production of H_2O_2 and Fe^{2+}cation, pH, density of current and concentration of electrolyte. Fenton oxidation consists of the reaction of the hydroxyl radicals on the alkyl chain of fatty acids of POME. The OH^-have strong affinity to destroy the aromatic ring attached with hydroxyl group in fatty acids. This results in the formation of water soluble compounds through removal of hydrogen and addition of oxygen atoms with the presence of ferric ions. An experiment using Fenton oxidation was conducted by Saeed et al. (2015) for the removal of COD and color from POME. The result obtained demonstrated that 85.1% COD and 92.1% color removal was achieved. However, the POME used for the study was obtained from the secondary treated effluent before discharge. Sono-Fenton method combines ultrasound and Fenton process for the treatment purposes. The combined process achieved 80% removal of COD (Taha and Ibrahim, 2014a). Aerated heterogeneous Fenton process also achieved 75% removal efficiency of COD from POME sample obtained from the treated effluent from the anaerobic pond (Taha and Ibrahim, 2014b). Ambient Fenton oxidation was used for the removal of COD and color using biologically treated POME for the investigation, the process reduced COD by 75.2% and color by 92.4%. The EF treatment process was used to study the changes in POME characteristics for 2 hours. It was observed that 46% removal efficiency of COD was obtained. Furthermore, the pH was observed to have increased from 5.3 to 7.4. This range of pH can be considered valuable for the effective reduction of COD, BOD, TOC, and TN (Babu et al., 2010).

The removal of organic pollutants from POME can be investigated under the influence of light by using conductor materials as catalyst. The process is known as photocatalysis. The photo catalytic degradation of POME for the reduction of COD was more effective using Ag/TiO_2 than with TiO_2 (Cheng et al., 2016). Also Ng and Cheng (2017) investigated COD removal from POME using photocatalysis. From their investigation, 55% removal of COD was achieved in the treatment process. The removal of COD, pH and color using Tungsten photo catalyst achieved 51.15% COD and 96.21% color removal in the study of Cheng et al. (2016).

The disadvantages of the conventional treatment methods include, large area required for the anaerobic ponds, high cost of routine maintenance is required for the reactors. Similarly, the methods are expensive and are not suitable for the treatment of POME at low concentration. In some other treatment processes using the physical and chemical methods such as coagulation and flocculation processes, there is the requirement of high dosage of adsorbent which may affect the pH of the treated POME and the use of membrane may be very expensive for the treatment of POME. Advanced oxidation such as Fenton oxidation and photocatalysis requires energy for the degradation of the POME.

However, the application of adsorption is a novel method of treatment of high strength wastewater such as POME, dyes and petrol chemical solutions. Adsorption process is affected by the nature of the adsorbate, adsorbent material, presence of other pollutants in solution and also atmospheric and experimental conditions.

1.4 Adsorption

Stability of emulsion is very important in terms of application and storage. Emulsion is an equilibrium system, but thermodynamically unfavorable systems, which tend to break down over time due to a number of physicochemical mechanisms, containing gravitational separation, flocculation, coalescence and Ostwald ripening (Dickinson, 1992; Friberg et al., 2004).

Adsorption is a valuable technology in the treatment of industrial wastewater. It is a very popular treatment method among other methods of industrial wastewater because of the simplicity of operation and efficiency of the treatment process. The method of adsorption has the capacity of treating high quality effluent with proper design consideration better than other chemical methods such as coagulation (Acero et al., 2016). Adsorption is the process through which substances present originally in one phase is removed from that phase by accumulation at the interface between the phase and a separate phase. Adsorption can take place at any solid liquid interface by the accumulation of a solute onto the adsorbent during the adsorption process. The adsorbate accumulates on the solid surface with interaction on the surface of the adsorbent. Such interaction is influenced by the pH of the medium, ion exchange, acid-base interactions, hydrogen bonding, hydrophobic and hydrophilic interactions and precipitation (Lalley et al., 2016).

Adsorption is used for the treatment of many industrial wastewaters because industrial wastewaters are toxic and difficult to remove through conventional secondary treatment methods. The substances are present in small concentrations and their removal becomes very difficult using other methods. The rate of adsorption is affected by the nature of the adsorbents, the concentration of adsorbate and the efficiency of the adsorption system. Adsorbents are materials that adsorbs wastewater pollutants while the adsorbate are the substances to be adsorbed. Pore structure of the adsorbate, pH of the solution, presence of inorganic salts, interacting solutes, temperature, pressure and the activation of the adsorbents are other factors that affect the adsorption of substances in water.

Adsorption can be classified into two types based on the nature and the interaction between the adsorbate molecules and the adsorbent surface. The classification could either be based on the process of physiosorption or chemisorption. Both types occur either in the gas−solid interface or liquid−solid interface due to the attractive forces at the surface adsorbent overcoming the activation energy of the adsorbent molecules (Ali, 2012).

1.4.1 Principles of Adsorption

Physiosorption is also called physical adsorption, which is a process whereby electronic arrangement of the molecules is affected by interactive forces called van der Waals forces. The van der Waals forces of the adsorption molecules occur as a result of the interaction between temporary and permanent electric dipoles. The adsorbed particles are very far from the surface plane but very active on the surface due to low binding energy. The desorption temperature is low as a result of the physiosorbates weak forces of interaction. In the process of physiosorption, single or multiple layers of the adsorbate molecules on the adsorbent surface occur as a result of low activation energy of adsorption (20−40 kJ) (Rouquerol et al., 2013). In the process of the interaction, there is spontaneous wetting of the surface of the adsorbent when in contact with the adsorbate, also there is tendency of the adsorbent material to dissolve in the adsorbate.

1.4.2 Chemisorption

The process of chemisorption of the adsorption system is otherwise referred to as chemical adsorption, which is defined as the process that occurs under the influence of chemical bond as forces of attraction between the adsorbed molecules and the adsorbent. Chemisorption occurs at very high temperature and the energy of adsorption exists between $(200-400 \text{ kJ/mol})$ (Rouquerol et al., 2013). The activation energy is usually very high at high pressure. The result obtained is a monolayer of the adsorbate attached to the surface of the adsorbent. Both physiosorption and chemisorption are affected by adsorption parameters such as pH, contact time, agitation speed, particle sizes, initial concentration, temperature and cationic exchange capacity (CEC). These parameters each have effect on adsorption process and determines the rate of adsorption of solutes onto the adsorbent. In addition, the pH of both the adsorbent and adsorbate is very significant to the process of adsorption. On the surface of the adsorbent media, the condition when the electric charge density on the surface is equal to zero is referred to as the point of zero charge (pzc). It is the pH value when the number of cations and anions on the surface of the adsorbent are equal. The pH_{pzc} is described in terms of the concentration of the solution. In cases whereby the pH of the solution is lower than the pH_{pzc}, the acidic medium donates more protons than the hydroxide group. Hence, the adsorbent surface is positively charged. The surface favours the adsorption of anions from the solution. On the other hand, when the pH of the solution is above pH_{pzc}, the surface of the adsorbent is negatively charged, in this instance, adsorption of cations from the solution is more favorable. An example of pH_{pzc} can be obtained by plotting a graph of final concentration pH of adsorbent against the initial concentration by adjusting the solution using HNO_3^- and NaOH (Fig. 1-2).

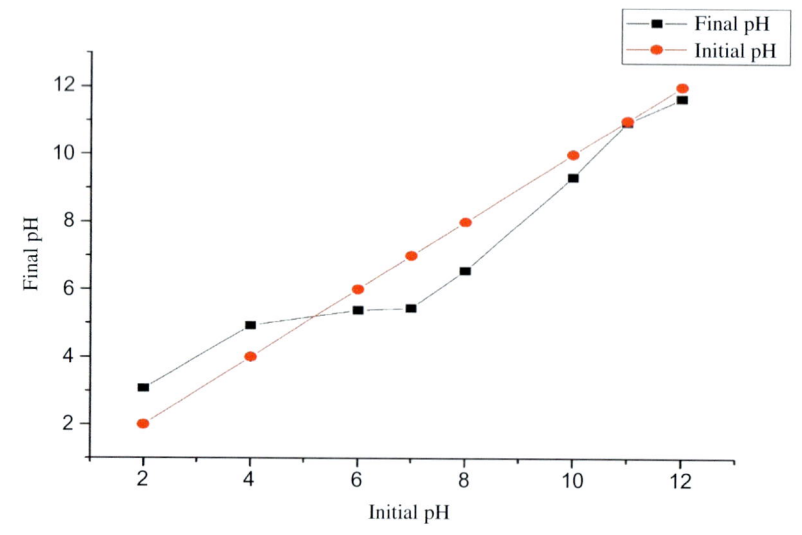

FIGURE 1-2 Point of zero charge.

The CEC is defined by the amount of negatively-charged sites that is available on the surface of the adsorbent, which has the capacity to retain positively charge ions, otherwise known as cations, such as Ca^{2+} Mg^{2+} and K^+ by the process of electrostatic attraction. The cations retained as a result of the electrostatic force are easily exchangeable with the cations in the wastewater. Adsorbents with higher CEC has the potential to engage in cationic exchange in solution than those with lower CEC.

1.4.3 Wetting and Fluid Adsorption

The concept of wetting is the response observed when a liquid is in contact with a solid surface (Bormashenko, 2013). According to Bornashenko, the liquid could spread spontaneously resulting in a film dependent on the mass of liquid available. When a liquid drop rests on the horizontal solid surface, an angle is formed by the liquid−solid interaction and the liquid vapor interface. The inclination to the surface is called contact angle, which is observed as the interfacial coexistence of the solid, liquid, and vapor phase. A contact angle less than 90 degrees shows favorable wettability of the liquid on the solid surface. In a case of contact angle greater than 90 degrees, it indicates low wettability of the liquid on the solid surface. When the contact angle is greater than 90 degrees, it signifies low wettability and the result is an unfavorable wetting on the surface. For a contact angle less than 90 degrees, the liquids drifts spontaneously into smaller pore spaces in contact, such condition is referred to as hydrophilic behavior. At higher contact angle, the liquid requires energy to be in contact with the solid surface, thus the liquid minimizes its contact forming a well-rounded liquid droplet. This property is referred to as hydrophobic behavior. When the contact angle is zero, the condition is known as complete wetting case. The droplet becomes flat with the solid surface. The spreading of the liquid develops films spontaneously on the surface. The surface tension of a liquid is used to determine the shape of the droplet of the liquid. The mechanism involves the molecules being pulled equally in all direction by adjacent liquid molecules leading to resultant net force of zero. The effect of this is not limited to the molecules at the surface thereby causing no interaction between adjacent molecules to provide a balanced net force. They are pulled inwardly by the adjacent molecule; this results to an internal pressure. The liquid contacts its surface area to maintain lowest surface free energy. Surface tension and contact angles are the results of intermolecular forces. The application of contact angle is to have the description of the behavior of the adsorbent when in contact fluid. It gives valuable understanding whether the adsorbent material have affinity and interaction with water. In other words, they can easily be covered by film of water, the behavior of such material is known as hydrophilic properties of the media. The application of contact angle also gives the description of the adsorbent properties from resisting wetting by the adsorbate. The property of materials showing such behavior is hydrophobic in nature.

1.4.4 Hydrophobicity of Adsorbent Material

When hydrophobic particles are suspended in water, they have tendency to interact with particles of their kinds rather than with water. Typical example is the oil droplet, because of this trend, hydrophobic particles suspended in water has the physical nature of voids in the bulk solution. The consequence of such behavior results in the expansion and breaking of hydrogen bonds between water particles. The effect of the breaking of the hydrogen bonding as a result of expansion brings together the hydrophobic particles immersed in water. This can be achieved through the following mechanisms. When two hydrophobic particles immersed in water and cluster, the surface area of the coalesced particle is smaller than the sum of the surface areas of the combining particles. As a result of this, the energy leading to the breaking down of water particles proportional to the surface area of the particles decreases as a result of coalescence of the particles. This makes the process leading to coalesces thermodynamically spontaneous. The interaction of the molecules is achieved at a higher state of entropy and this is the reason causing nonpolar molecules such as oil and grease, hexane, and organic solutes to clump together. They readily dissolve in nonpolar solvent. The process of clumping together of the nonpolar molecules reduces the surface area exposed to water, hence decreases the entropy of the system (Davies, 2012).

When two hydrophobic particles move together and combine, the free energy of the system favours the attachment of the particle to the surface (Mitik-Dineva et al., 2009). The more the surfaces are hydrophobic, the more strongly they cluster together (Lewandowski and Beyenal, 2013). For two molecules to be in contact, the attraction energy is about six times greater in water than the interaction of van der Waals in vacuum. There is an assumption by researchers that the small hydrophobic areas available on the surfaces of microbial cells are mainly responsible for the adherence of microorganisms to hydrophobic surfaces (Ferrara et al., 2013). The dissolved organic matter in an effluent is characterized by the hydrophobicity of the medium (Xia et al., 2015). Adsorption of organic molecules was effective using hydrophobic nano-gel in oil-in- water emulsion study (Wang et al., 2012). Activated carbon is widely known for its hydrophobic properties and also very effective as a catalytic support for the reduction of organic pollutants such as COD mg/L and other toxic compounds in wastewater. Increased hydrophobicity of carbon surface improves adsorbate-adsorbent interactions. The effectiveness of the removal by sorption depends on the hydrophobic nature of the aqueous organics (Pradhan et al., 2016). The ability of microorganism to adhere can be determined by the effectiveness of interacting forces such as Vander Waals, Coulomb, electrostatic as well as hydrophobic interactions. Among the noncovalent forces, hydrophobic interactions at nano level at medium range is considered as the most relevant in water and also electrostatic interactions in the short range (Hannig and Hannig, 2009). Hydrophobicity of a material is also dependent on the interactions with the spaces that result to changes in the membrane structure and morphology. The material membrane can be wetted by nonpolar phase such as nonpolar organics where the aqueous polar phase cannot penetrate into the pores (Drioli et al., 2011). The two phases are immiscible because the operating pressure is controlled. However, the pressure of the polar phase must be equal or greater than the pressure of the wetting phase. This will avoid the possibility of dispersion of drops from one phase to another. The interfacial area is established as the pore mouth if the penetration of the polar phase into the membrane pore is avoided. Hydrophobicity of material is not an absolute guarantee for keeping the pore spaces.

1.4.5 Hydrophilicity of Adsorbent Materials

When hydrophilic materials combine with water, a thermodynamic interaction is more favorable than the interactions with oil or other hydrophobes. The particles are charged-polarized and have strong ability to engage in hydrogen-bonding. This results in the adsorbent material to be soluble in water and other polar solvents. In hydrophilic medium, solid−liquid interface is more favorable resulting in contact angle to be less than 90 degrees (Nguyen, 2016). In the process of hydrophilic interaction, the water drop on the substrate aligns with the topography which results in the decrease of the contact angle. The moisture also spreads within the substrate and coexists with the solid filled with liquid. In a situation of partial wetting, the Wenzel contact angle is decreased but greater than zero (Grundke et al., 2015). Zeolite is often classified as hydrophilic materials which may be spherical or granular in nature. The rate of diffusion of inorganic materials is slow because of smaller pores. For macroporous structure, diffusion rate within the pores are higher than the rate of diffusion of ion in water. As the pore becomes large, there is transportation of the molecular ions through the pores by convection and diffusion. The importance of short diffusion can be achieved by developing composite materials with small particles forming into large particles (Andaç and Denizli, 2014). Hydrophilic materials are soluble in polar and have contact angle <90 degrees, also materials soluble in nonpolar solvents are hydrophobic and have contact angle >90 degrees (Shirtcliffe et al., 2010). Adsorption is readily and rapidly reversible on hydrophilic materials than hydrophobic surfaces (Park et al., 2015).

1.5 Treatment of POME Using Adsorption

Many conventional methods such as the membrane system, open ponding, anaerobic reactors are used in the treatment of POME, recent advancement in research on the treatment of industrial wastewater such as POME using adsorption is a novel technology for the removal of heavy metals and other pollutants contained in low concentration in POME. Experimental data to predict the efficiency of the adsorbent for the adsorption of solutes in wastewater are often fitted to empirical models in order to predict the pattern of adsorption. A study was conducted to investigate the removal efficiency of residual oil from POME using powder and flake chitosan. (Ahmad et al., 2005). The initial concentration of residual oil was 2 g/L contained in POME. The weight dosage, contact time and pH of chitosan both in powder and flake form were investigated to obtain the optimum condition for the adsorption of the residual oil from POME. It was observed from results that chitosan powder at dosage of 0.5 g/L, 15 minutes of contact time and pH value of 5 represented the most optimum condition for the adsorption of residual oil with a removal efficiency of 99% achieved. The powdered form of chitosan demonstrated better adsorption rate compared to the flake chitosan. Igwe et al. (2013) also studied the removal rate of residual oil from POME using boiler fly ash. The percentage residual oil uptake reduced from 80% to 5% at an increased contact time from 10 to 120 minutes. As the initial concentration of the residual oil increased from 0.04812 to 0.2406 mg/L, the adsorption rate increased from 2.74% to 72.98%.

The maximum mono layer adsorption capacity was 0.3476 mg/g. The result revealed that boiler fly ash was a very good adsorbent for the removal of residual oil from POME. In addition, Ahmad et al. (2005), adsorbed residual oil of POME using synthetic rubber, from their investigation, there was effective removal of oil and grease up to 88% reduction was obtained. This was achieved at a contact time of 3 hours, agitation speed of 150 rpm obtained at pH of 7. It was observed that the adsorption process fit in properly to the Freundlich isotherm model with R^2 value of 0.9721.

Similarly, low cost adsorbent using waste activated sludge (WAS) from POME treatment plant was used to adsorb ammonium from aqueous solution (Muttalib, 2012). The investigation was done in a batch test to determine the effect of initial concentration, temperature, pH and adsorbent dosage on the aqueous solution. The result of the investigation showed that the ammonium removal increased with increasing pH, initial concentration of ammonium and adsorbent dosage but decreased with temperature. The adsorption of ammonium agrees with Langmuir and Freundlich isotherm, which are the most applied empirical isotherm models to predict the nature of the adsorption process. Langmuir model predicts the adsorption of pollutants from the adsorbate on a monolayer surface while the adsorption on heterogeneous surface is best described by fitting experimental data to the Freundlich isotherm model. Other forms of isotherm model such as the Temkin model describes the heat of adsorption between the adsorbent and the adsorbate. Low cost adsorbent media can also be effective for the reduction of heavy metals in the wastewater. In the study conducted by Adeleke et al. (2017), the reduction of zinc from POME was investigated using coconut shells and cow bones under fixed condition of pH 7, 105 minutes contact time and 150 rpm shaking speed, it was observed that more than 90% removal efficiency was achieved for each of the adsorbents used. The experimental data fitted to the isotherm model revealed that the BET model is more suitable for the adsorption of zinc on coconut shell while the Langmuir model better expressed the pattern of the uptake of zinc ion from the experimental data using the activated cow bone powder. Lau et al. (2013) prepared palm shell activated carbon (PSAC) by steam activation for the removal of H_2S from biogas studying the initial concentration, adsorption temperature and space velocity. The effect of the parameters was studied to determine the rate of adsorption of H_2S from aqueous solution. The effect of temperature on adsorption did not have major effect on the adsorptive capacity of H_2S onto palm shell activated carbon.

The adsorption of POME using banana peel was investigated by modifying the carbonyl group of the peel by the method of esterification using acidic methanol at a carbonized temperature of 500 °C for 1 hour (Mohammed and Chong, 2014), the result of the investigation showed that BET surface area of 24.2572 m^2/g was achieved. WAS from palm oil mill effluent was used for the adsorption of methylene blue (MB) in a batch study (Gobi et al., 2011), the maximum mono layer adsorption capacity of WAS was found to be 66.23 mg/g at 30°C, the adsorption kinetic fitted well to the pseudo-second—order kinetic with R^2 of more than 0.95 recorded. The application of batch adsorption study for the removal of Cd was investigated in the study of Adeleke et al (2016) using activated cow bone powder as adsorbent, the result obtained revealed that optimum condition of adsorbent dosage was recorded at an average of 97.37% reduction achieved for the duplicate sample. It was observed that the least uptake of Cd was achieved using 30 g adsorbent dosage at 95.1% removal efficiency for both samples. The effect of palm oil mill fly ash (POFA) for the adsorption of Cd (II) and Cu (II) ions from aqueous solution through column studies was investigated by Aziz et al. (2014), the effect of the column dynamics and the break through curve was analyzed from the study. The result of the investigation showed that the highest bed capacity was recorded as 34.91 mg Cd (II)/g and 21.93 mg Cu (II)/g of POFA at 20 mg/L of influent metal concentrations, column bed depth of 20 mm and flow rate of 5 mL/min. The break through curve for both Cd (II) and Cu (II) fit properly to Thomas and Yoon-Nelson models, the initial breakthrough region was better illustrated using the BDST model. However, the result revealed that POFA (palm oil fuel ash) can be effectively utilized for the adsorption of Cd (II) and Cu (II) ions from aqueous solution in a fixed bed column.

The adsorptive capacity of natural zeolite for heavy metal ions removal was investigated by (Shavandi et al., 2012). The metals studied were Zn (II), Manganese (II) and Iron Fe (III). The initial concentration of dosage on the adsorption of the heavy metals as well as the contact time, agitation time, speed, pH was studied. The optimum adsorption was achieved at equilibrium contact time of 180 minutes. The sorption increased with pH and the rate of adsorption was in the range of 0.015 and 1.157 mg/g of zeolite. More than 50% of Zn (II) and Mn (II) and 60 % of Fe (III) were removed. The removal of oil from raw POME can be effective by using hydrophobic adsorbent material that has the ability to adsorb pollutants in the nonpolar phase (Adeleke et al., 2017). In the study conducted by Wahi et al. (2017), 80.23% oil was recovered from raw POME immediately after the milling process at optimum pH 4.18 at 24 hour contact time at 30°C. Similarly, Abdullahi et al. (2015) achieved 96%−99% removal of oil using structurally modified raw kapok fiber (Ceiba Pentandra). A summary is illustrated in Table 1-5.

Adsorption technique for the treatment of wastewater has shown great potentials above other treatment methods for the removal of organic pollutants. Adsorption has advantage over other treatment methods because of the simplicity of the design, it is cost effective and saves problem of land availability. The adsorption technology has successfully been proven as an effective treatment method for POME. The application of adsorption for the treatment of industrial wastewater has received a lot of attention from researchers. However, the search for low-cost adsorbent material with pollutant-binding potentials has increased over the years. Materials sourced from industrial, agricultural wastes and natural materials can be used as adsorbents for the treatment of industrial wastewater. The effectiveness of the sourced materials should be applied based on the potentials for the treatment of target pollutants. For example, the adsorption of polar and nonpolar solutes can be effective on the nature of the adsorbent material with pollutant binding effect (Adeleke et al., 2016). The method of adsorption also can be widely applied with other methods for effective treatment of high strength industrial wastewater. The combination of magnetic field and adsorption process has been reported to enhance adsorption process (Mohammed and Chong, 2014). The result of their findings demonstrated that magnetization process could accelerate the removal of color, TSS and COD in adsorption process, the percentage reduction of color, TSS

Table 1-5 Adsorption of Palm Oil Mill Effluent Using Local Adsorbent Materials

Adsorbent	Removal Efficiency	Limitation	Authors
Rubber powder	88% removal of residual oil	Residual oil (0.6%−0.7%) is not considered critical parameter of POME	Ahmad et al. (2005)
Flake chitosan	99% removal of oil and grease	Residual oil not a critical parameter of POME	Ahmad et al. (2005)
Natural zeolite	66.638%, 58.575% and 61.51% of Fe, Zn and Mn respectively	Treated POME from aerobic pond (secondary treatment)	Shavandi et al. (2012)
Banana peel	Color, TSS, BOD, Tanin and Lignin (95.96, 100, 100, 97.41 and 76.74%)	Treatment was done before discharge after final stage of pond treatment	Mohammed and Chong (2014)
Activated Carbon	COD 10 mg/L and SS 2 mg/ L	After secondary stage before discharge	Othman et al. (2014)
Granular activated carbon	71.26% suspended solids	Adsorption was used for pretreatment before ultrafiltration membrane	Azmi and Yunos (2014)
Raw kapok fibers	BOD 74%−98%, Total organic carbon 72−94% and 66−80% total nitrogen	Modification of the adsorbent surface	Ahmad et al. (2005)
Montimorillonite	>95% of COD, TSS and color	Secondary effluent used for the treatment	Said et al. (2016)
Oil palm leaves (OPL) and oil palm fronds (OPF)	83.74% oil and grease removal using OPL and 39.84% of oil and grease using OPF	The modified OPL more hrdophobic surface than OPF	Jahi et al. (2015)
Activated carbon and ultrasound cavitation	73.08% COD removed 98.33% TSS removed	Treatment done after final stage before discharge	Parthasarathy et al. (2016)
Bioadsorbent	69% COD removal 96% removal of SS	Biologically treated palm oil mill effleunt final discharge	Ibrahim et al. (2017)
Sago park fiber	80.23% oil removal achieved	Modification of surface through esterification process	Wahi et al. (2017)
Commelina Nudiflora	>40% COD removal after 9 h incubation	POME was investigated as secondary effluent before discharge	Kuppusamy et al. (2017)

and COD was observed as 39%, 61%, and 46% respectively. Membrane separator technology was combined with adsorption treatment method for the treatment of POME (Azmi and Yunos, 2014), the adsorption process was used as pretreatment and was achieved by stirring the raw POME with 0.20 g/L of palm kernel shell-based activated carbon at a contact time of 35.94 minutes and agitation speed of 39.82 rpm. There was a reduction of 71.26% of suspended solid, further treatment was achieved using the ultra-filtration technique. Adsorbent materials can also be combined with suitable adsorbent binder to form composite material. The combining adsorbent materials must be based on the target pollutants. In cases whereby the adsorption of polar solutes is required, hydrophilic adsorbent materials can be suitable (Adeleke et al., 2017). Activated carbon for adsorption of solutes are hydrophobic in nature and have the propensity of removing nonpolar solutes in the adsorbate. Hydrophobic particles have the tendency to interact with particles of the same kind rather than with water such as oil droplet. The process results in the breaking down of hydrogen bonding between the water particles. Similarly, hydrophilic adsorbents can be effective for the reduction of polar solutes in the wastewater. The combination of hydrophilic particles have a thermodynamic interaction which is more favorable than the interaction with hydrophobes such as oil. The degree of hydrophobic or hydrophilic particles is determined by the surface tension of the materials in the aqueous phase and the combination of the adsorbent materials can be very effective for the reduction of high strength waste water containing both polar and nonpolar solutes (Adeleke et al., 2016). The effectiveness of peat activated carbon composite adsorbent was applied for the reduction of SS, Color and Fe from landfill leachate (Rosli et al., 2017). At optimum condition of pH 7, 2 hours contact time and 200 rpm shaking speed, the optimum ratio of peat; activated carbon was achieved as 2:2 for color and 2.5:1.5 for Fe. A removal percentage 74.4% and 73.6% was achieved respectively. A batch adsorption study was conducted using composite adsorbent derived from activated coconut shell, activated cow bone and zeolite for the removal of COD and NH_3-N from POME using response surface methodology for the optimization of operational parameters (Adeleke et al., 2017). A central composite design (CCD) design expert 6.01 consisting of six independent variables was used in order to get the optimal conditions for the removal of COD and NH_3-N from POME. Eighty-five experimental design consisting of nine center points and eleven extra points were conducted in order to cover the possible effect of the combination of the operational factors. Batch adsorption experiments were performed in a random and with triplicate in order to reduce error percentage and the effects for the observed responses. The regression coefficients for reducing COD and NH_3-N in the POME at the end of the adsorption process were evaluated from the response surface quadratic model of the CCD to predict the reduction of COD and NH_3-N from POME on the composite. The result in Table 1-6 revealed that the regression model for the reduction of COD and NH_3-N was significant at a confidence level of 95% ($P < .05$) with determination coefficients

Table 1-6   Analysis of the Variance (ANOVA) of the Response Surface Quadratic Model for the Reduction of COD and NH_3-N from POME by Natural Composite

Source	Degree of Freedom	Sum of Squares		Mean Square		F Value		P Value	
		COD[a]	NH_3-N[b]	COD	NH_3-N	COD	NH_3-N	COD	NH_3-N
Model	53	5915.81	6647.89	111.62	125.43	14.09	6.50	<0.0001	<0.0001
Residual error	31	245.507	598.497	7.9195	19.3063				
Lack-of-fit	22	139.951	513.794	6.3614	23.3543	0.54	2.48	0.8836	0.0799
Pure error	9	105.555	84.7024	11.728	9.41138				
Total	84	6161.321	7246.391						

[a]$R^2 = 96.02\%$; R^2 (adj) $= 89.20\%$.
[b]$R^2 = 91.74\%$; R^2 (adj) $= 77.62\%$.

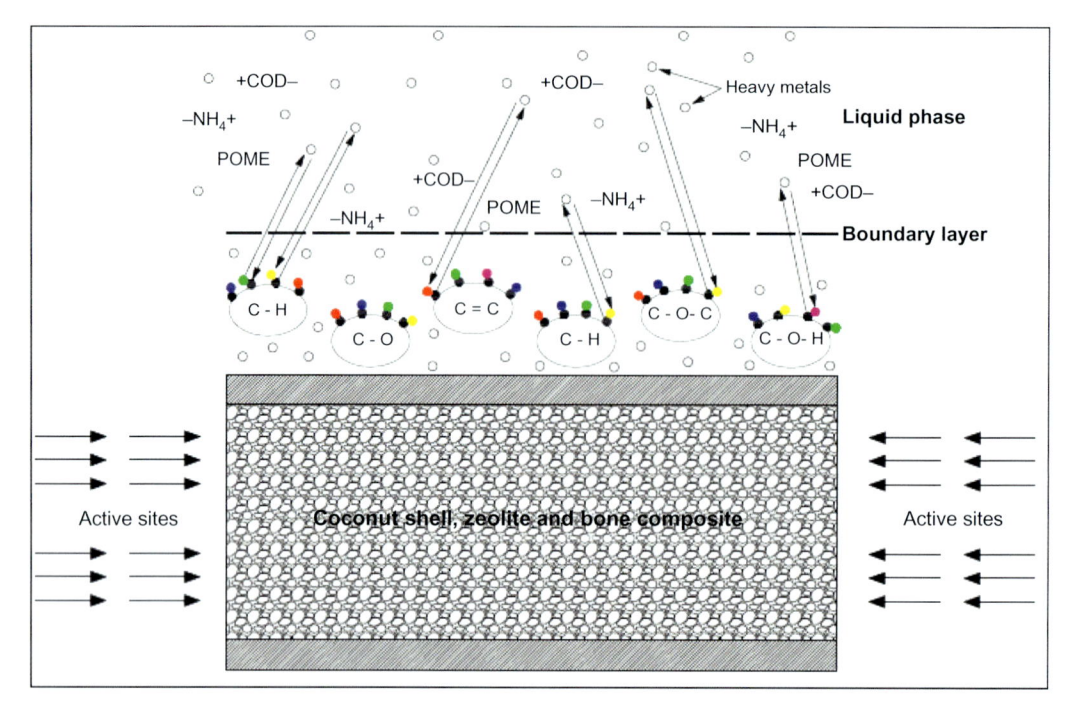

FIGURE 1-3 Ion exchange of pollutants and elements on the functional group on the active surface of the composite (Adeleke et al., 2017).

R^2 equal to 0.9602 and 0.9174 for COD and NH_3-N respectively, indicating the aptness of the model.

However, the application of composite adsorbent material from single adsorbent materials selected on the basis of target pollutants of wastewater has shown to have the potential and adsorption capacity than the individual adsorbent, although there could be the possibility of better removal capacity of pollutants using a single adsorbent material better than the use of composite depending on the nature of the adsorbent and the characteristics of the adsorbate. An illustration of the uptake of heavy metals from POME using composite adsorbent comprising of activated coconut shells activated cow bones and zeolite was illustrated in the works of Adeleke et al. (2017) is illustrated in Fig. 1-3.

The optimization of operational factors using regression model such as the central composite design can be more effective and more comprehensive optimization technique for the reduction of pollutants of wastewater than the conventional optimization method.

1.6 Conclusion

Open ponding systems, anaerobic reactors, membrane, and anaerobic systems are some of the conventional methods used for the treatment of palm oil mill effluent (POME). However, the major drawbacks of the conventional process is attributed to the cost of treatment, routine cost of maintenance and the effectiveness of the systems for the treatment of POME at low concentration. The future prospect of the effective treatment of high strength wastewater may not achieve the desired objective in terms of the reduction of pollutants of raw POME especially for critical parameters such as oil and grease and COD using conventional techniques. The application of locally-sourced adsorbent as activated carbon can be effective with respect to the target pollutants and has proven to be cost effective. Although, adsorption is effective as an alternative treatment method but the effectiveness of the treatment system is not wholly dependent on the nature of the adsorbent materials but also operational factors such as the pH, shaking speed, contact time, adsorbent dosage and initial concentration applicable to the adsorption system. The application low cost adsorbent is still a challenge due to the problem of the identification of proper adsorbent materials for the target pollutants. The combination of treatment methods such as membrane systems and adsorption has been applied for the reduction of the pollutants of POME. Additionally, the application of composite adsorbent has been applied and the optimization of the operational factors of POME using the composite adsorbent is still rare in the research phase for the treatment of POME.

References

Abdullah, M.A., Afzaal, M., Ismail, Z., Ahmad, A., Nazir, M.S., Bhat, A.H., 2015. Comparative study on structural modification of Ceiba pentandra for oil sorption and palm oil mill effluent treatment. Desalin. Water Treat. 54 (11), 3044−3053.

Acero, J.L., Benitez, F.J., Real, F.J., Teva, F., 2016. Micropollutants removal from retentates generated in ultrafiltration and nanofiltration treatments of municipal secondary effluents by means of coagulation, oxidation, and adsorption processes. Chem. Eng. J. 289, 48−58.

Adeleke, A.O., Al-Gheethi, A.A., Daud, Z., 2017. Optimization of operating parameters of novel composite adsorbent for organic pollutants removal from POME using response surface methodology. Chemosphere 174, 232−242.

Adeleke, A.R.O., Latiff, A., Aziz, A., Daud, Z., Ridzuan, B., Daud, M., et al., 2016. Remediation of raw wastewater of palm oil mill using activated cow bone powder through batch adsorption, Key Eng. Mater., 705. , pp. 380−384. Trans Tech Publications.

Adeleke, A.R.O., Latiff, A., Aziz, A., Daud, Z., Daud, M., Falilah, N., et al., 2017. Heavy metal removal from wastewater of palm oil mill using developed activated carbon from coconut shell and cow bones, Key Eng. Mater., 737. , pp. 428−432. Trans Tech Publications.

Ahmad, A.L., Chan, C., Abd Shukor, S., Mashitah, M., Sunarti, A., 2009. Isolation of carotenes from palm oil mill effluent and its use as a source of carotenes. Desalin. Water Treat. 7 (1-3), 251−256.

Ahmad, 2016. Renewable and sustainable bioenergy production from microalgal co-cultivation with palm oil mill effluent (POME): a review. Renew. Sustain. Energy Rev. 214−234.

Ahmad, A., Bhatia, S., Ibrahim, N., Sumathi, S., 2005. Adsorption of residual oil from palm oil mill effluent using rubber powder. Braz. J. Chem. Eng. 22 (3), 371−379.

Ali, A.A.M., 2012. Low cost adsorbents for the removal of organic pollutants from wastewater. J. Environ. Manage. 113, 170−183.

Andaç, M., Denizli, A., 2014. Affinity-recognition-based polymeric cryogels for protein depletion studies. RSC Adv. 4 (59), 31130−31141.

Aris, A., Ooi, B.S., Kon, S.K., Ujang, Z., 2008. Tertiary treatment of palm oil mill effluent using fenton oxidation. Malaysian J. Civ. Eng. 20 (1), 12−25.

Aziz, A.S.A., Manaf, L.A., Man, H.C., Kumar, N.S., 2014. Column dynamic studies and breakthrough curve analysis for Cd (II) and Cu (II) ions adsorption onto palm oil boiler mill fly ash (POFA). Environ. Sci. Pollut. Res. 21 (13), 7996−8005.

Azmi, N.S., Yunos, K.F.M., 2014. Wastewater treatment of palm oil mill effluent (POME) by ultrafiltration membrane separation technique coupled with adsorption treatment as pre-treatment. Agric. Agric. Sci. Proc. 2, 257−264.

Babu, B.R., Meera, K.S., Venkatesan, P., Sunandha, D., 2010. Removal of fatty acids from palm oil effluent by combined electro-fenton and biological oxidation process. Water, Air, Soil Pollut. 211 (1-4), 203−210.

Barhoumi, N., Oturan, N., Olvera-Vargas, H., Brillas, E., Gadri, A., Ammar, S., et al., 2016. Pyrite as a sustainable catalyst in electro-Fenton process for improving oxidation of sulfamethazine. Kinetics, mechanism and toxicity assessment. Water Res. 94, 52−61.

Bashir, M.J., Han, T.M., Wei, L.J., Aun, N.C., Amr, S.S.A., 2016. Polishing of treated palm oil mill effluent (POME) from ponding system by electrocoagulation process. Water Sci. Technol. 73 (11), 2704−2712.

Bashir, M.J., Wei, C.J., Aun, N.C., Amr, S.S.A., 2017. Electro persulphate oxidation for polishing of biologically treated palm oil mill effluent (POME). J. Enviro. Manage. 193, 458−469.

Bautista-Toledo, M.I., Rivera-Utrilla, J., Ocampo-Pérez, R., Carrasco-Marín, F., Sanchez-Polo, M., 2014. Cooperative adsorption of bisphenol-A and chromium (III) ions from water on activated carbons prepared from olive-mill waste. Carbon 73, 338−350.

Benhouria, A., Islam, M.A., Zaghouane-Boudiaf, H., Boutahala, M., Hameed, B., 2015. Calcium alginate−bentonite−activated carbon composite beads as highly effective adsorbent for methylene blue. Chem. Eng. J. 270, 621−630.

Bhatia, S., Othman, Z., Ahmad, A.L., 2007. Coagulation−flocculation process for POME treatment using Moringa oleifera seeds extract: optimization studies. Chem. Eng. J. 133 (1), 205−212.

Borja, R., Banks, C.J., Sánchez, E., 1996. Anaerobic treatment of palm oil mill effluent in a two-stage up-flow anaerobic sludge blanket (UASB) system. J. Biotechnol. 45 (2), 125−135.

Bormashenko, E.Y., 2013. Wetting of Real Surfaces (vol. 19). Walter de Gruyter.

Chan, Y.J., Chong, M.F., Law, C.L., 2012. An integrated anaerobic−aerobic bioreactor (IAAB) for the treatment of palm oil mill effluent (POME): start-up and steady state performance. Process Biochem. 47 (3), 485−495.

Cheng, C.K., Deraman, M.R., Ng, K.H., Khan, M.R., 2016. Preparation of titania doped argentum photocatalyst and its photoactivity towards palm oil mill effluent degradation. J. Cleaner Prod. 112, 1128−1135.

Cheng, Y.W., Chang, Y.S., Ng, K.H., Wu, T.Y., Cheng, C.K., 2017. Photocatalytic restoration of liquid effluent from oil palm agroindustry in Malaysia using tungsten oxides catalyst. J. Clean. Prod. 162, 205−219.

Chooi, C., 1984. Ponding system for palm oil mill effluent treatment [in Malaysia]. Paper presented at the Workshop on Review of Palm Oil Mill Effluent Technology vis-a-vis Department of Environment, Kuala Lumpur, 31 July, 1984.

Cooper, M., 2016. Conclusions: The Future of Sustainable Water Management Sustainable Water Management. Springer, pp. 175−185.

Davies, J.T., 2012. Interfacial Phenomena. Elsevier.

Drioli, E., Criscuoli, A., Curcio, E., 2011. Membrane Contactors: Fundamentals, Applications and Potentialities, Vol. 11. Elsevier.

Essington, M.E., 2015. Soil and Water Chemistry: An Integrative Approach. CRC press.

Faisal, M., Machdar, I., Gani, A., Daimon, H., 2016. The combination of air flotation and a membrane bioreactor for the treatment of palm oil mill effluent. Int. J. Technol 7 (5), 767−777.

Ferrara, N., Gerber, H.-P., Kowalski, J., Pisabarro, M.T., Sherman, D.E., 2013. Composition comprising and method of using angiopoietin-like protein 3 Angptl3: Google Patents.

Fiore, S., Ruffino, B., Campo, G., Roati, C., Zanetti, M., 2016. Scale-up evaluation of the anaerobic digestion of food-processing industrial wastes. Renew. Energy 96, 949−959.

Fulazzaky, M.A., Nuid, M., Aris, A., Muda, K., 2017. Kinetics and mass transfer studies on the biosorption of organic matter from palm oil mill effluent by aerobic granules before and after the addition of Serratia marcescens SA30 in a sequencing batch reactor. Process Saf. Environ. Protect. 107, 259−268.

Galimberti, C., Corti, I., Cressoni, M., Moretti, V.M., Menotta, S., Galli, U., et al., 2016. Evaluation of mercury, cadmium and lead levels in fish and fishery products imported by air in North Italy from extra-European Union Countries. Food Control 60, 329−337.

Gobi, K., Mashitah, M., Vadivelu, V., 2011. Adsorptive removal of methylene blue using novel adsorbent from palm oil mill effluent waste activated sludge: equilibrium, thermodynamics and kinetic studies. Chem. Eng. J. 171 (3), 1246−1252.

Grundke, K., Pöschel, K., Synytska, A., Frenzel, R., Drechsler, A., Nitschke, M., et al., 2015. Experimental studies of contact angle hysteresis phenomena on polymer surfaces—toward the understanding and control of wettability for different applications. Adv. Colloid and Interface Sci. 222, 350−376.

Guagliardi, I., Apollaro, C., Scarciglia, F., De Rosa, R., 2013. Influence of particle-size on geochemical distribution of stream sediments in the Lese river catchment, southern Italy. Biotechnol., Agronom., Soc. Environ. 17 (1), 43.

Haak, L., Roy, R., Pagilla, K., 2016. Toxicity and biogas production potential of refinery waste sludge for anaerobic digestion. Chemosphere 144, 1170−1176.

Habib, M., Yusoff, F., Phang, S., Ang, K., Mohamed, S., 1997. Nutritional values of chironomid larvae grown in palm oil mill effluent and algal culture. Aquaculture 158 (1-2), 95−105.

Hammed, 2003. Removal of residual oil from palm oil mill effuent using solvent extraction method. J. Teknol. 38, 33−42.

Hannig, C., Hannig, M., 2009. The oral cavity—a key system to understand substratum-dependent bioadhesion on solid surfaces in man. Clin. Oral Investig. 13 (2), 123−139.

Hazlan, 2006. Treatment of palm oil mill effluent (POME) using memebrane bioreactor. Faculty of Chemical and Natural Resources Engineering Technology University College of Engineering and Technology Malaysia. pp. 16−24.

Ho, K., Teow, Y., Ang, W., Mohammad, A., 2017. Novel GO/OMWCNTs mixed-matrix membrane with enhanced antifouling property for palm oil mill effluent treatment. Sep. Purif. Technol. 177, 337−349.

Hojjat, 2009. Optimization of POME anaerobic pond. Eur. J. Scient. 32 (4), 455−459.

Hossain, M.A., Rahman, G.K.M.M., Rahman, M.M., Molla, A.H., Rahman, M.M., Uddin, M.K., 2015. Impact of industrial effluent on growth and yield of rice (*Oryza sativa* L.) in silty clay loam soil. J. Environ. Sci. 30, 231−240.

Ibrahim, I., Hassan, M.A., Abd-Aziz, S., Shirai, Y., Andou, Y., Othman, M.R., Zakaria, M.R., 2017. Reduction of residual pollutants from biologically treated palm oil mill effluent final discharge by steam activated bioadsorbent from oil palm biomass. J. Clean. Prod. 141, 122−127.

Igwe, J.C., Arukwe, U., Anioke, S.N., 2013. Isotherm and kinetic studies of residual oil adsorption from palm oil mill effluent (Pome) using boiler fly ash. J. Environ. Eng. Manage. (EEMJ) 12 (3).

Jahi, N., Ling, E.S., Othaman, R., Ramli, S., 2015. Modification of oil palm plantation wastes as oil adsorbent for palm oil mill effluent (POME). Malaysian. J. Anal. Sci. 19 (1), 31−40.

James, R., Sampath, K., Alagurathinam, S., 1996. Effects of lead on respiratory enzyme activity, glycogen and blood sugar levels of the teleost Oreochromis mossambicus(Peters) during accumulation and depuration. Asian fisheries science. Metro Manila 9 (2), 87−100.

Kamarudin, K.F., Tao, D.G., Yaakob, Z., Takriff, M.S., Rahaman, M.S.A., Salihon, J., 2015. A review on wastewater treatment and microalgal by-product production with a prospect of palm oil mill effluent (POME) utilization for algae. Der. Pharma Chem. 7 (7), 73−89.

Khemkhao, M., Techkarnjanaruk, S., Phalakornkule, C., 2015. Simultaneous treatment of raw palm oil mill effluent and biodegradation of palm fiber in a high-rate CSTR. Bioresour. Technol. 177, 17−27.

Kuppusamy, P., Ilavenil, S., Srigopalram, S., Maniam, G.P., Yusoff, M.M., Govindan, N., et al., 2017. Treating of palm oil mill effluent using Commelina nudiflora mediated copper nanoparticles as a novel bio-control agent. J. Cleaner Prod. 141, 1023−1029.

Lalley, J., Han, C., Li, X., Dionysiou, D.D., Nadagouda, M.N., 2016. Phosphate adsorption using modified iron oxide-based sorbents in lake water: kinetics, equilibrium, and column tests. Chem. Eng. J. 284, 1386−1396.

Lau, L., Nor, N., Mohamed, A., Lee, K., 2013. Adsorption of hydrogen sulfide using palm shell activated carbon: an optimization study statistical analysis. Int. J. Res. Eng. Technol. 2, 302−311.

Lewandowski, Z., Beyenal, H., 2013. Fundamentals of Biofilm Research. CRC press.

Liew, W.L., Kassim, M.A., Muda, K., Loh, S.K., Affam, A.C., 2015. Conventional methods and emerging wastewater polishing technologies for palm oil mill effluent treatment: a review. J. Environ. Manage. 149, 222−235.

Malakahmad, A., Lahin, F.A., Yee, W., 2014. Biodegradation of high-strength palm oil mill effluent (POME) through anaerobes partitioning in an integrated baffled reactor inoculated with anaerobic pond sludge. Water Air Soil Pollut. 225 (3), 1.

Medellin-Castillo, N.A., Leyva-Ramos, R., Padilla-Ortega, E., Perez, R.O., Flores-Cano, J.V., Berber-Mendoza, M.S., 2014. Adsorption capacity of bone char for removing fluoride from water solution. Role of hydroxyapatite content, adsorption mechanism and competing anions. J. Ind. Eng. Chem. 20 (6), 4014−4021.

Mishra, V.K., Shukla, R., 2016. Aquatic Macrophytes for the Removal of Heavy Metals from Coal Mining Effluent Phytoremediation. Springer, pp. 143−156.

Mitik-Dineva, N., Wang, J., Truong, V.K., Stoddart, P., Malherbe, F., Crawford, R.J., et al., 2009. *Escherichia coli*, *Pseudomonas aeruginosa*, and *Staphylococcus aureus* attachment patterns on glass surfaces with nanoscale roughness. Curr. Microbiol. 58 (3), 268−273.

Mohammed, R.R., Chong, M.F., 2014. Treatment and decolorization of biologically treated palm oil mill effluent (POME) using banana peel as novel biosorbent. J. Environ. Manage. 132, 237−249.

Muhrizal, S., Shamsuddin, J., Fauziah, I., Husni, M., 2006. Changes in iron-poor acid sulfate soil upon submergence. Geoderma 131 (1), 110−122.

Mukherjee, I., Sovacool, B.K., 2014. Palm oil-based biofuels and sustainability in southeast Asia: a review of Indonesia, Malaysia, and Thailand. Renew. Sustain. Energy Rev. 37, 1−12.

Muttalib, N.A.A., 2012. Adsorption of ammonium using waste activated sludge from palm oil mill effluent treatment plant. Int. J. of Global Environ. Issues 256−268.

Nasrullah, M., Singh, L., Mohamad, Z., Norsita, S., Krishnan, S., Wahida, N., Zularisam, A., 2017. Treatment of palm oil mill effluent by electrocoagulation with presence of hydrogen peroxide as oxidizing agent and polialuminum chloride as coagulant-aid. Water Resour. Ind. 17, 7−10.

Neoh, C.H., Yung, P.Y., Noor, Z.Z., Razak, M.H., Aris, A., Din, M.F.M., et al., 2017. Correlation between microbial community structure and performances of membrane bioreactor for treatment of palm oil mill effluent. Chem. Eng. J. 308, 656−663.

Ng, K.H., Cheng, C.K., 2017. Photocatalytic degradation of palm oil mill effluent over ultraviolet-responsive titania: successive assessments of significance factors and process optimization. J. Cleaner Prod. 142, 2073−2083.

Nguyen, 2016. Comparison of solid substrates to differentiate the lubrication property of dairy fluids by tribological measurement. J. Food Eng. 185, 1−8.

Nor, M.H.M., Mubarak, M.F.M., Elmi, H.S.A., Ibrahim, N., Wahab, M.F.A., Ibrahim, Z., 2015. Bioelectricity generation in microbial fuel cell using natural microflora and isolated pure culture bacteria from anaerobic palm oil mill effluent sludge. Bioresour. Technol. 190, 458−465.

Nour, A.H., Nour, A.H., 2017. Production of Biogas and Performance Evaluation of Ultrasonic Membrane Anaerobic System (UMAS) for Palm Oil Mill Effluent Treatment (POME) Biological Wastewater Treatment and Resource Recovery: InTech. Chapter 13.

Ohimain, E.I., Izah, S.C., 2017. A review of biogas production from palm oil mill effluents using different configurations of bioreactors. Renew. Sustain. Energy Rev. 70, 242–253.

Othman, M.R., Hassan, M.A., Shirai, Y., Baharuddin, A.S., Ali, A.A.M., Idris, J., 2014. Treatment of effluents from palm oil mill process to achieve river water quality for reuse as recycled water in a zero emission system. J. Clean. Prod. 67, 58–61.

Park, S.Y., Chung, J.W., Kwak, S.-Y., 2015. Regenerable anti-fouling active PTFE membrane with thermo-reversible "peel-and-stick" hydrophilic layer. J. Membr. Sci. 491, 1–9.

Parthasarathy, S., Mohammed, R.R., Fong, C.M., Gomes, R.L., Manickam, S., 2016. A novel hybrid approach of activated carbon and ultrasound cavitation for the intensification of palm oil mill effluent (POME) polishing. J. Cleaner Prod. 112, 1218–1226.

Patel, S., Wei, S., Han, J., Gao, W., 2015. Transmission electron microscopy analysis of hydroxyapatite nanocrystals from cattle bones. Mater. Character. 109, 73–78.

Poh, P.E., Chong, M.F., 2009. Development of anaerobic digestion methods for palm oil mill effluent (POME) treatment. Bioresour. Technol. 100 (1), 1–9.

Poh, P.E., Chong, M.F., 2014. Upflow anaerobic sludge blanket-hollow centered packed bed (UASB-HCPB) reactor for thermophilic palm oil mill effluent (POME) treatment. Biomass Bioenergy 67, 231–242.

Poh, P.E., Tan, D.T., Chan, E.-S., Tey, B.T., 2015. Current Advances of Biogas Production via Anaerobic Digestion of Industrial Wastewater Advances in Bioprocess Technology. Springer, pp. 149–163.

Pradhan, S., Boernick, H., Kumar, P., Mehrotra, I., 2016. Removal of dissolved organic carbon by aquifer material: correlations between column parameters, sorption isotherms and octanol-water partition coefficient. J. Environ. Manage. 177, 36–44.

Rosli, M.A., Daud, Z., Awang, H., Zainorabidin, A., Halim, A.A., 2017. The effectiveness of peat-ac composite adsorbent in removing ss, colour and Fe from landfill leachate. Int. J. Integr. Eng. 9 (3).

Rossi, M., Gianazza, M., Alamprese, C., Stanga, F., 2003. The role of bleaching clays and synthetic silica in palm oil physical refining. Food Chem. 82 (2), 291–296.

Rouquerol, J., Rouquerol, F., Llewellyn, P., Maurin, G., Sing, K.S., 2013. Adsorption by Powders and Porous Solids: Principles, Methodology and Applications. Academic press.

Rupani, P.F., Singh, R.P., Ibrahim, M.H., Esa, N., 2010. Review of current palm oil mill effluent (POME) treatment methods: vermicomposting as a sustainable practice. World Appl. Sci. J. 11 (1), 70–81.

Saeed, M.O., Azizli, K., Isa, M.H., Bashir, M.J., 2015. Application of CCD in RSM to obtain optimize treatment of POME using Fenton oxidation process. J. Water Process Eng. 8, e7–e16.

Said, M., Abu Hasan, H., Mohd Nor, M.T., Mohammad, A.W., 2016. Removal of COD, TSS and colour from palm oil mill effluent (POME) using montmorillonite. Desalin. Water Treat. 57 (23), 10490–10497.

Saswattecha, K., Cuevas Romero, M., Hein, L., Jawjit, W., Kroeze, C., 2015. Non-CO_2 greenhouse gas emissions from palm oil production in Thailand. J. Integr. Environ. Sci. 12 (sup1), 67–85.

Shak, K.P.Y., Wu, T.Y., 2015. Optimized use of alum together with unmodified Cassia obtusifolia seed gum as a coagulant aid in treatment of palm oil mill effluent under natural pH of wastewater. Ind. Crops Products 76, 1169–1178.

Shavandi, M., Haddadian, Z., Ismail, M., Abdullah, N., Abidin, Z., 2012. Removal of Fe (III), Mn (II) and Zn (II) from palm oil mill effluent (POME) by natural zeolite. J. Taiwan Inst. Chem. Eng. 43 (5), 750–759.

Shirtcliffe, N.J., McHale, G., Atherton, S., Newton, M.I., 2010. An introduction to superhydrophobicity. Adv. Colloid Interface Sci. 161 (1), 124–138.

Stawiński, W., Węgrzyn, A., Dańko, T., Freitas, O., Figueiredo, S., Chmielarz, L., 2017. Acid-base treated vermiculite as high performance adsorbent: insights into the mechanism of cationic dyes adsorption, regeneration, recyclability and stability studies. Chemosphere 173, 107−115.

Subramaniam, M., Goh, P., Lau, W., Tan, Y., Ng, B., Ismail, A., 2017. Hydrophilic hollow fiber PVDF ultrafiltration membrane incorporated with titanate nanotubes for decolourization of aerobically-treated palm oil mill effluent. Chem. Eng. J. 316, 101−110.

Suwanno, S., Rakkan, T., Yunu, T., Paichid, N., Kimtun, P., Prasertsan, P., et al., 2017. The production of biodiesel using residual oil from palm oil mill effluent and crude lipase from oil palm fruit as an alternative substrate and catalyst. Fuel 195, 82−87.

Taha, M.R., Ibrahim, A., 2014a. Characterization of nano zero-valent iron (nZVI) and its application in sono-Fenton process to remove COD in palm oil mill effluent. J. Environ. Chem. Eng. 2 (1), 1−8.

Taha, M.R., Ibrahim, A., 2014b. COD removal from anaerobically treated palm oil mill effluent (AT-POME) via aerated heterogeneous Fenton process: optimization study. J. Water Process Eng. 1, 8−16.

Tee, P.-F., Abdullah, M.O., Tan, I.A.W., Amin, M.A.M., Nolasco-Hipolito, C., Bujang, K., 2016. Performance evaluation of a hybrid system for efficient palm oil mill effluent treatment via an air-cathode, tubular upflow microbial fuel cell coupled with a granular activated carbon adsorption. Bioresour. Technol. 216, 478−485.

Thangalazhy-Gopakumar, S., Al-Nadheri, W.M.A., Jegarajan, D., Sahu, J.N., Mubarak, N.M., Nizamuddin, S., 2015. Utilization of palm oil sludge through pyrolysis for bio-oil and bio-char production. Bioresour. Technol. 178, 65−69.

Tran, T., Da, G., Moreno-Santander, M.A., Vélez-Hernández, G.A., Giraldo-Toro, A., Piyachomkwan, K., et al., 2015. A comparison of energy use, water use and carbon footprint of cassava starch production in Thailand, Vietnam and Colombia. Resour., Conserv. Recycl. 100, 31−40.

Wahi, R., Abdullah, L.C., Mobarekeh, M.N., Ngaini, Z., Yaw, T.C.S., 2017. Utilization of esterified sago bark fibre waste for removal of oil from palm oil mill effluent. J. Environ. Chem. Eng. 5 (1), 170−177.

Wang, D., McLaughlin, E., Pfeffer, R., Lin, Y., 2012. Adsorption of oils from pure liquid and oil−water emulsion on hydrophobic silica aerogels. Sep. Purif. Technol. 99, 28−35.

Wang, J., Mahmood, Q., Qiu, J.-P., Li, Y.-S., Chang, Y.-S., Li, X.-D., 2015. Anaerobic treatment of palm oil mill effluent in pilot-scale anaerobic EGSB reactor. BioMed Res. Int. 2015.

Wang, H., Yuan, X., Wu, Y., Huang, H., Zeng, G., Liu, Y., et al., 2013. Adsorption characteristics and behaviors of graphene oxide for Zn (II) removal from aqueous solution. Appl. Surf. Sci. 279, 432−440.

Wei, H., Deng, S., Hu, B., Chen, Z., Wang, B., Huang, J., et al., 2012. Granular bamboo-derived activated carbon for high CO_2 adsorption: the dominant role of narrow micropores. ChemSusChem 5 (12), 2354−2360.

Wu, T., Mohammad, A.W., Jahim, J.M., Anuar, N., 2007. Palm oil mill effluent (POME) treatment and bioresources recovery using ultrafiltration membrane: effect of pressure on membrane fouling. Biochem. Eng. J. 35 (3), 309−317.

Xia, X., Li, H., Yang, Z., Zhang, X., Wang, H., 2015. How does predation affect the bioaccumulation of hydrophobic organic compounds in aquatic organisms? Environ. Sci. Technol. 49 (8), 4911−4920.

Yacob, S., Hassan, M.A., Shirai, Y., Wakisaka, M., Subash, S., 2006. Baseline study of methane emission from anaerobic ponds of palm oil mill effluent treatment. Sci. Tot. Environ. 366 (1), 187−196.

Zahrim, A., Nasimah, A., Hilal, N., 2014. Pollutants analysis during conventional palm oil mill effluent (POME) ponding system and decolourisation of anaerobically treated POME via calcium lactate-polyacrylamide. J. Water Process Eng. 4, 159−165.

Zinatizadeh, A.A., Ibrahim, S., Aghamohammadi, N., Mohamed, A.R., Zangeneh, H., Mohammadi, P., 2017. Polyacrylamide-induced coagulation process removing suspended solids from palm oil mill effluent. Sep. Sci. Technol. 52 (3), 520−527.

Further Reading

AbdulRahman, A., Latiff, A.A.A., Daud, Z., Ridzuan, M.B., Jagaba, A.H., 2016. Preparation and Characterization of Activated Cow Bone Powder for the Adsorption of Cadmium from Palm Oil Mill Effluent. In IOP Conference Series: Materials Science and Engineering (vol. 136, No. 1, p. 012045). IOP Publishing.

Ahmad, M., Lee, S.S., Yang, J.E., Ro, H.M., Lee, Y.H., Ok, Y.S., 2012. Effects of soil dilution and amendments (mussel shell, cow bone, and biochar) on Pb availability and phytotoxicity in military shooting range soil. Ecotoxicol. Environ. Saf. 79, 225–231.

Ahmad, Chan, C., Abd Shukor, S., Mashitah, M., Sunarti, A., 2009. Isolation of carotenes from palm oil mill effluent and its use as a source of carotenes. Desalin. Water Treat. 7 (1-3), 251–256.

Cheng, Chang, Y.S., Ng, K.H., Wu, T.Y., Cheng, C.K., 2017. Photocatalytic restoration of liquid effluent from oil palm agroindustry in Malaysia using tungsten oxides catalyst. J. Cleaner Prod. 162, 205–219.

Latiff, A., Aziz, A., Adeleke AbdulRahman, O., Daud, Z., Ridzuan, M.B., Daud, M., et al., 2015. Batch adsorption of manganese from palm oil mill effluent onto activated cow bone powder. ARPN J. Eng. Appl. Sci. 11 (4).

Oyekanmi, A.A., Daud, Z., Daud, N.M., Gani, P., 2017. Adsorption of heavy metal from palm oil mill effluent on the mixed media used for the preparation of composite adsorbent. MATEC Web Conf. 103, p. 06020). EDPSciences.

Zahrim, A.Y., Dexter, Z.D., 2016. Decolourisation of palm oil mill biogas plant wastewater using Poly-Diallyldimethyl Ammonium Chloride (polyDADMAC) and other chemical coagulants. In IOP Conference Series: Earth and Environmental Science (vol. 36, No. 1, p. 012025). IOP Publishing.

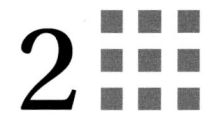

2

Locally Derived Activated Carbon From Domestic, Agricultural and Industrial Wastes for the Treatment of Palm Oil Mill Effluent

O. Abdulrahman Adeleke[1], Ab Aziz Abdul Latiff[2],
Mohammed Radin Saphira[2], Zawawi Daud[2], Norli Ismail[1],
Amimul Ahsan[3,4], N. Adila Ab Aziz[2], Mohammed Ndah[2],
Vicky Kumar[2], Adel Al-Gheethi[2], M. Arif Rosli[2], Mahmood Hijab[2]

[1]DIVISION OF ENVIRONMENTAL TECHNOLOGY, SCHOOL OF INDUSTRIAL TECHNOLOGY, UNIVERSITY SAINS MALAYSIA, PULAU PINANG, MALAYSIA [2]FACULTY OF CIVIL AND ENVIRONMENTAL ENGINEERING, UNIVERSITI TUN HUSSEIN ONN, MALAYSIA, UTHM, PARIT RAJA, BATU PAHAT, JOHOR, MALAYSIA [3]DEPARTMENT OF CIVIL AND CONSTRUCTION ENGINEERING, SWINBURNE UNIVERSITY OF TECHNOLOGY, MELBOURNE, VIC, AUSTRALIA [4]DEPARTMENT OF CIVIL ENGINEERING, UTTARA UNIVERSITY, DHAKA, BANGLADESH

2.1 Introduction

Activated carbon is the most effective adsorbent known for the treatment of domestic, municipal, agricultural and industrial wastewater. The use of activated carbon as adsorbent material is widely applicable due to readily available large surface for the attachment of solutes from wastewater. Others types of adsorbent materials are silica gel which is usually hard and granular made by precipitating solution of sodium silicate with acid. Activated alumina is made by activating aluminum oxide at very high temperature. The activated alumina is applicable for moisture adsorption. Aluminum silicates are the product of synthetic zeolite. They are mostly used in the separation process. Activated carbon is widely applied for adsorption because of its small particle sizes and good adsorption capacity (Pérez-González et al., 2012). It has been applied to adsorb heavy metals ions (Gabaldon et al., 2000), it was applied for the removal of phenol (Carrott et al., 2005; Mourao et al., 2006). It has been widely applied for the removal of dyes (Azad et al., 2016) for the removal of pesticides from water (Salman, 2014) and also used for the removal of poly chlorinated biphenyls PCBs (Antonetti et al., 2016). Activated carbon is available in two forms, either as powdered activated carbon (PAC) or granular activated carbon (GAC). Powdered activated carbon (PAC) comprises of organic materials having high content of carbon such as wood, lignite, and coal. PAC is widely applicable for wastewater treatment for the removal of taste, odor, and organic pollutants from water. PAC is added in the treatment process and removed subsequently by sedimentation or by backwashing using filter beds. The dosage of PAC is in the range of $1-100$ mg/L depending on the concentration of the pollutants. Dosage of 1 to 20 mg/L are usually applied for taste and odor control (O' Donnel et al., 2016). According to the International Union of Pure and Applied Chemistry (IUPAC), the pore size distribution determines the total volume of active sites available for the adsorption and desorption of solutes from the adsorbate. The pore of adsorbent can be classified as micropore if $d < 2$ nm, mesoporous adsorbent surface is in the range of $2-50$ nm while macropore is when $d > 50$ nm (Hesas et al., 2013). Granular activated carbon (GAC) is made from organic materials having high content of carbon. GAC is different from PAC because of the particle size diameter between 1.2 and 1.6 mm. The uniformity coefficient of GAC is usually large due to reduce desorption and the ability to aid stratification after backwashing (O' Donnel et al., 2016).

The contact volume of GAC is affected by the breakthrough time, empty bed contact time (EBCT) and design flow rate. The break through time is the time taken for the concentration of the contaminant in the effluent of the GAC unit to exceed the treatment requirement. As a rule of thumb, if the concentration of the effluent is more than the performance benchmark for three consecutive days, the GAC should be replaced or regenerated. The EBCT is usually calculated as ratio of the bed volume and the flow rate through the carbon. The EBCT can be achieved on the longer basis by increasing the bed volume or through the reduction of the flow rate of filter, lower EBCT makes the adsorption process to be slow because of less effectiveness of the diffusion process of the effluent solution (Song et al., 2016). The spent GAC can leach contaminants by percolation into soils in groundwater; therefore their regeneration is favoured over disposal. Despite its effectiveness and wide range applicability, its use for research is limited considering the rate at which water is polluted. The use is limited because of its high cost of production and problem of regeneration (Ahmaruzzaman, 2011). Materials of similar composition which are cheap and readily available are rather used for research for the treatment of wastewater.

2.2 Application of Locally Sourced Adsorbents Materials

The presence of heavy metals in water is a major health and environmental concern to wastewater managers and the ecosystem. The increasing level of contamination and deterioration of water quality is a major problem that needs attention by water users. This is because of the increasing need of water for domestic and industrial activities. The major cause of heavy metal pollution of water is the problem of discharge of effluents from industrial and agricultural sources without adequate treatment before discharge (Zaghden et al., 2016). Several types of treatments methods for the removal of heavy metals from water range from chemical treatments to low cost bio-sorption. The chemical treatment method for wastewater could not achieve the expected result in the wastewater treatment guidelines (Ismail et al., 2014). The application of conventional treatment methods need higher treatment cost of operation, the treatment method tends to have incomplete removal or residual of water contaminants causing further treatment problems (Chiban et al., 2012). The study of the removal of wastewater pollutants using inexpensive adsorbents is becoming very popular for the treatment of industrial wastewater. The low cost adsorbents used for adsorption purposes were waste products from industry, agriculture, and domestic wastes which are easily accessible and abundant in supply. Adsorbents are classified based on uses for gaseous phase application such as air purification and solvent extraction. The second classification of adsorbents is for liquid phase treatment. (Yang and Jiang, 2014).

The common adsorbents for the removal of water pollutants are agricultural wastes such as eggshells, agave bagasse, Maize cob, coconut shell (Foo and Hameed, 2010; Paranavithana et al., 2016; Velazquez-Jimenez et al., 2013). There are emerging adsorbents from mineral composition such as the treatment of heavy metals and organic pollutants using the activated cow bone powder (Latiff et al., 2015, Adeleke et al., 2016) The composition of most natural adsorbents for the treatment of wastewater contains calcium carbonate ($CaCO_3$), which can be very effective to adsorb heavy metals (Ismail et al., 2014). The use of agricultural waste products most especially materials containing lignocellose are very effective for the removal of metals in aqueous solution (Elhafez et al., 2016). Agricultural wastes such as rice husk, palm fruit shell, fruit seed, coconut shells and fruit peels can be used for adsorption purposes (Malik et al., 2016; Raval et al., 2016). These agricultural wastes are referred to as low-cost on the basis of their abundance in supply and the ease of getting the materials from the industry, agricultural areas and food processing production (Afroze et al., 2016). The application of composite adsorbent nowadays is recognized as modern and advanced technology in the adsorbent preparation techniques. The composite media can improve adsorption capacities significantly because of the physical and chemical characteristics of combining materials. They can also enhance the physical and chemical properties such as surface area, pore arrangement, functional group, and mechanical stability (Wang et al., 2014). The surface of composite adsorbent is proven to be effective for the removal of organic and inorganic compounds (Abbas et al., 2016; Orge et al., 2017). There are reported works on the use of composite adsorbent in wastewater treatment. Some of the researches include the use of sodium alginate/sodium-rectorate (SA/Na + REC) nano-composite for the adsorption of dyes and carbohydrate polymer. Hosseinzadeh and Khoshnood (2016) studied the removal of Congo red dye from aqueous solution (Sayğılı, 2015), the use of hydroxyapatite composite for the reduction of Co (II) and Eu (III) (Sheha et al., 2016), magnetic composite was applied for the removal of heavy metals (Reddy et al., 2013) while the use of zeolite-activated carbon as composite adsorbent was studied for the adsorption of acid orange 7 (Lim et al., 2013). Similarly, the application of bentonite bio-char composite for the removal of ammonium from fish tank was investigated by Ismadji et al. (2016), the use of carbon-mineral composite for the treatment of landfill leachate was studied by Halim et al. (2010a).

Wastewater treatment using adsorption is valuable for the treatment of varieties of wastewater with less economic implication, no requirement for land availability. It has no maintenance cost and produces quality effluent of treated wastewater. Adsorption is applicable for different categories of wastewater ranging from low, medium and high strength wastewater. Some of the commonly used adsorbent includes coconut shell adsorbents, zeolite and mineral enriched adsorbent materials.

2.2.1 Coconut Shell Adsorbent

Coconut shell is widely used for the production of GAC of high quality. It is widely used as an adsorbent because it is inexpensive and mostly available in developing countries such as Malaysia and Indonesia. Coconut shell activated carbon is usually prepared from coconut shell by heating at 700 °C, crushed coconut shells is modified with $ZnCl_2$. The resultant char is activated in steam or air at 900 °C. The carbon obtained has surface area of 800 m^2/g having very high adsorption capacities. The kinetics and equilibrium adsorption study of ammonium ion activated carbon prepared from coconut shell was studied by Boopathy et al. (2013), it was observed that the activated coconut shell carbon (ACSC) showed the highest removal efficiency at pH 9, temperature of 283k and contact time of 120 minutes. The optimum adsorption was obtained on a multilayer surface and intraparticle diffusion. The combination of heated coconut shell with acid can be effective for the adsorption purposes. Coconut shell treated with acid and coated with chitosan has better binding surface area for Zn (II) (Amuda et al., 2007). The adsorption efficiency of the coated coconut shell was well fitted in Langmuir and Freundlich isotherm (Amuda et al., 2007). Activated carbon coconut coir was used to adsorb Cr (VI) in the aqueous solution (Chaudhuri et al., 2010). The influence of contact time, adsorbent dosage and pH on the coconut coir activated carbon affected the removal of Cr (VI), Cu (II), and Cd (II). The application of 8 g/L of activated coconut coir carbon was used to adsorb 20 mg/L of concentration of Cr (VI) at 100% adsorption efficiency. The adsorption capacity of coconut coir activated carbon was proven to be better adsorbent than commercial activated carbon for the adsorption of Cu (II) and Cd (II), which fitted very well to Langmuir isotherm (Chaudhuri et al., 2010). The application of coconut husk in a study observed that the pH of solutions has great influence on the adsorption process (Abdulrasaq and Basiru, 2010). In their study, they observed that 50 mg/L removal of Fe (III), 50 mg/L Cu (II), and 10 mg/L Pb (II) were adsorbed at pH 5. The adsorption capacities were close to 90%. Also, coconut husk adsorbent for the removal of Fe (III) and Cu (II) were fitted to the Frendlich isotherm with $R^2 = 1$. Their conclusion was that Fe (III) and Cu (II) removal was due to their chemical bonding. Additionally, Pb (II) had better adsorption in Langmuir isotherm. The metals were also well-fitted to the second order kinetic due to chemisorption (Abdulrasaq and Basiru, 2010).

2.2.2 Bone Activated Carbon Adsorbent

The bone charcoal has the composition of carbon and minerals in the form of calcium carbonate ($CaCO_3$), it has the composition of hydrated calcium phosphate, hydroxyapatite (HAP) resulting in the formation of insoluble calcium phosphate when used as interface between the adsorbent and the adsorbate (AbdulRahman et al., 2016). This usually occurs at some certain temperature and is dependent on the nature of the adsorbate (Moreno-Piraján et al., 2010). However, the characteristics of the adsorbent have been explored in areas of wastewater treatment. Previous literature showed that activated carbon from cow bones is widely known to be used for the removal of color from sugar (Moreno-Piraján et al., 2010). The removal of organic solutes using cow bone carbon is still very rare in research phase in the treatment of POME. Cow bones were activated in the study of Adeleke et al. (2016) for the reduction of manganese and for the reduction of organic pollutants (Adeleke et al., 2016) in palm oil mill effluent (POME). The surface of activated cow bone powder is hydrophobic in nature and can be applied for the adsorption of organic solutes in high strength wastewater such as POME (Adeleke et al., 2017a,b). It has been reported that oxides of aluminum, calcium and phosphorus provide hydrophobic behavior to the composite material (Ito et al., 2013). These elements are the major components of cow bones. Calcium and phosphorus when added to anodized surfaces usually have contact angle in the range of 90 degrees, this indicates hydrophobicity (Yoshihara et al., 2010). When calcium oxide dissolves in water, it forms calcium carbonate which has tendency of forming an alkaline precipitate in the adsorbate (Adeleke et al., 2017a,b). Adsorbent enriched in CaO, raises the pH of the solution to at least 10, creating in the process a medium whereby excess calcium ions can react with phosphate to produce insoluble precipitate known as hydroxyapatite (Boanini et al., 2010). The characteristics of activated cow bone media for effective adsorption of heavy metals is due to presence of hydroxyapatite (Adeleke et al., 2016). Hydroxyapatite interface is recognized as good adsorbent for cationic exchange of various metals concentration in aqueous solution (Dong et al., 2010), which is known to have high degree of irreversible fixation of bone char through chemisorption by the precipitation of metal ions (Vareda et al., 2016). Cow bone is capable of removing heavy metals such as Cu (II), Mn (II), Fe (II), and Ni (II) and also organic compounds (Moreno-Piraján et al., 2010), and it was applied in the removal of lead ions (Cechinel and de Souza, 2014). A study using bone charcoal for the removal of Cr (III) in wastewater was conducted, the bone charcoal was obtained from bovine (Dahbi et al., 2002). The study on the bone charcoal showed that the removal of Hg (II) from the use of camel bone charcoal was due to the ion exchange of the mineral composition (calcium and phosphate components) (Hassan et al., 2008). The influence of pH on the metal removal affects the adsorption properties of the activated carbon. (Demirbas et al., 2004). The application of bone adsorbent as component of a composite material has the potential of showing the characteristics of activated carbon in the composite (Adeleke et al., 2017a,b). Materials are combined as composite adsorbent based on their effectiveness for the remediation of the target pollutants.

2.3 Composite Adsorbent

Several researches have been documented on the development of single adsorbents for range of pollutant removal in wastewater or in aqueous solution. There is increasing interest in regards to developing an adsorbent comprising of different single adsorbents with varying properties to form a composite adsorbent. Some of these compounds include zeolite, fly ash, bentonite, activated carbon, chitosan, montmorrillonite. The significance of the use of composite adsorbent is to enhance adsorption of wastewater. The materials used in developing the composite are identified based on their hydrophobic and hydrophilic properties (Kamarudin et al., 2015). The combination of different materials as composite has shown to possess better sorption of wastewater (Szychowski and Pacewska, 2012). The foremost research effort on carbon-mineral adsorbent was reported by (Leboda, 1992). In his findings, he concluded that mechanical mixing and carbonization of organic substances either physically or chemically bonded with the surface of mineral adsorbent. Hence, catalytic activities and adsorption of water pollutants were achieved. After his documented research effort, researchers began to explore different materials for developing composite adsorbent. Some of the researches conducted using composite adsorbent for wastewater treatment include the research effort of Velmurugan et al. (2015) on the use of magnesium and coconut shell activated carbon as composite for the removal of heavy metal ions in aqueous solution. The conclusion of their research was that the magnesium composite had a greater adsorption of Pb (II) > Cd (II) > Co (II) > Zn (II) and without the magnesium composite the removal was Zn (II) > Cu (II) > Co (II) > Pb (II) > Cd (II). Also, Ai and Jiang (2010), studied the effect of AC/ferrospinel composite for the removal of dyes. They observed adsorption of methyl orange (MO) and Basil fuchsin (BF) were well fitted to the Langmuir model, the adsorption capacities were recorded as 95.8 and 101.0 mg/g for MO and BF respectively. Sandeman. (2011) investigated and made comparative evaluation of the effectiveness of activated carbon, PVA hydrogels and PVA/AC composite for the removal of cationic methylene blue (MB), anionic methyl orange and Congo red (CR) from aqueous solution. Their conclusion was that the interactions with the ACs or PVA/AC available for adsorption as well as its structural and textural characteristics influenced adsorption capacity. Singh (2010) developed coconut shell carbon-iron oxide composite adsorbent, the optimum conditions of adsorption were achieved at concentration of 240 mg/L, temperature 50°C, pH 8.5 and adsorbent dosage of 1 g/L. The Langmuir model observed maximum adsorption efficiency of crystal violet (CV) at 81.70 mg/g. In the investigation of Adeleke et al., 2017a,b, composite adsorbent was developed from hydrophobic and hydrophilic adsorbent materials for the reduction of COD and NH_3-N from POME. Activated carbons are hydrophobic in nature in the aqueous solution and has the potential to adsorb non-polar solutes such as COD. Similarly, zeolite is a good example of hydrophilic adsorbent material and have the propensity to adsorb polar molecules such as NH_3-N. (Adeleke et al., 2016). However, the use of low cost adsorbents as a replacement for commercial activated carbon has brought new dimensions to the treatment of high strength wastewater such as POME. It has also provided greater opportunities in improving ways of water management. The removal efficiency of the pollutants on the adsorbent material is optimally reduced by combining the adequate ratio of the optimal conditions of adsorbent parameters. The effectiveness of a composite adsorbent for the reduction of solutes in wastewater is influenced by the optimal mixing ratio of the combining materials.

2.3.1 Optimization of Operating Conditions for the Adsorption Process

The optimization of operating parameters has been widely reported in the literature to determine the adsorptive capacity of the prepared adsorbent. The method of optimization of adsorption parameters can be achieved by combining the ratio of adsorbent of similar characteristics in terms of the removal efficiency of the target pollutants. In the study of the removal of high strength color from semiaerobic stabilized landfill leachate using limestone and activated carbon. The mixing ratio of 40 cm^3 of carbon-zeolite were used for the batch adsorption of leachate (Aziz et al., 2010). Similarly, adsorbent mixing ratio of 40:0, 35:5, 30:10, 25:15, 20:20, 15:25, 10:30, 5:35, 0:40 was adopted by volume of dosage of limestone and activated carbon in Aziz et al. (2011). The optimum removal efficiency of color (88%) was achieved using the mixture of limestone (35 cm^3) and AC (5 cm^3) (Aziz et al., 2011). The optimum ratio of 25:15 by volume of limestone- GAC was achieved at 58% removal efficiency of NH$_3$-N (Hussein et al., 2007). Similarly, the optimum mixing ratio for the removal of COD and NH$_3$-N from leachate was achieved at 150 rpm, pH 6 and 120 minutes contact time (Daud et al., 2017). Furthermore, the optimum mixing ratio of 150 rpm, pH 7 and 120 minutes was adopted for the adsorption of manganese from POME (Latiff et al., 2015). For effective adsorption of solutes, adsorbent materials are combined based on their behavior in the adsorbate. Halim et al. (2010a) classified adsorbents on the basis of their potentials to adsorb solutes in the polar and nonpolar phase of the adsorbate. They classified limestone and zeolite as hydrophilic materials and also activated carbon and rice husk carbon as hydrophobic materials. (Halim et al., 2010b). The description of the combining ratio can be illustrated in Table 2-1.

A parameter under fixed condition of other parameters can be used for optimization purposes. However, in real life application, investigating a parameter under fixed condition of other parameters may not comprehensively determine the nature of the adsorption process. The use of conventional approach in the study of adsorption, whereby a process is investigated by maintaining other parameters at a fixed condition may not be suitable to determine the combined effect of other parameters (Ravikumar et al., 2005). The operating parameters

Table 2-1   Combination of Adsorbent Material

Hydrophilic/Hydrophobic Ratio cm^3/Particle Size
0−40
5−35
10−30
15−25
20−20
25−15
30−10
35−5
40−0

to investigate the nature of adsorption can be suitably expressed by investigating the impact of multiple parameters simultaneously for adsorption purposes. The central composite design (CCD) of the response surface methodology (RSM) can achieve this purpose. RSM was used in the work of (Sahu et al., 2009) to establish the relationship of the adsorption of Cr (VI) using 25−30 designed experimental runs. The variables considered were represented in coded values as temperature (X_1), pH (X_2), initial concentration (X_3), and adsorbent dosage (X_4). Song et al. (2016) used 15−20 run of CCD experimental design to investigate the adsorption capacity of magnetic composite material under operating parameters of initial temperature, pH and initial concentration on the adsorption. In addition, Hu et al. (2013), investigated the effect of operating parameters of pH, initial concentration and temperature on the adsorption of Cu on graphene oxide composite using RSM. Debnath et al. (2015) analyzed the optimization of the adsorption of Congo red using lignocellulose composite adsorbent. The analysis of the effect of the operating parameters of pH and initial concentration combined was observed to be antagonistic. It was also observed that the effect of the combined operating parameters of initial concentration and temperature was synergistic while the effect of pH and temperature was reciprocal. In the study of Alkhatib et al. (2015), experiment of the optimization of the removal of color after the conventional treatment of POME was conducted using RSM. The operating variables used were the contact time, pH and the adsorbent dosage. The highest adsorption efficiency of 89.95% was obtained after the experimental output.

The application of RSM for multiple parameters makes the interpretation of the mechanism of adsorption, treatment process and application of models to be comprehensive. Hence, the use of RSM is a very valuable tool for optimization purposes.

2.3.2 Linear Optimization and the Preparation of a Composite Adsorbent

In the process of the preparation of the composite adsorbent, batch adsorption study under fixed conditions are usually conducted at different optimum ratio. In the study conducted by Adeleke et al. (2016), operating parameters at fixed condition of contact time of 105 minutes, 150 rpm shaking speed and pH 7 were used for the batch adsorption study to investigate the removal of pollutants of POME from the prepared adsorbent. Since the adsorbent materials were of different weight by dosage, to determine the right volume of the dosage by weight occupied in the POME, the dosage of each adsorbent used was evaluated using the bulk densities of the particle sizes of the starting materials (Adeleke et al., 2017a,b). This was obtained at predetermined volumes from density−mass relationship of the adsorbents.

The optimum condition of the hydrophobic-hydrophilic ratio was used to determine the optimum condition by percentage of the starting materials that will be mixed with the OPC. The optimum condition in this case is defined as the best mixed ratio that gives the best removal efficiency of the parameters at the lowest application of OPC, which served as the binder. All experiments to be optimized should be run in triplicate and the average recorded to reduce the unexpected variability of expected outcomes. The stages for the determination of the optimum condition used to prepare the composite are as follows: By the selection of the optimum particle size of the single adsorbents in a sequence batch study. By the determination of the optimum ratio of the hydrophobic media and the hydrophilic adsorbent under separate classification. The classification of adsorbent materials was based on the potential to adsorb nonpolar solutes in the adsorbate and also the ability to adsorb the polar solutes in the aqueous medium. This can be obtained using the contact angle experiment. The process of the preparation of composite adsorbent in the study of Adeleke et al., 2017a,b is presented in Figs. 2-1 and 2-2.

The volume of the adsorbent-binder ratio for the preparation of a composite adsorbent is defined by the adsorbent ratio that gives optimum reduction of the solutes in the adsorbate at minimum binder ratio. This is influenced by the porosity volume of the adsorbent is a very valuable parameter in fixed bed column adsorption for effective column design. An illustration of the fixed bed column. Other parameters that determines the nature of the

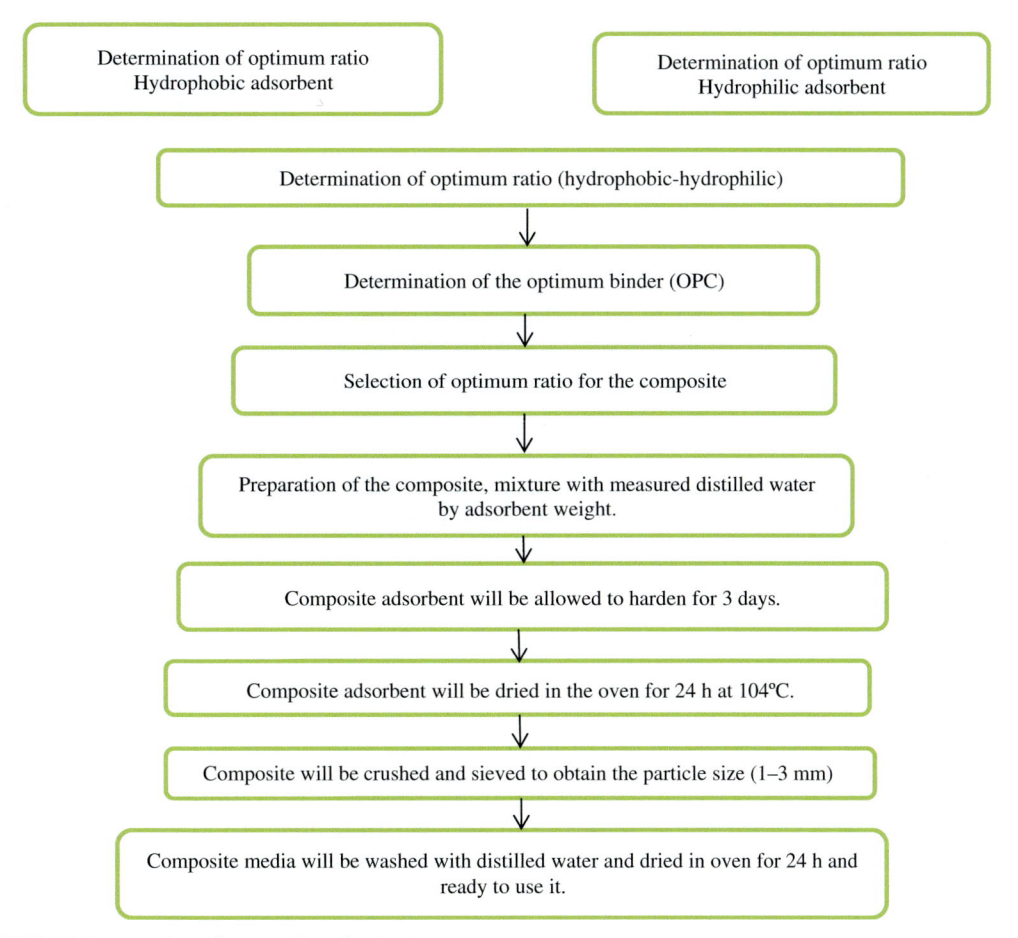

FIGURE 2-1 Preparation of composite adsorbent.

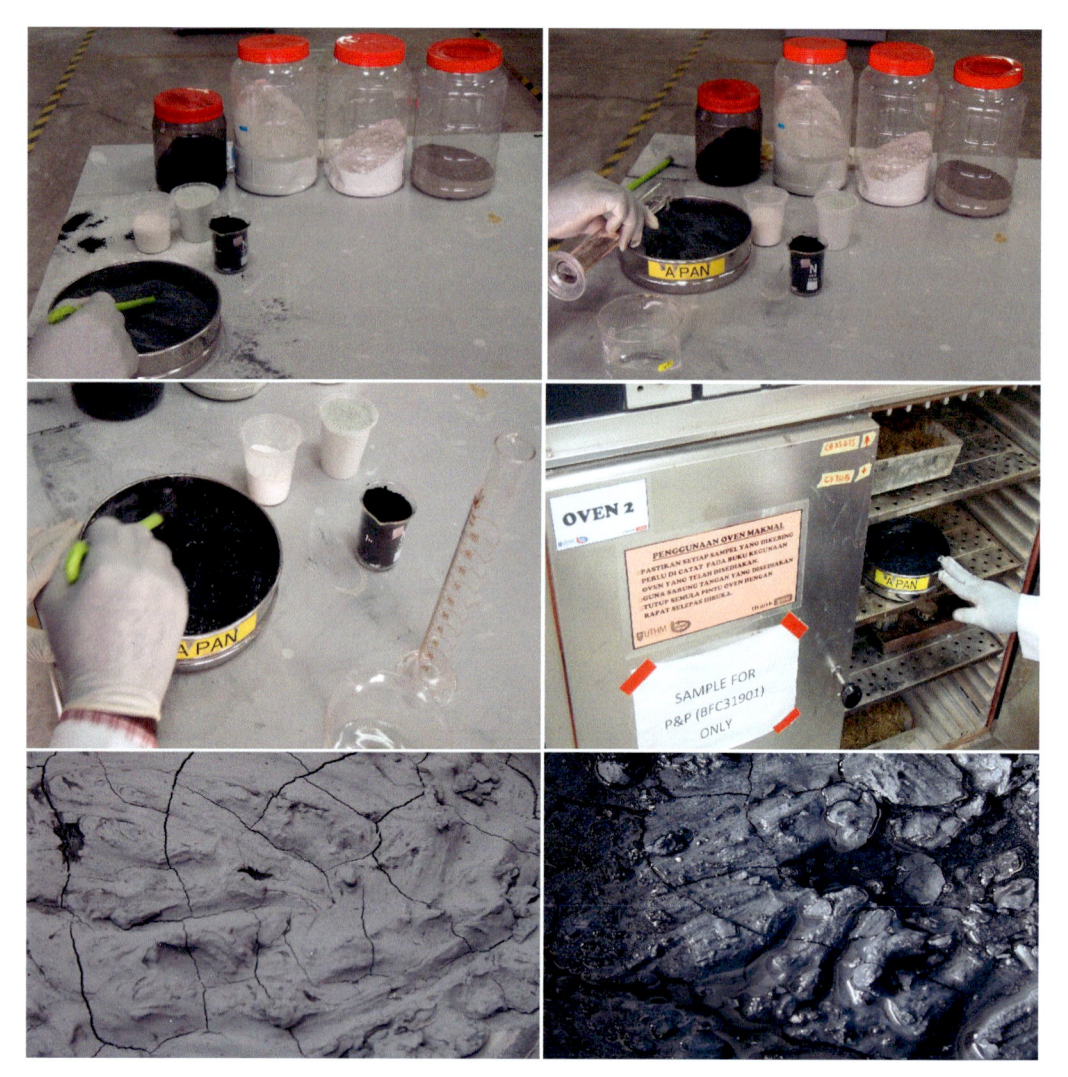

FIGURE 2-2 Preparation of composite adsorbent.

adsorption process includes the morphology of the adsorbent, the chemical composition of the adsorbent materials and also the point of which the adsorbent material is electrically neutral known as the point of zero charge of the adsorbent. The point of zero charge in relation to the pH of the adsorbate is used to determine whether the adsorbent surface is negative or positive. It is known that a negative surface has strong potential to engage in cationic adsorption of solutes in the adsorbate while a positive surface of adsorbent is known to be suitable for the adsorption of anionic adsorption in the adsorbate. The surface characteristics and pore size distribution of the adsorbent material is mostly determined by the Brunauer Emmet Teller (BET) analysis. The BET classifies adsorbent based on shape, pore, and sizes including both internal and external structures. The transport and diffusion mechanisms within the adsorbent pores and on the surface determines the nature of adsorption. This is determined by the nature of the concentration gradient. The chemical composition of an adsorbent is determined by the energy dispersive X-ray spectroscopy (EDX) or the application of the X-ray flourescence used for the characteristics of adsorbent material in the excited state to determine the elements and chemical compounds present in the adsorbent. The functional groups, cationic exchange capacity and the types of bonding of the solutes to the adsorbent are other chemical characteristics that defines the suitability of the adsorbent material for the attachment of solutes on the adsorbents. The description of some of the characteristics of adsorbent materials is described below.

2.3.3 Contact Angle Measurement

The contact angle measurements were conducted on the starting materials using Pendant drop penetration time test. The pendant drop test was used to investigate the interfacial tension on the starting materials when in contact with fluid. The curvature of the drop is known as Pendant drop penetration and is determined by the balance between the interface and the gravitational forces (Sarmadivaleh et al., 2015). This is expressed in Young−Laplace equation as:

$$\Delta PP = Y\left(\frac{1}{R1} + \frac{1}{R2}\right) \tag{2.1}$$

$$\Delta pk - \Delta PP = z\Delta pg \tag{2.2}$$

Where,

Y is defined as the interfacial tension
$R1$ and $R2$ are the radii of curvature of same point on the curved surfaces.
ΔPP is the pressure difference between the deionized water and the adsorbent at a random point p on the interface z.
Δpk is defined as the pressure difference between the deionized water and the adsorbent at height point of drop when $z = 0$.
Δp is the difference of the density between the deionized water and the adsorbent
g is the acceleration due to gravity.

In application, the deionized water is filled in to the cell through the needle at a constant flow rate using a high precision syringe pump. The deionized water forms a drop at the bottom of the needle and dropped on the surface of the adsorbent. Two additional drops are added to the surface of each starting material. The whole process is captured with a standby camera to measure the contact angle. The obtained contact angles of the materials are used to determine and classify the adsorbents on the basis of their hydrophobic or hydrophilic characteristics.

2.3.4 Bulk Density

Bulk density determines the weight of adsorbent by volume of the sample. Bulk density increases with decrease in porous surface of the adsorbent. Activated carbon is known to have a porous surface of around $3000 \, \text{m}^2/\text{g}$. Due to the large specific surface area and porous surface, activated carbon has lower bulk density. The measured dry weight is divided by the saturated volume to obtain the bulk density by the equation below.

$$\text{Bulk density} = \frac{\text{Dry weight of adsorbent (g)}}{\text{Volume of sample (mL).}} \tag{2.3}$$

2.3.5 Porosity Volume

Saturation method is mostly used to determine the porosity volume of the adsorbent. The porosity volume is much applicable for the determination of composite adsorbent in a continuous fixed column. In the case of fixed bed adsorption, water-saturated composite samples are introduced into the column until equilibrium concentration was achieved (Alonso et al., 2005). Saturation in a fixed column experiment is performed according to the packed column. An example of a fixed bed column adsorption is illustrated in Fig. 2-3.

In the study of Halim et al. (2010b), porosity volume experiment was conducted by filling the composite sample to maximum height of the column. The volume of the sample was

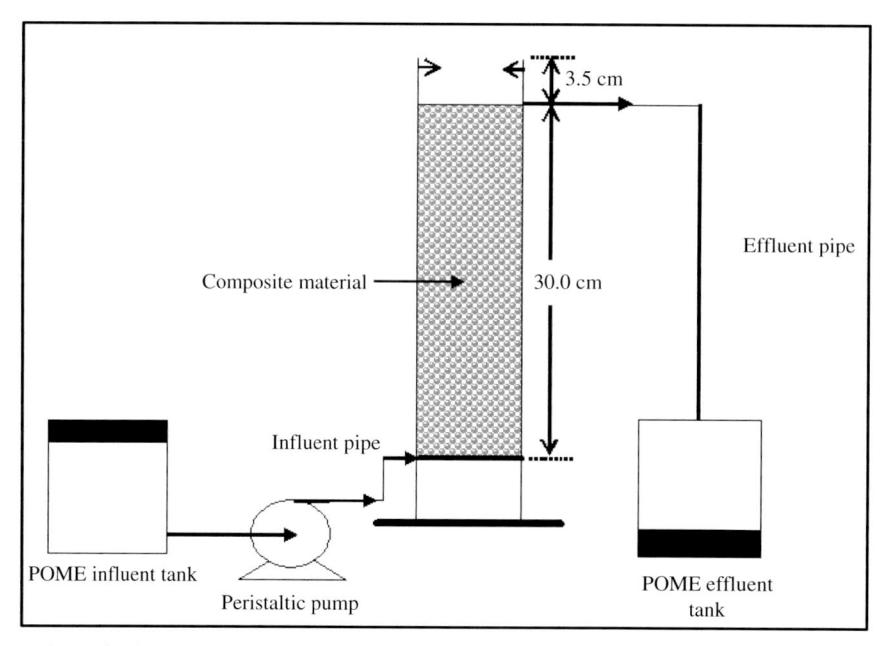

FIGURE 2-3 Column for fixed bed adsorption.

measured and recorded. A separate beaker was filled with deionized water containing more than the water needed to saturate the sample. The volume of the deionized water used at the saturation point was recorded. The volume of water used is the difference between the total volumes of water before saturation. This gives the pore volume of the sample. Therefore;

$$\text{Porosity } (Pt) = \frac{\text{Porevolume } (Vp)}{\text{Totalvolume } (Vt)} \tag{2.4}$$

Where total volume = solid volume (Vs) + pore volume (Vp)

2.3.6 Morphology Study

The surface morphology of the prepared adsorbent showing the textural and the pore size arrangement of the adsorbent are usually observed to determine the porosity and structure of adsorbent. The microstructure of the adsorbent is determined from the micrograph obtained from the scanning electron microscopy (SEM). In order to determine the rate of diffusion of the adsorbent and transport within the pores and on the surface of the adsorbents, the images of the adsorbent is obtained before and after the treatment of the batch adsorption. The images are obtained by scanning accelerated electron beam by the collection and recording of the secondary beam and backscatter electrons. To obtain a clear image, the micrograph is captured at high resolution and at a very clear and visible magnification. This is achieved by bombarding electrons on the surface of the sample particles.

The morphology of composite adsorbent for the treatment of POME in the study of Adeleke et al. (2017a,b) is illustrated in Fig. 2-4

The samples are usually coated with gold or platinum target to enhance the conductivity of the adsorbent materials before the bombardment.

2.3.7 Brunauer Emmet-Teller (BET) Surface Area

The BET surface analysis is conducted on the adsorbent to determine the surface area, pore volume and the pore size distribution of the adsorbent particle. The study is mostly determined by the N_2 adsorption/desorption using a micrometric ASAP 2000 system. The nitrogen adsorption-desorption on the sample is placed under vacuum at equilibrium time interval to remove moisture content from the solid surface. The analysis of the surface area, pore volume, and size of the composite is obtained from the plot of the relative pressure against the quantity adsorbed.

2.3.8 Chemical Characterization of Adsorbent

The chemical characterization focusses on the elemental composition and chemical compounds of the adsorbent materials and also the functional groups on the active sites of the adsorbent. Acidic surface have the potential for cationic exchange due to the presence of carboxylic compounds on the surface of the adsorbent. Elemental analysis of the prepared

(A)

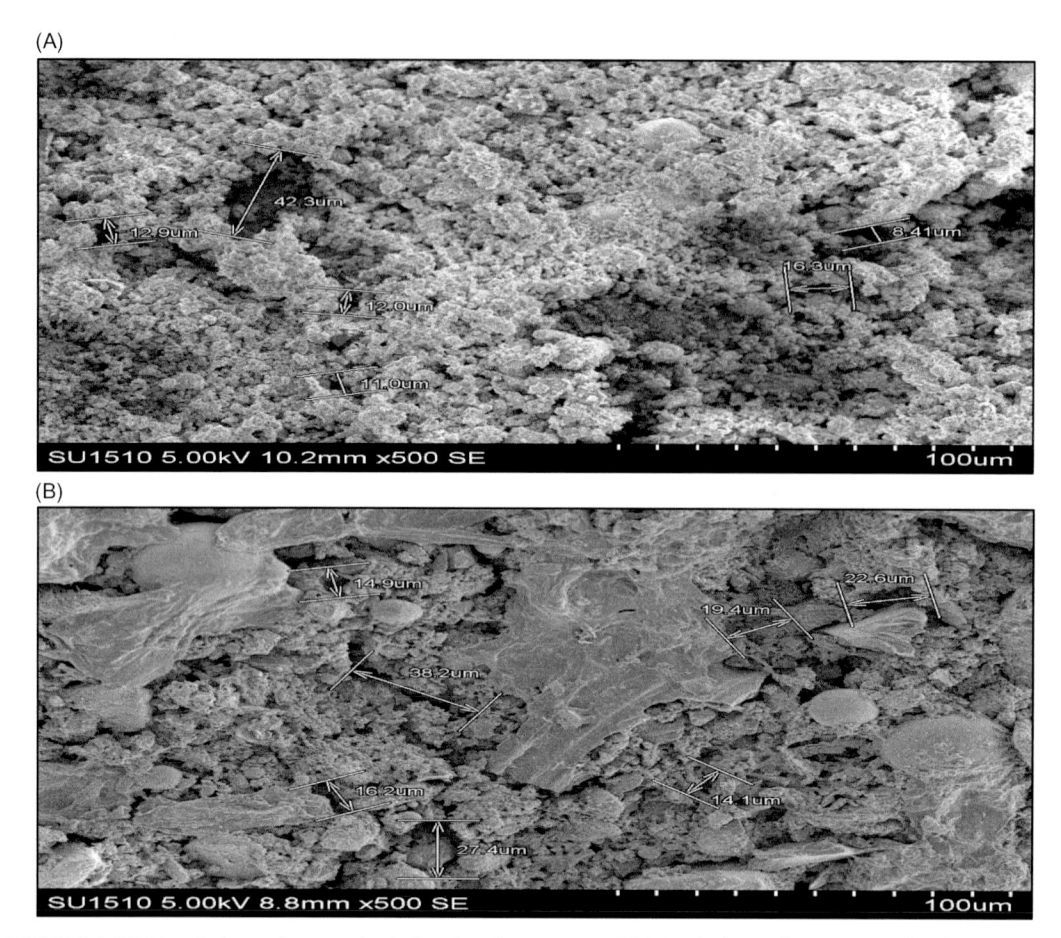

(B)

FIGURE 2-4 (A) Morphology of composite before batch treatment. (B) Morphology of composite after batch adsorption.

adsorbent materials can be achieved using the energy dispersive X-ray (EDX) incorporated in SEM (JOEL-JSM 6400). Prior to the investigation, adsorbent materials are coated to enhance the conductivity of the adsorbent material in order to achieve better resolution of the microstructure of the adsorbent and to avoid the accumulation of electrostatic charge. Some of the materials used for coating to enhance the conductivity of the sample includes gold, platinum, chromium, and tungsten. The coating material on the surface is usually conducted at low-vacuum sputter coating or by high vacuum evaporation. The images produced by SEM is as a result of the interaction between electron beams of the samples obtained at various depths. The images are transmitted in form of signals. Some of the most common types of signals produced by SEM includes the secondary electrons, back-scattered electrons and transmitted electrons. Nonconductive sample to be analyzed by the SEM obtain charges when scanned by the electron beam can result to improper scanning of the material. The elemental compositions of the adsorbents were achieved as percentage atomic contents in the adsorbents. This was used to determine the heavy metals before and after the batch adsorption study. The emission of secondary X-rays from an excited material involves the bombardment with high-energy X-rays. This is achieved by the radiation of the element at different electronic orbitals by the removal of the electron inside the orbitals by an energetic photon which is provided by a primary source of radiation. For the excitation of the atoms, a source of radiation with sufficient energy is needed to expel tightly held inner electrons. X-ray transmitters in the range of $20-60$ kV are used to obtain atomic information of the excited material.

2.3.9 Functional Group on Adsorbent

Fourier transform infrared irradiation (FT-IR) is an analytical procedure of obtaining the functional groups on the surface of an adsorbent material which is obtained at high spectra resolution. The effectiveness of the FT-IR spectroscopy is determined by the extent at which adsorbent material adsorbs light at a specified wavelength. FT-IR is used for the identification of the chemical bonds in a molecule by the production of infrared absorption spectrometer. FTIR is applicable to analyze the functional groups of the irradiated adsorbent materials. The powdered form of the adsorbent materials is dried in the oven at 105°C usually with silicon carbide. The dried sample is placed on the translucent crystal plate, the manual bench-press is adjusted and placed on the sample to ensure light compaction until the spectrometer detected the functional groups on the adsorbents. The scan of the adsorbent is done for within a few seconds to achieve better sensitivity. The spectra are mostly in the range of $400-4000$ cm^{-1} (Model Perkin Elmer FT-IR-2000, US).

2.3.10 Point of Zero Charge

The point of zero charge (pH_{pzc}) determines the point at which the interfacial region of a material is electrically neutral. When the pH of the system is lower than pH_{pzc}, the medium is said to be acidic because acidic water donates protons more than the hydroxide group. In the process, the surface becomes positively charged, which implies that there is high tendency for the attraction of ions. Conversely, when the pH of the system is higher than the pH_{pzc}, the surface tends to be negatively charged occasioned by the attraction of cations and repulsion of the anions. The pH of the Pzc is mostly obtained by the acid-base titration. Another common technique is described in the study of Cechinel and de Souza (2014). The experiment to determine the pH_{pzc} of activated cow bone adsorbent was carried out using the 11 point pH measurement as adopted by Cechinel and de Souza (2014). 0.025 g of adsorbent weight of material was placed in a 250 mL conical flask, 25 mL of deionized water was added to each of the flasks at different initial pH conditions. The pH was adjusted using 0.1 M NaOH and HNO$_3$. The samples were agitated at 120 rpm for 24 hours at 25°C. At the end of the contact time, the final pH of each of the samples was measured. A plot of the final pH against the initial pH was obtained from the obtained values. The point at which the plot of the final pH intersects with the plot of the initial pH value was recorded as the pH_{pzc}.

2.3.11 Analysis of Cation Exchange Capacity

The ammonium acetate saturation (AMAS) method is widely conducted to determine the CEC of the adsorbent media which is often used to explain the mechanism of adsorption. This was achieved by the saturation of the ammonium acetate (NH_4OAc) solution. About $5-10$ g of each of the adsorbents are put in a 250 mL flask. 100 mL of NH_4OAc is added to each of the conical flask and shaken thoroughly for 1 hour after which the sample is shaken at 150 rpm for 24 hours. The samples are filtered with 2.5 μm (Whatman no. 42) using the vacuum pump by applying light suction until a clear filtrate is obtained and discarded. The sample is washed with ethanol to remove excess saturating solution. The solution is allowed to completely filter through the flask with light suction after which the supernatant is discarded and cleaned from the flask. Each of the samples is washed with 25 mL of 1 M KCl. At this stage, the sample is discarded and the filtrate transferred to a 250 mL volumetric flask. A Nessler colorimetric test is conducted to test for NH_4-N in the filtrate at 420 nm. A blank sample is prepared according to the Nessler method to adjust the spectrophotometer in order to obtain the reading of the sample for the evaluation of the CEC.

The CEC is evaluated as:

$$\text{CEC (meq/g)} = \frac{0.1\,\text{L}}{14} \times \frac{1}{\text{weight of sample (g)}} \times \text{conc } NH_4^+ (mg/L) \tag{2.5}$$

2.4 Optimization of Process Parameters

The optimization of process parameters in adsorption process can be processed either by the conventional method by the optimization of operational parameters under fixed condition in terms of other parameters. The other method involves the optimization of multiple variables for the prediction of response parameter. One of the commonly used method is by applying response surface methodology (RSM). The two forms of optimization method was adopted in the study of (Latiff et al., 2015, Adeleke et al., 2017a,b). Linear optimization, which is conventionally used in literature, was adopted for the determination of the optimum condition of the particle size and ratio of adsorbent used for the preparation of the composite adsorbent in terms of the removal efficiencies of COD and NH_3-N (Latiff et al., 2015). A typical example of the combining ratio of the adsorbent parameters using linear optimization is illustrated in Table 2-1. The other form of optimization of the process parameters was achieved using the RSM, to investigate various variables and factors of adsorption and the efficiency of the prepared composite for the treatment of the pollutants of POME (Adeleke et al., 2017a,b). Two responses of COD and NH_3-N were also used for the optimization purpose. This was adopted because COD is the most critical parameter of raw POME (40,000-100,000) mg/L (Cheng et al., 2015). Also, the concentration of NH_3-N in water bodies may result to the eutrophication effect in the flora and fauna (Boyd, 1947). The linear optimization is conducted under fixed condition of pH 7, contact time 105 minutes, shaking speed 150 rpm. The factors used for the optimization is determined under fixed condition. Batch adsorption study or fixed bed column adsorption can be for the optimization of experimental factors.

2.5 Batch Adsorption Experiment

Batch adsorption experiments is conducted to obtain information about the uptake of parameters. Adsorption experiments is investigated using 250 mL Erlenmeyer flask with a stopper. 100 mL of the POME at different weight of adsorbent dosage are put in the flasks and placed in the isothermal water-bath shaker (Haake Wia Model, Japan) for the predetermined period of contact time. The water bath has a temperature controller adjustable from 25 to 100 ± 1 rpm. The adsorbent dosage used for the adsorption study can either be quantified in terms of weight or volume of the adsorbent. In the case of volume of adsorbent, it is evaluated from the bulk density of the adsorbent material. Prior to the investigation of the adsorption efficiency of the adsorbent for the uptake of the parameter, the solutions is filtered using 0.45 μm Whattman filter paper to obtain the residual concentration of the treated effluent. The analysis of COD from the residual concentration of the filtered sample is known as soluble COD. The analysis of each of the parameter of the wastewater characteristics is based on the standard methods. The amount of the adsorption of the adsorbate at equilibrium contact time qe (mg/g) is evaluated according to Eq. (2.6) and the removal efficiency is measured according to Eq. (2.7).

$$qe = \frac{(Co - Ce)v}{w} \tag{2.6}$$

$$\text{Adsorption efficiency} = \left(\frac{Co - Ce}{Co}\right) \times 100 \tag{2.7}$$

Where,

  Co is the initial liquid−phase concentration of adsorbate (mg/L)
  Ce is the final liquid−phase concentration of adsorbate (mg/L)
  v is the volume of solution (L)
  w is the weight of the adsorbent (g)

2.6 Optimization of the Process Parameters Using Central Composite Design

The optimization of process parameters is used to determine the best condition of the operational factors that can be suitable to achieve the effective treatment efficiency of the adsorbent materials. One of the most comprehensive method of optimizing the parameters of adsorption is by the use of RSM, which is widely applied in varieties of wastewater such as the domestic, agricultural, municipal and industrial effluents. RSM was applied to the study of Adeleke et al. (2017a,b) for the treatment of organic pollutants of POME. The effect of six factors (independent variables) which included pH (x_1), shaking speed (x_2), contact time (x_3), particle size (x_4), initial concentration (x_5) and adsorbent dosage (x_6) were investigated for the removal of COD (y_1) and NH$_3$-N (y_2) which were the response parameters. The central composite design (CCD, expert 6.0.10) method was used to determine the number of experimental analysis to be evaluated for the optimization of the variables and responses. This was conducted using the Design Expert software to analyze the evaluated data for the first order response surface equations of the model. The significance of the six variables on the adsorption process was analyzed using the analysis of variance (ANOVA, $P < .05$). The significance of the adjusted coefficient of determination (R^2 adj) was for the checking of the fitness of the linear model. The interactions between the factors and their effects on the reduction of both organic solutes could have a synergistic or antagonistic effect on the investigated responses. The results of the prediction of the interaction of the variable factors for the reduction of the responses can be presented and visualized using three-dimensional graphical representation of the system behavior, called RSM.

2.6.1 Experimental Design Using the Central Composite Design

The central composite design (CCD, expert 6.0.10) method is used to determine the number of experiments to be evaluated for the optimization of the variables and responses. The minimum, intermediate, and maximum values of each variables are labeled as -1, 0, and $+1$, respectively. This is illustrated in Table 2-2.

In the study of Adeleke et al. (2017a,b), optimization of process parameters of composite adsorbent was conducted using 85 experimental analyses considering 6 operational variables to investigate the optimum condition that gave better removal efficiency of the response factors, a central composite design (CCD) design expert 6.01 consisting of six independent variables was used in order to get the optimal conditions for the removal of COD and NH_3-N from POME by a composite adsorbent using response surface methodology. Eighty-five runs [nine (9) center points and eleven (11) extra points)] were conducted in order to cover the possible combinations of operational factors and their effect on the reduction of the investigated responses. Batch adsorption experiments were performed in a random and with triplicate in order to reduce the unexpected variability effects for the observed responses. The results revealed that the investigated factors significantly influenced the reduction of COD and NH_3-N and was better than the conventional method of optimization. The effective of reduction of the organic parameters of POME would lead to improve characteristics of the treated effluent before discharge.

Table 2-2 Experimental Design for the Optimization of the Response Factors

Factor	Symbol	Level		
		Low (-1)	Middle (0)	High ($+1$)
pH	x_1	4	7	10
Shaking speed (rpm)	x_2	50	175	300
Contact time (h)	x_3	2	13	24
Particle size of composite (mm)	x_4	1	2	3
Initial concentration of composite (dilution factor)	x_5	100	300	500
Adsorbent dosage (g/L)	x_6	65	95	125

Moreover, in the optimization process, the optimum operation for the reduction percentage of parameters are usually varied with each factor separately and as combined factors. The combination of factors could either be synergistic by influencing the reduction of the response parameter or antagonistic by reducing the efficiency of the treatment process. To determine the effectiveness of the treatment process, the efficiency of the adsorbent materials is evaluated using the reduction percentage from the relationship between the initial concentration and the residual concentration of the investigated response parameter. The effectiveness of the adsorption process is determined from the Eq. 2.7.

2.6.2 Development of Regression Model for the Response Parameters of the Adsorbent

The regression coefficients is the numerical index used to quantify the reduction percentage of the investigated parameter with the consideration of the interaction of the variable factors and the effect on the adsorption process. The variable factors for the prediction of the response parameter can be quantified from the linear, quadratic and cubic function of the CCD, the coefficient of the model in terms of the operating parameters is used to predict the regression equation of the prediction of the model. The F-value is used to determine the suitability of the model for the prediction of the response factors while the *P*-value ($P < .05$) shows the significance of the model at confidence level of 95%. The response is satisfactorily predicted when there is high coefficient of *F*-value in terms of the standard error. The ability to determine the aptness of the model is usually with respect to the standard error, a low value of standard error in relation to the *F*-value signifies that the model is sound and effective for the prediction of the response parameters. However, a low value of the standard error observed in Table 3-5 in relation to the *F*-value signified that the model was effective for the prediction of the reduction of the COD and NH_3-N in terms of the operating conditions (Adeleke et al., 2017a,b). A typical result of regression coefficient and their significance of the linear, quadratic and cubic model for the reduction of parameters of POME is illustrated in Table 2-3

The optimum operation for the reduction percentage was varied with each factor separately and also as a combined factor. The reduction percentage which is the dependent variables was calculated according to Eq. (3.4). The evaluated data was analyzed using the analysis of variance (ANOVA, $P < .05$). The independent factors which included pH values (4−10), shaking speed (50−300 rpm), contact time (2−24 h), particle size (1−3 mm), initial concentration of POME (100−500 dilution factor) and adsorbent dosage (65−125 g/L) were selected in order to examine the comprehensive factors which might have effects on the reduction of COD and NH_3-N. The calculation of the regression coefficients of the linear and quadratic, as well as the interaction between the factors in the model, was conducted to explain the variability of the experimental data. The effect of each independent factor was considered significant at *P*-value $<.05$ which is equivalent to 95% of the confidence level. The results found that five examined factors have a positive significant linear effect on the reduction of COD, while time has a negative significant effect which means that the increase of time affects or reduces the efficiency of the adsorption process. In contrast, three factors including pH and adsorbent dosage and initial concentration have positive effect on NH_3-N. pH and initial concentration was observed not significant on the reduction of NH_3-N except the adsorbent dosage which positive effect reduced the NH_3-N concentration in the POME. Also two factors of contact time and particle size of the adsorbent have a negative significant effect on the removal of NH_3-N. As the contact time increases, adsorption decreases depending on the behavior of the surface.

Also, the size of adsorbent is dependent on the rate of adsorption. The solute molecule is expected to readily penetrate pores of the smaller diameter than larger pore spaces. As a result, adsorption capacity increases with a decrease in particle diameter (Li et al., 2016). The quadratic effect of the pH is significant but the quadratic effect of particle size of the composite is not significant on the reduction of both COD and NH_3-N. Also, the quadratic

Table 2-3 Regression Coefficient and Their Significance of the Linear, Quadratic, and Cubic Model for the Reduction of COD and NH_3-N From POME by Using Central Composite Design

Term	Coefficient		Standard Error		F value		P-value	
	COD	NH_3-N	COD	NH_3-N	COD	NH_3-N	COD	NH_3-N
Model	**68.923**	**52.166**	**0.890**	**1.389**	**14.094**	**6.497**	**<0.0001**	**<0.0001**
X_1	1.682	0.636	0.414	0.647	16.469	0.966	0.000	0.333
X_2	1.013	-0.700	0.414	0.647	5.975	1.170	0.020	0.288
X_3	-2.399	-1.804	0.414	0.647	33.503	7.775	<0.0001	0.009
X_4	1.872	-1.794	0.414	0.647	20.414	7.691	<0.0001	0.009
X_5	1.850	0.357	0.511	0.797	13.129	0.200	0.001	0.658
X_6	1.948	2.246	0.414	0.647	22.088	12.050	<0.0001	0.002
X_1^2	0.889	1.667	0.272	0.425	10.645	15.349	0.003	0.001
X_2^2	-3.313	3.691	0.272	0.425	147.821	75.274	<0.0001	<0.0001
X_3^2	-2.349	-1.292	0.272	0.425	74.326	9.223	<0.0001	0.005
X_4^2	0.955	0.914	0.272	0.425	12.274	4.612	0.001	0.040
X_5^2	-0.541	5.418	0.738	1.152	0.537	22.118	0.469	<0.0001
$X_1 X_5$	1.725	0.893	0.352	0.549	24.038	2.645	<0.0001	0.114
$X_2 X_3$	2.902	4.897	0.352	0.549	68.051	79.496	<0.0001	<0.0001
$X_2 X_4$	-1.426	-0.059	0.352	0.549	16.439	0.012	0.000	0.915
$X_2 X_5$	-0.860	-1.355	0.352	0.549	5.981	6.088	0.020	0.019
$X_2 X_6$	-1.251	-1.391	0.352	0.549	12.652	6.418	0.001	0.017
$X_3 X_4$	-3.104	-0.289	0.352	0.549	77.849	0.276	<0.0001	0.603
$X_3 X_6$	-0.686	1.881	0.352	0.549	3.799	11.734	0.060	0.002
$X_4 X_5$	-0.954	0.417	0.352	0.549	7.356	0.577	0.011	0.453
X_5^3	0.096	-1.300	0.319	0.497	0.091	6.831	0.766	0.014
X_6^3	-0.055	-0.351	0.112	0.175	0.243	4.013	0.626	0.054
$X_1 X_2 X_4$	-0.258	-1.273	0.352	0.549	0.540	5.374	0.468	0.027
$X_1 X_2 X_6$	-1.022	0.128	0.352	0.549	8.444	0.054	0.007	0.817
$X_3 X_5 X_6$	-0.768	-0.086	0.352	0.549	4.760	0.024	0.037	0.877
$X_4 X_5 X_6$	0.829	-0.737	0.352	0.549	5.550	1.799	0.025	0.190

effect of speed and time is not significant for the reduction of COD, but the quadratic effect of the shaking speed and the initial concentration is significant to the reduction of NH_3-N. The synergistic effects of the examined factors investigated in the study was based on ANOVA analysis. In the case of COD reduction, the results of the analysis revealed that pH factor has a positive significant synergic effect with an initial concentration of composite ($P < .05$) but not significant for the reduction of NH_3-N. The shaking speed factor exhibited a positive significant synergic effect with time which improved the reduction of both organic pollutants investigated in the POME. Also a negative significant effect of the synergic of initial concentration and particle size of the composite and also the shaking speed and initial concentration have negative impact on the reduction of both pollutants. Meanwhile, the negative effect of the synergic between the shaking speed and initial concentration is not significant to the reduction of both COD and NH_3-N similar to the cubic effect, the effect of the shaking speed and the adsorbent particle size is not significant for the reduction of COD but influenced the removal of NH_3-N. In the case of the contact time and particle size, a negative effect for the reduction of COD concentration was observed but significant for NH_3-N. A case of contrast was observed for the contact time and adsorbent dosage where there was negative effect on COD but positive effect on NH_3-N to the adsorption process. The combined effect of pH, shaking speed and particle size was significant for the reduction of COD but significantly decreased the efficiency of NH_3-N. Both the negative effect of pH, speed and adsorbent dosage on COD and positive effect on NH_3-N is not suitable for the adsorption process. The negative effect of the contact time, initial concentration and the adsorbent dosage is not suitable for the reduction of COD but suitable for the reduction of NH_3-N concentration in the POME. The effect of the particle size, initial concentration and adsorbent dosage is significant for the reduction of COD but not significant for the reduction of NH_3-N. The prediction of the reduction of the investigated pollutants of POME can be represented through the numerical coefficient of the regression model.

The second-order model describes the significant relationship between the reduction of COD and NH_3-N and selected factors in terms of coded factors as given by Eqs. (2.8) and (2.29).

$$y_1 = +68.92 + 1.68x_1 + 1.01x_2 - 2.40x_3 + 1.87x_4 + 1.85x_5 + 1.95x_6 + 0.89x_1^2 - 3.31x_2^2 - 2.35x_3^2$$
$$+ 0.95x_4^2 + 1.72x_1x_5 + 2.90x_2x_3 - 1.43x_2x_4 - 0.86x_2x_5 - 1.25x_2x_6 - 3.10x_3x_4 \qquad (2.8)$$
$$- 0.69x_3x_6 - 0.95x_4x_5 - 1.02x_1x_2x_6 - 0.77x_3x_5x_6 + 0.83x_4x_5x_6$$

$$y_2 = +52.17 - 1.80x_3 - 1.79x_4 + 2.25x_6 + 1.67x_1^2 + 3.69x_2^2 - 1.29x_3^2 + 0.91x_4^2 + 5.42x_5^2 + 4.90x_2x_3$$
$$- 1.36x_2x_5 - 1.39x_2x_6 + 1.88x_3x_6 - 1.30x_5^3 - 0.35x_6^3 - 1.27x_1x_2x_4 \qquad (2.9)$$

Where y_1 and y_2 are the responses for the reduction percentage of COD and NH_3-N respectively. The summary of the analysis of variance (ANOVA) for the quadratic model represents the significance of the prediction of the model, It was noted that the regression model for the reduction of response parameters was significant at a confidence level of 95% ($P < .05$) with determination coefficients (R^2 adj) equal to 0.892 and 0.776 for COD and NH_3-N respectively, indicating the aptness of the model. The ANOVA of the response surface quadratic model

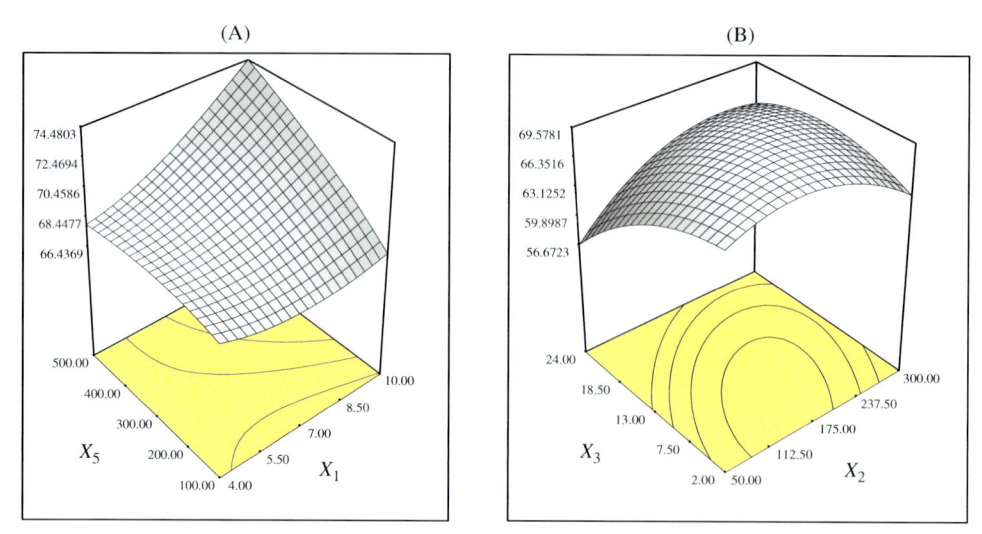

FIGURE 2-5 (A) Interaction between pH and initial concentration of composite. (B) Interaction between shaking speed and time.

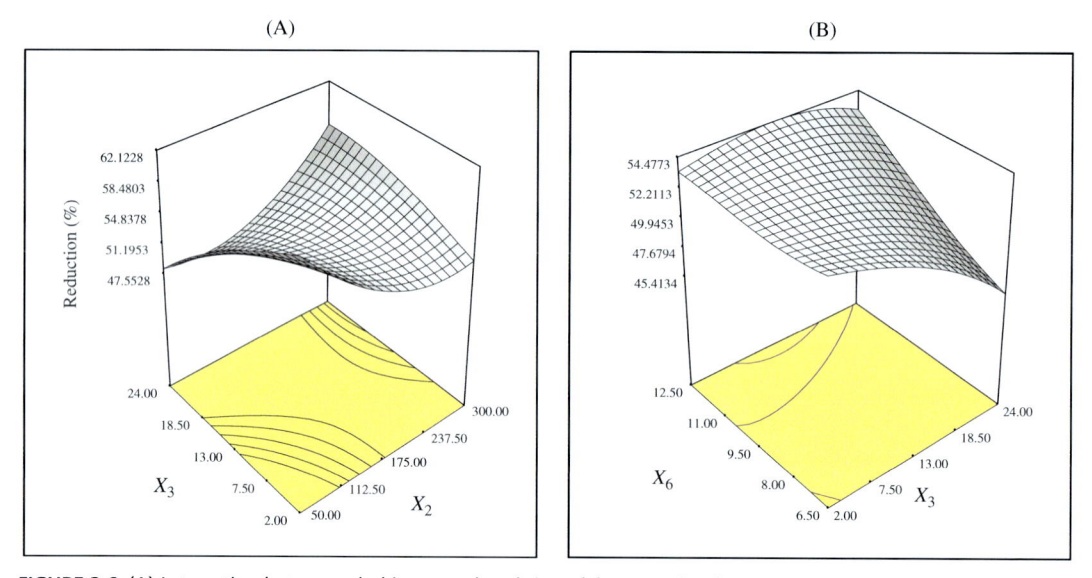

FIGURE 2-6 (A) Interaction between shaking speed and time. (B) Interaction between time and adsorbent dosage.

for the reduction of response parameters from POME on natural composite The strong significant ($P < .01$) interaction between shaking speed and time as well as between time and adsorbent dosage strongly enhanced the reduction of COD and NH_3-N from POME by 74.48% and 69.56% (Fig. 2-3A and B) and 62.12% and 54.48 % (Figs. 2-5A and B, 2-6).

From the result obtained from the optimization process, the use of a single parameter in the adsorption, in terms of other fixed parameters, may not be suitable to optimize operational conditions. The ability to reduce experimental trials that are required to express multiple trials and interactions makes RSM a better optimization technique. The experimental design tool such as RSM has been proven to be very effective with a statistical model in understanding the interaction among the optimized parameters. The use of RSM has been widely applied in the literature, for example in the study conducted on cationic dye by (Azzaz et al., 2016), shoe waste by (Iqbal et al., 2016) and peat (Rosli et al., 2017). The condition for the best operation of adsorption process was performed based on the results obtained from the screening of independent factors which revealed the possible direction for maximizing the adsorption process of the adsorbent material. The optimal operation parameters of the adsorption process for the reduction of the organic pollutants of POME was recorded at pH 10, 50 rpm of shaking speed for 2 h and by using 3 mm of composite particle size and 125 g/L of the adsorbent dosage of the treated sample of POME (Adeleke et al., 2017a,b). The use of activated carbon-peat composite have promising potential for the treatment of POME (Rosli et al., 2017). The observed and predicted reduction of COD and NH_3-N were 89.60% versus 85.01% and 75.61% versus 74.04%, respectively. Under these conditions, the independent factors exhibited strong interactions with 95% confidence level for both COD and NH_3-N reduction.

2.7 Conclusion

Conventional methods for the treatment of POME may not be effective for low concentration of the effluent. Similarly, the cost of the treatment process and routine maintenance of bioreactor is not economical including open ponding system which requires large land mass. However, adsorption is more applicable for the treatment of effluent at low concentration although the efficiency of the treatment process depends on effectiveness of the adsorbent material. The effectiveness of the adsorption process is influenced by the nature of the adsorbent materials and the characteristics of the adsorbents such as the pore size, functional group, surface morphology, bulk density including the hydrophobic and hydrophilic behavior. Low cost adsorbents are synthesized as activated carbon for the treatment of POME as a replacement for the commercial carbon. The combining ratio of adsorbent materials determines the reduction of POME characteristics, adsorbent parameters are more comprehensive and better optimized to achieve effective reduction of response factors using response surface methodology. The application of regression model from the central composite design can perform optimization of operational parameters of wastewater better than the conventional linear optimization techniques.

References

Abbas, A., Al-Amer, A.M., Laoui, T., Al-Marri, M.J., Nasser, M.S., Khraisheh, M., et al., 2016. Heavy metal removal from aqueous solution by advanced carbon nanotubes: critical review of adsorption applications. Sep. Purif. Technol. 157, 141−161.

AbdulRahman, A., Latiff, A.A.A., Daud, Z., Ridzuan, M.B., Jagaba, A.H., 2016. Preparation and Characterization of Activated Cow Bone Powder for the Adsorption of Cadmium from Palm Oil Mill Effluent. In IOP Conference Series: Materials Science and Engineering (vol. 136, No. 1, p. 012045). IOP Publishing.

Abdulrasaq, O.O., Basiru, O.G., 2010. Removal of copper (II), iron (III) and lead (II) ions from mono-component simulated waste effluent by adsorption on coconut husk. Afr. J. Environ. Sci. Technol. 4 (6).

Adeleke, A.O., Al-Gheethi, A.A., Daud, Z., 2017a. Optimization of operating parameters of novel composite adsorbent for organic pollutants removal from POME using response surface methodology. Chemosphere 174, 232−242.

Adeleke, A.R.O., Latiff, A., Aziz, A., Daud, Z., Ridzuan, B., Daud, M., et al., 2016. Remediation of raw wastewater of palm oil mill using activated cow bone powder through batch adsorption, Key Eng. Mater., Vol. 705. Trans Tech Publications, pp. 380−384.

Adeleke, A.R.O., Latiff, A., Aziz, A., Daud, Z., Daud, M., Falilah, N., et al., 2017b. Heavy metal removal from wastewater of palm oil mill using developed activated carbon from coconut shell and cow bones, Key Eng. Mater., Vol. 737. Trans Tech Publications, pp. 428−432.

Afroze, S., Sen, T.K., Ang, H.M., 2016. Adsorption removal of zinc (II) from aqueous phase by raw and base modified Eucalyptus sheathiana bark: Kinetics, mechanism and equilibrium study. Process Saf. Environ. Protect. 102, 336−352.

Ahmaruzzaman, M., 2011. Industrial wastes as low-cost potential adsorbents for the treatment of wastewater laden with heavy metals. Adv. Colloid Interface Sci. 166 (1), 36−59.

Ai, L., Jiang, J., 2010. Fast removal of organic dyes from aqueous solutions by AC/ferrospinel composite. Desalination 262 (1), 134−140.

Alkhatib, M., Mamun, A.A., Akbar, I., 2015. Application of response surface methodology (RSM) for optimization of color removal from POME by granular activated carbon. Int. J. Environ. Sci. Technol. 12 (4), 1295−1302.

Alonso, E., Alcoverro, J., Coste, F., Malinsky, L., Merrien-Soukatchoff, V., Kadiri, I., et al., 2005. The FEBEX benchmark test: case definition and comparison of modelling approaches. Int. J. Rock Mech. Min. Sci. 42 (5), 611−638.

Amuda, O., Giwa, A., Bello, I., 2007. Removal of heavy metal from industrial wastewater using modified activated coconut shell carbon. Biochem. Eng. J. 36 (2), 174−181.

Antonetti, C., Licursi, D., Galletti, A.M.R., Martinelli, M., Tellini, F., Valentini, G., et al., 2016. Application of microwave irradiation for the removal of polychlorinated biphenyls from siloxane transformer and hydrocarbon engine oils. Chemosphere 159, 72−79.

Azad, F.N., Ghaedi, M., Dashtian, K., Hajati, S., Pezeshkpour, V., 2016. Ultrasonically assisted hydrothermal synthesis of activated carbon−HKUST-1-MOF hybrid for efficient simultaneous ultrasound-assisted removal of ternary organic dyes and antibacterial investigation: Taguchi optimization. Ultrasonics Sonochem. 31, 383−393.

Aziz, H.A., Foul, A.A., Isa, M.H., Huang, Y.T., 2010. Physico-chemical treatment of anaerobic landfill leachate using activated carbon and zeolite: batch and column studies. Int. J. Environ. Waste Manage. 5 (3-4), 269−285.

Aziz, H.A., Hin, L.T., Adlan, M.N., Zahari, M.S., Alias, S., AM, A., et al., 2011. Removal of high strength colour from semi-aerobic stabilzed landfill leachate via adsorption on limestone and activated carbon mixture. Res. J. Chem. Sci. 1 (6), 1−7.

Azzaz, A.A., Jellali, S., Bousselmi, L., March, 2016). Optimization of a cationic dye adsorption onto a chemically modified agriculture by-product using response surface methodology. In Renewable Energy Congress (IREC), 2016 7th International (pp. 1−5). IEEE.

Boanini, E., Gazzano, M., Bigi, A., 2010. Ionic substitutions in calcium phosphates synthesized at low temperature. Acta Biomaterial. 6 (6), 1882−1894.

Boopathy, R., Karthikeyan, S., Mandal, A.B., Sekaran, G., 2013. Adsorption of ammonium ion by coconut shell-activated carbon from aqueous solution: kinetic, isotherm, and thermodynamic studies. Environ. Sci. Pollut. Res. 20 (1), 533−542.

Boyd, G.E., Adamson, A.W., Myers Jr, L.S., 1947. The exchange adsorption of ions from aqueous solutions by organic zeolites. II. Kinetics1. J. Am. Chem. Soc. 69 (11), 2836−2848.

Carrott, P., Mourao, P., Ribeiro Carrott, M., Gonçalves, E., 2005. Separating surface and solvent effects and the notion of critical adsorption energy in the adsorption of phenolic compounds by activated carbons. Langmuir 21 (25), 11863−11869.

Cechinel, M.A.P., de Souza, A.A.U., 2014. Study of lead (II) adsorption onto activated carbon originating from cow bone. J. Cleaner Prod. 65, 342−349.

Chaudhuri, M., Kutty, S.R.M., Yusop, S.H., 2010. Copper and cadmium adsorption by activated carbon prepared from coconut coir. Nat. Environ. Pollut. Technol. 9 (1), 25−28.

Cheng, C.K., Derahman, M.R., Khan, M.R., 2015. Evaluation of the photocatalytic degradation of pre-treated palm oil mill effluent (POME) over Pt-loaded titania. J. Environ. Chem. Eng. 3 (1), 261−270.

Chiban, M., Zerbet, M., Carja, G., Sinan, F., 2012. Application of low-cost adsorbents for arsenic removal: a review. J. Environ. Chem. Ecotoxicol. 4 (5), 91−102.

Dahbi, S., Azzi, M., Saib, N., De la Guardia, M., Faure, R., Durand, R., 2002. Removal of trivalent chromium from tannery waste waters using bone charcoal. Anal. Bioanal. Chem. 374 (3), 540−546.

Daud, Z., Abubakar, H.M., Kadir, A.,N., Abdul Latiff, A., Awang, H., Abdul Halim, A., et al., 2017. Batch study on COD and ammonia nitrogen removal using granular activated carbon and cockle shells. Int. J. Eng. 30, 937−944.

Demirbas, E., Kobya, M., Senturk, E., Ozkan, T., 2004. Adsorption kinetics for the removal of chromium (VI) from aqueous solutions on the activated carbons prepared from agricultural wastes. Water Sa 30 (4), 533−539.

Dong, L., Zhu, Z., Qiu, Y., Zhao, J., 2010. Removal of lead from aqueous solution by hydroxyapatite/magnetite composite adsorbent. Chem. Eng. J. 165 (3), 827−834.

Elhafez, S.A., Hamad, H., Zaatout, A., Malash, G., 2016. Management of agricultural waste for removal of heavy metals from aqueous solution: adsorption behaviors, adsorption mechanisms, environmental protection, and techno-economic analysis. Environ. Sci. Pollut. Res. 1−19.

Foo, K., Hameed, B., 2010. Insights into the modeling of adsorption isotherm systems. Chem. Eng. J. 156 (1), 2−10.

Gabaldon, C., Marzal, P., Seco, A., Gonzalez, J.A., 2000. Cadmium and copper removal by a granular activated carbon in laboratory column systems. Sep. Sci. Technol. 35 (7), 1039−1053.

Halim, A.A., Aziz, H.A., Johari, M.A.M., Ariffin, K.S., 2010a. Comparison study of ammonia and COD adsorption on zeolite, activated carbon and composite materials in landfill leachate treatment. Desalination 262 (1-2), 31−35.

Halim, A.A., Aziz, H.A., Johari, M.A.M., Ariffin, K.S., Adlan, M.N., 2010b. Ammoniacal nitrogen and COD removal from semi-aerobic landfill leachate using a composite adsorbent: fixed bed column adsorption performance. J. Hazard. Mater. 175 (1−3), 960−964.

Hassan, S.S., Awwad, N.S., Aboterika, A.H., 2008. Removal of mercury (II) from wastewater using camel bone charcoal. J. Hazard. Mater. 154 (1), 992−997.

Hesas, R.H., Arami-Niya, A., Daud, W.M.A.W., Sahu, J.N., 2013. Preparation and characterization of activated carbon from apple waste by microwave-assisted phosphoric acid activation: application in methylene blue adsorption. BioResources 8 (2), 2950−2966.

Hosseinzadeh, H., Khoshnood, N., 2016. Removal of cationic dyes by poly (AA-co-AMPS)/montmorillonite nanocomposite hydrogel. Desalin. Water Treat. 57 (14), 6372−6383.

Hu, X.-j, Liu, Y.-g, Wang, H., Chen, A.W., Zeng, G.M., Liu, S.M., et al., 2013. Removal of Cu (II) ions from aqueous solution using sulfonated magnetic graphene oxide composite. Sep. Purif. Technol. 108, 189−195.

Hussain, S., Aziz, H.A., Isa, M.H., Adlan, M.N., Asaari, F.A., 2007. Physico-chemical method for ammonia removal from synthetic wastewater using limestone and GAC in batch and column studies. Bioresour. Technol. 98 (4), 874−880.

Iqbal, M., Iqbal, N., Bhatti, I.A., Ahmad, N., Zahid, M., 2016. Response surface methodology application in optimization of cadmium adsorption by shoe waste: a good option of waste mitigation by waste. Ecol. Eng. 88, 265−275.

Ismadji, S., Tong, D.S., Soetaredjo, F.E., Ayucitra, A., Yu, W.H., Zhou, C.H., 2016. Bentonite hydrochar composite for removal of ammonium from Koi fish tank. Appl. Clay Sci. 119, 146−154.

Ismail, F.A., Aris, A.Z., Latif, P.A., 2014. Dynamic behaviour of Cd2 + adsorption in equilibrium batch studies by caco3 − -rich Corbicula fluminea shell. Environ. Sci. Pollut. Res. 21 (1), 344−354.

Ito, T., Takemasa, M., Makino, K., Otsuka, M., 2013. Preparation of calcium phosphate nanocapsules including simvastatin/deoxycholic acid assembly, and their therapeutic effect in osteoporosis model mice. J. Pharmacy Pharmacol. 65 (4), 494−502.

Kamarudin, K.F., Tao, D.G., Yaakob, Z., Takriff, M.S., Rahaman, M.S.A., Salihon, J., 2015. A review on wastewater treatment and microalgal by-product production with a prospect of palm oil mill effluent (POME) utilization for algae. Der Pharma Chem. 7 (7), 73−89.

Latiff, A., Aziz, A., Adeleke AbdulRahman, O., Daud, Z., Ridzuan, M.B., Daud, M., et al., 2015. Batch adsorption of manganese from palm oil mill effluent onto activated cow bone powder. ARPN J. Eng. Appl. Sci. 11 (4).

Leboda, R., 1992. Carbon-mineral adsorbents—new type of sorbents? Part I. The methods of preparation. Mater. Chem. Phys. 31 (3), 243−255.

Li, Z., Hu, X., Xiong, D., Li, B., Wang, H., Li, Q., 2016. Facile synthesis of bicontinuous microporous/mesoporous carbon foam with ultrahigh specific surface area for supercapacitor application. Electrochim. Acta 219, 339−349.

Lim, C.K., Bay, H.H., Neoh, C.H., Aris, A., Majid, Z.A., Ibrahim, Z., 2013. Application of zeolite-activated carbon macrocomposite for the adsorption of Acid Orange 7: isotherm, kinetic and thermodynamic studies. Environ. Sci. Pollut. Res. 20 (10), 7243−7255.

Malik, D., Jain, C., Yadav, A.K., 2016. Removal of heavy metals from emerging cellulosic low-cost adsorbents: a review. Appl. Water Sci. 1−24.

Moreno-Piraján, J., Gómez-Cruz, R., García-Cuello, V., Giraldo, L., 2010. Binary system Cu (II)/Pb (II) adsorption on activated carbon obtained by pyrolysis of cow bone study. J. Anal. Appl. Pyrol. 89 (1), 122−128.

Mourao, P., Carrott, P., Carrott, M.R., 2006. Application of different equations to adsorption isotherms of phenolic compounds on activated carbons prepared from cork. Carbon 44 (12), 2422−2429.

O' Donnel, A.J., Lytle, D.A., Harmon, S., Vu, K., Chait, H., Dionysio, D.D., 2016. Removal of strontium from drinking water by conventional treatment and lime softening in bench-scale studies. Water Res. 319−333.

Orge, C.A., Faria, J.L., Pereira, M.F.R., 2017. Photocatalytic ozonation of aniline with TiO_2-carbon composite materials. J. Environ. Manage. 195 (Pt 2), 208−215.

Paranavithana, G., Kawamoto, K., Inoue, Y., Saito, T., Vithanage, M., Kalpage, C., et al., 2016. Adsorption of Cd2 + and Pb2 + onto coconut shell biochar and biochar-mixed soil. Environ. Earth Sci. 75 (6), 1−12.

Pérez-González, A., Urtiaga, A., Ibáñez, R., Ortiz, I., 2012. State of the art and review on the treatment technologies of water reverse osmosis concentrates. Water Res. 46 (2), 267−283.

Raval, N.P., Shah, P.U., Shah, N.K., 2016. Adsorptive amputation of hazardous azo dye Congo red from wastewater: a critical review. Environ. Sci. Pollut. Res. 23 (15), 14810−14853.

Ravikumar, K., Pakshirajan, K., Swaminathan, T., Balu, K., 2005. Optimization of batch process parameters using response surface methodology for dye removal by a novel adsorbent. Chem. Eng. J. 105 (3), 131−138.

Reddy, K., McDonald, K., King, H., 2013. A novel arsenic removal process for water using cupric oxide nanoparticles. J. Colloid Interf. Sci. 397, 96−102.

Rosli, M.A., Daud, Z., Ab Aziz, A., Sharil, E., Adeleke, A.O., Awang, H., et al., 2017. The Effectiveness of Peat-AC Composite Adsorbent in Removing SS, Colour and Fe from Landfill Leachate. Inter. J. Integrated Eng. 9 (3).

Sahu, J., Acharya, J., Meikap, B., 2009. Response surface modeling and optimization of chromium (VI) removal from aqueous solution using Tamarind wood activated carbon in batch process. J. Hazard. Mater. 172 (2), 818−825.

Salman, J., 2014. Optimization of preparation conditions for activated carbon from palm oil fronds using response surface methodology on removal of pesticides from aqueous solution. Arabian J. Chem. 7 (1), 101−108.

Sandeman., 2011. Adsorption of anionic and cationic dyes by activated carbons, PVA hydrogels, and PVA/AC composite 358, 582−592.

Sarmadivaleh, M., Al-Yaseri, A.Z., Iglauer, S., 2015. Influence of temperature and pressure on quartz−water−CO 2 contact angle and CO 2−water interfacial tension. Journal of colloid and interface science 441, 59−64.

Sayğılı, G.A., 2015. Synthesis, characterization and adsorption properties of a novel biomagnetic composite for the removal of Congo red from aqueous medium. J. Mol. Liquids. 211, 515−526.

Sheha, R., Moussa, S., Attia, M., Sadeek, S., Someda, H., 2016. Novel substituted hydroxyapatite nanoparticles as a solid phase for removal of Co (II) and Eu (III) ions from aqueous solutions. J. Environ. Chem. Eng. 4 (4), 4808−4816.

Song, S.-T., Hau, Y.-F., Saman, N., Johari, K., Cheu, S.-C., Kong, H., et al., 2016. Process analysis of mercury adsorption onto chemically modified rice straw in a fixed-bed adsorber. J. Environ. Chem. Eng. 4 (2), 1685−1697.

Szychowski, D., Pacewska, B., 2012. Methods of preparation and properties of mineral-carbon sorbents obtained from coal-tar pitch-polymer compositions. J. Thermal Anal. Calorimetry 109 (2), 789−795.

Vareda, J.P., Valente, A.J., Durães, L., 2016. Heavy metals in Iberian soils: removal by current adsorbents/ amendments and prospective for aerogels. Adv. Colloid Interface Sci. 237, 28−42.

Velazquez-Jimenez, L.H., Pavlick, A., Rangel-Mendez, J.R., 2013. Chemical characterization of raw and treated agave bagasse and its potential as adsorbent of metal cations from water. Ind. Crops Products 43, 200−206.

Velmurugan, G., Ahamed, K.R., Azarudeen, R.S., 2015. A novel comparative study: synthesis, characterization and thermal degradation kinetics of a terpolymer and its composite for the removal of heavy metals. Iran. Polym. J. 24 (3), 229−242.

Wang, R., Shin, C.-H., Park, S., Park, J.-S., Kim, D., Cui, L., et al., 2014. Removal of lead (II) from aqueous stream by chemically enhanced kapok fiber adsorption. Environ. Earth Sci. 72 (12), 5221−5227.

Yang, G.X., Jiang, H., 2014. Amino modification of biochar for enhanced adsorption of copper ions from synthetic wastewater. Water Res. 48, 396−405.

Yoshihara, K., Yoshida, Y., Nagaoka, N., Fukegawa, D., Hayakawa, S., Mine, A., et al., 2010. Nano-controlled molecular interaction at adhesive interfaces for hard tissue reconstruction. Acta Biomater. 6 (9), 3573−3582.

Zaghden, H., Serbaji, M.M., Saliot, A., Sayadi, S., 2016. The Tunisian Mediterranean coastline: potential threats from urban discharges Sfax-Tunisian Mediterranean coasts. Desalin. Water Treat. 57 (52), 24765−24777.

Further Reading

Oyekanmi, A.A., Daud, Z., Daud, N.M., Gani, P., 2017. Adsorption of heavy metal from palm oil mill effluent on the mixed media used for the preparation of composite adsorbent. MATEC Web Conf. 103, 06020. EDPSciences.

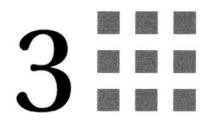

3

Nanoparticles as a Powerful Tool for Membrane Pore Size Determination and Mercury Removal

E. Arkhangelsky[1], V. Inglezakis[2], V. Gitis[3], A.V. Korobeinyk[2,4]

[1]DEPARTMENT OF CIVIL AND ENVIRONMENTAL ENGINEERING, ENVIRONMENTAL SCIENCE & TECHNOLOGY GROUP (ESTG), NAZARBAYEV UNIVERSITY, ASTANA, REPUBLIC OF KAZAKHSTAN [2]DEPARTMENT OF CHEMICAL ENGINEERING, ENVIRONMENTAL SCIENCE & TECHNOLOGY GROUP (ESTG), NAZARBAYEV UNIVERSITY, ASTANA, REPUBLIC OF KAZAKHSTAN [3]UNIT OF ENERGY ENGINEERING, BEN-GURION UNIVERSITY OF THE NEGEV, BEER-SHEVA, ISRAEL [4]O.O. CHUIKO INSTITUTE OF SURFACE CHEMISTRY, NAS OF UKRAINE, KYIV, UKRAINE

3.1 Introduction

Pathogenic microorganisms removal by ultrafiltration (UF)/microfiltration (MF) membranes is mainly based on the size exclusion mechanism (Cheng et al., 2016; ElHadidy et al., 2013), and therefore, membrane pore size must be precisely determined. Several physical methods—that is, liquid and gas displacement methods (Jakobs and Koros, 1997; Li, 2007; Mulder, 2000; Hernández et al., 1996), permeation method (Li, 2007), microscopic methods, gas adsorption-desorption (Burggraaf and Cot, 1996; Mulder, 2000), permporometry (Mulder, 2000), thermoporometry (Cuperus et al., 1992), and nuclear magnetic resonance spectroscopy (Hernández et al., 1998)—exist for determining membrane pore size distribution. However, these physical methods cannot be conducted during continuous membrane operation; and it is impossible to relate the measured pore size to membrane retention efficiency. The standard pore size determination is therefore based on permeation and rejection performance using reference molecules (Aimar and Meireles, 2010; Lebleu et al., 2010; Langlet et al., 2009; Latulippe et al., 2007). Classic tests are performed with polyethylene glycols (PEGs), polyvinylpyrrolidones, dextrans, and proteins (Arkhangelsky et al., 2008). The polymers are nonionic, relatively monodisperse, easily detectable, and they are available in a range of molecular weights. Due to the linear structure, some of the polymer will always penetrate the membrane. It is therefore hard to determine the ultimate (100%) membrane removal capability. Instead, a concept of 90% retention is used. This parameter alone,

Nanotechnology in Water and Wastewater Treatment. DOI: https://doi.org/10.1016/B978-0-12-813902-8.00003-4

however, does not reflect an accurate estimate of membrane retention capability due to non-uniform pore sizes. Moreover, an absence of international standards for determining membrane pore size makes a comparison of membranes from different manufacturers very difficult.

The work presented here describes a development of nanotool and its application to various pristine and cleaned membranes. In contrast to the concept of 90% retention, the developed nanotool enables to find the actual retention value for a particle of a given size and detect the increase of pore diameters in a chemically cleaned membrane. Here highly monodisperse gold nanoparticle (NP) solutions, with 3−45 nm size range were transferred through a ceramic and a number of polymeric membranes. The obtained data were analyzed and compared to results received with other membrane pore size evaluation tests such as microscopy analysis and polymers' permeation. Experiments showed that the NP based test offers a wide range of advantages for the membrane separation industry including the ability to estimate the actual pore size distributions of pristine and cleaned membranes in the final assembled filter systems; examination of actual pore size distribution of both categories of membranes such as depth and screen; obtaining the absolute retention in addition to pore size distribution by the same method; better prediction of the retention of viruses. The NP based method is a valuable tool. It could provide a means to confirm the compliance of membrane systems with the stringent regulatory requirements of the wastewater treatment and the drinking water industries.

3.2 Monodisperse Nanoparticle Solutions for Membrane Pore Size Determination

Prior to synthesis and application of a set of gold NPs to membranes, membrane characteristics were measured in a series of parallel experiments. Four commercially available UF membranes (Sterlitech Corporation, Kent, WA, United States) made of polyvinylidene fluoride, ceramics, and polycarbonate were employed in the study. General membrane parameters are summarized in Table 3-1.

To demonstrate ability of the NPs to detect alterations in the pore size distributions of chemically cleaned membranes, the nanoparticles were applied to cleaned PVDF30 membranes. Cleaning was performed at 20°C with 0.7% nitric acid for 20 hours followed by 48 hours cleaning with 1% hypochlorite.

Table 3-1   Membrane Characteristics

Membrane Name	Configuration	Membrane Material	Filtration Mechanism	Molecular Weight Cut-Off (kDa)	Pore Size (nm)	Flux (L/m^2-hour)
PVDF30	Flat sheet	Polyvinylidene fluoride	Depth	30	−	276
C50	Disc	Ceramics	Depth	50	−	222
PC30	Disc	Polycarbonate	Screen	−	30	120
PC100	Disc	Polycarbonate	Screen	−	100	1500

Pore size distribution was plotted using a log-normal probability density function in a form (Ren et al., 2006):

$$f(d) = \frac{1}{d_{pore}\, ln\sigma\, \sqrt{2\pi}} exp\left[-\frac{1}{2}\left(\frac{ln\left(d_{pore}/d_{50}\right)}{ln\sigma} \right)^2 \right] \tag{3.1}$$

where d_{pore} is the pore diameter; f(d) is probability density function; d_{50} is a diameter corresponding to f(d) = 50%; and σ is a geometric standard deviation of the membrane, which was calculated as $\sigma = d_{90}/d_{50}$.

Zeta potential (Pontié et al., 1998), contact angle (Kuzmenko et al., 2005), and scanning electron microscopy (SEM) analysis and PEG pore size test (feed solution contained 0.3 g/L of a PEG; concentrations were determined by a combustion-infrared method using a total organic carbon analyzer, Apollo 9000 TOC analyzer, Tekmar Company, Cincinnati, OH, United States) were used to determine charge, hydrophobicity, and pore size of the membranes, respectively. The prepared gold NPs were characterized in terms of size, shape, monodispersity, and zeta potential. Transmission electron microscopy (TEM) analysis was performed with a JEM-1230 equipped with a TemCam-F214 (TVIPS Company, Germany) camera. Zeta potential and size distribution were measured with a ZetaPlus (Brookhaven Instruments Corporation, Holtsville, NY, United States). Measurements of gold concentrations consisted of two steps, that is, dissolution of gold NPs in aqua regia and atomic absorption spectroscopy analysis at Perkin Elmer AAnalyst200, Waltham, MA, United States.

Membrane zeta potential values are depicted in Fig. 3-1.

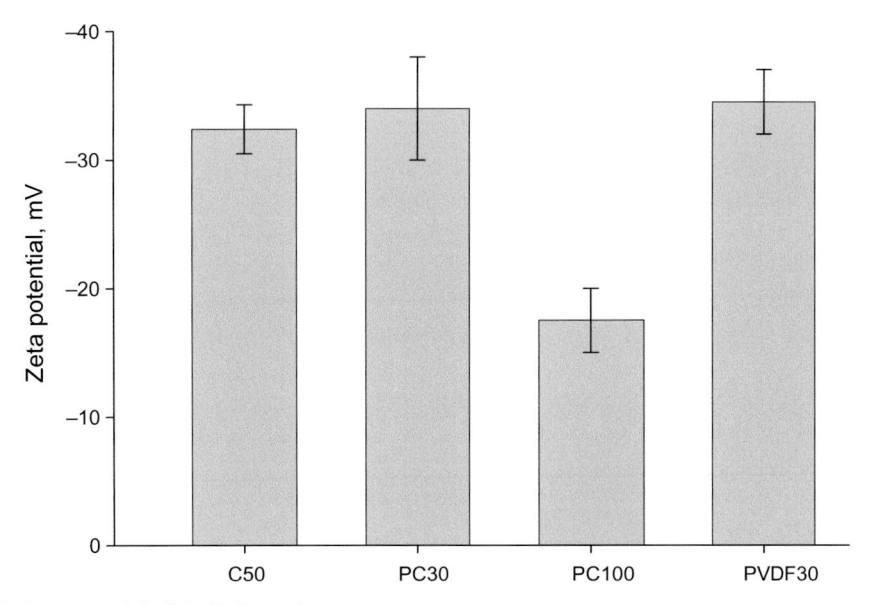

FIGURE 3-1 Zeta potential of studied membranes.

The zeta potentials of C50 and PVDF30 membranes were −32 and −35 mV, respectively. Similar values for ceramic membranes were reported by (Benfer et al., 2004). The zeta potential of the 100-nm PC membrane was almost two times lower than that of the 30-nm membrane: −17.5 vs −34.5 mV (Keesom et al., 1988). This is because the porosity of the PC30 membrane $(6 \times 10^8 \text{ cm}^{-2})$ was 1.5 times higher than that of the PC100 membrane $(4 \times 10^8 \text{ cm}^{-2})$.

Results of contact angle measurements are presented in Fig. 3-2.

The contact angles of the C50, PC30, PC100, and PVDF30 membranes were 35, 45, 45, and 75 degrees, respectively. C50 was the most hydrophilic and PVDF30 the most hydrophobic membranes.

PEG test results are provided in Fig. 3-3.

As the molecular weights of PEGs increased from 0.2 to 35 kDa, their retention levels fluctuated slightly, in the ranges of 0%−7% and 0%−14% for PVDF30 and C50 membranes, respectively. The retention of high molecular weight PEGs (100, 200, and 600 kDa) was significantly higher: 69%−59% for PVDF30 and 70%−95% for C50 membranes. No 90% retention was found for the PVDF30 membrane (Fig. 3-3A). The mode pore diameter of the C50 membrane was 4.06 nm.

SEM was used to study membrane surface. The micrographs are presented in Fig. 3-4.

The SEM micrographs show that membrane pores are round. However, numerous irregularly shaped pores were found in PC100 membranes. The approximate pore sizes of the PC30 and PC100 membranes were 35 and 116 nm, respectively.

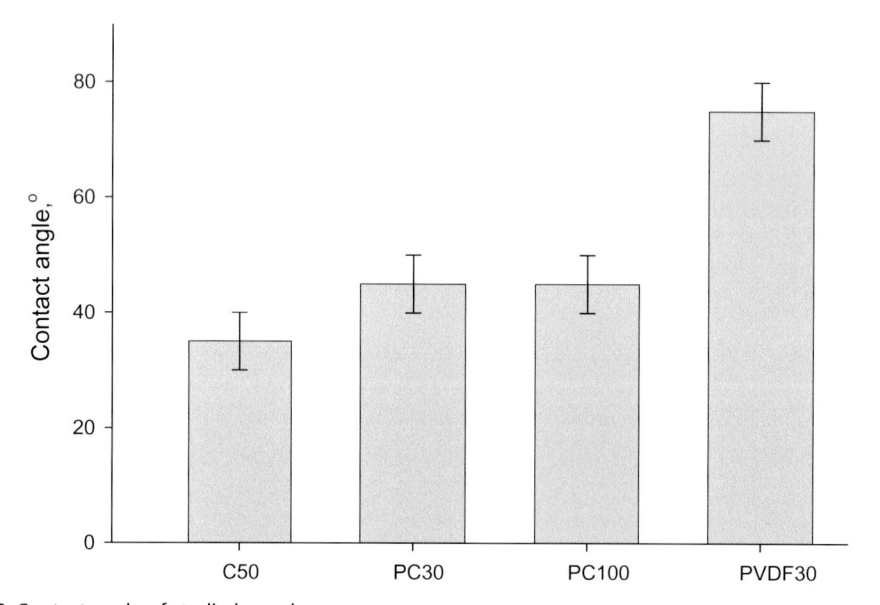

FIGURE 3-2 Contact angle of studied membranes.

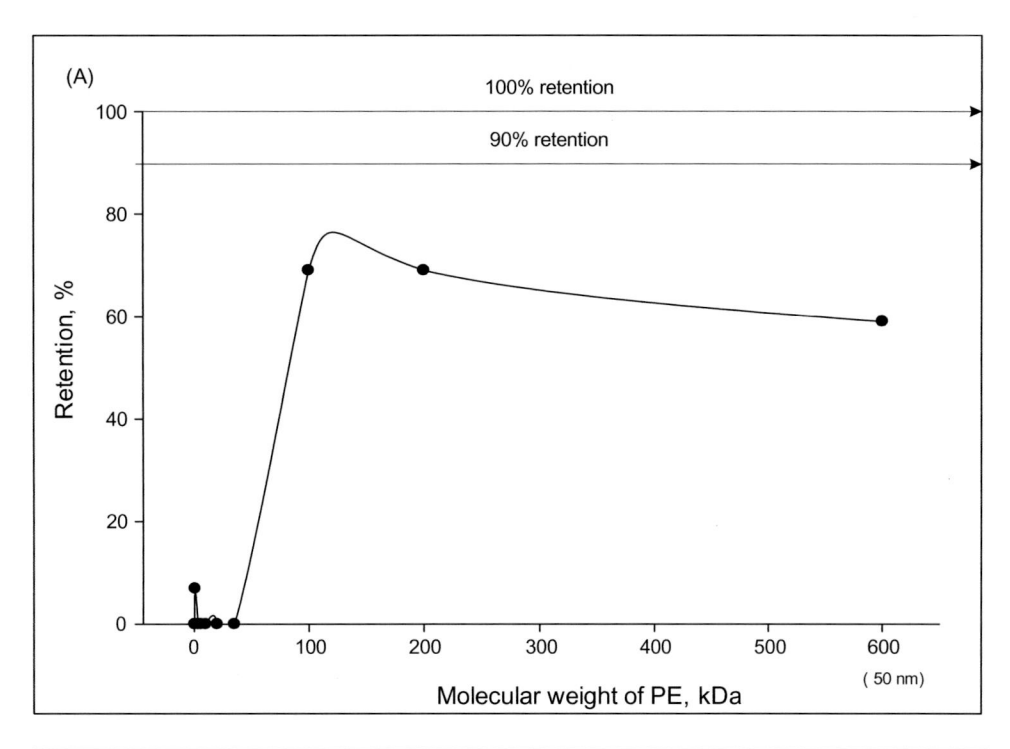

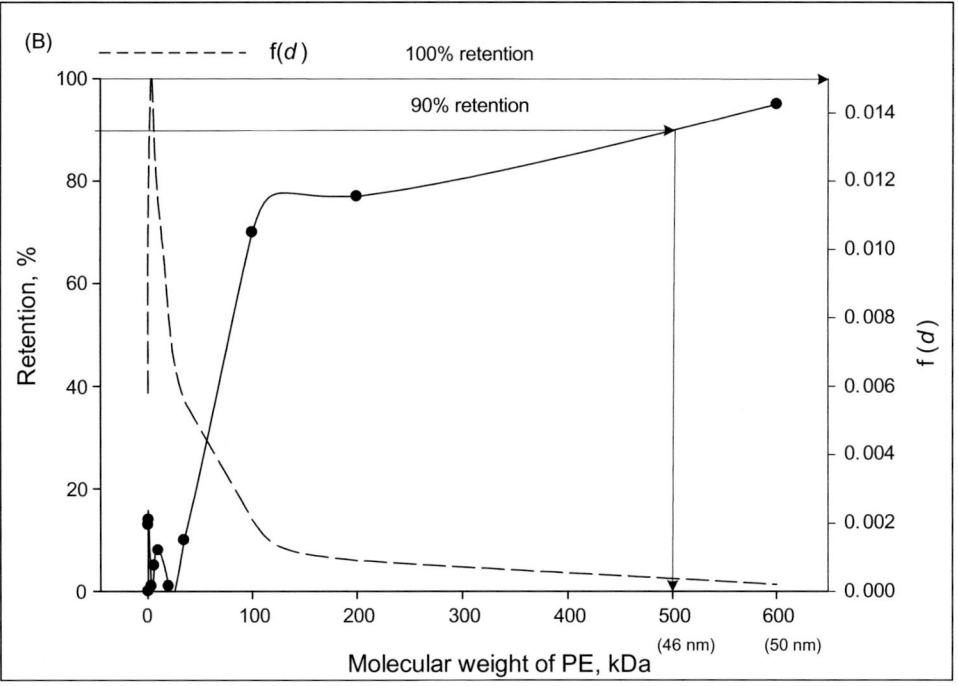

FIGURE 3-3 PEG curves for PVDF30 (A) and C50 (B) membranes.

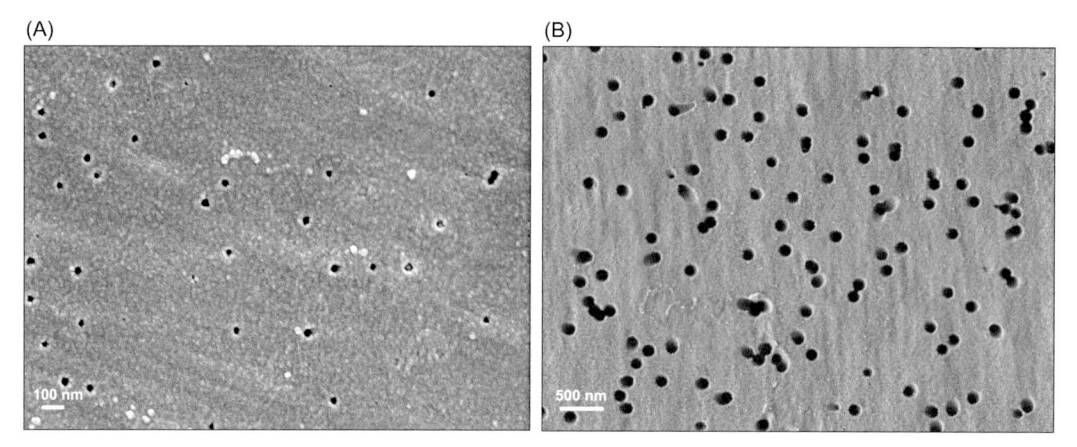

FIGURE 3-4 SEM images of PC30 (A) and PC100 (B) membranes.

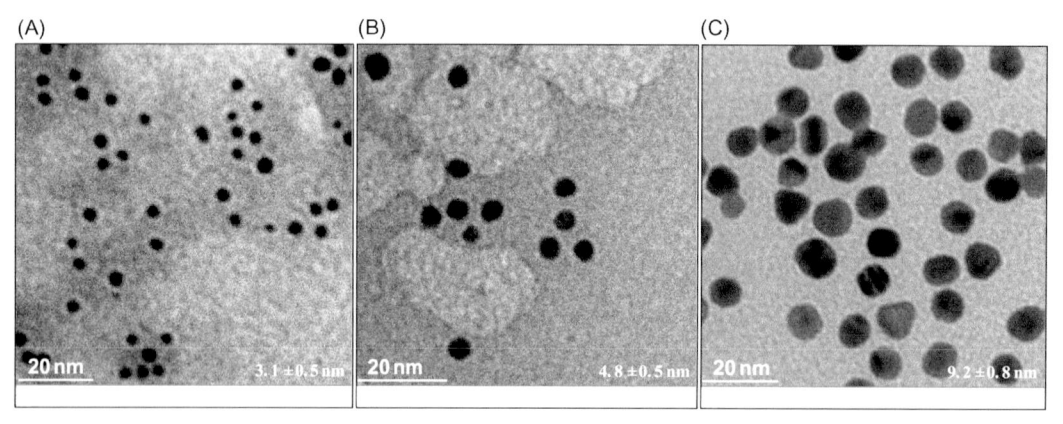

FIGURE 3-5 TEM images of gold nanoparticles—solution **a**: 3.1 ± 0.5 nm (A); solution **b**: 4.8 ± 0.5 nm (B); solution **c**: 9.2 ± 0.8 nm (C).

Colloidal gold particles were prepared in accordance with both original and modified protocols (Slot and Geuze, 1985; Turkevich et al., 1951). TEM images of the colloids are presented in Fig. 3-5. Size distribution diagrams are plotted in Fig. 3-6.

The NPs did not form aggregates, were spherically shaped, and exhibited high monodispersity.

Suspensions having a measured zeta potential above 30 mV or below -30 mV are considered stable (Gibson et al., 2009). Solutions of 28, 37, and 45 nm were therefore assumed stable, and solutions of 3, 5, 9, and 18 nm were assumed highly stable (Fig. 3-7).

The typically higher absolute zeta potential values of small colloids can be explained by their higher surface charge density (Basu et al., 2008; Sonavane et al., 2008; Kim et al., 2005), which was due to the higher concentration of citrate ions used in the reduction/stabilization

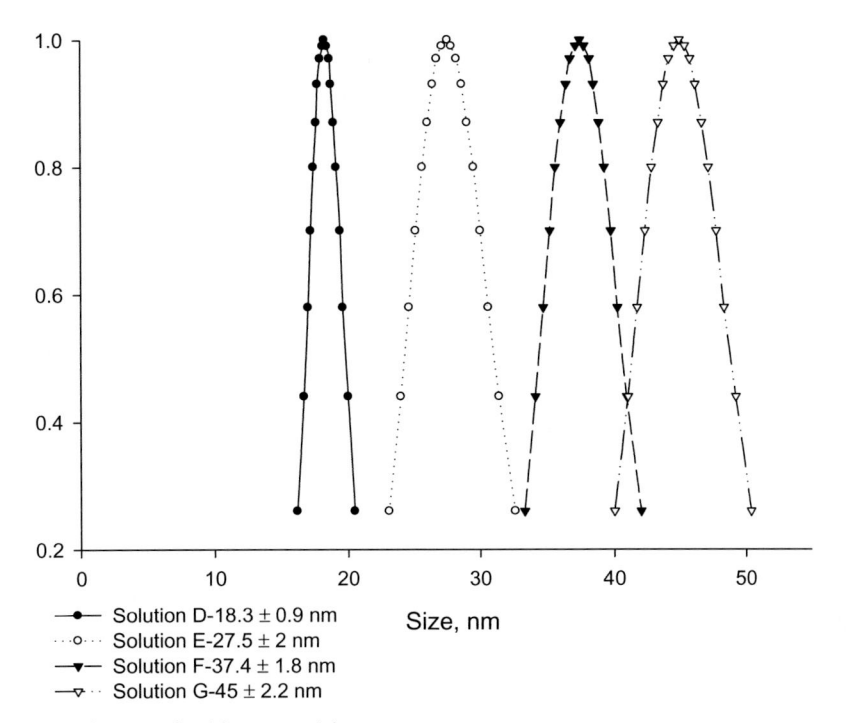

FIGURE 3-6 Size distribution of gold nanoparticles.

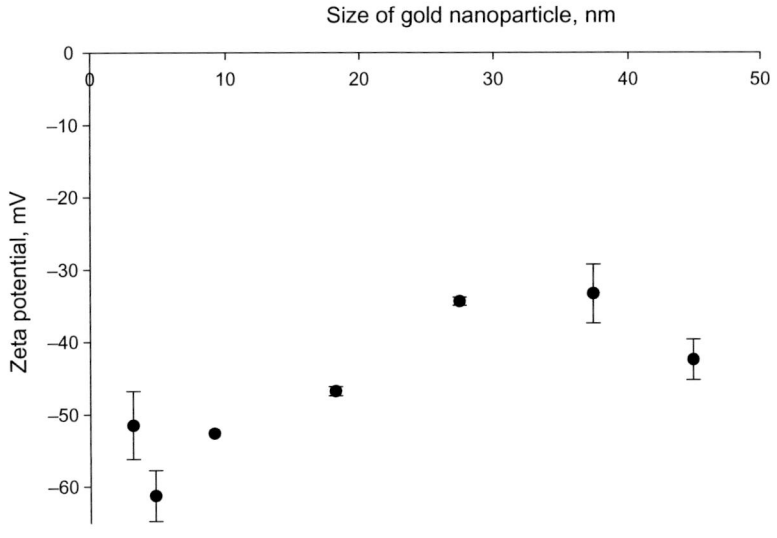

FIGURE 3-7 Zeta potentials of gold nanoparticles.

of the smaller colloids. Because high concentrations of citrate were used for the synthesis of the small colloids, residual citrate ion concentrations in these solutions was also high. As a result, increased numbers of residual citrate ions adsorbed to the particle, thereby increasing the colloid's surface charge density and zeta potential.

Membranes pore sizes were revalidated in tests with gold NPs. The results are depicted in Figs. 3-8 and 3-9.

In contrast to the filtration results of PEGs, here a clear trend was observed. Higher levels of retention were observed in the filtration of larger gold NPs. Moreover, 100% retention was observed for all membranes except PC100. The log-normal probability density function shows that mode pore diameters of the PVDF30, C50, PC30, and PC100 membranes were 4.79, 3.14, 3.14, and >45 nm, respectively.

The degree of membrane fouling during tests with the NPs was evaluated in filtration experiments with particles of 18 nm size through a PVDF30 membrane. The filtration mode was as follows: water—colloid solution—water. The results are presented in Fig. 3-10. The results obtained for the clean membrane demonstrated a 30%—40% decrease in water flux. The decrease is attributed to compaction of the membrane caused by its creep deformation over time (Kallioinen et al., 2007; Roberts Alley, 2007; Bohonak and Zydney, 2005). Membrane compaction occurs immediately after pressure is applied, which is followed by further gradual reduction in the compaction coefficient (Tarnawski and Jelen, 1986). No significant changes in water flux due to the test were observed.

The primary function of membrane systems in the treatment of drinking water is to remove pathogens. Therefore, it is very important that new reference material and pathogens (bacteriophage X174 and MS2 have been used herein this work) will demonstrate similar removal ratios. Retention values of X174 (26 nm) and of 27.5-nm gold NPs by PVDF30 were 87% and 90%, respectively. The retention of MS2 (30 nm), however, was 99.9%. The difference was explained by the higher hydrophobicity of MS2 (log removal value (LRV) of 6.33 by hydrophobic resin). Both X174 (LRV of 0.22) and gold NPs (LRV of 0.46) are more hydrophilic. The observed differences in retention values suggest that the prediction is somewhat limited to particles of similar characteristics. The correct prediction is based not only on size, but also on hydrophobicity and surface charge. However, recent developments in nanoparticle synthesis methods offer the promise of almost limitless possibilities conferred by the ability to fine tune nanoparticle characteristics, that is, to adjust size, shape, charge, and hydrophilicity of gold nanoparticles to match the parameters of certain viruses (Kango et al., 2013; Boyer et al., 2010; Tréguer-Delapierre et al., 2008; Hu et al., 2007; Liu et al., 2007; Kimling et al., 2006; dos Santos et al., 2005).

A comparison of PEG permeation through pristine (Fig. 3-3A) and cleaned (Fig. 3-11A) PVDF30 membranes showed a significant decrease in polymer permeation after chemical cleaning. For example, the retentions of 0.2—35-kDa PEGs increased from 0%—7% to 0%—48% and of 100—600-kDa PEGs from 69%—59% to 87%—99%.

The straightforward conclusion reached in response to the observed tendency is that the HNO_3-NaOCl cleaning treatment significantly narrowed membrane pores. However, experiments performed with gold NPs demonstrated an opposite trend, that is, a reduced retention

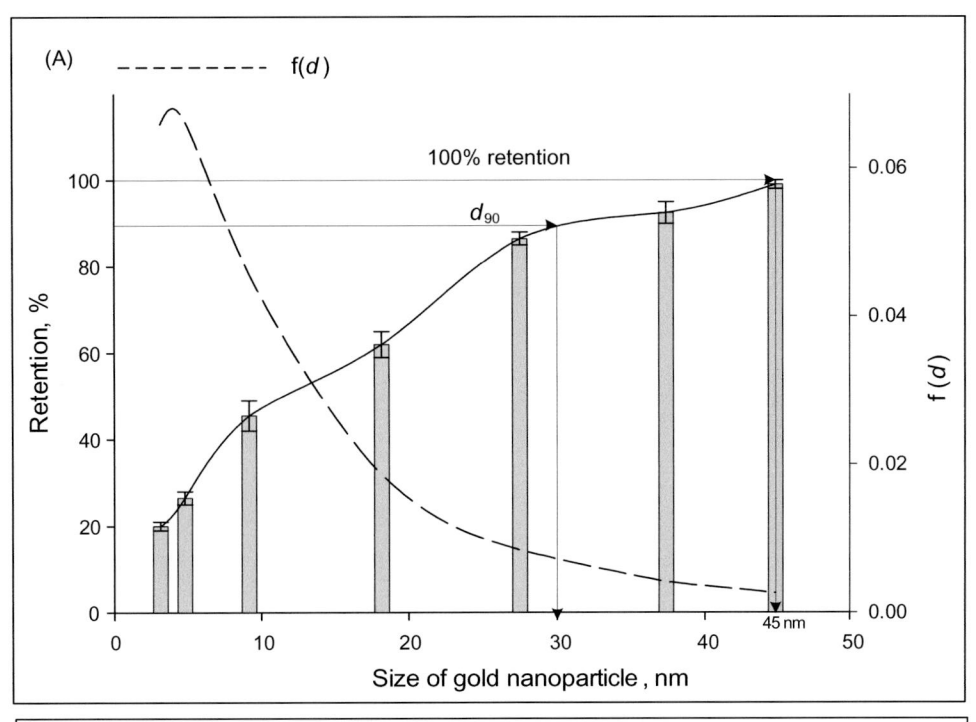

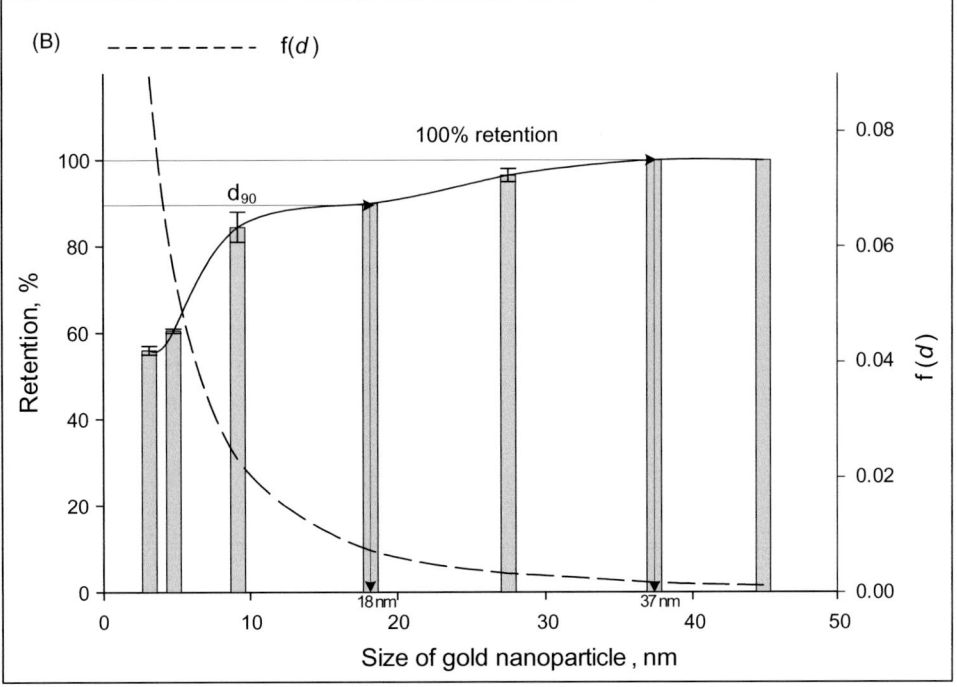

FIGURE 3-8 Gold nanoparticle separation and pore size distribution curves: PVDF30 (A) and C50 (B).

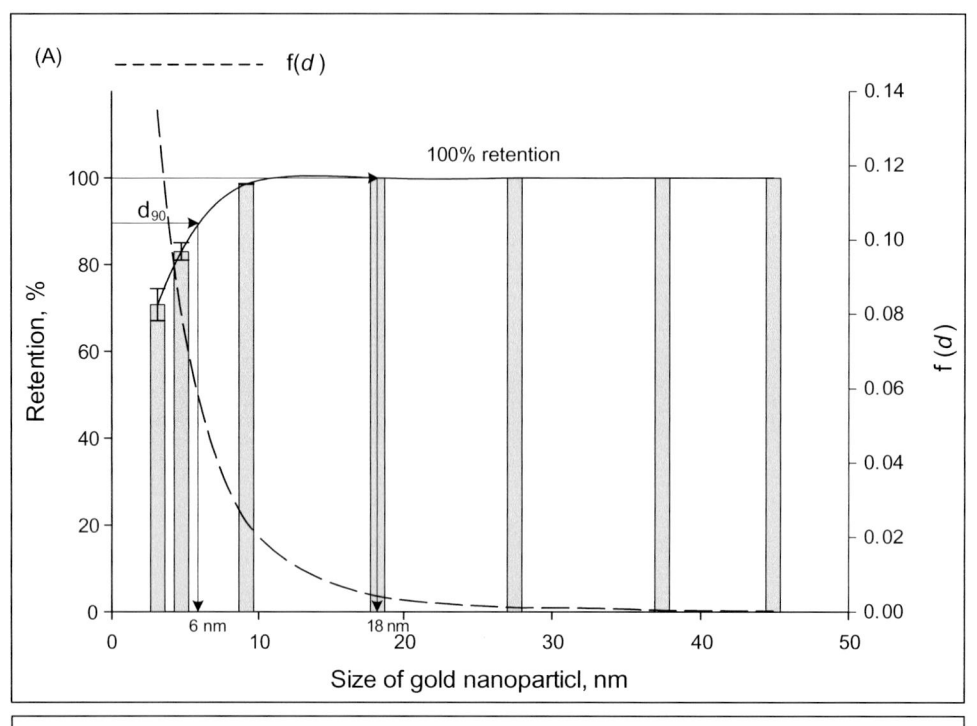

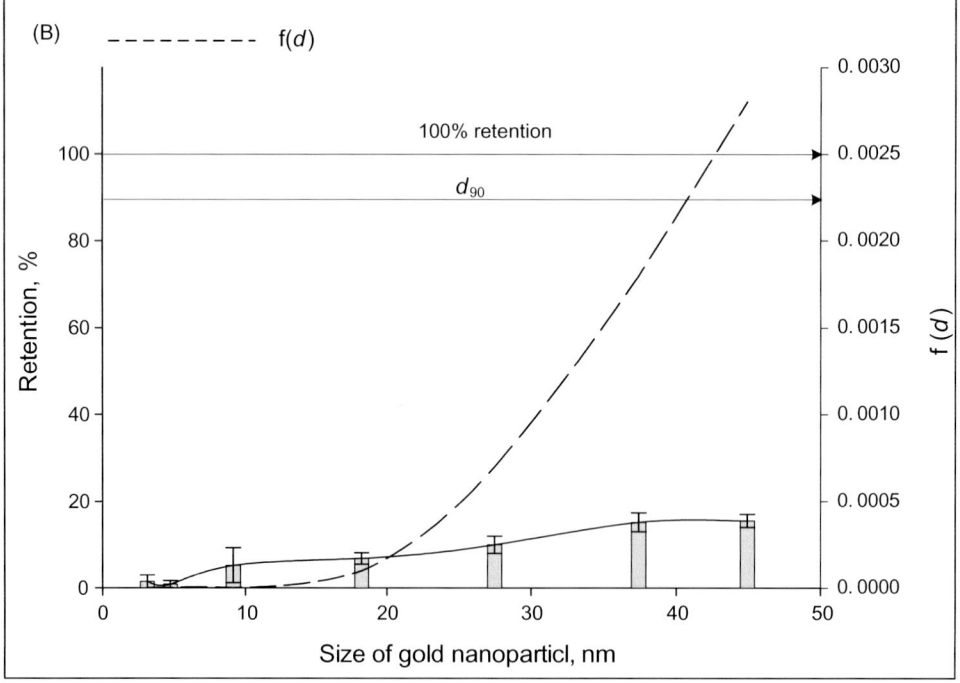

FIGURE 3-9 Gold nanoparticle separation and pore size distribution curves: PC30 (A) and PC100 (B).

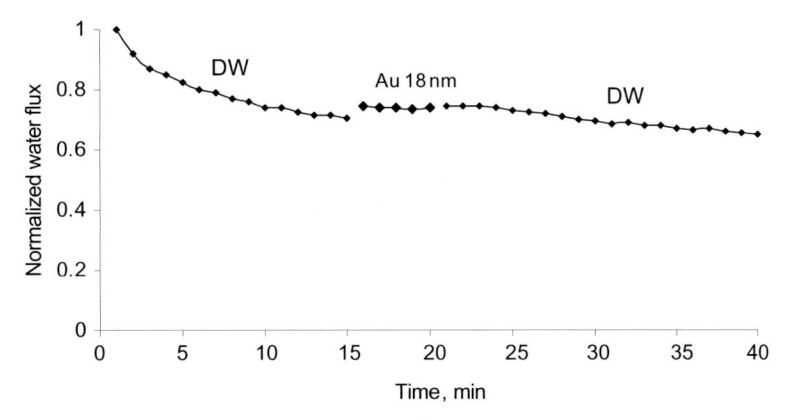

FIGURE 3-10 Influence of gold nanoparticles on membrane flux.

of colloids by a membrane treated with HNO_3-NaOCl (Figs. 3-8A and 3-11B). A comparison of the results obtained for pristine and cleaned PVDF30 membranes show that the retention levels of 4.8, 9.2, 18.3, 27.5, and 37.4-nm particles by pristine membrane were higher than those by the cleaned membrane. According to the log-normal probability density function, after cleaning the PVDF30 membrane, its mode pore diameter increased 4.5 times to 21.6 nm. In an effort to explain the increased retention of PEGs by the cleaned membrane, the zeta potential and contact angle of a chemically treated membrane were measured. Treatment with HNO_3-NaOCl did not affect PVDF membrane charge, but its contact angle increased from 75 to 105 degrees (Arkhangelsky et al., 2007; Puspitasari et al., 2010). It is reasonable to assume that polyvinylpyrrolidone, which was used in the initial membrane casting as a pore former and hydrophilizer, prevented the adsorption of PEGs onto the membrane surface. Cleaning with HNO_3-NaOCl, meanwhile, dislodged the polyvinylpyrrolidone that decreased membrane hydrophilicity, thus leading to the increased affinity of the polymer to the membrane (Arkhangelsky et al., 2007; Puspitasari et al., 2010).

Table 3-2 demonstrates the pore sizes of different membranes as claimed by the manufacturer, as calculated by the formula of (Ioan et al., 2000):

$$d_{pore} = 0.11(MW)^{0.46} \tag{3.2}$$

(where *MW* is molecular weight of polymer), and as found from SEM analysis and PEG and gold nanoparticle tests.

Pore sizes of the PVDF30 and C50 membranes were transformed to 12.6 and 16 nm, respectively, using Eq. (3.2). The PEG test results were >50 nm and 46 nm for PVDF30 and C50, respectively. The pore sizes of PC30 and PC100 membranes were 35 and 116 nm, higher than the values of 30 and 100 nm provided by the manufacturer.

Figs. 3-8 and 3-9 showed that in opposition to the 90% retention concept and SEM analysis, gold NPs were able to provide precise data about absolute membrane pore sizes. For example, the absolute and mode pore sizes of the screen filter PC30 were 18.2 and 3.1 nm,

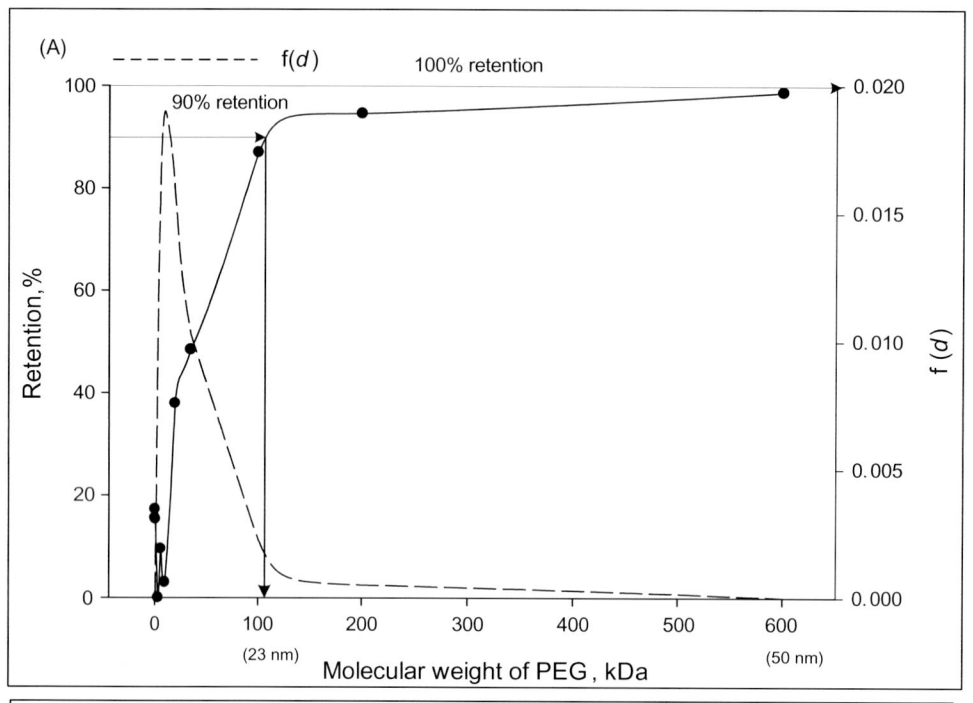

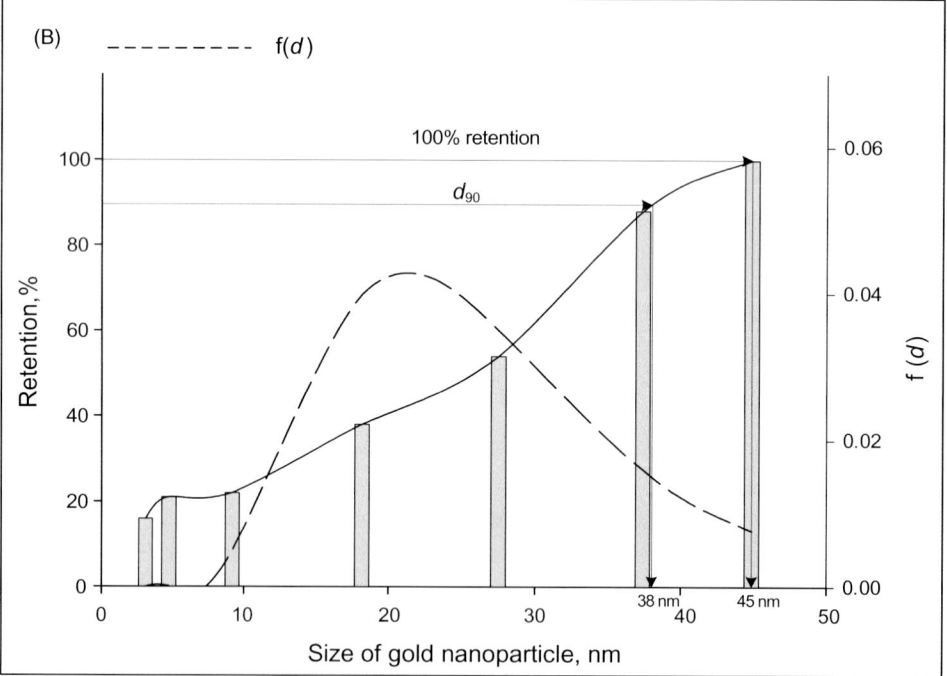

FIGURE 3-11 PEG (A) and gold nanoparticle (B) tests performed on a HNO_3-NaOCl cleaned PVDF30 membrane.

Table 3-2   Membrane Pore Sizes

			PEGs/SEM			Gold NPs		
Membrane	Pore Size Provided by Manufacturer	Pore Size Transferred From Eq. (3.2) (nm)	90% Retention/ Surface Pore Size (nm)	100% Retention (nm)	Mode Pore Diameter (nm)	d_{90} (nm)	100% Retention (nm)	Mode Pore Diameter (nm)
PVDF30	30 kDa	12.6	>50	>50		30	≈ 45	4.8
C50	50 kDa	16	46	>50	4.6	18	37	3.1
PC30	30 nm		35			6.3	18.2	3.1
PC100	100 nm		116			>45	>45	>45
cleaned PVDF30			22.5	≈ 50	7.6	38.4	≈ 45	21.6

respectively. A comparison of these results with an SEM image of a PC30 membrane shows that the pore size obtained with gold NPs was 5.6 times lower than that obtained by microscope analysis. This phenomenon can be attributed to the tapered shape of the pore channels in track-etch membranes, as was described previously by several research groups (Kozlovskiy et al., 2017; Apel et al., 2006; Lettmann et al., 1999; Ferain and Legras, 1997; Kim and Stevens, 1997; Schönenberger et al., 1997). As such, we assumed that a 3—45-nm gold nanoparticle set would also allow us to determine pore size distribution of the PC100 membrane. However, all gold solutions freely penetrated the membrane, thereby indicating that the absolute and mode pore diameters of this membrane were higher than 45 nm.

In contrast to the 90% removal concept, a set of gold NPs can be used to find the actual retention value for a particle of a given size and vice versa. For example, C50 and PC30 membranes will retain 90% of 18- and 6.3-nm particles, and 100% of 37- and 18.2-nm particles, respectively (Table 3-2). In addition, experiments demonstrated another interesting trend, that is, the pore size of pristine depth filters determined by PEGs was higher than that determined by gold NPs. Such behavior is related to the presence/absence of the rigid outer envelope that significantly affects particle retention value.

3.3 Mercury Pollution and Overview of Available Treatment Methods

As has been mentioned above the second part of this chapter deals with the purification of mercury contaminated aqueous solutions by employing silver NPs. As a global pollutant mercury affects human health (Ballester et al., 2018) and ecosystem (Unep, 2013). The first health effects awareness of Hg was manifested in the early 1950s in Minatama (Japan) after Hg contaminated fish were consumed as a main food source (Kurland et al., 1960). Mercury exposure even at low concentrations (0.3 mg/kg body weight per day as oral reference dose) is responsible for many fatal diseases (Cadet and Bolla, 2007). The lowest observed adverse effect level (LOAEL) for inorganic mercury at oral intake was registered as 0.016 mg/kg-day (Druet et al., 1978). Several adsorbents can be used in the extraction of this pollutant from aqueous solutions such as sulfurized activated carbon (Asasian and Kaghazchi, 2015), mineral zeolites (Azizi et al., 2013), Hg-specific resins (Monier et al., 2015) polymer membranes (Zhang et al., 2018), and etc. Nevertheless, utilization of those filter materials in a field where concentration of Hg is high is limited due to high cost, bio/mechanical filter clogging or low filter capacity (Newcombe et al., 1997; Weisener et al., 2005; Mudasir et al., 2016). Mercury as a liquid metal can be found as solid in supercooled condition or as an amalgam compound. Amalgamation not only a unique term referring to mercury ability "dissolve" some metal (Harika et al., 2018) it also can reduce the mercury toxicity via formation of Hg-amalgam with noble metals (Poulston et al., 2007). The effectiveness of mercury removal with silver was reported for having very little mercury release from corresponding amalgam (Henderson et al., 2001). However, an application of bulk silver is limited due to its small contact area. This problem can be solved with the silver nanoparticles (AgNPs) development.

Numerous approaches and methods have been reported for the synthesis of AgNPs using eco-friendly, biological, photochemical, physical, and chemical routes. Each method produces nanoparticles for narrow range of applications due common problems being particle size/shape and scalability (Hsien-Hsueh et al., 2005; Agnihotri et al., 2014; Wakuda et al., 2008). Physical routes to prepare silver NPs need expensive equipment and high vacuum/temperature conditions (Siegel et al., 2017; Kawasaki and Nishimura, 2006; Malekzadeh and Halali, 2011). Current approaches in a noble metal nanoparticles generation involve the utilization of reducing and stabilizing agents (i.e., ethylene glycol (Leng et al., 2018), sodium borohydride (Tyliszczak et al., 2017), citrate (Khatoon et al., 2017), and etc.) in the aqueous and nonaqueous dispersions, which result in nanoparticle surface contamination by those agents' residuals and thus limiting the final nanoparticle reactivity (Loran et al., 2018).

Direct synthesis of solid supported AgNPs (Lazić et al., 2018) is attracting increasing interest in terms of practical applications and synthetic challenges. Activated carbon (Arivizhivendhan et al., 2018), graphene (Donini et al., 2018), cellulose (Volova et al., 2018), and polyelectrolytes (Kruk et al., 2016; Oćwieja et al., 2015; Kawada et al., 2014) were successfully utilized as supporting materials for noble metal generation via chemical (Ferraris et al., 2017) and physical (Loran et al., 2018; Chen and Yu, 2017; Ye et al., 2018) in situ methods. In situ generation of AgNPs on the solid substrate from solution improves the nanoparticle dispersion and prevents their aggregation. Recently was reported that supporting silver NPs on silica improves the photocatalytic properties of metal (Chen et al., 2010). Fumed silica (SiO_2) has been thoroughly investigated due to chemical and thermal stability, biocompatibility and sorption properties (1995a). Because of that this inert material has widespread application, for example, in catalysis, nano-liquid chromatography, wastewater treatment, etc. (Bloisi et al., 2016; Coz et al., 2009; Aydoğan, 2016). Surface of commercial available fumed silica contains only siloxane or silanol groups (1995b) and to improve reactivity of silica surface, prior to any use, modification of silica surface must be performed (Mora-Barrantes et al., 2011). To control size, surface purity and discreet distribution of the noble nanoparticles they were generated via in situ reduction on the silica surface modified by reductive silicon hydride ($\equiv SiH$) groups (Katok et al., 2012). An average metal NP diameter in a resulting silica/noble metal NPs composite is easily controllable by adjusting the synthetic procedure.

A newly developed method, presented in this chapter, involves in situ reduction of the noble metal on the silica surface containing the reductive silicon hydride groups, which control the NP size, surface purity, and discreet distribution. The approach allows easy synthesis and size control of 17−90 nm silica supported noble metal NPs. Adsorption tests of the synthesized NPs demonstrate mercury adsorption increase with reduction in the NP average diameter. Also the 17 nm NPs solutions showed hyperstoichiometry effects (interaction more than the stoichiometric amount of atoms) in mercury extraction. The proposed NP synthesis route demonstrate the main advantages of hyperstoichiometric rate of mercury extraction, environmentally friendly synthesis pathway, have no impact on the environment as chemically attached NPs are fixed on the support surface, and low cost due to the usage of low amounts of the noble metal in preparation.

3.4 Silver NPs in Mercury Removal From Aqueous Solutions

The chemicals $AgNO_3$ (>99.9%), SiO_2 (300 m^2/g), Triethoxysilane (TEOS, 95%), $HgCl_2$ (≥99.5%), acetic acid (glacial), acetone and ethanol (all from Sigma-Aldrich) were employed in the study. Ultrapure water with resistivity of 18.3 M$\Omega \cdot$ cm was obtained by reversed osmosis followed by ion-exchange and filtration (Millipore). The silver NP/SiO_2 nanocomposites were prepared by in situ redox reaction method. Silica surface was modified with TEOS in glacial acetic acid as described elsewhere (Katok et al., 2012) in order to create redox-active silicon-hydride functional groups. The common procedure for silver NPs/SiO_2 preparation was as follows: 1.0 g of TEOS-modified SiO_2 sample was immersed in aqueous solution of silver nitrate of 0.1 M and appropriate volume in order to create silver NPs loading of 0.05, 0.2, 0.4 and 0.5 mmol/g of silica. The mixture was stirred in the dark for 1 hour to the reaction completion. The mixture was then filtered and dried in a bench oven at 105°C for 8 hours. The derived materials have been characterized by TEM, which was carried out using a JEOL-JEM2100 instrument operated at 200 keV. All samples were ultrasonically dispersed in isopropyl alcohol at room temperature, and then a portion of the solution was placed onto a copper grid for TEM measurements. Wide-angle X-ray diffraction (XRD) was used to characterize the noble metal NP diffraction patterns. All measurements were collected over the 2θ range from 10 to 70 degrees on a Rigaku SmartLab diffractometer equipped with a CuKα radiation source (0.1549 nm). The adsorption of Hg^{2+} on AgNP/SiO_2 nanocomposite was studied by batch technique. All batch adsorption experiments were performed by mixing a 1.0 g of nanocomposite sorbent into a $HgCl_2$ solution (8.5 mg/L, 100 mL) with stirring (120 r/min) at room temperature for 120 minutes. Then, the supernatant was separated from the mixture solution by filtration with 0.22 μm MF membrane. The concentration of Hg^{2+} ions remaining in solution was measured by mercury analyzer (RA-915 +, Lumex). The amount of Hg^{2+} ions adsorbed by nanocomposite q_t (mg/g) is calculated using the following Eq. (3.3):

$$q_t = \frac{(C_0 - C_t) \cdot V}{m} \tag{3.3}$$

where C_0 and C_t are the initial and a time t liquid-phase Hg^{2+} concentrations (mg/L), V is the volume of the solution (mL), and m is the mass of dry adsorbent used (g).

The chemical composition of the modified silica surface was determined by FTIR analysis and iodometric titration. Presence of silicone hydride groups is confirmed by absorption band at 2240 cm^{-1} (Fig. 3-12 spectrum 1) (Bellamy, 1975). Iodometric titration results (not shown here) revealed the presence and total number of silicone hydride groups (0.8 mmol/g). Disappearance of the 2240 cm^{-1} band after reaction with Ag^{+} (Fig. 3-12 spectrum 2) confirms consumption of those groups in reaction of silver reduction.

XRD patterns of AgNPs/silica nanocomposites are recorded in Fig. 3-13. A large halo at 22.5 2θ indicates the reflection of amorphous structure of silica. Well resolved pattern of five bands at 2θ values 38.05, 44.3, 64.5, 77.6, and 81.5 degrees that corresponds to (111), (200), (220), (311), and (222) planes of cubic nanocrystalline silver (JCPDS, File No. 4-0783) (Okitsu et al., 1996) and indicates that silver nanoparticles in the nanocomposite structure are well crystallized. No peaks of other impurity crystalline phases have been detected. Increase in the intrinsic intensity of XRD peaks is represents an increase in silver particle size.

The formation of metal nanoparticles on the silica support was confirmed by TEM observations. Figs. 3-14−3-17 shows TEM images of metal nanoparticles within the SiO$_2$ matrix. The presence of nanoparticles is clearly seen at silver loading: 0.05 (Fig. 3-14), 0.2 (Fig. 3-15), 0.4 (Fig. 3-16) and 0.5 mmol/g (Fig. 3-17). At large magnification the crystalline lines within the nanoparticle structure of silver NPs are clearly seen. The results showed that the smaller

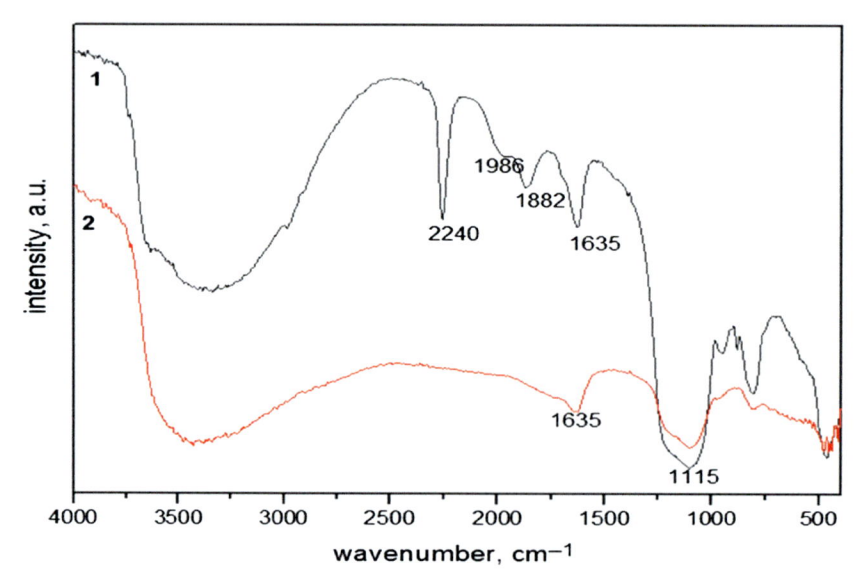

FIGURE 3-12 IR Fourier spectra: 1—silica modified with triethoxysilane and 2—sample after interaction with silver ions.

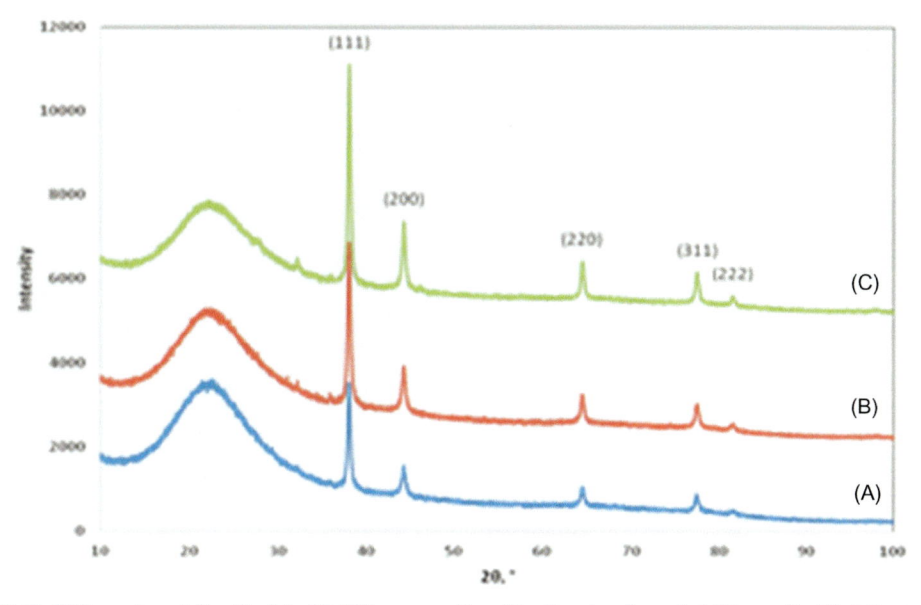

FIGURE 3-13 XRD spectra of (A–C) of AgNPs/SiO$_2$ composite with silver loading of: 0.05 mmol/g (A), 0.2 mmol/g (B) and 0.5 mmol/g (C).

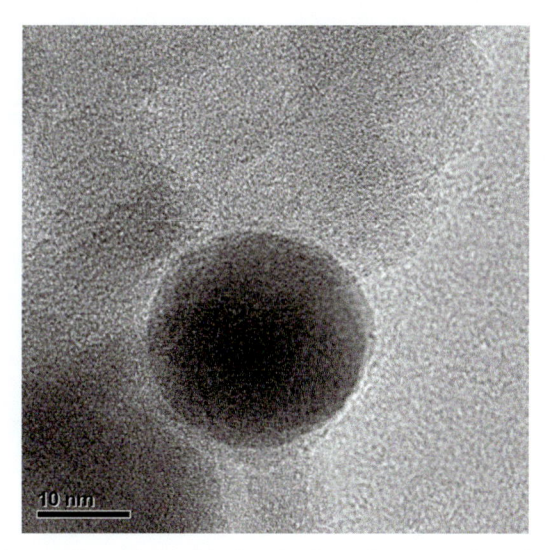

FIGURE 3-14 TEM image of AgNPs/SiO$_2$ sample with silver loading of 0.05 mmol/g at high magnification.

the metal/ modified silica ratio, the smaller the nanoparticle. Thus, the loading of metal ion may play a more important role than its redox potential.

According to the adsorption experiments, nanocomposite adsorbent samples reach a mercury saturation in 120 minutes (in whole range of silver loading). Therefore the time

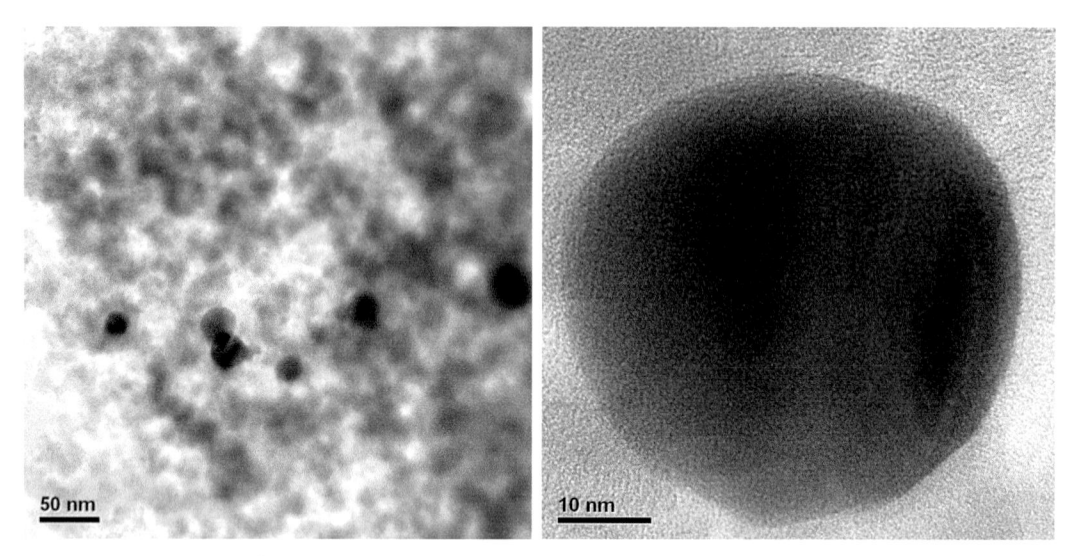

FIGURE 3-15 TEM images of AgNPs/SiO$_2$ sample with silver loading of 0.2 mmol/g at low (left) and high (right) magnification.

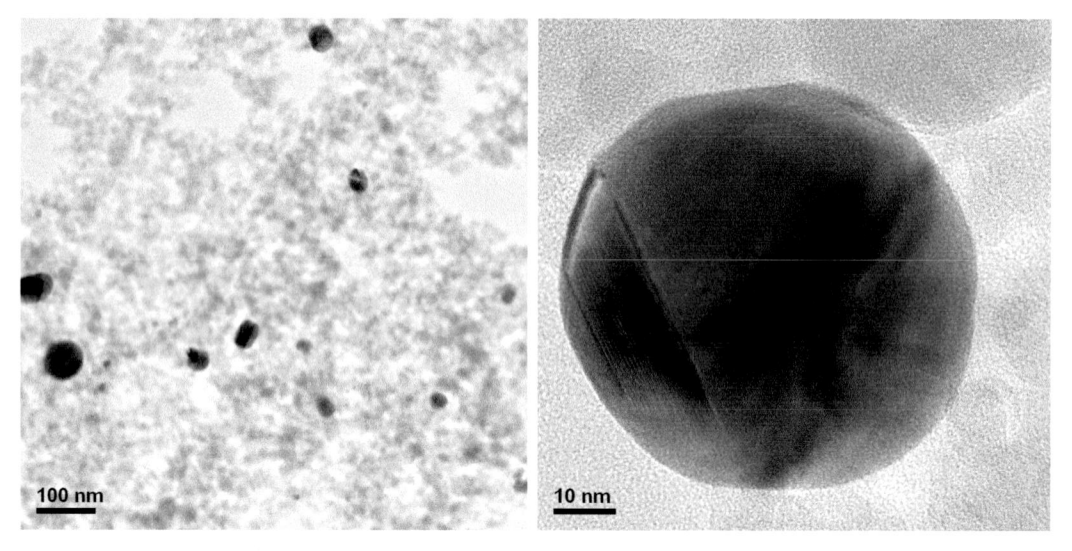

FIGURE 3-16 TEM images of AgNPs/SiO$_2$ sample with silver loading of 0.4 mmol/g at low (left) and high (right) magnification.

allowed for establishing equilibrium was 120 minutes with constant rotation. The uptake of Hg^{2+} (Table 3-3) increased with the increase in silver content in composite, however if calculate silver/mercury ratio the composites with smallest AgNPs diameters (AgNPs/SiO$_2$_0.05) were found to have greater adsorption capacities than silica's with higher silver content under the experimental condition used.

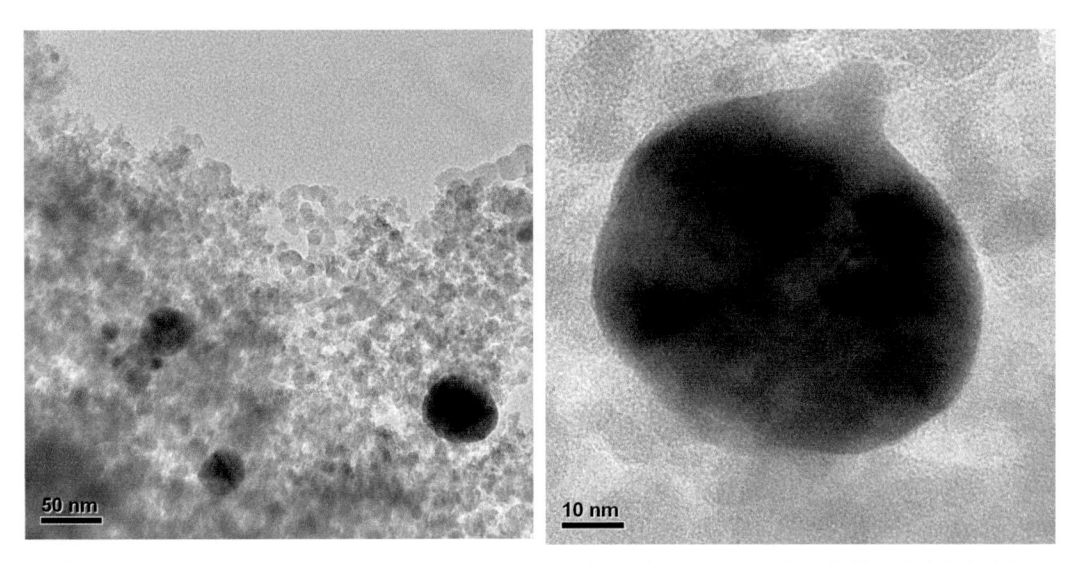

FIGURE 3-17 TEM images of AgNPs/SiO$_2$ sample with silver loading of 0.5 mmol/g at low (left) and high (right) magnification.

Table 3-3   Mercury Adsorption Results on Different AgNPs/SiO$_2$ Samples

Sample	Ag Content (mmol of Ag/g SiO$_2$)	Average AgNPs Diameter (nm)	Maximum Hg Loading (mmol of Hg/g of Composite)
AgNPs/SiO$_2$_0.05	0.05	17	0.1
AgNPs/SiO$_2$_0.2	0.2	24	0.17
AgNPs/SiO$_2$_0.4	0.4	63	0.3
AgNPs/SiO$_2$_0.5	0.5	90	0.4

In our study the excess of mercury adsorption (deviation from stoichiometry, i.e., hyperstoichiometry) was observed. These findings demonstrate the need for further investigation on mercury adsorption mechanism in order to assess potential application in mercury removal from the aqueous solution.

Acknowledgements

This work has been financially supported by Nazarbayev University; "HyperActiv" ORAU project (Nazarbayev University); "NanoMed" project, Horizon 2020 MSCA-RISE-2016, European Commission; projects IAPP/1516/46 and IAPP/1516/, the Royal Academy of Engineering (United Kingdom). Authors also express their gratitude to Dr. Alex Jass, Serikbayev Ust' Kamenogorsk State University, Kazakhstan, for TEM experiments and discussion.

References

Agnihotri, S., Mukherji, S., Mukherji, S., 2014. Size-controlled silver nanoparticles synthesized over the range 5-100 nm using the same protocol and their antibacterial efficacy. RSC Adv. 4, 3974–3983.

Aimar, P., Meireles, M., 2010. Calibration of ultrafiltration membranes against size exclusion chromatography columns. J. Memb. Sci. 346, 233–239.

Apel, P.Y., Blonskaya, I.V., Dmitriev, S.N., Orelovitch, O.L., Sartowska, B., 2006. Structure of polycarbonate track-etch membranes: origin of the "paradoxical" pore shape. J. Memb. Sci. 282, 393–400.

Arivizhivendhan, V., Mahesh, M., Boopathy, R., Karthikeyan, S., Mary, R.R., Sekaran, G., 2018. Functioned silver nanoparticle loaded activated carbon for the recovery of bioactive molecule from bacterial fermenter for its bactericidal activity. Appl. Surf. Sci. 427, 813–824.

Arkhangelsky, E., Kuzmenko, D., Gitis, V., 2007. Impact of chemical treatment on properties and functioning of polyethersulfone membranes. J. Memb. Sci. 305, 176–184.

Arkhangelsky, E., Levitsky, I., Gitis, V., 2008. Electrostatic repulsion as a mechanism in fouling of ultrafiltration membranes. Water Sci. Technol. 58, 1955–1961.

Asasian, N., Kaghazchi, T., 2015. Sulfurized activated carbons and their mercury adsorption/desorption behavior in aqueous phase. Int. J. Environ. Sci. Technol. 12, 2511–2522.

Aydoğan, C., 2016. Boronic acid-fumed silica nanoparticles incorporated large surface area monoliths for protein separation by nano-liquid chromatography. Anal. Bioanal. Chem. 408, 8457–8466.

Azizi, S.N., Dehnavi, A.R., Joorabdoozha, A., 2013. Synthesis and characterization of LTA nanozeolite using barley husk silica: Mercury removal from standard and real solutions. Mater. Res. Bull. 48, 1753–1759.

Ballester, F., Iñiguez, C., Murcia, M., Guxens, M., Basterretxea, M., Rebagliato, M., et al., 2018. Prenatal exposure to mercury and longitudinally assessed fetal growth: relation and effect modifiers. Environ. Res. 160, 97–106.

Basu, S., Pande, S., Jana, S., Bolisetty, S., Pal, T., 2008. Controlled interparticle spacing for surface-modified gold nanoparticle aggregates. Langmuir 24, 5562–5568.

Bellamy, A.L.J., 1975. The Infra-Red Spectra of Complex Molecules. Chapman and Hall, London.

Benfer, S., Árki, P., Tomandl, G., 2004. Ceramic membranes for filtration applications — preparation and characterization. Adv. Eng. Mater. 6, 495–500.

Bloisi, F., Califano, V., Perretta, G., Nasti, L., Aronne, A., Girolamo, D.I., et al., 2016. Lipase immobilization for catalytic applications obtained using fumed silica deposited with MAPLE technique. Appl. Surf. Sci. 374, 346–352.

Bohonak, D.M., Zydney, A.L., 2005. Compaction and permeability effects with virus filtration membranes. J. Memb. Sci. 254, 71–79.

Boyer, C., Whittaker, M.R., Chuah, K., Liu, J., Davis, T.P., 2010. Modulation of the surface charge on polymer-stabilized gold nanoparticles by the application of an external stimulus. Langmuir 26, 2721–2730.

Burggraaf, A., Cot, L., 1996. Fundamentals of Inorganic Membrane Science and Technology. Elsevier Science B. V, Amsterdam.

Cadet, J.L., Bolla, K.I., 2007. Chapter 111 - environmental toxins and disorders of the nervous system a2 - Schapira, Anthony H.V. In: Byrne, E.A., Dimauro, S., Frackowiak, R.S.J., Johnson, R.T., Mizuno, Y., Samuels, M.A., Silberstein, S.D., Wszolek, Z.K. (Eds.), Neurology and Clinical Neuroscience. Mosby, Philadelphia.

Chen, X., Zheng, Z., Ke, X., Jaatinen, E., Xie, T., Wang, D., et al., 2010. Supported silver nanoparticles as photocatalysts under ultraviolet and visible light irradiation, Green Chem., 12. pp. 414–419.

Chen, Y.-C., Yu, K.-P., 2017. Enhanced antimicrobial efficacy of thermal-reduced silver nanoparticles supported by titanium dioxide. Colloids Surf. B 154, 195–202.

Cheng, X., Liang, H., Ding, A., Qu, F., Shao, S., Liu, B., et al., 2016. Effects of pre-ozonation on the ultra-filtration of different natural organic matter (NOM) fractions: membrane fouling mitigation, prediction and mechanism. J. Memb. Sci. 505, 15−25.

Coz, A., Andrés, A., Soriano, S., Viguri, J.R., Ruiz, M.C., Irabien, J.A., 2009. Influence of commercial and residual sorbents and silicates as additives on the stabilisation/solidification of organic and inorganic industrial waste. J. Hazard. Mater. 164, 755−761.

Cuperus, F.P., Bargeman, D., Smolders, C.A., 1992. Critical points in the analysis of membrane pore structures by thermoporometry. J. Memb. Sci. 66, 45−53.

Donini, C.A., Da Silva, M.K.L., Simões, R.P., Cesarino, I., 2018. Reduced graphene oxide modified with silver nanoparticles for the electrochemical detection of estriol. J. Electroanal. Chem. 809, 67−73.

Dos Santos, J.D.S., Alvarez-Puebla, R.A., Oliveira, J.O.N., Aroca, R.F., 2005. Controlling the size and shape of gold nanoparticles in fulvic acid colloidal solutions and their optical characterization using SERS. J. Mater. Chem. 15, 3045−3049.

Druet, P., Druet, E., Potdevin, F., Sapin, C., 1978. Immune type glomerulonephritis induced by $HgCl_2$ in the Brown Norway rat. Ann. d'immunol. 129 C, 777−792.

Elhadidy, A.M., Peldszus, S., Van Dyke, M.I., 2013. An evaluation of virus removal mechanisms by ultrafiltration membranes using MS2 and φX174 bacteriophage. Sep. Purif. Technol. 120, 215−223.

Ferain, E., Legras, R., 1997. Characterisation of nanoporous particle track etched membrane. Nucl. Instrum. Methods Phys. Res. Sect. B 131, 97−102.

Ferraris, S., Miola, M., Cochis, A., Azzimonti, B., Rimondini, L., Prenesti, E., et al., 2017. In situ reduction of antibacterial silver ions to metallic silver nanoparticles on bioactive glasses functionalized with polyphenols. Appl. Surf. Sci. 396, 461−470.

Gibson, N., Shenderova, O., Luo, T.J.M., Moseenkov, S., Bondar, V., Puzyr, A., et al., 2009. Colloidal stability of modified nanodiamond particles. Diamond Relat. Mater. 18, 620−626.

Harika, V.K., Kumar, V.B., Gedanken, A., 2018. One-pot sonochemical synthesis of Hg−Ag alloy microspheres from liquid mercury. Ultrason. Sonochem. 40, 157−165.

Henderson, D.C., Clifford, R., Young, D.M., 2001. Mercury-reactive lymphocytes in peripheral blood are not a marker for dental amalgam associated disease. J. Dent. 29, 469−474.

Hernández, A., Calvo, J.I., Prádanos, P., Tejerina, F., 1996. Pore size distributions in microporous membranes. A critical analysis of the bubble point extended method. J. Memb. Sci. 112, 1−12.

Hernández, A., Calvo, J.I., Prádanos, P., Tejerina, F., 1998. Pore size distributions of track-etched membranes; comparison of surface and bulk porosities. Colloids and Surfaces A: Physicochemical and Engineering Aspects 138, 391−401.

Hsien-Hsueh, L., Kan-Sen, C., Kuo-Cheng, H., 2005. Inkjet printing of nanosized silver colloids. Nanotechnology. 16, 2436−2441.

Hu, J., Wang, Z., Li, J., 2007. Gold nanoparticles with special shapes: controlled synthesis, surface-enhanced raman scattering, and the application in biodetection. Sensors (Basel) 7, 3299−3311.

Ioan, C.E., Aberle, T., Burchard, W., 2000. Structure properties of dextran. 2. Dilute solution. Macromolecules 33, 5730−5739.

Jakobs, E., Koros, W.J., 1997. Ceramic membrane characterization via the bubble point technique. J. Memb. Sci. 124, 149−159.

Kallioinen, M., Pekkarinen, M., Mänttäri, M., Nuortila-Jokinen, J., Nyström, M., 2007. Comparison of the performance of two different regenerated cellulose ultrafiltration membranes at high filtration pressure. J. Memb. Sci. 294, 93−102.

Kango, S., Kalia, S., Celli, A., Njuguna, J., Habibi, Y., Kumar, R., 2013. Surface modification of inorganic nanoparticles for development of organic−inorganic nanocomposites—a review. Progr. Polym. Sci. 38, 1232−1261.

Katok, K.V., Whitby, R.L.D., Fukuda, T., Maekawa, T., Bezverkhyy, I., Mikhalovsky, S.V., et al., 2012. Hyperstoichiometric interaction between silver and mercury at the nanoscale. Angew. Chem. Int. Ed. 51, 2632−2635.

Kawada, S., Saeki, D., Matsuyama, H., 2014. Development of ultrafiltration membrane by stacking of silver nanoparticles stabilized with oppositely charged polyelectrolytes. Colloids Surf. A 451, 33−37.

Kawasaki, M., Nishimura, N., 2006. 1064-nm laser fragmentation of thin Au and Ag flakes in acetone for highly productive pathway to stable metal nanoparticles. Appl. Surf. Sci. 253, 2208−2216.

Keesom, W.H., Zelenka, R.L., Radke, C.J., 1988. A zeta-potential model for ionic surfactant adsorption on an ionogenic hydrophobic surface. J. Colloid. Interface. Sci. 125, 575−585.

Khatoon, U.T., Nageswara, R.A.O., Mohan, G.V.S., Ramanaviciene, K.M., Ramanavicius, A., 2017. Antibacterial and antifungal activity of silver nanospheres synthesized by tri-sodium citrate assisted chemical approach. Vacuum 146, 259−265.

Kim, K.J., Stevens, P.V., 1997. Hydraulic and surface characteristics of membranes with parallel cylindrical pores. J. Memb. Sci. 123, 303−314.

Kim, T., Lee, K., Gong, M.-S., Joo, S.-W., 2005. Control of gold nanoparticle aggregates by manipulation of interparticle interaction. Langmuir 21, 9524−9528.

Kimling, J., Maier, M., Okenve, B., Kotaidis, V., Ballot, H., Plech, A., 2006. Turkevich method for gold nanoparticle synthesis revisited. J. Phys. Chem. B 110, 15700−15707.

Kozlovskiy, A., Zdorovets, M., Arkhangelsky, E., 2017. Track-etch membranes: the kazakh experience. Desalin. Water Treat. 76, 143−147.

Kruk, T., Szczepanowicz, K., Kręgiel, D., Szyk-Warszyńska, L., Warszyński, P., 2016. Nanostructured multilayer polyelectrolyte films with silver nanoparticles as antibacterial coatings. Colloids Surf. B 137, 158−166.

Kurland, L.T., Faro, S.N., Siedler, H., 1960. Minamata disease. The outbreak of a neurologic disorder in Minamata, Japan, and its relationship to the ingestion of seafood contaminated by mercuric compounds. World Neurol. 1, 370−395.

Kuzmenko, D., Arkhangelsky, E., Belfer, S., Freger, V., Gitis, V., 2005. Chemical cleaning of UF membranes fouled by BSA. Desalination 179, 323−333.

Langlet, J., Ogorzaly, L., Schrotter, J.-C., Machinal, C., Gaboriaud, F., Duval, J.F.L., et al., 2009. Efficiency of MS2 phage and Qβ phage removal by membrane filtration in water treatment: applicability of real-time RT-PCR method. J. Memb. Sci. 326, 111−116.

Latulippe, D.R., Ager, K., Zydney, A.L., 2007. Flux-dependent transmission of supercoiled plasmid DNA through ultrafiltration membranes. J. Memb. Sci. 294, 169−177.

Lazić, V., Smičiklas, I., Marković, J., Lončarević, D., Dostanić, J., Ahrenkiel, S.P., et al., 2018. Antibacterial ability of supported silver nanoparticles by functionalized hydroxyapatite with 5-aminosalicylic acid. Vacuum 148, 62−68.

Lebleu, N., Roques, C., Aimar, P., Causserand, C., 2010. Effects of membrane alterations on bacterial retention. J. Memb. Sci. 348, 56−65.

Leng, Z., Wu, D., Yang, Q., Zeng, S., Xia, W., 2018. Facile and one-step liquid phase synthesis of uniform silver nanoparticles reduction by ethylene glycol. Optik−Int. J. Light Electron Opt. 154, 33−40.

Lettmann, C., Möckel, D., Staude, E., 1999. Permeation and tangential flow streaming potential measurements for electrokinetic characterization of track-etched microfiltration membranes. J. Memb. Sci. 159, 243−251.

Li, K., 2007. Ceramic Membranes for Separation and Reaction. Chichester, John Wiley & Sons, LTD.

Liu, Y., Shipton, M.K., Ryan, J., Kaufman, E.D., Franzen, S., Feldheim, D.L., 2007. Synthesis, stability, and cellular internalization of gold nanoparticles containing mixed peptide − poly(ethylene glycol) monolayers. Anal. Chem. 79, 2221−2229.

Loran, S., Yelon, A., Sacher, E., 2018. Short communication: unexpected findings on the physicochemical characterization of the silver nanoparticle surface. Appl. Surf. Sci. 428, 1079–1081.

Malekzadeh, M., Halali, M., 2011. Production of silver nanoparticles by electromagnetic levitation gas condensation. Chem. Eng. J. 168, 441–445.

Monier, M., Elsayed, N.H., Abdel-Latif, D.A., 2015. Synthesis and application of ion-imprinted resin based on modified melamine–thiourea for selective removal of Hg(II). Polym. Int. 64, 1465–1474.

Mora-Barrantes, I., Rodriguez, A., Ibarra, L., Gonzalez, L., Valentin, J.L., 2011. Overcoming the disadvantages of fumed silica as filler in elastomer composites. J. Mater. Chem. 21, 7381–7392.

Mudasir, M., Karelius, K., Aprilita, N.H., Wahyuni, E.T., 2016. Adsorption of mercury(II) on dithizone-immobilized natural zeolite. J. Environ. Chem. Eng. 4, 1839–1849.

Mulder, M., 2000. Basic Principles of Membrane Technology. Kluwer Academic Publishers, Dordrecht.

Newcombe, G., Drikas, M., Hayes, R., 1997. Influence of characterised natural organic material on activated carbon adsorption: II. Effect on pore volume distribution and adsorption of 2-methylisoborneol. Water Res. 31, 1065–1073.

Oćwieja, M., Adamczyk, Z., Morga, M., Kubiak, K., 2015. Influence of supporting polyelectrolyte layers on the coverage and stability of silver nanoparticle coatings. J. Colloid. Interface. Sci. 445, 205–212.

Okitsu, K., Mizukoshi, Y., Bandow, H., Maeda, Y., Yamamoto, T., Nagata, Y., 1996. Formation of noble metal particles by ultrasonic irradiation. Ultrason. Sonochem. 3, S249–S251.

Pontié, M., Durand-Bourlier, L., Lemordant, D., Lainé, J.M., 1998. Control fouling and cleaning procedures of UF membranes by a streaming potential method. Sep. Purif. Technol. 14, 1–11.

Poulston, S., Granite, E.J., Pennline, H.W., Myers, C.R., Stanko, D.P., Hamilton, H., et al., 2007. Metal sorbents for high temperature mercury capture from fuel gas. Fuel 86, 2201–2203.

Puspitasari, V., Granville, A., Le-Clech, P., Chen, V., 2010. Cleaning and ageing effect of sodium hypochlorite on polyvinylidene fluoride (PVDF) membrane. Sep. Purif. Technol. 72, 301–308.

Ren, J., Li, Z., Wong, F.-S., 2006. A new method for the prediction of pore size distribution and MWCO of ultrafiltration membranes. J. Memb. Sci. 279, 558–569.

Roberts Alley, E., 2007. Water Quality Control Handbook. McGraw-Hill, New-York.

Schönenberger, C., Van Der Zande, B.M.I., Fokkink, L.G.J., Henny, M., Schmid, C., Krüger, M., et al., 1997. Template synthesis of nanowires in porous polycarbonate membranes: electrochemistry and morphology. J. Phys. Chem. B 101, 5497–5505.

Siegel, J., Lyutakov, O., Polívková, M., Staszek, M., Hubáček, T., Švorčík, V., 2017. Laser-assisted immobilization of colloid silver nanoparticles on polyethyleneterephthalate. Appl. Surf. Sci. 420, 661–668.

Slot, J.W., Geuze, H.J., 1985. A new method of preparing gold probes for multiple-labeling cytochemistry. Eur. J. Cell Biol. 38, 87–93.

Sonavane, G., Tomoda, K., Sano, A., Ohshima, H., Terada, H., Makino, K., 2008. In vitro permeation of gold nanoparticles through rat skin and rat intestine: effect of particle size. Colloids Surf. B 65, 1–10.

Tarnawski, V.R., Jelen, P., 1986. Estimation of compaction and fouling effects during membrane processing of cottage cheese whey. J. Food Eng. 5, 75–90.

Tréguer-Delapierre, M., Majimel, J., Mornet, S., Duguet, E., Ravaine, S., 2008. Synthesis of non-spherical gold nanoparticles. Gold Bull. 41, 195–207.

Turkevich, J., Stevenson, P.C., Hillier, J., 1951. A study of the nucleation and growth processes in the synthesis of colloidal gold. Discuss. Faraday Soc. 11, 55–75.

Tyliszczak, B., Drabczyk, A., Kudøacik-Kramarczyk, S., Bialik-Wås, K., Kijkowska, R., Sobczak-Kupiec, A., 2017. Preparation and cytotoxicity of chitosan-based hydrogels modified with silver nanoparticles. Colloids Surf. B 160, 325–330.

Unep, 2013. Global MercuryAssessment 2013: Sources, Emissions, Releases, and Environmental Transport. *UNEP*.

Chapter 1 Silica: preparation and properties. In: Vansant, E.F., Van Der Voort, P., Vrancken, K.C. (Eds.), Studies in Surface Science and Catalysis. Elsevier.

Chapter 3 The surface chemistry of silica. In: Vansant, E.F., Van Der Voort, P., Vrancken, K.C. (Eds.), Studies in Surface Science and Catalysis. Elsevier.

Volova, T.G., Shumilova, A.A., Shidlovskiy, I.P., Nikolaeva, E.D., Sukovatiy, A.G., Vasiliev, A.D., et al., 2018. Antibacterial properties of films of cellulose composites with silver nanoparticles and antibiotics. Polym. Test. 65, 54−68.

Wakuda, D., Kim, K.-S., Suganuma, K., 2008. Room temperature sintering of Ag nanoparticles by drying solvent. Scr. Mater. 59, 649−652.

Weisener, C.G., Sale, K.S., Smyth, D.J.A., Blowes, D.W., 2005. Field column study using zerovalent iron for mercury removal from contaminated groundwater. Environ. Sci. Technol. 39, 6306−6312.

Ye, M., Wang, R., Shao, Y., Tian, C., Zheng, Z., Gu, X., et al., 2018. Silver nanoparticles/graphitic carbon nitride nanosheets for improved visible-light-driven photocatalytic performance. J. Photochem. Photobiol. A 351, 145−153.

Zhang, Q., Liu, N., Cao, Y., Zhang, W., Wei, Y., Feng, L., et al., 2018. A facile method to prepare dual-functional membrane for efficient oil removal and in situ reversible mercury ions adsorption from wastewater. Appl. Surf. Sci. 434, 57−62.

4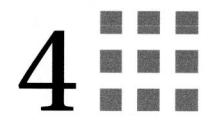

Overview of Potential Applications of Nano-Biotechnology in Wastewater and Effluent Treatment

Rama Rao Karri[1], Shahriar Shams[2], J.N. Sahu[3]

[1]PETROLEUM AND CHEMICAL ENGINEERING, FACULTY OF ENGINEERING, UNIVERSITI TEKNOLOGI BRUNEI, GADONG, BRUNEI [2]CIVIL ENGINEERING PROGRAMME AREA, FACULTY OF ENGINEERING, UNIVERSITI TEKNOLOGI BRUNEI, GADONG, BRUNEI [3]FACULTY OF CHEMISTRY, INSTITUTE OF CHEMICAL TECHNOLOGY, UNIVERSITY OF STUTTGART, STUTTGART, GERMANY

4.1 Introduction

As we are aware, the total saltwater oceans make up to 71% of the Earth's surface, whereby the balance 29% is made up of Earth's continents and islands. Even though three-quarters of the Earth is made up of water, the demand for potable water is increasing due to enormous rise in the population and the rapid growth in the industrial sector. The main threat for this increase in water demand would be to provide clean water to the required sector. Besides household usage as drinking water and for cooking, clean water also a demanding feedstock for growing industries such as food, pharmaceuticals, electronics and medical. On the other hand, the attainable supply of fresh water is going through a complication in which the total amount of freshwater is reducing due to global warming, population growth and more drastic regulation based on health. On that note, 2 million people die yearly due to diarrheal disease and lack of improved sanitation facility. Hence, a continuous supply of clean water is essential for our daily life usage. Besides that, many manufacturing companies that produce paper, plastic, and metal depends on the supply of clean water for production. Hence, one can conserve the usage of water by further understanding the use and reuse of products that are being used to produce a clean and potable water. An increase in the demand for clean water accelerates the demand in the reuse of wastewater and freshwater for agriculture and aquaculture as well. To meet this demand, more novel techniques and framework for water treatment technology are required as the current ones are reaching their final stages of providing adequate water treatment to fit into daily human and environmental needs.

Generally, water is polluted due to the effluent of the industrial sector as well as due to lack of improved sanitation facility. Inconsiderate human behavior as deploying rubbish into any water resources such as river or lake also contribute to water pollution. The traditional

Nanotechnology in Water and Wastewater Treatment. DOI: https://doi.org/10.1016/B978-0-12-813902-8.00004-6

method of water treatment, water distribution and discharge practices are no longer sustainable as it relies highly on a centralized system. Reliable and sustainable water supply is one of the most fundamental humanitarian needs and yet remains a significant challenge to meet the 21st century's global demand. Rising cost of potable water driven by growing populations influenced by a variety of climatic and environmental issues (Adeleye et al., 2016) is driving cause for failing to people to provide adequate drinking water. Today, 1.1 billion people lack a supply of adequate drinking water (Organization, 2015). With rapid industrial growth, urbanization and increased population, improved living standard, expansion of agriculture, a considerable amount of effluents are generated and released into waterbodies and contaminating the water, which are posing a threat both to human and aquatic lives. Wastewater and effluents from various small and medium-sized enterprises (SME's) and process industries like metal plating, battery manufacturing firms, mining operations, are the primary sources of contaminants in which heavy metals are the primary pollutants. These heavy metals which are not biodegradable pose a severe threat to plants, animals and human beings upon consumption of these contaminated water (El-Latif et al., 2013). Efforts have been made the to harness water quantity through the development of various hydraulic structures such as dams, channels, reservoirs whereas water conservation, recycling and desalination of water has been at the forefront of technological options particularly in water-stressed countries (Sa and Premalatha, 2016). These technological options are often expensive and not cost-effective; hence efforts are given on treatment of wastewater as there is a growing realization that water is finite (Rao et al., 2010; Karri et al., 2017a; Karri et al., 2017b; Lingamdinne et al., 2018). The need for technological innovation to enable sustainable and integrated water management is a step forward for achieving water security.

"Nano" are typically defined as materials one billionth of a meter (10^{-9}) and even smaller. At this nanoscale, materials are categorized by its physical, chemical and biological properties rather than their equivalent standard sizes (Davies, 2006). For example, materials such as metal oxides, polymers, metals, and carbon derivatives have a higher proportion of surface area to the particle size at the nanoscale level. These nanoscale particles display different mechanical, electrical, magnetic and optical properties, which is usually distinctive from the properties exhibited at a macroscopic scale (Theodore and Kunz, 2005; Lingamdinne et al., 2018). This distinguishing characteristic can be due to the fact that the higher the number of molecules atoms leads to decrease in particle size. At this scale, materials often possess size dependent properties which are different from their counterparts at macroscale.

Recently, research has been done by utilizing nano-biotechnology to replace the current water treatment technology. Qu et al. (Qu et al., 2012; Lingamdinne et al., 2018) explained through his research that nanotechnology, which was enabled by an efficient and multifunctional process is anticipated to offer a satisfaction water treatment solution which doesn't rely on any significant and expensive infrastructures (Qu et al., 2012). Hence, this astounding nanotechnology can be utilized with other water treatment methods as adsorption, coagulation, membrane technology, and photocatalysis establish a water treatment methods with higher efficiency, lower cost and environmentally acceptable (Qu et al., 2013; Karri and Sahu, 2017; Karri et al., 2017b). This article provides a brief summary of the extensive utilization of nano-biotechnology based nanomaterials based on its application in water treatment and collaboration with other treatment techniques. The primary objective of this paper would be to discuss the opportunities and challenges of utilizing nanomaterials cojointly with biotechnology for a different type of water treatment approach.

4.2 Current Nano-Based Methodologies in Wastewater and Effluent Treatments

4.2.1 Carbon-Based Nano-Adsorbents

The removal of both organic and inorganic pollutants in the treatment of wastewater is majorily accomplished through adsorption (Fig. 4-1). Carbon-based nano absorbents are evolving through the development of carbonaceous nanomaterials (CNMs) which includes carbon nanoparticles, carbon nano sheets and carbon nanotubes (CNT's). CNT's are more preferred as compared to activated carbons due to higher adsorption efficiency of numerous organic chemicals (Pan and Xing, 2008). Its outstanding chemical resistance, high specific surface

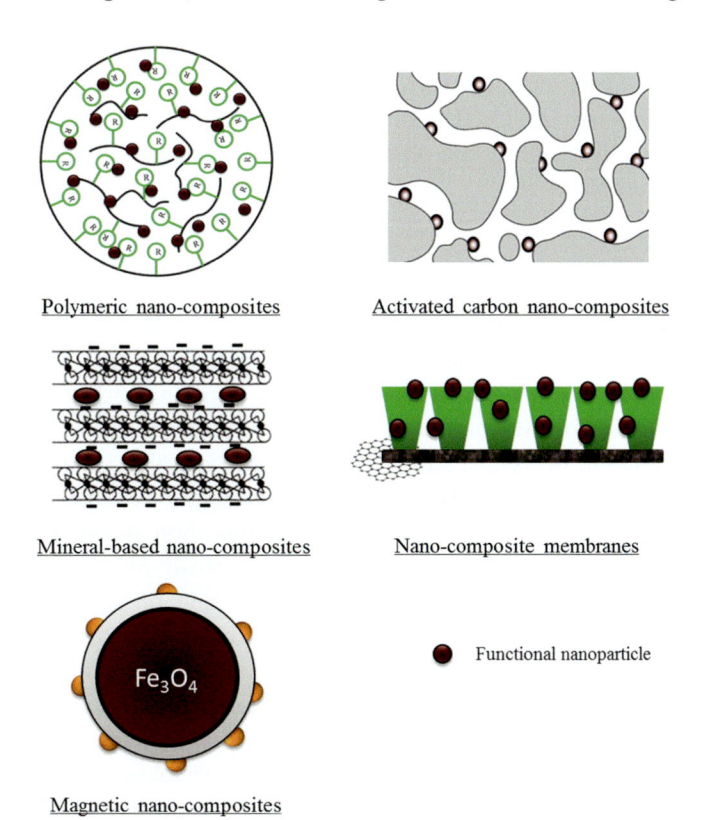

Polymeric nano-composites   Activated carbon nano-composites

Mineral-based nano-composites   Nano-composite membranes

● Functional nanoparticle

Fe_3O_4

Magnetic nano-composites

FIGURE 4-1 Various applications of nano-composites used in water treatments. Source: *Zhang, Y., Wu, B., Xu, H., Liu, H., Wang, M., He, Y., et al., 2016. Nanomaterials-enabled water and wastewater treatment. NanoImpact 3-4, 22–39.*

area, excellent adsorption capacity, and mechanical strength diverse CNT applications (Lee et al., 2012). CNT's are particularly useful in removing heavy metals such as nickel ions (Ni^{2+}) from water (Ren et al., 2011). The surface characteristics of CNT's can be improved further through grafting. The grafting comprises of functional groups through various processes such as microwave, plasma technique, and chemical modification (Zhang et al., 2012; Chen et al., 2012).

CNT transforms into aggregates in aqueous condition due to hydrophobic effect of graphitic surfaces. Interstitial spaces and grooves within the aggregates form high absorption energy for trapping organic molecules. Notably, bulky organic molecules are stuck due to large pores owing to high adsorption capacity of CNT's. Covalent bonding and electrostatic interactions, hydrogen bonding (for compounds with $-OH$, $-NH_2$, $-COOH$, functional groups), pi$-$pi interactions (for polar aromatic compounds, polycyclic aromatic hydrocarbons) are other reasons for trapping organic components present in wastewater. Chemical bonding and electrostatic attraction create CNTs to adsorption metal ions due to surface functional groups (carboxyl, hydroxyl, and phenol) (Rao et al., 2007). CNT's-based composite filter can remove heavy metal ions from water very efficiently (Parham et al., 2013). One of the significant advantages of CNT nano absorbents is the removal of Zn^{2+} efficiently through continuous regenerated and reused CNT nano absorbents. Besides these, multiwall carbon nanotubes (MWCNT) (Tang et al., 2012; Tarigh and Shemirani, 2013) has been also used to remove heavy metals such as Cu(II) (Tang et al., 2012), Pb(II), and Mn(II) (Tarigh and Shemirani, 2013), thus overcoming the obstacles of poor diffusion ability, difficulty in separation and small particles size.

4.2.2 Metal-Based Nano-Adsorbents

These metal-based nano absorbents such as metal oxides (MO) are lower in cost and efficient adsorbents for removing heavy metals and radionuclides (Sharma et al., 2009). In Particular, MO-based nanomaterials have exhibited superior performance over the traditional activated carbon form (Mayo et al., 2007). The sorption in these adsorbents is significant due to a combination of oxygen in metal oxides and dissolved metals. This approach is a two-step process as shown in Fig. 4-2. In this approach, the external surface of the metal-based nano absorbents adsorbs metal ions quickly and then intraparticle diffusion takes place in micropore walls (Trivedi and Axe, 2000) because of minimum environmental impacts, the higher specific surface area, less solubility, and no residual pollutants (Gupta et al., 2015). The higher number of surface (edges and corners) reaction sites and diffusion distance results due to shorter intraparticle distance. The adsorption capacity of arsenic increases over 100 times if nano-magnetite particle size decreases from 300 to 11 nm (Yean et al., 2005). Currently, various disinfection processes involve the comprehensive application of TiO_2 as they can activate as photocatalytic even during visible sunlight. Other unique features of TiO_2 are nontoxicity if by ingesting. Besides, it is widely applicable for low-cost. Oxides of magnesium, cerium, and manganese are also very capable for removal of heavy metals (Liu et al., 2008; Huang and Chen, 2009). The performance of adsorption of nano-based metal oxides depends on shape and size. Facile synthetic methods (physical and chemical) as shown in Table 4-1 have used widely to obtain the desired shape. The applications of different metal-based nano absorbents used for removing contaminants from wastewater a given Table 4-2 below.

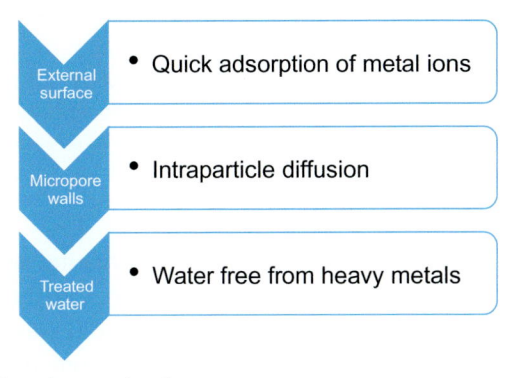

FIGURE 4-2 Process for metal-based nano absorbents.

Table 4-1   Methods Applied for Controlling Shape of Nano-Based Oxides

Physical Method	Chemical Method
Ultrasound shot peening	Chemical vapor condensation
Inert gas condensation	Pulse electrode position
Plastic deformation	Controlled chemical coprecipitation
High-energy ball milling	Liquid-phase reduction, liquid flame spray

Table 4-2   Application of Metal-Based Nano Absorbents

Metal-Based Nano Absorbent	Method of Synthesis	Particles Size	Types of Heavy Metals Removed	Adsorption Capacity (mg/g)	References
ZnO	Hydrothermal	1 μm	Pb(II)	6.7	Shang et al. (2007)
Hematite (α-Fe$_2$O$_4$)	Coprecipitation	74 nm	Cu(II)	84.46	Grossl et al. (1994)
Hydrous Al oxides	Precipitation	1.9 nm	Pb(II)	6.7	Hiraide et al. (1994)
MnO$_2$	Precipitation	2.1 nm	Pb(II)	78.74	Babel and Kurniawan (2004)
TiO$_2$	Hydrolysis	2.1 nm	Zn(II), Cd(II)	15.3 for Zn(II), 7.9 for Cd(II)	Engates and Shipley (2010)
Fe$_2$O$_4$-MnO$_2$	Hydrothermal	15.15 nm	Ni (II)	55.63	Zhao et al. (2016)
γ-Al$_2$O$_3$	Precipitation	7.5 nm	Ni(II)	176	Ghaedi et al. (2008)

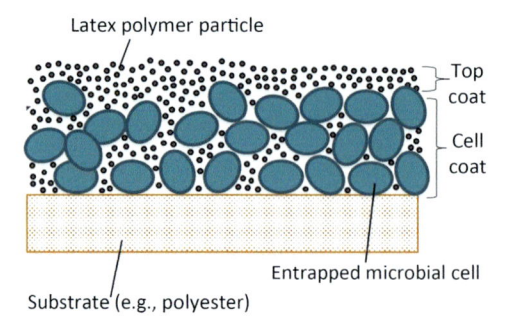

FIGURE 4-3 Latex polymer particle entrapping microbial cell. Source: *Cortez, S., Nicolau, A., Flickinger, M.C., Mota, M., 2017. Biocoatings: a new challenge for environmental biotechnology. Biochem. Eng. J. 121, 25−37.*

4.2.3 Polymeric Nano-Adsorbents

These nano absorbents are proficient in removing both organics and heavy metals through specific cavities for removal of pollutants. Heavy metals are absorbed due to the presence of hydroxyl at the external surface, while organic pollutants from wastewater are removed due to sorption by the hydrophobic action at its inner surface. A polymeric nano absorbent such as dendrimer-ultrafiltration system has efficiently removed metal ions from aqueous solutions (Diallo et al., 1999). Synthetic polymers such as molecularly imprinted polymers (MIP's) are gaining a reputation for wastewater treatment by mimicking natural detection and identification, demonstrating high affinity and selectivity for its structural analogs (Advincula, 2011). There is an excellent potential for wastewater treatment using MIP's-based composite materials. MIPs can reserve their prominent selectivity and high adsorption capability through the use of composite materials such quantum dots, magnetic, and nanoparticles. Polymeric nanoparticles obtained through synthesis of Fe- and Al-doped, activated micron (~ 0.8 mm), and nano (~ 100 nm) sized porous adsorbents has successfully removed arsenic (V) (~ 40 mg/g) and fluoride (~ 100 mg/g) ions from effluents and wastewater (Kumar et al., 2011). Adhesive biocatalytic coatings having nanoporous microstructure created by partly-combined latex polymer particles (Fig. 4-3) capture highly concentrated microbial cells (harmful pathogenic bacteria) in a dry state stabilized by carbohydrate osmoprotectants during wastewater treatment (Cortez et al., 2017).

4.2.4 Magnetic Nanoparticles

These magnetic based nanoparticles comprise of two components, namely a magnetic material (nickel, iron, cobalt) and a chemical component that has characteristics for removal of heavy metals. The significant advantages of using magnetic nanoparticles are faster and efficient method in comparison to filtration methods (Ambashta and Sillanpää, 2010). For example capacity of arsenic, adsorption increased greater than 100 times when magnetic nanoparticles size is decreased by 30 times (Yean et al., 2005). Superparamagnetic nanoparticles (maghemite, magnetite) hold much potential for wastewater treatment since they

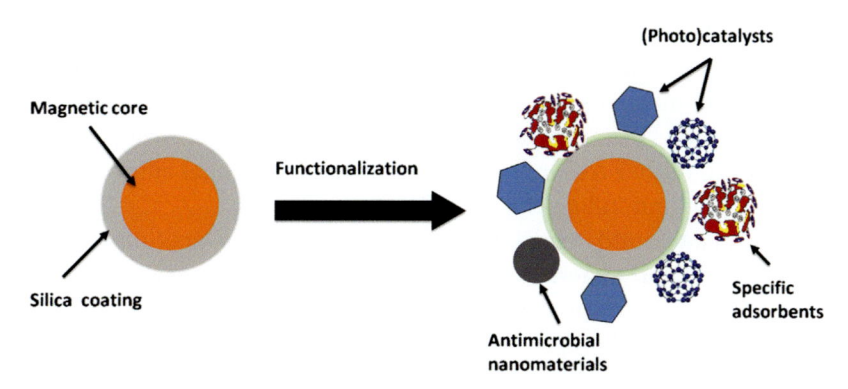

FIGURE 4-4 Multifunctional magnetic nanoparticles. Source*: Qu, X., Alvarez, P.J.J., Li, Q., 2013. Applications of nanotechnology in water and wastewater treatment. Water Res. 47, 3931−3946.*

express excellent biocompatibility which concerning the impacts of the material on the environment is an advantage compared to metallic nanoparticles. Magnetic particles become superparamagnetic when the size decreases (<40 nm) which allow quick split-up and repossession by a low-gradient magnetic field due to losing permanent magnetic moments attracted by external magnetic field. Magnetic nanoparticle structure comprises a core-shell coated with silica (Fig. 4-4) in which the outer shell provides the required tasks for the magnetic core to accomplish the magnetic separation (Qu et al., 2013).

Maghemite nanoparticles are used for its simple, cost-effective, space-saving and environmentally friendly approach through a combination of magnetic separation and adsorption process for recovering/removing chromium from wastewater (Hu et al., 2005). Original metal removing the capacity of nanoscale maghemite was retained after six adsorptions (Optimal pH value 2.5) -desorption cycles. Carbonaceous materials, ferrite with Ba, Co, and Mn, and maghemite are also used for removal of several types of pollutants from wastewater, wherein iron oxide is utilized as Fenton catalysts (Shahwan et al., 2011; Sun and Lemley, 2011). Zn^{2+} and Cd^{2+} are removed from effluents through the synthesis of magnetic hydroxyapatite nanoparticles (MNHAP) adsorbents (Feng et al., 2010). Fe_3O_4-magnetite is used for coating with humic acid for successful and productive removal of heavy metals (Liu et al., 2008).

4.3 Reuse and Retention of Nanomaterials

In the outline of nano technology-based devices, the preservation and reuse of nanomaterials play a vital role in dictating the efficiency of the process as well as the operation cost of the process. This is customarily done by immobilizing the nanomaterials in the treatment approach. Membrane filtration was observed to be extremely viable in the different detachment process, because of notable highlights like minimal chemical utilization and can be utilized in the commercial process where the operation is continuous. Polymeric and ceramic membranes have found wide applications in the treatment of water (Tewari et al., 2010;

Padaki et al., 2015; Aba et al., 2015), whereas the main drawback of this process is that the suspended particles in the effluent water are held by the film, thus reducing the effectiveness of the membrane. Nanomaterials can be immobilized on membranes and resins to avoid further separation. However, the present immobilization procedures result in massive loss of efficiency. Henceforth, further research is needed to create minimal effort methods to immobilize nanomaterials without influencing its efficiency.

In the present digital world, very little research available in open-literature has investigated on the release of nano materials and its strategies from nanotechnology-based applications. The measure of nanomaterial discharge relies upon the immobilization method and the procedure utilized for the separation. The extraction of nanomaterial covered on the material surface is simple and very quick unless there is an application of downstream separation process. While the extraction of nanomaterials which is embedded in a solid matrix with the treatment material is extremely troublesome and insignificant discharge can be expected before the material is discarded. For nanomaterials that release metal ions, their breaking down ought to be painstakingly controlled by streamlining the shape, size, and coating thickness. For hazard evaluation, the identification of nanomaterial release is a noteworthy pragmatic obstacle and remains challenging. The methods which can distinguish nanomaterials in complex effluents are expensive and sophisticated, and have numerous pragmatic constraints (da Silva et al., 2011). It will be a great need for the society if there is any quick and precise investigative method to identify particular nanomaterials.

4.4 Benefits of Nano-Biotechnology Based Applications for Water Sustainability

Nano-biotechnology was found many applications due to its immense benefits in addressing challenges related to food, water, and biodiversity. Due to its promising capability to enhance the quality of water resources for human utilization, it finds eminence for sustainability. The importance and alliance of nanoscience, nanomaterials, biotechnology, and its associated fields, scientists and researchers are continuously striving in developing and fabricating innovative and functional biological nano-systems. Above all, the controlled exploration of engineered bio-coatings leads to the low-cost biochemical transformation of effluent and wastewaters with enhanced efficiency and intensity.

Membrane separation technology has gained prominence in water treatment process and recently the thin-film composite (TFC) membrane with imparted antibacterial material has found many applications (Kwak et al., 2001; Kim et al., 2003; Ong et al., 2016). This TFC membrane enhances the antimicrobial activity of the membrane. The benefits of this TFC membrane is that it results in the more significant active surface area and thus increases the membrane performance (Kwak et al., 2001; Ong et al., 2016). Kim et al. (2003) performed experiments where TiO_2 was immobilized onto the surface and found that TiO_2 self-assembled hybrid membrane possessed a remarkably higher photocatalytic bactericidal efficiency.

Stoller (2009) analyzed the molecule measure dispersion in the effluent at the outlet of each stage and underlined the advantageous impact made by auxiliary flocculation in case of using the AS flocculant through the accumulation of the derived salts nearby the membrane layer. This helped carry away the particles by the tangential turbulence and reduced fouling. Similar results were observed by Stoller and Bravi (2010) upon applying additional UV photocatalysis (PC) with TiO_2 nanoparticles or aerobic digestion. All pretreatment processes provided final treated streams apt for irrigation, but in particular UV/TiO_2 photocatalysis ensured the optimum membrane performance and implied the shortest residence time (24 hours). The most important advantages of TiO_2/UV light photocatalysis are related to the very high ratio of surface/volume of TiO_2 nanoparticles, which enhances the mass transfer when used as photocatalytic material, even at ambient temperature conditions, the possibility of doping these nanoparticles in order to habilitate their activation with sun irradiation, and their resistance to photo-corrosion (Chatterjee and Dasgupta, 2005; Chen et al., 2007). Nogueira et al. (2015) investigated the treatment of Olive mill wastewater through photocatalytic oxidation with TiO_2 and Fe_2O_3 nanocatalysts combined with an ulterior biological process with *Pleurotus sajor caju* and *Phanerochaete chrysosporium* fungi. Results showed that nano-$TiO_2/H_2O_2/UV$ ensured up to 43%, 14%, 38% and 31% removal values for color, aromatic compounds content, chemical oxygen demand (COD), and total phenols, respectively, but no toxicity decrease was attained. However, the integration of a biological treatment helped reduce the COD and total phenols concentration as well as the toxicity. The treatment with *P. chrysosporium* promoted the highest reduction in toxicity, but *P. sajor caju* yielded the highest COD and total phenols abatement. Furthermore, the effectiveness of the biological treatment stage was enhanced by pretreatment with hydrogen peroxide.

Dwivedi et al. (2017) used Cellulose Nanofiber (CNF) composite; four CNF derivatives are designed by introducing the mussel-inspired conjugation of dopamine (a natural catecholamine of water habitat mussels) by exploiting stable complexation between catechol and metal ions, and iron-based cross-linkers with nano-cellulose. The CNF derivatives demonstrated efficient uptake of carcinogenic arsenic and chromium from the mixed contaminated waters, a distinct improvement compared to commercially available materials. Table 4-3 shows the application of various treatment process for diverse wastewater sources along with the scale of the experiment and achievements.

4.5 Conclusions

Due to rapid industrial growth, urbanization and increased population, a massive amount of effluents are generated and released into waterbodies which contaminate the water. In this regard, there is a major need for scientific innovation to enable sustainable and integrated water management, which would be a step forward in achieving water security. Owing to the multiple benefits of nanotechnology, these methodologies are overtaking other conventional water treatment methods such as adsorption, coagulation, membrane technology, and photocatalysis. This approach not only results in higher contaminant removal, it is also lower

Table 4-3 Application of Various Treatment Process for Diverse Wastewater Sources

References	Wastewater/ Contaminated Water Source	Treatment Process	Scale (Pilot/ Lab)	Country	Achieved Standards
Stoller and Bravi (2010)	Olive mill	UV photocatalysis along with nanometric TiO_2 and aerobic digestion	Pilot	Italy	Final RO permeate streams with COD equal to 456, 242, and 385 mg/L for AS, AD + AS, and PC, respectively
Němeček et al. (2016)	Groundwater	Injecting Zn(VI) suspensions and subsequently an organic substrate	Pilot	Czech Republic	Indicated that Zn(VI) was efficient in removal of Cr(VI) from the groundwater (LOQ 0.05 mg/L) and the subsequent application of organic substrate resulted in a high removal (97%) of chlorinated ethenes
Nogueira et al. (2015)	Olive mill	PC oxidation with TiO_2 and Fe_2O_3 nanocatalysts plus biological process	Lab	Portugal	Nano-TiO_2/UV/H_2O_2 ensured up to 43%, 38%, 31%, and 14% removal of color, COD, total phenols, and aromatic compounds, respectively.
Costa and Alves (2013)	Olive mill	PC nano-TiO_2	Lab	Portugal	$90.8 \pm 2.7\%$ removal of the phenolic content
Feng et al. (2010)	Synthetic wastewater contain Zn^{2+} and Cd^{2+}	Adsorption: magnetic hydroxyapatite nanoparticles	Lab scale	China	Maximum adsorption capacities towards Zn^{2+} and Cd^{2+} ions are 2.151 and 1.964 mmol/g, respectively
Zhou et al. (2014)	Synthetic wastewater contain Pb^{2+} ions	Adsorption: carboxylated cellulose nano fibrils-filled magnetic chitosan hydrogel beads	Lab scale	China	Proven effective in eliminating Pb (II) ions
Dwivedi et al. (2017)	Synthetic wastewater contain As (V) and Cr(VI) ions	Adsorption: CNF composite	Lab scale	South Korea	The CNF derivatives demonstrated efficient uptake of carcinogenic arsenic and chromium from the mixed contaminated waters, a distinct improvement compared to commercially available materials
Ahmad et al. (2017)	Coking wastewater	Adsorption and biodegradation with the help of a modified biocarrier	Lab scale	China	Simultaneous nitrification/denitrification were reached with the removal of up to 322 mg/L (98%) NH_4, 311 mg/L (99%) NO_2, and 633 mg/L (97%) total nitrogen

in cost, as well as environmentally friendly. Many researchers are continuously striving to improve and develop novel nanomaterials which further enhance the usage of nanoparticles for treating wastewater and effluents at a large scale so that they increase the selectivity, affinity, capacity, and capability to work at any operating conditions. Nano-biotechnology where the benefits of biotechnology is integrated with nano materials has found many applications and due to its promising capability to enhance the quality of water resources for human utilization, it finds eminence for sustainability. As the technology is growing and stringent environmental regulations are in place, scientists and researchers are continuously striving in developing and fabricating innovative and functional biological nano-systems. Above all, the controlled exploration of engineered bio-coatings leads to the low-cost biochemical transformation of effluent and wastewaters with enhanced efficiency and intensity. Looking at the current nano-based technologies for the treatment of wastewater, bio based nano materials seems to the viable and efficient. Above all, the controlled exploration of engineered bio-coatings leads to the low-cost biochemical transformation of effluent and wastewaters with enhanced efficiency and intensity.

References

Aba, N.F.D., Chong, J.Y., Wang, B., Mattevi, C., Li, K., 2015. Graphene oxide membranes on ceramic hollow fibers—microstructural stability and nanofiltration performance. J. Memb. Sci. 484, 87—94.

Adeleye, A.S., Conway, J.R., Garner, K., Huang, Y., Su, Y., Keller, A.A., 2016. Engineered nanomaterials for water treatment and remediation: costs, benefits, and applicability. Chem. Eng. J. 286, 640—662.

Advincula, R.C., 2011. Engineering molecularly imprinted polymer (MIP) materials: developments and challenges for sensing and separation technologies. Kor. J. Chem. Eng. 28, 1313—1321.

Ahmad, M., Liu, S., Mahmood, N., Mahmood, A., Ali, M., Zheng, M., et al., 2017. Synergic adsorption—biodegradation by an advanced carrier for enhanced removal of high-strength nitrogen and refractory organics. ACS Appl. Mater. Interfaces 9, 13188—13200.

Ambashta, R.D., Sillanpää, M., 2010. Water purification using magnetic assistance: a review. J. Hazard. Mater. 180, 38—49.

Babel, S., Kurniawan, T.A., 2004. Cr (VI) removal from synthetic wastewater using coconut shell charcoal and commercial activated carbon modified with oxidizing agents and/or chitosan. Chemosphere 54, 951—967.

Chatterjee, D., Dasgupta, S., 2005. Visible light induced photocatalytic degradation of organic pollutants. J. Photochem. Photobiol. C 6, 186—205.

Chen, D., Jiang, Z., Geng, J., Wang, Q., Yang, D., 2007. Carbon and nitrogen Co-doped TiO_2 with enhanced visible-light photocatalytic activity. Ind. Eng. Chem. Res. 46, 2741—2746.

Chen, H., Li, J., Shao, D., Ren, X., Wang, X., 2012. Poly(acrylic acid) grafted multiwall carbon nanotubes by plasma techniques for Co(II) removal from aqueous solution. Chem. Eng. J. 210, 475—481.

Cortez, S., Nicolau, A., Flickinger, M.C., Mota, M., 2017. Biocoatings: a new challenge for environmental biotechnology. Biochem. Eng. J. 121, 25—37.

Costa, J.C., Alves, M.M., 2013. Posttreatment of olive mill wastewater by immobilized TiO_2 photocatalysis. Photochem. Photobiol. 89, 545—551.

Da Silva, B.F., Pérez, S., Gardinalli, P., Singhal, R., Mozeto, A.A., Barceló, D., 2011. Analytical chemistry of metallic nanoparticles in natural environments. TrAC Trends Anal. Chem. 30, 528—540.

Davies, J.C., 2006. Managing the effects of nanotechnology.

Diallo, M.S., Balogh, L., Shafagati, A., Johnson, J.H., Goddard, W.A., Tomalia, D.A., 1999. Poly(amidoamine) dendrimers: a new class of high capacity chelating agents for Cu(II) ions. Environ. Sci. Technol. 33, 820−824.

Dwivedi, A.D., Sanandiya, N.D., Singh, J.P., Husnain, S.M., Chae, K.H., Hwang, D.S., et al., 2017. Tuning and characterizing nanocellulose interface for enhanced removal of dual-sorbate (AsV and CrVI) from water matrices. ACS Sustain. Chem. Eng. 5, 518−528.

El-Latif, M.M.A., Ibrahim, A.M., Showman, M.S., Hamide, R.R.A., 2013. Alumina/iron oxide nano composite for cadmium ions removal from aqueous solutions. Int. J. Nonferrous Metall. 2, 47.

Engates, K.E., Shipley, H.J., 2010. Adsorption of Pb, Cd, Cu, Zn, and Ni to titanium dioxide nanoparticles: effect of particle size, solid concentration, and exhaustion. Environ. Sci. Pollut. Res. 18, 386−395.

Feng, Y., Gong, J.-L., Zeng, G.-M., Niu, Q.-Y., Zhang, H.-Y., Niu, C.-G., et al., 2010. Adsorption of Cd (II) and Zn (II) from aqueous solutions using magnetic hydroxyapatite nanoparticles as adsorbents. Chem. Eng. J. 162, 487−494.

Ghaedi, M., Niknam, K., Shokrollahi, A., Niknam, E., Rajabi, H.R., Soylak, M., 2008. Flame atomic absorption spectrometric determination of trace amounts of heavy metal ions after solid phase extraction using modified sodium dodecyl sulfate coated on alumina. J. Hazard. Mater. 155, 121−127.

Grossl, P.R., Sparks, D.L., Ainsworth, C.C., 1994. Rapid kinetics of Cu(II) adsorption/desorption on goethite. Environ. Sci. Technol. 28, 1422−1429.

Gupta, V.K., Tyagi, I., Sadegh, H., Ghoshekand, R.S., Makhlouf, A.S.H., Maazinejad, B., 2015. Nanoparticles as adsorbent; a positive approach for removal of noxious metal ions: a review. Sci., Technol. Dev. 34, 195−214.

Hiraide, M., Sorouradin, M.-H., Kawaguchi, H., 1994. Immobilization of dithizone on surfactant-coated alumina for preconcentration of metal ions. Anal. Sci. 10, 125−127.

Hu, J., Chen, G., Lo, I.M.C., 2005. Removal and recovery of Cr(VI) from wastewater by maghemite nanoparticles. Water Res. 39, 4528−4536.

Huang, S.-H., Chen, D.-H., 2009. Rapid removal of heavy metal cations and anions from aqueous solutions by an amino-functionalized magnetic nano-adsorbent. J. Hazard. Mater. 163, 174−179.

Karri, R.R., Sahu, J.N., 2017. Modeling and optimization by particle swarm embedded neural network for adsorption of zinc (II) by palm kernel shell based activated carbon from aqueous environment. J. Environ. Manage. 206, 178−191.

Karri, R.R., Jayakumar, N., Sahu, J., 2017a. Modelling of fluidised-bed reactor by differential evolution optimization for phenol removal using coconut shells based activated carbon. J. Mol. Liq. 231, 249−262.

Karri, R.R., Sahu, J.N., Jayakumar, N.S., 2017b. Optimal isotherm parameters for phenol adsorption from aqueous solutions onto coconut shell based activated carbon: Error analysis of linear and non-linear methods. J. Taiwan Inst. Chem. Eng. 80, 472−487.

Kim, S.H., Kwak, S.-Y., Sohn, B.-H., Park, T.H., 2003. Design of TiO_2 nanoparticle self-assembled aromatic polyamide thin-film-composite (TFC) membrane as an approach to solve biofouling problem. J. Memb. Sci. 211, 157−165.

Kumar, V., Talreja, N., Deva, D., Sankararamakrishnan, N., Sharma, A., Verma, N., 2011. Development of bi-metal doped micro- and nano multi-functional polymeric adsorbents for the removal of fluoride and arsenic(V) from wastewater. Desalination 282, 27−38.

Kwak, S.-Y., Kim, S.H., Kim, S.S., 2001. Hybrid organic/inorganic reverse osmosis (RO) membrane for bactericidal anti-fouling. 1. Preparation and characterization of TiO_2 nanoparticle self-assembled aromatic polyamide thin-film-composite (TFC) membrane. Environ. Sci. Technol. 35, 2388−2394.

Lee, X.J., Lee, L.Y., Foo, L.P.Y., Tan, K.W., Hassell, D.G., 2012. Evaluation of carbon-based nanosorbents synthesised by ethylene decomposition on stainless steel substrates as potential sequestrating materials for nickel ions in aqueous solution. J. Environ. Sci. 24, 1559−1568.

Lingamdinne, L.P., Koduru, J.R., Chang, Y.-Y., Karri, R.R., 2018. Process optimization and adsorption modeling of Pb(II) on nickel ferrite-reduced graphene oxide nano-composite. J. Mol. Liq. 250, 202−211.

Liu, J.-F., Zhao, Z.-S., Jiang, G.-B., 2008. Coating Fe_3O_4 magnetic nanoparticles with humic acid for high efficient removal of heavy metals in water. Environ. Sci. Technol. 42, 6949−6954.

Mayo, J.T., Yavuz, C., Yean, S., Cong, L., Shiple, H., Yu, W., et al., 2007. The effect of nanocrystalline magnetite size on arsenic removal. Sci. Technol. Adv. Mater. 8, 71−75.

Nogueira, V., Lopes, I., Freitas, A.C., Rocha-Santos, T.A.P., Gonçalves, F., Duarte, A.C., et al., 2015. Biological treatment with fungi of olive mill wastewater pre-treated by photocatalytic oxidation with nanomaterials. Ecotoxicol. Environ. Saf. 115, 234−242.

Němeček, J., Pokorný, P., Lhotský, O., Knytl, V., Najmanová, P., Steinová, J., et al., 2016. Combined nano-biotechnology for in-situ remediation of mixed contamination of groundwater by hexavalent chromium and chlorinated solvents. Sci. Total Environ. 563-564, 822−834.

Ong, C.S., Goh, P., Lau, W., Misdan, N., Ismail, A.F., 2016. Nanomaterials for biofouling and scaling mitigation of thin film composite membrane: a review. Desalination 393, 2−15.

Organization, W.H., 2015. Progress on Sanitation and Drinking Water: 2015 Update and MDG Assessment. World Health Organization.

Padaki, M., Murali, R.S., Abdullah, M.S., Misdan, N., Moslehyani, A., Kassim, M., et al., 2015. Membrane technology enhancement in oil−water separation. A review. Desalination 357, 197−207.

Pan, B., Xing, B., 2008. Adsorption mechanisms of organic chemicals on carbon nanotubes. Environ. Sci. Technol. 42, 9005−9013.

Parham, H., Bates, S., Xia, Y., Zhu, Y., 2013. A highly efficient and versatile carbon nanotube/ceramic composite filter. Carbon. N. Y. 54, 215−223.

Qu, X., Brame, J., Li, Q., Alvarez, P.J.J., 2012. Nanotechnology for a safe and sustainable water supply: enabling integrated water treatment and reuse. Acc. Chem. Res. 46, 834−843.

Qu, X., Alvarez, P.J.J., Li, Q., 2013. Applications of nanotechnology in water and wastewater treatment. Water Res. 47, 3931−3946.

Rao, G., Lu, C., Su, F., 2007. Sorption of divalent metal ions from aqueous solution by carbon nanotubes: a review. Sep. Purif. Technol. 58, 224−231.

Rao, K.R., Srinivasan, T., Venkateswarlu, C., 2010. Mathematical and kinetic modeling of biofilm reactor based on ant colony optimization. Process Biochem. 45, 961−972.

Ren, X., Chen, C., Nagatsu, M., Wang, X., 2011. Carbon nanotubes as adsorbents in environmental pollution management: a review. Chem. Eng. J. 170, 395−410.

Sa, R.M., Premalatha, M., 2016. Applications of nanotechnology in waste water treatment: a review. Imp. J. Interdiscip. Res. 2.

Shahwan, T., Abu Sirriah, S., Nairat, M., Boyacı, E., Eroğlu, A.E., Scott, T.B., et al., 2011. Green synthesis of iron nanoparticles and their application as a Fenton-like catalyst for the degradation of aqueous cationic and anionic dyes. Chem. Eng. J. 172, 258−266.

Shang, T.-M., Sun, J.-H., Zhou, Q.-F., Guan, M.-Y., 2007. Controlled synthesis of various morphologies of nanostructured zinc oxide: flower, nanoplate, and urchin. Cryst. Res. Technol. 42, 1002−1006.

Sharma, Y.C., Srivastava, V., Singh, V.K., Kaul, S.N., Weng, C.H., 2009. Nano-adsorbents for the removal of metallic pollutants from water and wastewater. Environ. Technol. 30, 583−609.

Stoller, M., 2009. On the effect of flocculation as pretreatment process and particle size distribution for membrane fouling reduction. Desalination 240, 209−217.

Stoller, M., Bravi, M., 2010. Critical flux analyses on differently pretreated olive vegetation waste water streams: some case studies. Desalination 250, 578−582.

Sun, S.-P., Lemley, A.T., 2011. p-Nitrophenol degradation by a heterogeneous Fenton-like reaction on nano-magnetite: process optimization, kinetics, and degradation pathways. J. Mol. Catal. A 349, 71−79.

Tang, W.-W., Zeng, G.-M., Gong, J.-L., Liu, Y., Wang, X.-Y., Liu, Y.-Y., et al., 2012. Simultaneous adsorption of atrazine and Cu (II) from wastewater by magnetic multi-walled carbon nanotube. Chem. Eng. J. 211-212, 470−478.

Tarigh, G.D., Shemirani, F., 2013. Magnetic multi-wall carbon nanotube nanocomposite as an adsorbent for preconcentration and determination of lead (II) and manganese (II) in various matrices. Talanta 115, 744−750.

Tewari, P., Singh, R., Batra, V., Balakrishnan, M., 2010. Membrane bioreactor (MBR) for wastewater treatment: filtration performance evaluation of low cost polymeric and ceramic membranes. Sep. Purif. Technol. 71, 200−204.

Theodore, L., Kunz, R.G., 2005. Nanotechnology: turning basic science into reality. Nanotechnol.: Environ. Implic. Solutions 61−107.

Trivedi, P., Axe, L., 2000. Modeling Cd and Zn sorption to hydrous metal oxides. Environ. Sci. Technol. 34, 2215−2223.

Yean, S., Cong, L., Yavuz, C.T., Mayo, J.T., Yu, W.W., Kan, A.T., et al., 2005. Effect of magnetite particle size on adsorption and desorption of arsenite and arsenate. J. Mater. Res. 20, 3255−3264.

Zhang, C., Sui, J., Li, J., Tang, Y., Cai, W., 2012. Efficient removal of heavy metal ions by thiol-functionalized superparamagnetic carbon nanotubes. Chem. Eng. J. 210, 45−52.

Zhang, Y., Wu, B., Xu, H., Liu, H., Wang, M., He, Y., et al., 2016. Nanomaterials-enabled water and wastewater treatment. NanoImpact 3-4, 22−39.

Zhao, J., Liu, J., Li, N., Wang, W., Nan, J., Zhao, Z., et al., 2016. Highly efficient removal of bivalent heavy metals from aqueous systems by magnetic porous Fe_3O_4-MnO_2: adsorption behavior and process study. Chem. Eng. J. 304, 737−746.

Zhou, Y., Fu, S., Zhang, L., Zhan, H., Levit, M.V., 2014. Use of carboxylated cellulose nanofibrils-filled magnetic chitosan hydrogel beads as adsorbents for Pb(II). Carbohydr. Polym. 101, 75−82.

5

Applications of Emerging Nanomaterials for Oily Wastewater Treatment

P.S. Goh[1], C.S. Ong[2], B.C. Ng[1], Ahmad Fauzi Ismail[1]

[1]ADVANCED MEMBRANE TECHNOLOGY RESEARCH CENTRE (AMTEC), SCHOOL OF CHEMICAL AND ENERGY ENGINEERING, FACULTY OF ENGINEERING, UNIVERSITI TEKNOLOGI MALAYSIA, JOHOR BAHRU, MALAYSIA [2]WATER DESALINATION AND REUSE CENTER, DIVISION OF BIOLOGICAL AND ENVIRONMENTAL SCIENCE AND ENGINEERING, KING ABDULLAH UNIVERSITY OF SCIENCE AND TECHNOLOGY, THUWAL, SAUDI ARABIA

5.1 Introduction

Water is a critical resource used for economic, social and cultural development. However, with the increase of population and the developments brought by the refineries, petrochemical and transportation industrial revolution, water pollution is an inevitable issue faced by the mankind in this century (Jamaly et al., 2015). Currently, one of the most significant waste found in the water sources is oil discharge from oil and gas industries. In general, oily wastewater refers to the wastewater that has mixed with oil with a broad range of concentrations. The oil compounds found in oily wastewater may consist of fats and hydrocarbons as well as petroleum fractions. Many industries produce a huge amount of oily wastewater which have significant adverse impacts and threats to the surrounding environment and human beings due to the presence of hazardous contents in the wastewater. For instance, high consumption of oil and gas in conventional petrochemical refinery industries poses critical environmental issues in the waste disposal management. Oily wastewater which collected from different industrial source may vary in terms of its chemical composition, physical characteristics. Hence, different treatment designs have been reported based on the required specifications, wastewater's characteristics and pollution parameters. Produced water generated from oil wells is known to be one of the most hazardous waste streams if the wastewater is disposed directly without treatment as it contains large quantities of oily pollutants, such as dissolved oil, grease, suspended particles, gases and minerals, as well as some insoluble organic substances. The evaporation of these contents may cause air pollution and the penetration of produced water into underground water resources. These pollutants also resulted in several problems in process equipment such as tubular corrosion and scaling formation in

Table 5-1 Physical Classification of Oil and Grease Compounds in Aqueous Medium (Pintor et al., 2016)

Classification	Diameter Range	Characteristics
Free oil	>150 μm	Droplets that float on the surface of aqueous medium due to the density difference of oil and water
Dispersed oil	20–150 μm	Droplets stabilized by electric charges
Emulsified oil	<20 μm	Droplets stabilized by the chemical action of surface active agent
Soluble oil	<5 μm	Very fine droplets that dissolve in aqueous medium

heat exchanger. Besides that, the discharge of these oily components also known to affect crop production and destructing the natural landscape.

An appropriate wastewater strategy is required to treat the produced water prior to their disposal or reuse for other purposes. The appropriateness of an identified technology is normally dictated by the characteristics of the oil components in water. Particularly, it is known that the degree of dispersion and the oil droplet stability in water has a strong influence in the readiness of separation. The physical characteristics of oil components and their diameter range are tabulated in Table 5-1. Different treatment methods have been utilized to remove the oil impurities. The treatment processes normally involved the removal of dispersal oil and grease; soluble organic; suspended solid removal, dissolved gas such as carbon dioxide; desalination for salt removal and disinfection of microorganisms. Currently, a wide selection of treatment methods has been established for the removal of the oil impurities in order to minimize or prevent the negative impacts of oily wastewater on our environments. Some of the prevailing technologies include membrane filtration, electrochemical treatment, adsorption, floatation, and chemical coagulations as well as the hybrid technologies that integrate two or more approaches mentioned above. These technologies are known to serve different purposes in oily wastewater treatment. The primary treatment that consist of gravitational separation and sedimentation normally aims to remove oil and grease components in free oil form and the settlement of some unstable colloidal particles.

The emergence of a wide range of nanomaterials rendered with astonishing properties has promised innovative and novel ways to treat wastewater, hence minimizing the negative impact of the oily wastewater discharge into the water sources (Yu et al., 2017). In terms of nanomaterial selection, several kinds of surface wetting properties have been considered to deliver efficient oily wastewater treatment. These surface properties include: hydrophobic and oleophilic; hydrophilic and oleophobic; superhydrophilic and superoleophobic, and responsive wetting properties. For example, the superoleophobic nanomaterials normally exhibit very high water affinity but extremely low underwater oil adhesion force. As a result, higher water permeation can be obtained and excellent oil repellent can be achieved under water. The surface properties of nanomaterials are known to be affected by their chemical composition and surface roughness. It is also known that superwetting surface can be created by choosing the suitable materials and through various physical or/and chemical treatment (Hou et al., 2017). Hence, various strategies have been developed to synthesize or modify the nanomaterial surfaces in order to obtain desired features like high permeability, self-cleaning ability, catalytic reactivity and ability to mitigate fouling. To date, many efforts have been paid to design and fabricate superwetting nanomaterials for adsorption and membrane-based oil/water separation.

In this chapter, the applications of innovative and novel nanomaterials for oily wastewater treatment are presented. The oily wastewater treatment technologies are first reviewed to provide a general overview on the three major approaches: adsorption, electrochemical and membrane separation. Next, the roles of emerging nanomaterials in reducing the adverse effect of oily wastewater through different approaches are highlighted. Finally, the challenges and opportunities of this particular field for future development are also briefly discussed.

5.2 Contemporary Technologies for Oily Wastewater Treatment

5.2.1 Adsorption

One widely applied technique to remove dissolved organic substances in water is adsorption where the substances are separated from one phase to the surface of another. Adsorption is one of the most frequently used separation processes for the treatment of waste in industry scale. It is a process that involves the adhesion of pollutants onto the surface of a substance without creating a huge amount of hazardous sludge. The adsorbing medium is generally known as the adsorbent whereas the material adsorbed on the surface is known as the adsorbate. Some of the important criteria for the selection of nanomaterials are their accessibility, cost effectiveness and ability to reduce environmental impact. In terms of the mechanisms involved, adsorption can be divided into two, i.e. chemical and physical adsorption. The former mechanism involves the chemical reaction where adsorbate is chemically reacted with the surface. On the other hand, the latter involves the attachment of waste substances through physical forces such as hydrogen ponding and dipole-dipole interaction (Kausar et al., 2018). A list of materials has been applied for water remediation. Some naturally found mineral or organic adsorbent shown benefits such as low cost, environmentally friendliness and high oil removal efficiency. The typically required characteristics of adsorbent are high surface area, charge density and hydrophobicity. In terms of surface feature, corrugations and hair-like features are also desired to promote oil uptake through capillary forces (Pintor et al., 2016).

Among these materials, activated carbon has been widely used owing to its unique adsorbing properties such as highly porous and large surface area. It has been used for the treatment of petroleum contaminated ground-water and it is known that powdered activated carbon (PAC) is generally more effective than granular activated carbon (GAC). However, despite the efficiency in biochemical oxygen demand (BOD) and COD removal, adsorption based on activated carbon is largely limited by its expensive cost. Furthermore, the high oil concentrations may result in pore clogging and deteriorate the adsorption performance. Clay is another class of material that exhibits promising adsorptive behavior to remove organic pollutants. The utilization of different clays as adsorbents for the removal of oily components from wastewater has received attentions mainly due to the eco-friendly nature of clay materials. It has been reported that bentonite organo-clay can serve as an effective adsorbent to remove oil from oily waters in a column system in which the capability is 5−7 times better than that of activated carbon. Over the last few decades, the application of natural fibers for oily wastewater treatment has also been reported (Wahi et al., 2013). Some commonly used natural oil adsorbents are rice husk, sugarcane bagasse and wood residues which are normally formed into sheets, filers or fiber assemblies. These natural adsorbents are derived from the waste hence chemical-free and highlight biodegradable.

Recently, many nanoparticles have been studied for their potential as adsorbents. The nanosized adsorbents are featured by two main properties, i.e. innate surface and external functionalization (Anjum et al., 2016). It is well agreed that the physical and chemical properties of these nanomaterials are also strongly associated to their extrinsic properties such as surface structure and apparent size. The effectiveness of nanoparticles that used for adsorption are influenced by their size and shape, surface chemistry, solubility and dispersion state, chemical composition as well as crystallinity. However, for oily wastewater treatment, it is generally required that these nanoparticles should be nontoxic, high adsorption capacity, able to adsorb pollutants in concentration as low as ppb level, easy desorption process and can be recycled for several times without much affecting the overall effectiveness.

5.2.2 Electrochemical Separation

Electrochemical separation of oily waste involves the destabilization the oil emulsion in wastewater through the application of electrical current (Jamaly et al., 2015). In fact, electrochemical method has been favorably used in many types of wastewater treatments due to its environmentally friendliness that produces zero or very less secondary pollution. This technique features many interesting properties for oily wastewater treatment such as highly efficient, simple post processing, stable and small footprint (Yang et al., 2015). Electrochemical methods are based on electrochemical oxidation processes using several electrodes. As such, the properties of electrode materials are of crucial importance to determine the efficiency of electrochemical separation. Different types of conductive materials such as iron, aluminum and boron doped diamond have been widely used as the electrode materials. Electroflotation and electrocoagulation are the two common electrochemical methods used for oily wastewater. The principle of electrofloatation lies on the dissociation of water to produce H_2 and O_2 with low density small bubble particle size that ranges from 20 to 50 μm. These micro-sized bubbldes can quickly absorb emulsified oil droplet and other suspended waste to form bigger lumps that float to the water surface. Their strong adsorption capacity and floating load capacity of these bubbles can rapidly result in fast flotation separation and effective oil removal (Yang et al., 2015). On the other hand, electrocogulation relies on the production of in situ coagulants from an electrode through the electric current applied to the electrodes. The production of ions is subsequently followed by the electrophoretic concentration that takes place at the anode. These ions are attracted by the colloidal particles and their charges are neutralized to allow coagulation to take place. The hydrogen gas formed at the cathode is then interact with the particles to facilitate the removal of the unwanted materials (Andrade and Costa Marques, 2012).

5.2.3 Membrane Technology

Membrane technology is a promising solution for oily wastewater treatment due to its advantages such as cost effectiveness, free from chemical additives, modular installation and operation at ambient temperature compared to those conventional methods (Padaki et al., 2015). In a broad term, membrane filtration involves the physical separation of the unwanted impurities from the bulk solutions through a semipermeable membrane. Depending on the principles of the membrane technology, hydraulic pressure or osmotic pressure is required for the operation of this technology. Different types of pressure driven membrane-based processes such as microfiltration (MF), ultrafiltration (UF), nanofiltration (NF), reverse osmosis (RO) have been widely used to treat oily waste water. Recently, forward osmosis (FO) has also been explored for this purpose. Owing to the modular operations of membrane separation, some of these processes have been integrated to attain higher removal efficiencies. Extensive research in membrane filtration technologies for oily wastewater treatment has been performed over the past 20 years. In general, two broad categories of commercial and lab scaled membranes have been applied for oil removal, that is, ceramic membranes and polymeric membranes. Polymeric membranes have been substantially used for oily wastewater treatment due to its low cost and easy handling. However, compared to the polymeric membranes, ceramic membranes are known to have greater mechanical and chemical stability as well as better resistance towards membrane fouling.

Regardless of the types of membranes used, constant efforts have been made to improve the membrane and membrane system in terms of cost and affordability, energy consumption and sustainability. One promising strategy to achieve these improvement is through the development of advanced membrane materials. The application of nanomaterials in membrane technologies to revolutionize the membrane materials in order to tackle the problems related to oily wastewater has become the major focus of the research community in this field (Ahmadi et al., 2017). One of the most significant breakthrough in polymeric membrane development is the fabrication of mixed matrix membrane (MMM) or thin film nanocomposite (TFN). These nano-enabled membranes are formed by incorporating nanomaterials which act as nanofiller into the polymeric substrate (Lee et al., 2011). In MMM, the nanomaterials are commonly embedded and randomly distributed within the polymer matrix through physical mixing prior to the formation of MMM through phase inversion technique. On the other hand, in the formation of TFN, the nanomaterials can be selectively introduced into the polymeric supportive substrate or/and within the polyamide selective layer. As such, TFN usually allows higher degree of freedom in terms of the nanocomposite membrane design. Fig. 5-1 illustrates the typical procedure used to prepare TFN that is incorporated

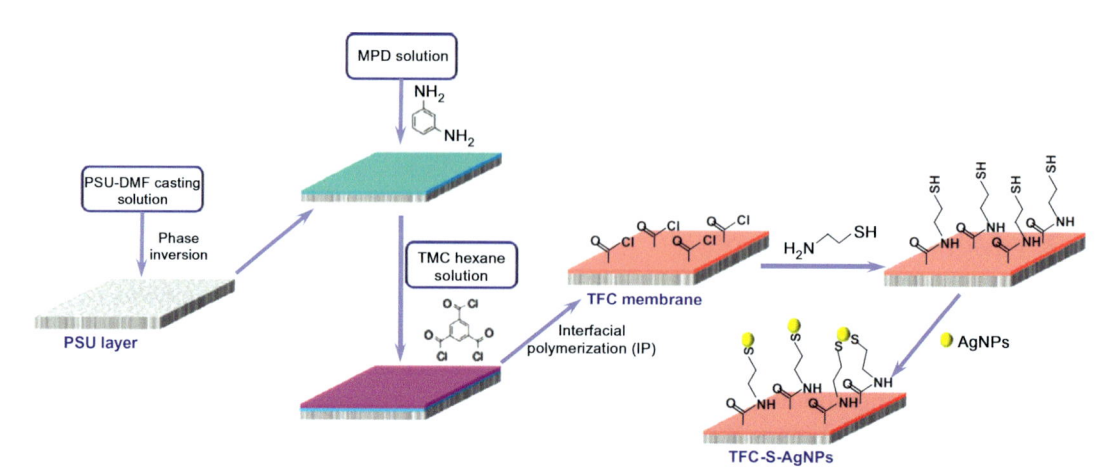

FIGURE 5-1 Schematic diagram of TFN that is incorporated with silver nanoparticles at the polyamide layer (Yin et al., 2013).

with silver nanoparticles at the polyamide layer through interfacial polymerization (Yin et al., 2013).

To date, different types of nanomaterials have been attempted for this purpose. To name a few, carbon-based nanomaterials such as carbon nanotubes (CNTs) and graphene oxide (GO), nanosized metal oxides such as titanium dioxide and silica, silicate-based nanomaterials such as zeolites have been widely applied for the formation of MMM and TFN. The addition of nanosized materials into the membrane matrix has shown to improve the flux, rejection and also adsorption capacity of these membranes and the usage of these membranes are scattered around the field of water treatment and also gas separation and gas adsorption. It also changes the morphology of the membrane surface and cross section. In general, these nanomaterials possess desired properties such as hydrophilicity, surface charge and mechanical strength to heighten the performance of the resultant nanocomposite membranes. These properties are crucial to resolve some intrinsic membrane problems, that is, concentration polarization and fouling. A relatively new class of membrane known as photocatalytic membranes have been recently developed to address the issues related to oily wastewater treatment. Photocatalytic membrane is the integration of photocatalysis and membrane technology which can degrade organic pollutants into harmless by-products. This versatile technique has combine the advantages of both photocatalysis and filtration to achieve synergetic effects for oily wastewater treatment. Polymeric membranes such as polysulfone (PSf) and polyvinylidene difluoride (PVDF), ceramic membranes such as alumina and zeolite membranes are commonly used as the host for photocatalyst. Conventional photocatalyst such as Degussa P25 titanium dioxide has been commonly incorporated into the MMM to form photocatalytic membranes with ability to degrade the oil molecules hence reduce the fouling tendency. Moreover, when the photocatalyst is introduced onto the ceramic support or dispersed into the membrane matrix, they can be physically retained and

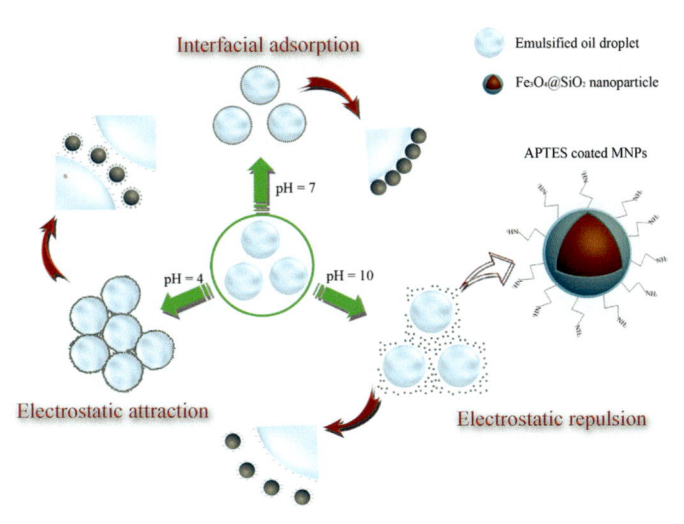

FIGURE 5-2 Schematic illustration of the interaction mechanism between the silica coated magnetic nanoparticles and emulsified oil droplets (Lü et al., 2017).

eliminate the step of photocatalyst recovery from the aqueous suspension (Subramaniam et al., 2016).

5.3 Roles of Nanomaterials in Oily Wastewater Treatment

5.3.1 Adsorption

Zhang et al. prepared Fe_3O_4 magnetic nanoparticles by a coprecipitation method and followed by surface coating with silica and 3-aminopropyltriethoxysilane (APTES). The resultant nanocomposite adsorbent was then supported onto quaternized chitosan (QC) for adsorption of oil. QC is known to demonstrate permanent cationic charges hence could establish electrostatic interaction with negatively charged oil droplets (Zhang et al., 2017). Owing to the excellent adsorptive properties, the adsorbent exhibited promising oil removal capability at various pH conditions where under both neutral and acidic conditions, the water transmittance of 98% was achieved at 34 mg/L and under alkaline conditions, water transmittance of 98% was achieved with dosage of 38 mg/L. Importantly, the magnetic adsorbent shown good reusability for practical application where it still exhibited god performance of more than 90% water transmittance after 8 cycles of operations. Lu et al. also synthesized pH-sensitive and recyclable silica-coated Fe_3O_4 magnetic nanoparticles using dense liquid silica coating method (Lü et al., 2017). The oilwater separation mechanism proposed by the authors pointed out that both electrostatic interaction and interfacial activity has significant roles in the oil–water separation. As shown in Fig. 5-2, under acidic condition, the magnetic nanoparticles adsorbed to the negatively charged oil droplets hence facilitate the flocculation of the oil droplets via electrostatic interaction thus can be easily separated through physical magnetic separation. On the other hand, under the neutral condition, the magnetic nanoparticles served as non-ionic surfactant and accumulated at the interface of oil−water whereas under alkaline condition, the silica-coated nanoparticles were detached from each other due to the electrostatic repulsion. These interesting characteristics of the magnetic nanoparticles have allowed the recycling of this material based on its pH sensitivity where it can be used as an effective adsorbent to remove emulsified oil droplets under acidic and neutral conditions, and can be recycled in alkaline solution through simple rinsing.

In a recent study conducted by Song et al., bismuth vanadate ($BiVO_4$) nanoparticle with sunlight responsive self-cleaning properties has been applied for the synthesis of underwater superoleophobic mesh (Song et al., 2017). Compared to conventionally used TiO_2, the unique optical and electrical properties of $BiVO_4$ allow the photocatalytic activity to be activated under visible light. Owing to the excellent underwater superoleophobicity, the water in the oil/water mixture can penetrate the mesh easily hence oil droplets can be efficiently removed. A variety of oil/water mixtures such as hexane, diesel, dichloroethane and chloroform have been successfully separated. Furthermore, the adsorption capability can be easily recovered under the irradiation of visible light due to the photocatalytic activity demonstrated by $BiVO_4$ to degrade the organic compounds.

5.3.2 Membrane Technology

One of the most critical challenges in membrane processes, particularly the pressure driven processes, is the membrane fouling issue caused by the deposition of oil droplets and soluble organics onto the membrane surfaces. The plugging of the membrane pores is directly associated to the deterioration of membrane performance in in term of the water productivity. Commonly, membrane cleaning is necessary to recover the membrane flux hence impose additional chemical cost to the overall process. Membrane fouling also unfavorably results in the shortening of membranes' life span. In order to combat this primary challenge, many efforts have been devoted to modify and improve the properties of the membrane materials. Different approaches and strategies and have developed to mitigate fouling during operation. These include cross-linking, polymer blending and incorporation of inorganic nanomaterial fillers. Among these approaches, incorporation of hydrophilic nanomaterials has attracted tremendous attentions of the research community due to its effectiveness and feasibility to enhance the performance. In general, upon the introduction of hydrophilic nanomaterials, the resultant nanocomposite membranes demonstrate much improved hydrophilicity hence can easily reject the hydrophobic oil droplet found in the oily wastewater based on the repulsive force formed between the oil molecules and membrane surfaces.

Kusworo et al. compared the performance of PES membrane incorporated with ZnO and SiO2 for produced water treatment and it was found that the former was better in terms of the permeate flux (Kusworo et al., 2017). Permeate flux improvement up to 200% and pollutant rejection efficiency of 16%−18% were observed in the nanocomposite membrane. The oil droplet removal efficiency was much better compared to that of neat PES due to the improved hydrophilicity of the membrane, as evidenced from the decrease in surface contact

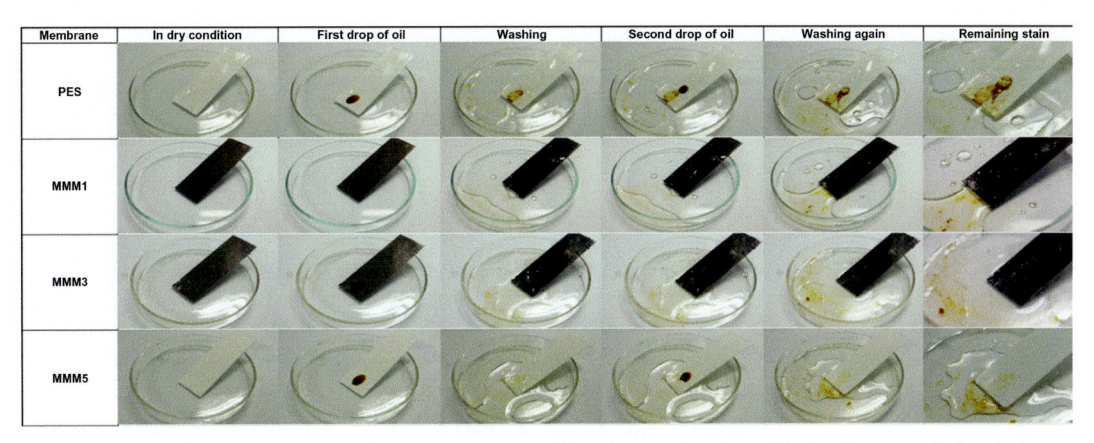

FIGURE 5-3 Antifouling studies of the MMMs that involved membrane cleaning with simple water rinsing (Lai et al., 2017).

angle. Additionally, lesser foulant deposition was observed on the membrane surface compared to that of neat PES membrane. Saadati et al. prepared PSf/pebax MMMs with different loadings of functionalized multi-walled CNTs (MWCNT) for the separation of oil/water emulsion (Saadati and Pakizeh, 2017). By incorporating 2 wt% of MWCNT into the polymeric matrix, oil rejection has been improved by 8% at transmembrane pressure of 10 bar compared to that of neat PSf/pebax membranes. The increment was mainly ascribed to the increase in hydrophilicity upon the addition of functionalized MWCNT to increase the water permeation through the membrane. Additionally, the reduction in membrane pore size and porosity upon the addition of MWCNT with loading of more than 0.5 wt% also led to favorably increased oil rejection.

MMMs consisted of dual-nanofillers of hydrous manganese oxide (HMO) and titanium dioxide (TiO_2) at different have also been reported to improve the properties of polyethersulfone (PES) membrane for oil−water separation (Lai et al., 2017). It was observed that the incorporation of the dual-nanofiller, particularly with higher amount of HMO used, has enhanced the surface hydrophilicity of the membranes. This improvement can be attributed to the presence of huge amount -OH functional groups found in HMO compared to TiO_2. The incorporation of hydrophilic nanomaterials has affected the solvent-non solvent exchange rate during the formation of membrane through phase inversion technique, hence prompted the formation of more long finger-like structure. The favorable changes in the cross-sectional morphology and surface hydrophilicity have shown positive effect on the water flux enhancement as lower transport resistance for water molecules was anticipated. In this study, the authors reported the two best MMMs that consisted of HMO and TiO_2 in the ratio of 0.75:0.25 and 0.25:0.75. These membranes have achieved 31.73% and 26.41% higher water flux, respectively as compared to the neat PES membrane without compromising the oil removal performance. Additionally, lower flux decline was observed due to the improved surface resistance against fouling. Consequently, these samples can be used for longer life span and hold good potential for long-term operation. Fig. 5-3 displays the

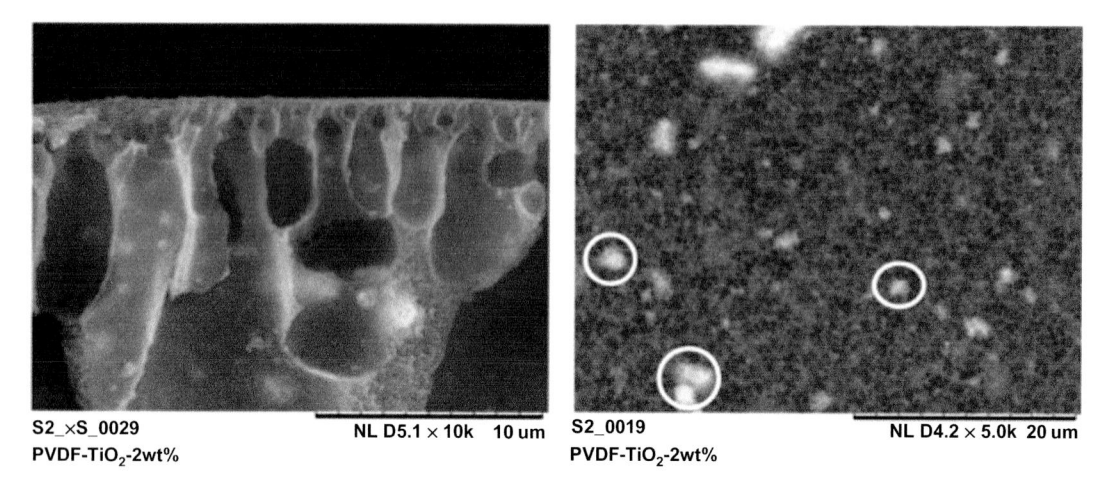

S2_xS_0029   NL D5.1 × 10k   10 um   S2_0019   NL D4.2 × 5.0k 20 um
PVDF-TiO$_2$-2wt%   PVDF-TiO$_2$-2wt%

FIGURE 5-4 SEM images (cross-section and outer surface) of PVDF membranes with 2 wt% of TiO$_2$ (Ong et al., 2015).

antifouling properties of membrane by subjecting the membrane to oil contamination and followed by simple water cleaning for two cycles. It is found that the oil stain was easily removed from the membrane surface in comparison to the control membrane due to the presence of hydrophilic nanofillers.

In terms of photocatalytic membranes, study done by Ong et al. showed that TiO$_2$ incorporated PVDF membranes were able to treat cutting oil effluent effectively, where a treatment efficiency of more than 90% was achieved (Ong et al., 2015). They found that the addition of TiO$_2$ with loading of 2 wt% has contributed to the enhanced membrane properties by increasing the surface hydrophilicity, pore size and surface roughness of the resultant nanocomposite membranes. However, when the loading of TiO$_2$ was further increased above 2 wt%, agglomeration of nanoparticles took place and deteriorated the performance of the membranes. As shown in the scanning electron microscope (SEM) images in Fig. 5-4, the surface and cross-sectional morphology of the resultant MMM hollow fiber membranes was not significantly influenced by the addition of the photocatalyst. However, at higher concentration, agglomeration (white circle) was observed and resulted in the formation of voids which tend to compromise the oil rejection properties. Another study on oily wastewater done by Rusli et al. (2016) showed an oil removal efficiency of 60% within 6 hours of treatment using a hybrid PVDF/TiO$_2$ membrane where the photocatalyst was hot pressed onto a PVDF flat sheet membrane for better adhesion.

5.3.3 Hybrid Technology

Despite the maturity of conventionally used technologies for oily wastewater treatment, the feasibility of these technologies for large scale treatment is always restricted by their limitations such as high energy consumption, cost and low efficiency. To counter these shortages, combination of conventional processes such as coagulation, adsorption and ion exchange (hybrid processes) with membrane technology demonstrate some benefits and improvements, such as improving quality of the treated water, minimizing fouling of the membranes, energy savings hence reduce the operating cost of the treatment plants. Rasouli et al. performed oily wastewater treatment by hybrid adsorption-MF process using powdered activated carbon (PAC), natural zeolite powder and ceramic membranes (Rasouli et al., 2017). In their study, PAC and natural zeolite powder by concentrations of 100−800 mg/L have been used as adsorbing agent in in-line adsorption-MF process and they reported the increase in total organic carbon rejection in all concentration of PAC for all membranes when compared to MF process alone.

5.4 Conclusion and Future Outlook

Oily wastewater is a severe issue before it is disposed to the water sources in a manner that does not harm to the mankind and environments. Due to the challenges encountered in oily wastewater treatment, many techniques as well as the combinations of different techniques have been developed to remediate the negative impacts caused by this waste. The development and application of advanced functional nanomaterials for oily wastewater treatment has attracted significant interest. In order to efficiently treat stable oil−water emulsion and separate relatively small oil droplets from their bulk solution, numerous studies have been dedicated to design novel functional nanomaterials for oily wastewater separation. Recent advances in the development of nanomaterials have allowed some critical underlying issues related to the existing technologies to be resolved. For instance, the severe membrane fouling caused by the surfactant adsorption and/or pore plugging by oil droplets can be tackled by incorporating functional nanomaterials to obtained the desired properties for oily wastewater treatment. Nevertheless, despite the promising results showed by these nanomaterials in heightening the performances of various commonly used treatment technologies, the complicated issues regarding the oil recovery and the recycling of nanomaterials are a subject to be studied in depth in the economic point of view. In term of adsorption process, it is expected that the development of magnetic nanomaterials can serve as a good option due to its recyclability. Besides that, the embedment of nanomaterials within polymeric matrix also promises the sustainability of this material as the leaching of nanomaterials can be prevented.

In term of adsorption, it is anticipated that nanomaterials with switchable wettability will be the future trend of the research. Nanomaterials with responsive wettability that can be altered from wetting to antiwetting or vice versa can offer a significant improvement for controllable oily wastewater separation, particularly in the harsh and complex environmental condition. Currently, some efforts have been made to alter the superwettability of the nanomaterial such as CNT film using UV irradiation and dark storage, the recovery time of the switching requires one day. Hence, smarter nanomaterials need to be explored to accelerate the switching period for practical usage (Yang et al., 2017). In view of the advantages and limitations of the standalone technologies mentioned above, the hybrid technology that consists of pretreatment techniques such as photocatalytic reactor, gravitational separation, flocculation and adsorption prior to membrane technology will be advantageous to resolve the challenging treatment process. Undeniably, owing to their unique and unprecedented properties, various nanomaterials have found their important roles in different oily wastewater treatment technologies. With more advances and innovations made in this field, it is expected that more emerging and potential nanomaterials will be developed to deal with a large variety of industrial oily wastewater.

Acknowledgment

The authors would like to acknowledge the financial support provided by the Ministry of Higher Education under HiCOE Grant (4J182) and Universiti Teknologi Malaysia under Research University Grant (18H35).

References

Ahmadi, A., Qanati, O., Seyed Dorraji, M.S., Rasoulifard, M.H., Vatanpour, V., 2017. Investigation of antifouling performance a novel nanofibrous S-PVDF/PVDF and S-PVDF/PVDF/GO membranes against negatively charged oily foulants. J. Membr. Sci. 536 (1), 86−97.

Andrade, A., Costa Marques, M.R. da, 2012. Electrolytic treatment of wastewater in the oil industry. New Technol. Oil Gas Ind. 3−28.

Anjum, M., Miandad, R., Waqas, M., Gehany, F., Barakat, M.A., 2016. Remediation of wastewater using various nano-materials. Arab. J. Chem.

Hou, K., Zeng, Y., Zhou, C., Chen, J., Wen, X., Xu, S., et al., 2017. Durable underwater superoleophobic PDDA/halloysite nanotubes decorated stainless steel mesh for efficient oil−water separation. Appl. Surf. Sci. 416, 344−352.

Jamaly, S., Giwa, A., Hasan, S.W., 2015. Recent improvements in oily wastewater treatment: progress, challenges, and future opportunities, J. Environ. Sci., 37. pp. 15−30.

Kausar, A., Iqbal, M., Javed, A., Aftab, K., Nazli, Z.-H., Bhatti, H.N., et al., 2018. Dyes adsorption using clay and modified clay: a review. J. Mol. Liq. 256, 395−407.

Kusworo, T.D., Qudratun, Utomo, D.P., 2017. Performance evaluation of double stage process using nano hybrid PES/SiO$_2$-PES membrane and PES/ZnO-PES membranes for oily waste water treatment to clean water. J. Environ. Chem. Eng. 5 (6), 6077−6086.

Lai, G.S., Yusob, M.H.M., Lau, W.J., Gohari, R.J., Emadzadeh, D., Ismail, A.F., et al., 2017. Novel mixed matrix membranes incorporated with dual-nanofillers for enhanced oil-water separation. Sep. Purif. Technol. 178.

Lee, K.P., Arnot, T.C., Mattia, D., 2011. A review of reverse osmosis membrane materials for desalination—development to date and future potential. J. Membr. Sci. 370 (1−2), 1−22.

Lü, T., Zhang, S., Qi, D., Zhang, D., Vance, G.F., Zhao, H., 2017. Synthesis of pH-sensitive and recyclable magnetic nanoparticles for efficient separation of emulsified oil from aqueous environments. Appl. Surf. Sci. 396, 1604−1612.

Ong, C.S., Lau, W.J., Goh, P.S., Ng, B.C., Ismail, A.F., 2015. Preparation and characterization of PVDF−PVP−TiO$_2$ composite hollow fiber membranes for oily wastewater treatment using submerged membrane system. Desalin. Water Treat. 53 (5).

Padaki, M., Murali, R.S., Abdullah, M.S., Misdan, N., Moslehyani, A., Kassim, M.A., et al., 2015. Membrane technology enhancement in oil − water separation. A review. Design 357, 197−207.

Pintor, A.M.A., Vilar, V.J.P., Botelho, C.M.S., Boaventura, R.A.R., 2016. Oil and grease removal from wastewaters: sorption treatment as an alternative to state-of-the-art technologies. A critical review. Chem. Eng. J. 297, 229−255.

Rasouli, Y., Abbasi, M., Hashemifard, S.A., 2017. Oily wastewater treatment by adsorption-membrane filtration hybrid process using powdered activated carbon, natural zeolite powder and low cost ceramic membranes. Water Sci. Technol. 76 (4), 895−908.

Rusli, U.N., Alias, N.H., Shahruddin, M.Z., Othman, N.H., 2016. Photocatalytic degradation of oil using polyvinylidene fluoride/titanium dioxide composite membrane for oily wastewater treatment. MATEC Web Conf. 69.

Saadati, J., Pakizeh, M., 2017. Separation of oil/water emulsion using a new PSf/pebax/F-MWCNT nanocomposite membrane. J. Taiwan Inst. Chem. Eng. 71, 265−276.

Song, S., Yang, H., Zhou, C., Cheng, J., Jiang, Z., Lu, Z., et al., 2017. Underwater superoleophobic mesh based on $BiVO_4$ nanoparticles with sunlight-driven self-cleaning property for oil/water separation. Chem. Eng. J. 320, 342−351.

Subramaniam, M.N., Goh, P.S., Ismail, A.F., Lau, W.J., 2016. Effect of titania nanotubes on the flux and separation performance of polyethersulfone membranes. IOP Conference Series: Earth and Environmental Science.

Wahi, R., Abdullah, L., Shean, T., Choong, Y., Ngaini, Z., 2013. Oil removal from aqueous state by natural fibrous sorbent: an overview.pdf. Sep. Purif. Technol. 113, 51−63.

Yang, M., Li, Q., Yin, X., Zhai, L., Jing, B., 2015. Field Test of Electrochemical Degradation and Oil-Removal Technology Used the Treatment of Wastewater Containing Polymer in Offshore Oilfield. Sci. Res 1409−1415.

Yang, W., Li, J., Zhou, P., Zhu, L., Tang, H., 2017. Superhydrophobic copper coating: Switchable wettability, on-demand oil-water separation, and antifouling. Chem. Eng. J. 327, 849−854.

Yin, J., Yang, Y., Hu, Z., Deng, B., 2013. Attachment of silver nanoparticles (AgNPs) onto thin- fi lm composite (TFC) membranes through covalent bonding to reduce membrane biofouling. J. Membr. Sci. 441, 73−82.

Yu, L., Ruan, S., Xu, X., Zou, R., Hu, J., 2017. One-dimensional nanomaterial-assembled macroscopic membranes for water treatment. Nano Today 17, 79−95.

Zhang, S., Lü, T., Qi, D., Cao, Z., Zhang, D., Zhao, H., 2017. Synthesis of quaternized chitosan-coated magnetic nanoparticles for oil-water separation. Mater. Lett. 191, 128−131.

Comparative Impact Assessment of TiO$_2$ and ZnO Nanoparticles to Rocket (*Eruca sativa* L) Plant

Sevinç Adiloğlu[1], Yusuf Solmaz[1], Deniz İzlen Çifçi[2], Süreyya Meriç[2]

[1]DEPARTMENT OF SOIL AND PLANT PROTECTION AND NUTRITION, AGRICULTURAL FACULTY, NAMIK KEMAL UNIVERSITY, TEKIRDAĞ, TURKEY [2]DEPARTMENT OF ENVIRONMENTAL ENGINEERING, FACULTY OF ÇORLU ENGINEERING, NAMIK KEMAL UNIVERSITY, TEKIRDAĞ, TURKEY

6.1 Introduction

In recent years, an explosion in nanoparticles (NPs) production and use has taken place. These particles of less than 100 nm have unique physical and chemical properties, making them extremely attractive for applications in many consumer products (fuel additives, coating, catalysis, cosmetics, etc.), or for military or medical purposes. Titanium dioxide (TiO$_2$) is a photocatalyst that has been used in solar cells, paints, and coatings. Zinc oxide (ZnO) and TiO$_2$ are finding extensive application in sunscreens, cosmetics, and bottle coatings because of their ultraviolet blocking ability and the visible transparency of nanoparticulate forms. It is very likely that a non-negligible part of ENPs, directly released during their life cycle, enters the environment (Keller et al., 2010). In the case of soil compartment, key processes relating to transformation and potential risk from manufactured NPs of (1) dissolution; (2) sorption/ aggregation; (3) plant bioaccumulation; (4) invertebrate accumulation and toxicity; (5) microbial toxicity; (6) direct particle uptake/ toxicity and (7) particle migration are coming out (Klaine et al., 2008). Although there has been numerous studies such as bioavailability (Johnston et al., 2010), ecotoxicity on aquatic organisms (Federici et al., 2007; Heinlaan et al., 2008; Peng et al., 2011; Clément et al., 2013) and on terrestrial organisms (Nations et al., 2011; Zhang et al., 2012) there are still limited number of studies on plants (Lin and Xing, 2011; Menard et al., 2011).

Rocket (*Eruca vesicaria subsp.sativa* (Mill)) is a member of Brassicaceae (Cruciferae) family (Siedemann, 2005). It is commonly known as "garden rocket" and consumed as a salad vegetable in Southern Europe, forms an oilseed in India and is used for its fodder. Becoming a popular vegetable for salad in Western Europe, it is widely distributed across Southern Europe, North Africa, Western Asia and India (Dixon, 2007). Rocket remains very much valued currently in many countries of the Mediterranean region such as Italy, Greece and Turkey where it is consumed mainly in salads (Eryilmaz Acikgoz, 2011). There has been reported only one study investigating the effect of Zn on Roket plant (*Eruca sativa* L.) by Urlic et al. (2014) during 30 days incubation in green house. Rocket seedlings were grown for 30 days at three Zn (1, 75, and 150 μm). They found out that Zn affected the rooth of the plant more than green parts. The Zn concentration in the rooth was measured higher than that of measured in the leaves. Plants grown at 75 and 150 μm Zn in nutrient solution accumulated Zn at 620 and 850 mg/kg in leaves, and 3500 and 6000 mg/kg in roots, respectively although concentration of Zn > 300 mg/kg leaf dry weight taken as a threshold for toxicity in most species.

To the best of our knowledge, this work is the first initiative to evaluate the ecotoxicity of TiO_2 and ZnO ENMs to Rocket plant (*Eruca sativa* L.).

6.2 Materials and Methods

6.2.1 Chemicals

TiO_2 powder was obtained from Sigma Aldrich (CAS 13463-67-7) with specific surface area of 35-65 m^2/g (BET) and mean particle size of 21 nm. While ZnO was provided from Merck (CAS 1314-13-2, catalog no 8849) with specific surface area of 47 m^2/g and mean particle size of 230 nm (Hilal et al., 2010; Alizadeh et al., 2007).

6.2.2 Phytoxicity Experiments

A series of 50, 100, 200, 400, 600 ppm concentrations of Ti and Zn were added into each pot containing 4 kg dried soil and mixed well before the seeds were germinated in the test pots as triplicate on 3rd January 2015 considering "Randomly block design." Three controls were run in paralled for each concentration too (Fig. 6-1A). When the plants were germinated and grown enough (after 10 days from seedling) only three plants were left in each pot (Fig. 6-1B). The plants were irrigated by distilled water. Urea, triple superphosphate and potassium sulfate were added into pots at the necessary amounts during vegetation period of 60 days (Fig. 6-1C). After that, plants were harvested and were submitted to many biological parameters analyses such as length and diameter of the plant (cm), length of leafs (cm), diameter of leafs (cm), root length (cm), weight of the plant (g) and number of leafs to understand the mechanism of the toxicity (Eryilmaz Acikgoz, 2011). Toxicity results were expressed as average of triplicate experiments. All experiments were run at room temperature (25°C).

6.2.3 Analysis

Soil samples used in the experiments were air-dried and sieved from 2 mm and submitted to various analysis before experiment (pH, $CaCO_3$, electrical conductivity (EC), organic matter amount, available phosphorus, extractable K and available Fe, Zn, Cu and Mn) contents

FIGURE 6-1 View of containers (A) for planting Rocket plant (B) and just before of harvesting (C).

(Jackson, 1967; Lindsay and Norvell, 1978; Olsen and Sommers, 1982). Soil texture analysis before toxicity experiments was also determined using Bouyoucos hydrometer method (Bouyoucos, 1955). Soil was dried and sieved through 2 mm pore size.

Harvested Rocket plants were washed with distilled water, then were dried for 48 hours at 68°C to define dry weights. After that, the plants were prepared and microwave-assisted extracted for metal analysis using Inductively Couple Plasma Spectrophotometer (ICP-OES, spectro/spectro blue (marka/model). Limit of detection (LOD) of Ti and Zn were 1.566 and 10 ppb Ti respectively while and limit of quantification (LOQ) of Ti and Zn were 1.203 and 10 ppb respectively. The soil was also analyzed for Ti and Zn contents prior and after harvesting.

6.2.4 Statistics

Data were processed using PASW Statistics 18 for Windows. Varians (ANOVA) analysis was used to test the differences among the data groups. Duncan multivariate test was used for the significant data groups ($P < .05$).

6.3 Results and Discussion

6.3.1 Chemical Analysis of Soil and Plant

Physical and chemical composition of the soil samples used in the experiments are shown in Table 6-1. Texture class of the soil was defined to be clay (Bouyoucos, 1955).

When Rocket plants were harvested after 60 days of vegetation, plants were submitted to metal analyses to understand the plant uptake percentages of Zn and Ti. Table 6-2 displays the Zn and Ti accumulated in the Rocket plant samples. According to statistical analyses of the data shown in Table 6-2, increasing ZnO doses affected the plant uptake at 5% statistically significant level compared to the control experiment. Additionally, the plant uptake increased *versus* increasing ZnO doses. The same results were obtained in the TiO_2 experiments. Considering the Duncan test results there was a significant difference among the Zn doses applied in comparison with the control tests (shown to be a, b, c, d, e, f in Table 6-2).

Table 6-1 Physical and Chemical Composition of the Soil Used in the Experiments

Parameter	Unit	Value
pH	(1: 2.5)	7.01
ECx10^6	dS/m	150
Texture class	—	Clay
Organic matter	(%)	0.98
P_2O_5	(kg/da)	12.95
K_2O	(ppm)	65.52
$Ca(OH)_2$	(%)	6.95
Fe	(ppm)	8.63
Zn	(ppm)	1.90
Cu	(ppm)	1.51
Mn	(ppm)	9.10
Ti	(ppm)	1.203

Table 6-2 Average Concentrations of Zn and Ti in the Green Parts of the Rocket Plant ($P < .5\%$)

Initial metal oxide doses (ppm)	Zn (ppm)	Ti (ppm)
Control	43.86 ± 1.64**a**	4.63 ± 0.12**a**
I	46.34 ± 0.68**b**	11.34 ± 0.62**b**
II	54.97 ± 0.01**c**	18.66 ± 0.64**c**
III	72.10 ± 0.54**d**	19.49 ± 0.63**c**
VI	114.23 ± 0.04**e**	20.05 ± 0.57**c**
V	145.58 ± 1.14 **f**	43.03 ± 0.72 **d**

a, b, c, d, e, f: Values in the same column with different letters are statistically different at the 5% significance level.

Table 6-3 Variance Analysis for Ti ve Zn Concentrations Measured in the Green Parts of the Rocket Plant After Harvesting

SV	(df)	Ti			Zn		
		Total of squares	Average of squares	*F* value	Total of squares	Average of squares	*F* value
General	17	2558.39			25811.99		
Doses	5	2548.29	509.66	605.82*	25797.04	5159.40	4140.5[a]
Error	12	10.09	0.84		14.95	1.24	

[a]$P \le .05$; SV, source of variance.

On the other hand, the significant difference defined by Duncan test among the initial TiO$_2$ doses was only observed at 50 ve 600 ppm doses for TiO$_2$ tests (shown as c in Table 6-2).

As seen in Table 6-3, when variance analysis was performed between metal concentrations measured in plant samples, a 5% significance level was observed to indicate the solubility of the metals affecting their uptake by the Rocket plant as confirmed by the variance test results shown in Table 6-3.

6.3.2 Phyto-Toxicity Experiments

The effects of Zn and Ti on biological parameters were observed as seen in Figs. 6-2 and 6-3 respectively. Zinc (Zn) is an important plant micronutrient involved in many metabolic reactions and is required for photosynthesis, membrane maintenance, and N metabolism. In most crops, the typical Zn concentration in leaves is 15–50 mg Zn/kg dry weight (Marschner 1996) and can become toxic above these levels. Some plants can accumulate up to a few thousand mg Zn/kg if excess Zn is available in growing medium (Roosens et al., 2008). Although Rocket plant up took Zn at higher level than Ti while their increasing initial doses (Table 6-2) in this study, the biological factors affected by Zn were found to be statistically insignificant except for the internal leaf length. The reasons for this finding can be attributed to (1) sufficient bioavailable Zn level in the soil used or (2) low solubility of Zn at the pH of the soil as Ma et al. (2013) reported that solubility to ionic zinc represents an important mechanism of ZnO NPs toxicity. In fact, Hernandez-Viezcas et al. (2011) reported that ZnO NPs toxicity to springtails in soil can be explained from Zn dissolution but not from particle size. Similarly, in the study by Kool et al. (2011) ZnO NPs and non-nano ZnO were found to be equally toxic to *Folsomia candida* in soil. ZnO NP toxicity in soil was most probably due to Zn dissolution from the NPs. In a recent study (Bandyopadhyayd et al., 2015), ZnO NPs, bulk ZnO, and ZnCl$_2$ were exposed to the symbiotic alfalfa (*Medicago sativa* L.)–*Sinorhizobium meliloti* association at concentrations ranging from 0 to 750 mg/kg soil. Plant growth, Zn bioaccumulation, dry biomass, leaf area, total protein, and catalase (CAT) activity were measured in 30 day-old plants. ZnO reduced root and shoot biomass by 80% and 25%, respectively compared to control at 750 mg/kg dose. STEM-EDX imaging revealed the presence of ZnO particles in the root, stem, leaf, and nodule tissues. Seed germination and root

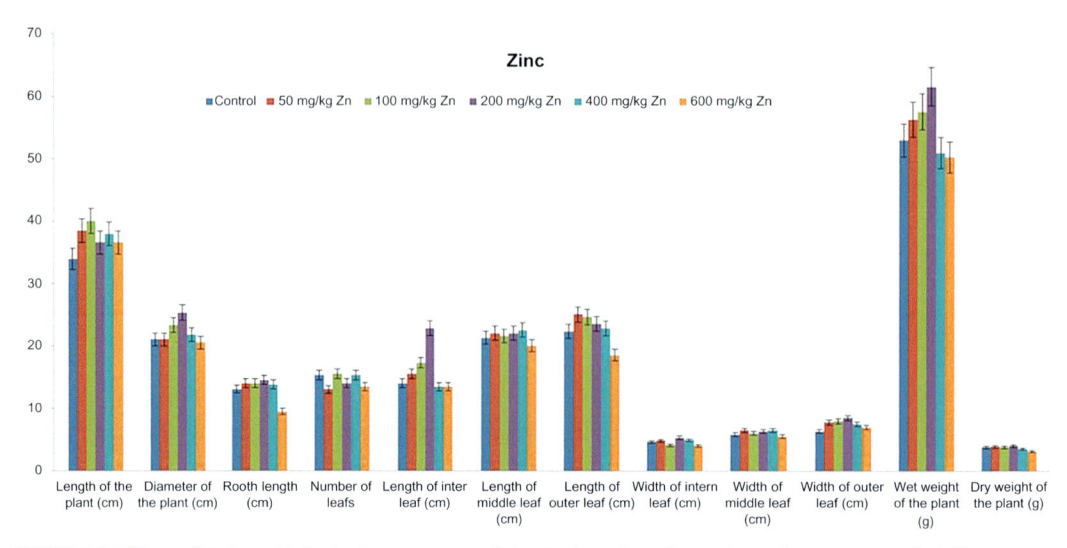

FIGURE 6-2 Effect of ZnO on biological parameters of the Rocket plant (Data show the averages of triplicate experiments).

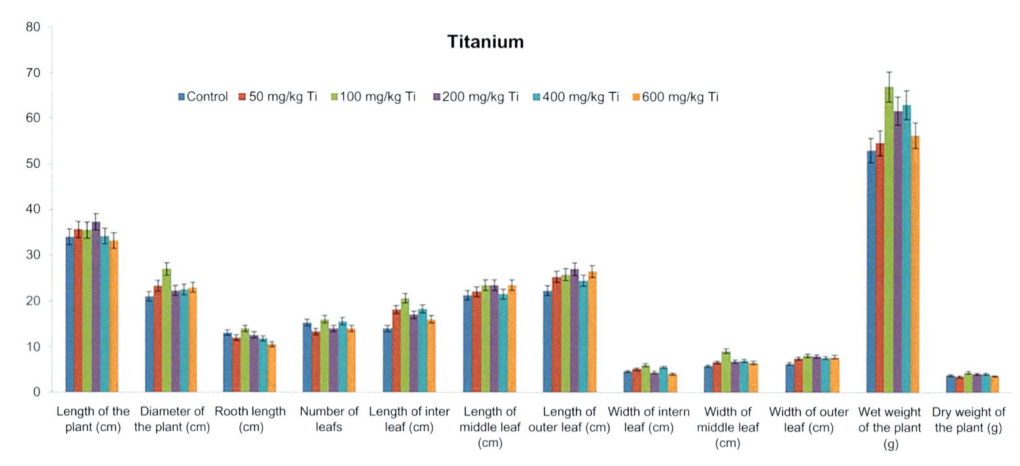

FIGURE 6-3 Effect of TiO_2 on biological parameters of the Rocket plant (Data show the averages of triplicate experiments).

growth of six higher plant species (radish, rape, ryegrass, lettuce, corn, and cucumber) exposed to Zn were investigated by Lin et al. (2011). Seed germination was not affected except for the inhibition of nanoscale zinc (nano-Zn) on ryegrass and zinc oxide (nano-ZnO) on corn at 2000 mg/L. Suspensions of 2000 mg/L nano-Zn or nano-ZnO practically terminated root elongation of the tested plant species. Fifty percent inhibitory concentrations (IC_{50}) of nano-Zn and nano-ZnO were estimated to be near 50 mg/L for radish, and about 20 mg/L for rape and ryegrass.

As seen in Fig. 6-2, the root length was also significantly inhibited by the increased Zn concentration of 600 mg/kg as it slightly occurred in case of the same dose of Ti application as Ti was classified as a beneficial element for plants (Tlustos et al., 2005; Kleibler, 2017). A 10 ppm concentration of Ti was found to be beneficial for different plants and phytotoxic was observed at the concentrations >50 ppm (Kuzel et al., 2007). Levels of Ti in plants vary rather considerably within the range of 0.15–80 ppm (Kabata-Pendias and Pendias, 2001). Wallace et al. (1977) described Ti toxicity symptoms such as the chlorotic and necrotic sport that were observed on leaves of bush bean that contained Ti at a concentration of about 200 ppm. According to statistical analysis, increasing Ti doses did not significantly ($P < .05$) affect the aspects of length and diameter of leaves, rooth length, number of leaves, length and width of the middle leaf, and wet weight of the Rocket plant whereas Ti did significantly ($P < .05$) affect the intern and outer length and width of the leaves and dry weight of Rocket plant. These low toxicity results can be attributed to the low level of extractable Ti levels due to the texture of the soil (clay) used in the study in accordance with the results reported by Zheng et al. (2005) that nano-TiO₂ significantly increased seed germination and plant growth of spinach at low concentrations, but those parameters decreased at high Ti concentrations. Nanoanatase TiO₂ promoted photosynthesis, improve spinach (*Spinacia oleracea*) growth, promote the vigor of aged seeds, and chlorophyll biosynthesis in spinach (Hong et al., 2005).

6.4 Conclusion

Plant toxicity is important as they are in direct contact with both soil and water quality. The number of toxicity studies of TiO₂ and ZnO has been increasing, mostly focusing on bacterial inactivation. This study is a first comparative example to examine the toxic effects of TiO₂ and ZnO to the Rocket plant which is a favored vegetable food too. The results evidenced that both NPs exhibited slightly significant (5%) effects on biological parameters of the plant. The uptake of Zn was higher than TiO₂, which is attributed to the texture of the soil while less number of biological parameters were affected by ZnO NPs, which was found in accordance with the literature. Further studies such as varying soil texture (its pH is also important for the solubility of those metal oxides.), increasing initial doses of Ti and Zn, searching physiological characteristics such as enzyme activities, oxidative stress to better understand the bioavailability and effects of those NPs would be necessary to explain the mechanisms toxicity to the Rocket plant.

Acknowledgments

The authors would like to Namik Kemal University Research Fund (NKUBAP.00.17.AR.14.17 and NKUBAP.00.24.AR.14.03 projects) for the support.

References

Alizadeh, R., Jamshidi, E., Ale Ebrahim, H., 2007. Catalytic effect of zinc oxide on the reduction of barium sulfate by methane. Thermochim. Acta. 460, 44−49.

Bandyopadhyayd, S., Plascencia-Villa, G., Mukherjee, A., Rico, C.M., José-Yacamán, M., Peralta-Videa, J.R., et al., 2015. Comparative phytotoxicity of ZnO NPs, bulk ZnO, and ionic zinc onto the alfalfa plants symbiotically associated with Sinorhizobium meliloti in soil. Sci. Total Environ. 515-516, 60−69.

Bouyoucos, G.J., 1955. A recalibration of the hydrometer method for making mechanical analaysis of the soils. Agron. J. 4 (9), 434.

Clément, L., Hurel, C., Marmier, N., 2013. Toxicity of TiO_2 nanoparticles to cladocerans, algae, rotifers and plants-effects of size and crystalline structure. Chemosphere 90, 1083−1090.

Dixon, G.R., 2007. Vegetables Brassica and Related Crucifers. CABI Publishing, Wallingford, UK9780851993959.

Eryilmaz Acikgoz, F., 2011. The effects of different sowing time practises on vitamin C and mineral material content for rocket (*Eruca vesicaria subsp. sativa* (Mill)). Sc. Res. Essays. 6 (15), 3127−3131.

Federici, G., Shaw, B.J., Handy, R.D., 2007. Toxicity of titanium dioxide nanoparticles to rainbow trout (*Oncorhynchus mykiss*): gill injury, oxidative stress, and other physiological effects. Aquat. Toxicol. 84 (4), 415−430.

Heinlaan, M., Ivask, A., Blinova, I., Dubourguier, H.C., Kahru, A., 2008. Toxicology of nanosized and bulk ZnO, CuO and TiO2 to bacteria *Vibrio fisheri* and crustaceans *Daphnia magna* and *Thamnocephalus platyurus*. Chemosphere 71 (7), 1308−1316.

Hernandez-Viezcas, J.A., Castillo-Michel, H., Servin, A.D., Peralta-Videa, J.R., Gardea-Torresdey, J.L., 2011. Spectroscopic verification of zinc absorption and distribution in the desert plant *Prosopis juliflora-velutina (velvet mesquite)* treated with ZnO nanoparticles. Chem. Eng. J. 170 (1-3), 346−352.

Hilal, H.S., Al-Nour, G.Y.M., Zyoud, A., Helal, M.H., Saadeddin, I., 2010. Pristine and supported ZnO-based catalysts for phenazopyridine degradation with direct solar light. Solid State Sci. 12 (4), 578−586.

Hong, F., Zhou, J., Liu, C., Yang, F., Wu, C., Zheng, L., Yang, P., 2005. Effect of nano-TiO_2 on photochemical reaction of chloroplasts of spinach. Biol. Trace Elem. Res. 105, 269−280.

Jackson, M.C., 1967. Soil Chemical Analaysis. Prentice Hall of India Private' Limited, New Delhi, Indian.

Johnston, B.D., Scown, T.M., Moger, J., Cumberland, S.A., Baalousha, M., Linge, K., et al., 2010. Bioavailability of nanoscale metal oxides TiO_2, CeO_2, and ZnO to fish. Environ. Sci. Technol. 44 (3), 1144−1151.

Kabata-Pendias, A., Pendias, H., 2001. Trace Elements in Soils and Plants, third ed. CRC Press, Inc., Florida, ABD.

Keller, A.A., Wang, H., Zhou, D., Lenihan, H.S., Cherr, G., Cardinale, B.J., et al., 2010. Stability and aggregation of metal oxide nanoparticles in natural aqueous matrices. Environ. Sci. Technol. 44, 1962−1967.

Klaine, S.J., Alvarez, P.J.J., Batley, G.E., Fernandes, T.F., Handy, R.D., Lyon, D.Y., et al., 2008. Nanomaterials in the environment: behavior, fate, bioavailability, and effects. Environ. Toxicol. Chem. 27 (9), 1825−1851.

Kleibler, T., 2017. Effect of titanium application on lecttuce growth under Mn. Stress. J. Elementol. 22 (1), 329−337.

Kool, P.L., Ortiz, M.D., van Gestel, C.A.M., 2011. Chronic toxicity of ZnO nanoparticles, non-nano ZnO and $ZnCl_2$ to Folsomia candida (Collembola) in relation to bioavailability in soil. Environ. Pollut. 159 (10), 2713−2719.

Kuzel, S., Cigler, P., Hruby, M., Vydra, J., Pavlikova, D., Tlustos, P., 2007. The effect of simultaneous magnesium application on the biological effects of titanium. Plant Soil Environ. 53 (1), 16−23.

Lin, D., Xing, B., 2011. Phytotoxicity of nanoparticles: Inhibition of seed germination and root growth. Environ. Pollut. 150, 243−250.

Lindsay, W.L., Norvell, W.A., 1978. Development of a DTPA soil test for zinc, iron, manganase and copper. Soil Sci. Soc. Am. J. 42, 421−428.

Ma, H., Williams, P.L., Diamond, S.A., 2013. Ecotoxicity of manufactured ZnO nanoparticles - A review. Environ. Pollut. 172, 76−85.

Marschner, H., 1996. Mineral Nutrition of Higher Plants, second ed Academic Press, New York.

Menard, A., Drobne, M., Jemec, A., 2011. Ecotoxicity of nanosized TiO$_2$. Review of in vivo data. Environ. Pollut. 159, 677−684.

Nations, S., Long, M., Wages, M., Canas, J., Maul, J.D., Theodorakis, C., et al., 2011. Effects of ZnO nanomaterials on Xenopus laevis growth and development. Ecotoxicol. Environ. Saf. 74, 203−210.

Olsen, S.R., Sommers, L.E., 1982. Phosphorus. In: Page, A.L., Miller, R.H., Keeney, D.R. (Eds.), Methods of Soil Analysis. Part II. Chemical and Microbiological Properties, second ed ASA SSSA Publisher, Agronomy, Madison, WI, pp. 403−427.

Peng, X., Palma, S., Fisher, N.S., Wong, S.S., 2011. Effect of morphology of ZnO nanostructures on their toxicity to marine algae. Aquat. Toxicol. 102, 186−196.

Roosens, N.H.C.J., Willems, G., Saumitou-Laprade, P., 2008. Using *Arabidopsis* to explore zinc tolerance and hyperaccumulation. Trends Plant Sci. 13, 208−215.

Siedemann, J., 2005. World Spice Plants (Economic Usage, Botany, Taxonomy). Springer-Verlag, Berlin Heidelberg, Germany.

Tlustos, P., Cigler, P., Hruby, M., Kuzel, S., Szakova, J., Balık, J., 2005. The role of titanium in biomass production and its influence on essential elements contents in field growing crops. Plant Soil Environ. 51 (1), 19−25.

Urlic, B., Dumiçiç, G., Ban, S.M., 2014. Zinc and sulfur effects on growth and nutrient concentrations in rocket (*Eruca sativa L.* Commun. Soil Sci. Plant Anal. 45, 1831−1839.

Wallace, A., Alexander, G.V., Chaudhry, F.M., 1977. Phytotoxicity of cobalt, vanadium, titanium, silver and chromium. Commun. Soil Sci. Plant Anal. 8, 751−752.

Zhang, J., Wages, M., Cox, S.B., Maul, J.D., Li, Y., Barnes, M., et al., 2012. Effect of titanium dioxide nanomaterials and ultraviolet light coexposure on African clawed frogs (*Xenopus laevis*). Environ. Toxicol. Chem. 31 (1), 176−183.

Zheng, L., Hong, F., Lu, S., Liu, C., 2005. Effect of nano-TiO$_2$ on strength of naturally aged seeds and growth of spinach. Biol. Trace Elem. Res. 104, 83−92.

7

Photocatalytic Decolorization of Two Remazol Dyes Using TiO$_2$ Impregnated Pumice Composite as Catalyst

Deniz İzlen Çifçi, Sema Terzi, Süreyya Meriç

DEPARTMENT OF ENVIRONMENTAL ENGINEERING, FACULTY OF ÇORLU ENGINEERING, NAMIK KEMAL UNIVERSITY, TEKIRDAĞ, TURKEY

7.1 Introduction

Dyes are widely used in textiles, pulp mills, leather, dye synthesis, printing, painting, photography, food and plastics industries and most of them display resistance to removal with biological treatment systems (Çifçi and Meriç, 2015). There have been more than 100,000 commercial dyes and more than 7×10^5 tons of paints registered per year in the world (Malakootian et al., 2014). Most of the dyes are known to be toxic or carcinogenic, and they reduce light transmittance in the aquatic environment, thus, advanced treatment methods are required to remove color and toxic content before discharge (Kanakaraju et al., 2015; Mahmoun et al., 2013). Photocatalysis is a promising method among the advanced oxidation processes (AOPs) for the removal of the toxic dyes to carbondioxide and water (Çifçi and Meriç, 2015). Because TiO$_2$ is not a good adsorbent, the use of TiO$_2$ in combination with a porous adsorbent, which is owning the high surface area that improves the photocatalytic activitiy by preventing electron/hole recombination, enhances the aggregation of TiO$_2$ (Li et al., 2014; Zabihi-Mobarake and Nezamzadeh-Ejhieh, 2015). In addition, other studies have shown that photocatalytic activity increases with catalysts such as TiO$_2$ supported chitosan, activated carbon and TiO$_2$ supported zeolite (Wang et al., 2015; Afzal et al., 2017; Bagheri et al., 2016; Gomez et al., 2013; Omri et al., 2014). However, there have any studies with TiO$_2$ supported pumice up to date. Pumice, which is a volcanic stone, is an intensive porous material with an average porosity of 90% (Calabrò et al., 2012). As a cheap adsorbent, pumice has become popularly used in the adsorption of heavy metals and dyes (Çifçi and Meriç, 2016). However, there are still few studies on its use as catalyst, in particular, there is no study reported on TiO$_2$ impregnated pumice yet. This study is a first attempt to evaluate the decolorization of photocatalysis using TiO$_2$ impregnated pumice composite as catalyst under

UV-A illumination. The reusability and stability of composite synthesized was evaluated for three times. Toxicity of oxidation by-products to *Daphnia magna* was tested following the standardized test.

7.2 Materials and Methods

7.2.1 Materials

Titanium isopropoxide (TTIP, 97% CAS#546-68-9) was purchased from Sigma Aldrich. Ethanol (CAS#102371), HNO_3 (CAS#100456) and NaOH (CAS#106462) were purchased from Merck and used as analytical grade. Distilled water was used in all the experiments. The pumice was obtained from a company working in Cappadocia region (Nevşehir city, Turkey), where a large pumice reserve as is present. The particle size of pumice powder ranged from nano to micron ($0-125$ μm). Remazol red (RR) and remazol brilliant blue (RB) were purchased from a dye producer industry in Turkey.

7.2.2 Synthesis of TiO_2 Impregnated Pumice Composite

TiO_2 supported pumice composite was synthesized following the literature with the application of some modifications (Kanakaraju et al., 2015; Afzal et al., 2017). Briefly, 10 mL TTIP was dropped slowly into the 100 mL anhydrous ethanol under continuous stirring at room temperature. Meanwhile, concentrated nitric acid was dilluted with distilled water with a volume ratio of 1:5. This solution was added dropwise into the TTIP solution and stirred for 1 hour. After that, pumice was added to the solution at a mass ratio of 2% between TiO_2 and pumice. Than, the solution of pH was adjusted to 4 using concentrated nitric acid and stirred for 24 hours for the ageing step. Latter, it was dried at $105°C$ for 24 hours and calcinated at $550°C$ for 2 hours.

7.2.3 Photocatalytic Degradation Experiments

A photoreactor equipped with sixteen UV-A light lamps (Philips, 8 W, 350 nm wavelength) was used during photocatalytic experiments carried out in 200 mL sample volume at $30°C$ constant temperature. pH was adjusted to the desired values by dosing 1 M NaOH and 1 M HNO_3. The effect of the composite amount ($0.5-2.5$ g/L) and the effect of pH ($3-11$) on both dye removal were evaluated in this study. Experiments were run for 2 hours and a 3 mL of sample were collected at different time intervals. The samples were centrifuged for 5 minutes at 4000 rpm speed before analysis. The experiments were repeated at optimized conditions to collect samples for toxicity tests. To determine the reusability and stability the composite was centrifuged, washed with distilled water and dried at $105°C$ each time before reuse experiments run for three times. During reuse studies, a solution containing 50 mg/L of dye solution was submitted to photocatalysis at pH 3 using a 2.0 g/L of composite amount for RR and 2.5 g/L for RB dye. The color removal of RR and RB dyes was calculated on the basis of their peak absorbance values measured at 520 nm (Mahmoodia and Aramib, 2009) and at 610 nm (Dong et al., 2011) respectively prior and post photocatalytic oxidation.

7.2.4 Analysis

The synthesized composite was submitted to electron scanning microscopy (SEM)-energy dispersive X-ray analyzer (EDX) (Fei-Quanta FEG 250) and Fourier transform infrared spectroscopy (FTIR) (Bruker Vertex 70 ATR) analysis to obtain its surface properties. SEM-EDX was operated at 5 kV accelerating, 60 Pa pressure and 3 spot at 20000 $\times$ and 40000 $\times$ magnification. FTIR was obtained in the range of 400 to 4000 cm^{-1}. Absorbance of the dyes was scanned using a spectrophotometer (Shimadzu UV-2401).

7.2.5 Acute Toxicity

Newborn *Daphnia magna* ($<$24 hours) was exposed to dye solutions for 24 and 48 hours. at 50% and 25% dilution ratios (ISO 6341, 2012). *D. magna* were inoculated in a temperature constant vivarium at NKU Environmental Engineering Department's Laboratory. The pH of the feeding and test solutions was kept at 8.0. Toxicity tests were performed in the dark at temperature constant room (20°C). Experiments were performed as quadruplicate and 5 daphnids were tested in each replicate. Acute toxicity was assessed by means of immobilization percentage determined by dividing the total number of immobilized organisms in four replicate to total number of test organisms to be 20.

7.3 Results and Discussion

7.3.1 Characterization of TiO$_2$ Impregnated Pumice Composite

SEM analysis illustrating surface morphology of TiO_2 impregnated pumice composite showed that pumice has an irregular shape with a porous surface that is heterogeneous (Fig. 7-1). While the TiO_2 film layer is mostly coated on the external surface of the pumice, it can be also coated on the inner surfaces due to the pore structure (Zhang et al., 2016; Nagarjuna et al., 2015). This leads to increase the active surface area of TiO_2 thin layer but it is, in any case, limited by the pore size (Pires et al., 2015; Liu et al., 2015). Chemical composition of pumice is mostly composed of silisium and oxygen. It numerically contains 23.81% O, 2.00% Na, 9.74% Al, 64.45% Si and trace amount of Fe. After TiO_2 impregnation the composition of pumice was characterized by 1.6% Ti as well as 22.85% O, 1.72% Na, 10.22% Al, 63.61% Si and trace amount of Fe.

As seen in Fig. 7-2, the FTIR spectrum of TiO_2 impregnated pumice composite showed a peak near 600 cm^{-1} representing the stretching of Ti-O bond which is different from the FTIR spectrum of pumice alone (Singh et al., 2016; Bagheri et al., 2016; Farhadian and Kazemzad, 2016; Zhang et al., 2012). The vibrations at 1650 and 3400 cm^{-1} assigned to $^-$OH vibration. The peaks appeared at 450, 785, and 1030 cm^{-1} are the characteristic bands of silisiums as Si-O, Si-O-Si or O-Si-O bonds (Jardim et al., 2017). The absorption band at 450 cm^{-1} could be attributed to bending of Si$-$O vibration and the intensity of this band for TiO_2 impregnated pumice composite is higher than the band of pumice alone. The peak which appears at 1030 cm^{-1} slightly shifted to 1058 cm^{-1} for TiO_2 impregnated pumice composite.

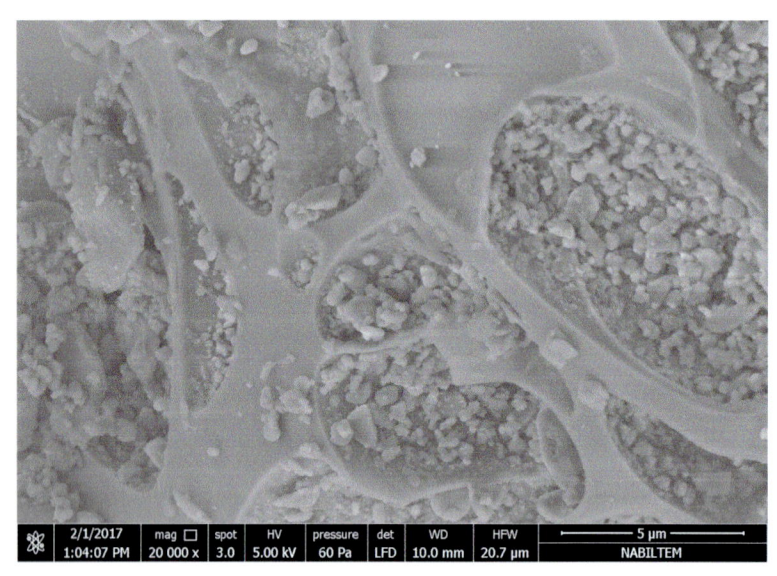

FIGURE 7-1 SEM image of TiO$_2$ impregnated pumice composite.

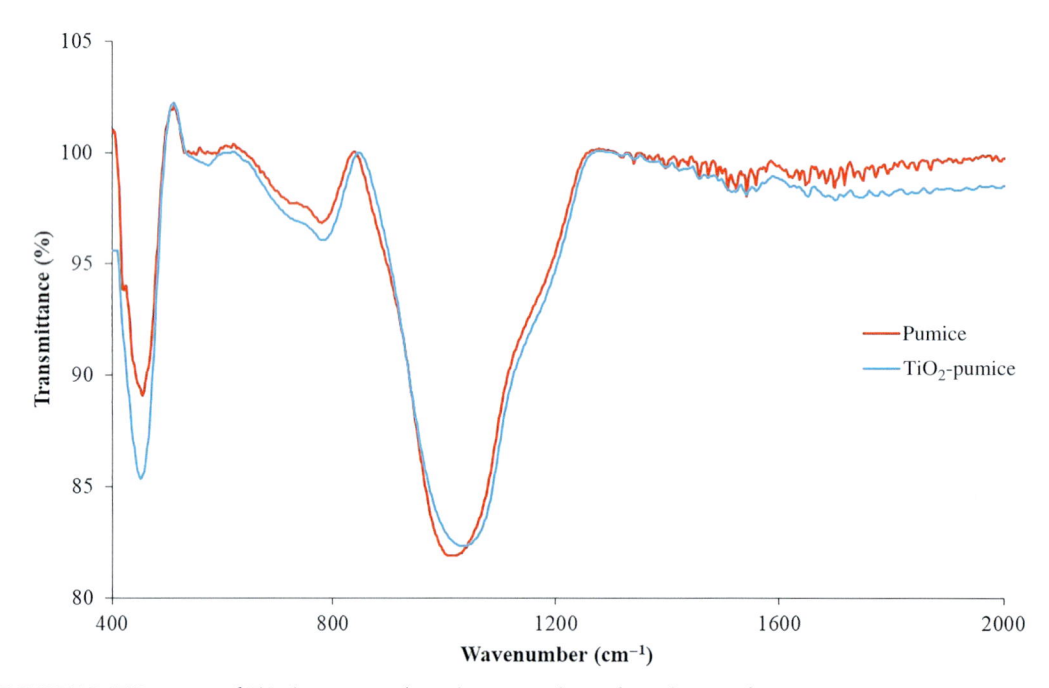

FIGURE 7-2 FTIR spectra of TiO$_2$ impregnated pumice composite and pumice powder.

7.3.2 Effect of the Amount of TiO$_2$ Impregnated Pumice Composite on Decolorization

The effect of the amount of TiO$_2$ impregnated pumice composite used as a catalyst during photocatalytic oxidation on the decolorization of RB and RR dyes is given in Fig. 7-3. Increasing the amount of the catalyst enhanced decolorization of RB and RR dyes due to the presence of TiO$_2$ nanoparticles promoting the formation of OH radicals. As the amount of the catalyst increased from 0 to 2.5 g/L, the decolorization of RB shifted from 36.1% to

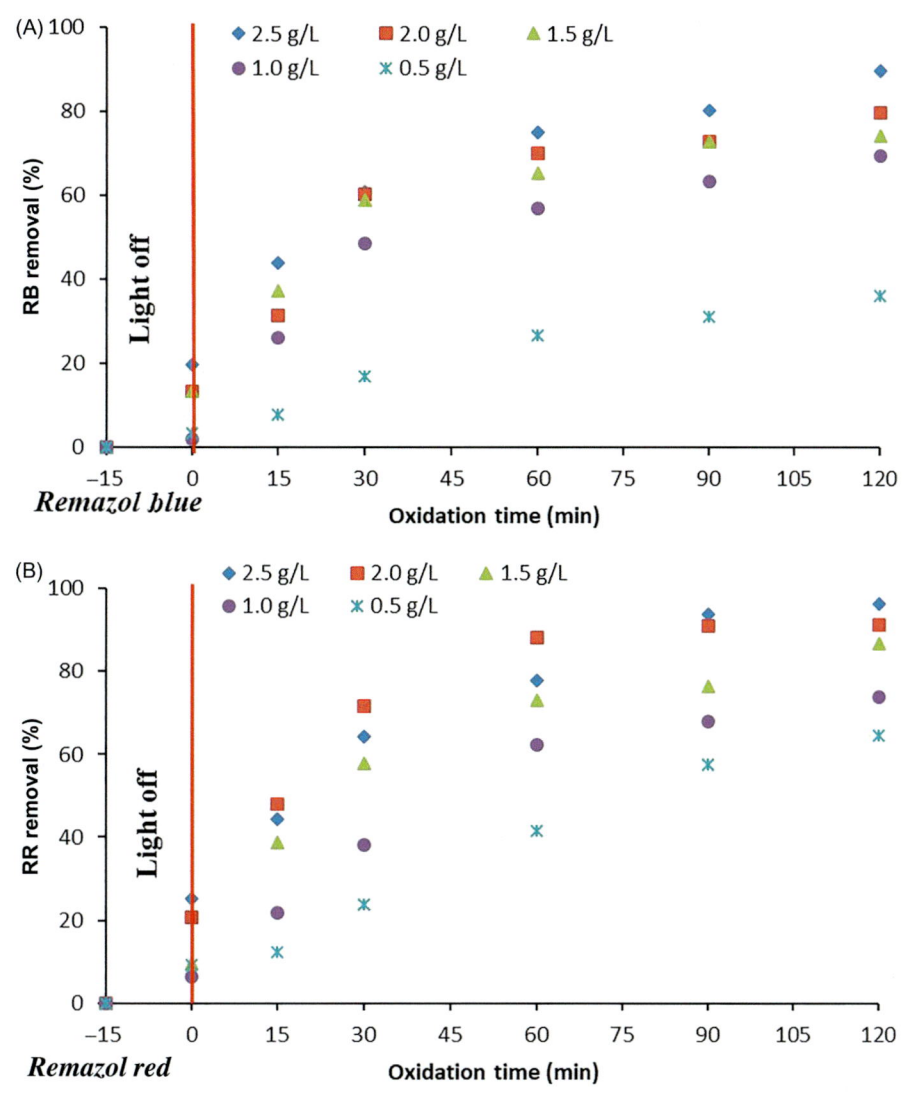

FIGURE 7-3 Decolorization of the dyes versus oxidation time using different concentrations of TiO$_2$ impregnated pumice composite as catalyst (pH:7).

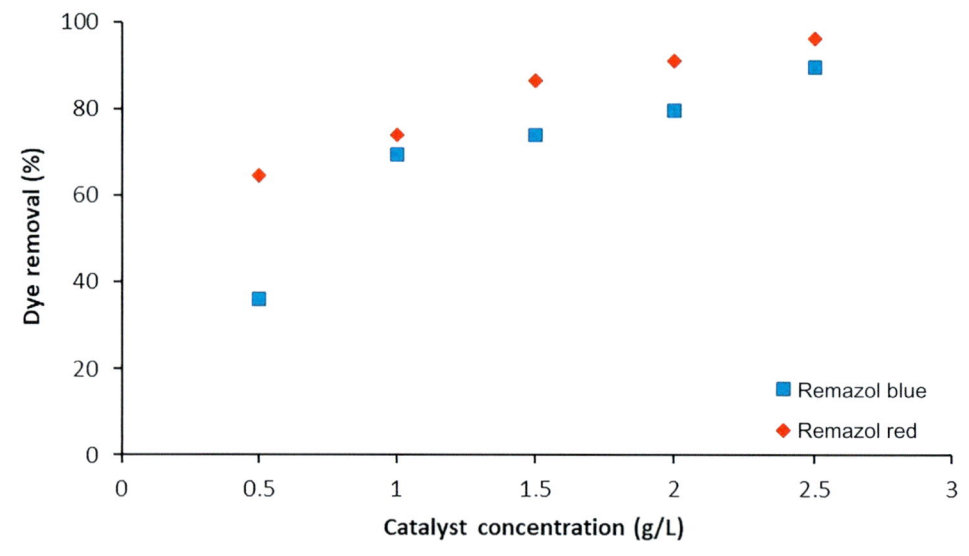

FIGURE 7-4 Effect of TiO_2 impregnated pumice composite dose on the decolorization of the dyes (pH:7, oxidation time: 120 minutes).

89.7%, while decolorization of RR dyes increased from 64.7% to 96.2% (Fig. 7-4). The degradation of RR dye did not significantly change during 90 and 120 minutes oxidation. While the degradation of RB dye continued during 120 minutes illumination. Maximum degradation of RB was obtained using 2.5 g/L catalyst, while the removal of RR did not significantly change when a 2.0 and 2.5 g/L of catalyst was used. Pseudo first (Eq. 7.1) order kinetics fitted the data as given in the following the equation:

$$\ln C/C_0 = -kt \tag{7.1}$$

Where C is the initial dye concentration (mg/L), C_0 is the concentration of the dye at the end of the reaction (mg/L), t (min) indicates the reaction occurring at any time, $k1$ (1/min) is the equilibrium rate constant of pseudo first order adsorption. Accordingly, the reaction rate constant ($k1$) is calculated from the intercept of the plot of $-\ln (C/C_0)$ against time t. The decolorization rates of dyes using the synthesized catalyst are given in Table 7-1. The kinetic constant of RB dye increased with increasing catalyst dose up to 2.5 g/L. However, the kinetic constant of RR decreased as the amount of the catalyst increased from 2.0 g/L to 2.5 g/L. Highest kinetic constant was calculated to be $17.28.10^{-3}$ min^{-1} at 2.5 g/L catalyst dose for RB and $27.19.10^{-3}$ min^{-1} at 2.0 g/L catalyst dose for RR dye.

7.3.3 Effect of pH on the Color Removal

The pH is important for the color removal of dyes as the adsorption of dye molecules on the catalyst surface is affected by the change of surface charge with pH varying (Gupta et al., 2012). The effect of pH on the color removal of both RB and RR dyes is given in Fig. 7-5.

Table 7-1 Pseudo First Order Kinetic Rates for Photocatalytic Decolorization of Dyes Studied Using Different Amounts of the Composite (pH:7)

Composite Concentration (g/L)	RB			RR		
	Removal %	k_1 (min^{-1}) ($*10^{-3}$)	R^2	Removal %	k_1 (min^{-1}) ($*10^{-3}$)	R^2
1.0	69.4	11.07	0.8297	73.9	11.75	0.9527
1.5	74.2	12.16	0.7705	86.7	16.60	0.9308
2.0	79.7	13.54	0.9490	91.3	27.19	0.9472
2.5	89.7	17.28	0.9686	96.2	25.02	0.9808

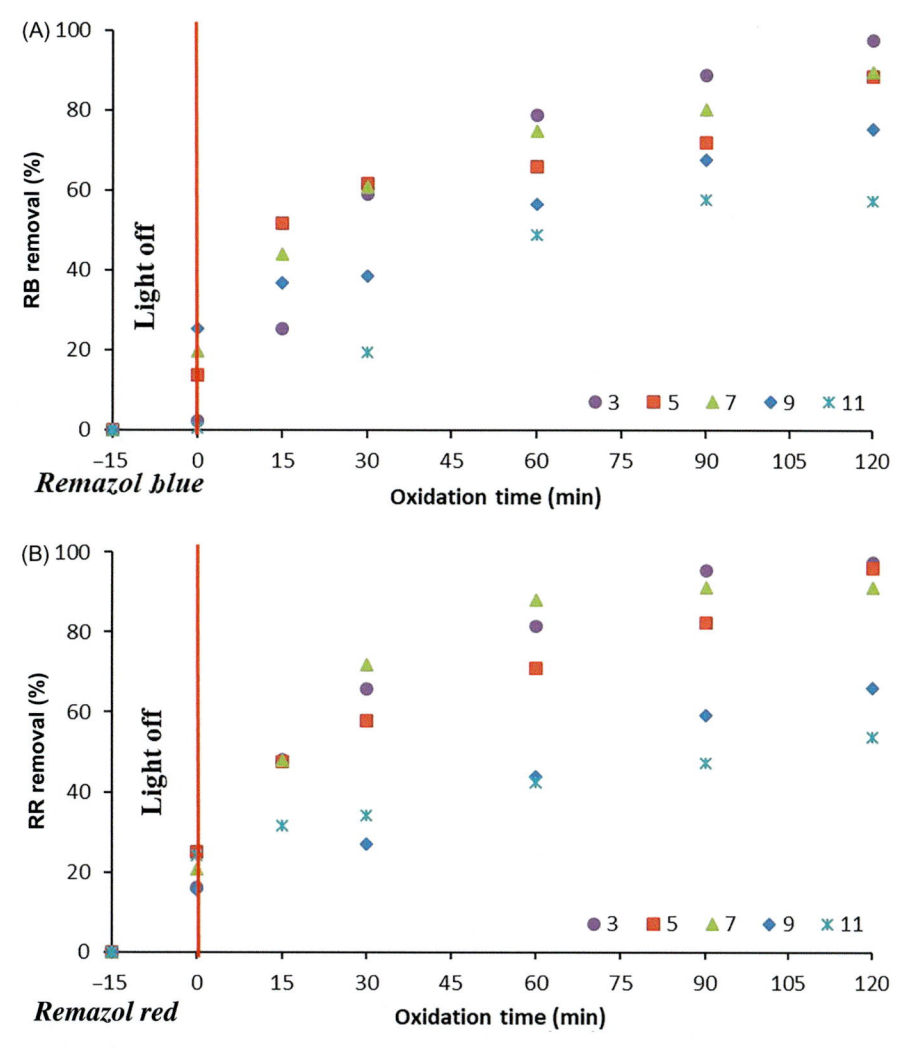

FIGURE 7-5 Effect of pH on the decolorization of the dyes (catalyst concentration: 2.0 g/L for RR and 2.5 g/L for RB, oxidation time: 120 minutes).

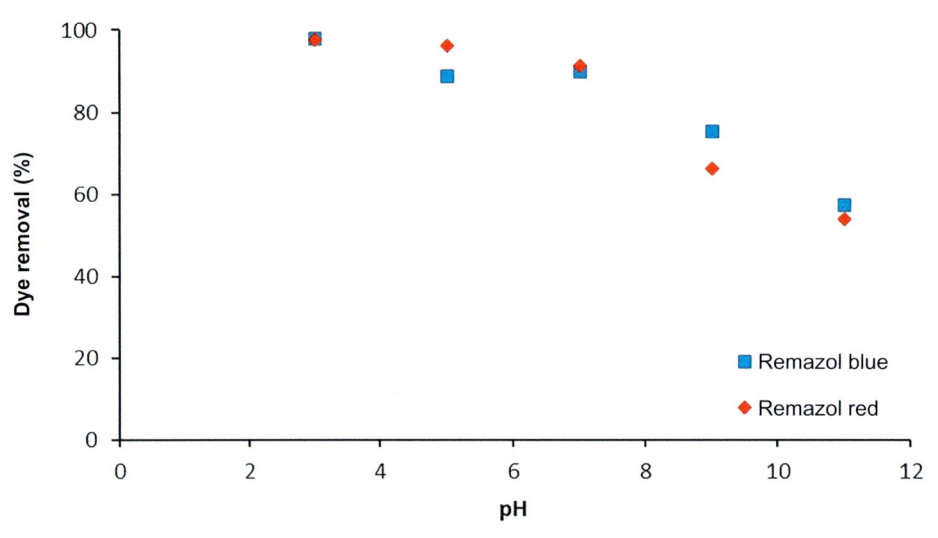

FIGURE 7-6 Effect of pH on the decolorization of the dyes (catalyst concentration: 2.0 g/L for RR and 2.5 g/L for RB, oxidation time: 120 minutes).

Table 7-2 Pseudo First Order Kinetics for Photocatalytic Decolorization of Dyes at Varying pH Values (Composite Amount: 2.0 g/L for RR and 2.5 g/L for RB)

	Remazol Blue		Remazol Red	
pH	k_1 (min^{-1}) ($^{*}10^{-3}$)	R^2	k_1 (min^{-1}) ($^{*}10^{-3}$)	R^2
3.00	28.29	0.9652	29.84	0.9866
5.00	15.74	0.9116	20.92	0.9092
7.00	17.28	0.9686	27.19	0.9472

Higher catalytic activity obtained at acidic condition for both dyes and the removal efficiency decreased by the increase of pH of the solutions. Specifically, decolorization of RB and RR dyes were found to be 97.8% and 97.6% at pH 3, respectively whereas those color removal percentages decreased to 57.4% and 53.9%, respectively at pH 11 after 2 hours of photocatalytic oxidation (Fig. 7-6). When pH was adjusted from 5 to 3, the $k1$ enhanced from $15.74.10^{-3}$ to $28.29.10^{-3}$ min^{-1} for RB dye and from $20.92.10^{-3}$ to $29.84.10^{-3}$ min^{-1} for RR dye (Table 7-2).

7.3.4 Toxicity of Photocatalytic Oxidation By-Products

As it was reported for the other advanced oxidation processes, oxidation by products can form in photocatalysis process that can also cause more toxic effluent than prior to the treatment (Lofrano et al., 2016; Ozkal et al., 2016). In this study, photocatalytic treated all samples resulted in 100% immobilization of *Daphnia magna* when they were tested without dilution.

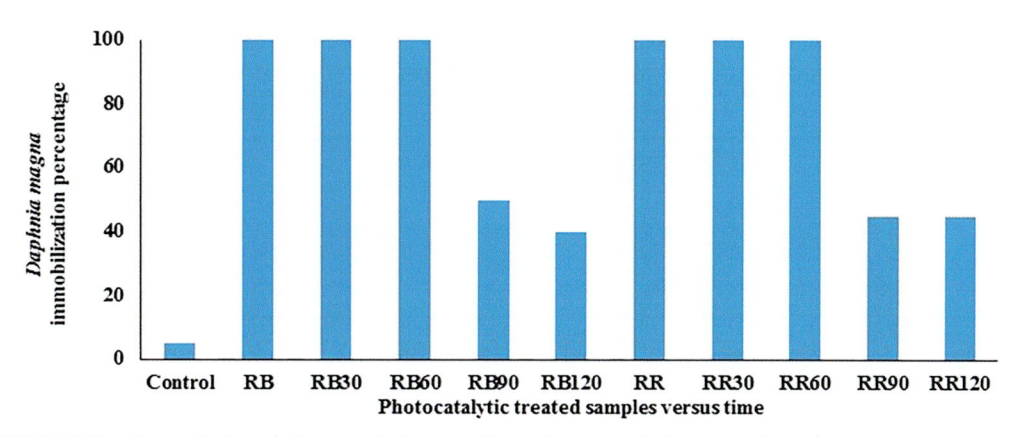

FIGURE 7-7 Toxicity evolution of photocatalytic treated samples on *Daphnia magna* for 24 h exposure time and tested at 75% dilution ratio (catalyst concentration: 2.0 g/L for RR and 2.5 g/L for RB, pH:3).

The treated samples were also tested after dilution at which the samples resulted in 100% toxic at 50% dilution ratio after 24 hours exposure time. As a further dilution, the samples were tested at 75% dilution at which both photocatalytic treated dye solutions resulted in 100% toxic after 30 and 60 minutes of irradiation. Toxicity to *Daphnia magna* decreased below 50% immobilization after 90 and 120 minutes of irradiation for both dyes (Fig. 7-7). This decrease is found promising for the effectiveness of photocatalytic treatment, however there is an urgent need to prolong the treatment duration and/or for increasing catalyst dose to completely remove toxicity in the undiluted samples.

7.3.5 Reuse and Stability of TiO$_2$ Impregnated Pumice Composite

A slight decrease in the photocatalytic color removal was observed for both dyes when TiO$_2$ impregnated pumice composite was reused. This reduction of removal efficiency was probably due to the adsorption of dyes on the surface of the composite (Yener et al., 2017). Dye molecules accumulate in the pores of pumice leads to inhibition of the photocatalytic degradation. The degradation efficiencies of RB dye were found to be 97.8%, 82.7%, and 75.3% whereas a 97.6%, 85.3%, and 84.0% degradation of RR were observed for the first, second, and third cycles of reusing composite, respectively. Besides, $k1$ decreased from $28.29.10^{-3}$ to $11.01.10^{-3}$ min^{-1} for RB and from $29.84.10^{-3}$ to $14.03.10^{-3}$ min^{-1} for RR dye (Table 7-3). Similar reductions in removal efficiency for reuse of catalyst was reported in the literature (Kanakaraju et al., 2015; Yener et al., 2017; Hadjltaief et al., 2016; Gomez et al., 2013). However, Gomez et al. (2013) found that the calcination of catalyst is important to regenerate the catalyst for reuse. On the other hand, Zabihi-Mobarake and Nezamzadeh-Ejhieh (2015) reported that the calcination at 450°C for used catalyst was more efficient than at 25°C and 200°C. The slight decline of degradation observed in our study could occur since calcination is not applied for each cycle.

Table 7-3 Pseudo First Order Kinetics Changes for Photocatalytic Decolorization of Dyes Reusing Composite (Composite Amount: 2.0 g/L for RR and 2.5 g/L for RB, pH:3)

Reuse of Composite	RB			RR		
	Removal (%)	k_1 (min^{-1}) ($^*10^{-3}$)	R^2	Removal (%)	k_1 (min^{-1}) ($^*10^{-3}$)	R^2
1st	97.8	28.29	0.9652	97.6	29.84	0.9866
2nd	82.7	15.71	0.9754	85.3	15.00	0.9766
3rd	75.3	11.01	0.9924	84.0	14.03	0.9624

R, regression coefficient.

7.4 Conclusion

This study, as a first attempt, aimed to evaluate photocatalytic decolorization of of two remazol dyes (blue and red) using TiO_2 impregnated pumice composite catalyst. The SEM-EDX analysis showed that TiO_2 nanoparticules were mostly impregnated on the pores of pumice with a ratio of 1.60% Ti. A 97.8% and 97.6% of color removal was achieved for RB and RR at pH 3, respectively, whereas those color removal percentages decreased to 57.4% and 53.9%, respectively, at pH 11 after 2 hours illumination. The color removal efficiencies of 97.8%, 82.7%, and 75.3% for RB dye and of 97.6%, 85.3%, and 84.0% for RR dye were observed after first, second, and third cycles reuse of the synthesized catalyst. Toxicity of photocatalytic treated samples of both dyes decreased below 50% after 90 and 120 minutes irradiation when they were diluted at 75%. This result indicate there is an urgent need to prolong the irradiation time and/or for increasing catalyst dose to completely remove toxicity in the undiluted samples.

Acknowledgments

The authors thank Soylu Ltd for providing pumice. This study was inspired from the EU COST Actions ES1202 and supported by Namik Kemal University Research funding Office (NKUBAP.06.GA.17.094).

References

Afzal, S., Samsudin, E.M., Mun, L.K., Julkapli, N.M., Abd Hamid, S.B., 2017. Room temperature synthesis of TiO_2 supported chitosan photocatalyst: study on physicochemical and adsorption photo-decolorization properties. Mater. Res. Bull. 86, 24—29.

Bagheri, S., Hir, Z.A.M., Yousefi, A.T., Hamid, S.B.A., 2016. Photocatalytic performance of activated carbon-supported mesoporous titanium dioxide. Desalin. Water Treat. 57 (23), 10859—10865.

Calabrò, P.S., Moraci, N., Suraci, P., 2012. Estimate of the optimum weight ratio in zero-valent iron/pumice granular mixtures used in permeable reactive barriers for the remediation of nickel contaminated groundwater. J. Hazard. Mater. 207-208, 111—116.

Çifçi, D.I., Meriç, S., 2015. Optimization of suspended photocatalytic treatment of two biologically treated textile effluents using TiO_2 and ZnO catalysts. Global NEST J. 17 (4), 653—663.

Çifçi, D.I., Meriç, S., 2016. A review on pumice for water and wastewater treatment. Desalin. Water Treat. 57 (39), 18131–18143.

Dong, Y., Dong, W., Cao, Y., Han, Z., Ding, Z., 2011. Preparation and catalytic activity of Fe alginate gel beads for oxidative degradation of azo dyes under visible light irradiation. Catal. Today 175 (1), 346–1355.

Farhadian, M., Kazemzad, M., 2016. Photocatalytic degradation of malachite green by magnetic photocatalyst. Synth. React. Inorg. M. 46 (3), 458–463.

Gomez, S., Marchena, C.L., Pizzio, L., Pierella, L., 2013. Preparation and characterization of TiO_2/HZSM-11 zeolite for photodegradation of dichlorvos in aqueous solution. J. Hazard. Mater. 258-259, 19–26.

Gupta, V.K., Jain, R., Mittal, A., Saleh, T.A., Nayak, A., Agarwal, S., et al., 2012. Photo-catalytic degradation of toxic dye amaranth on TiO_2/UV in aqueous suspensions. Mater. Sci. Eng. C. 32, 12–17.

Hadjltaief, H.B., Zina, M.B., Galvez, M.E., Da Costa, P., 2016. Photocatalytic degradation of methyl green dye in aqueous solution over natural clay-supported $ZnO-TiO_2$ catalysts. J. Photochem. Photobiol. A Chem. 315, 25–33.

ISO 6341, 2012. International Organization for Standardisation Water Quality: Determination of the Inhibition of the Mobility of *Daphnia magna* Straus (Cladocera, Crustacea)-Acute Toxicity Test. Geneva, Switzerland.

Jardim, A.A.M.L.F., Bacani, R., Gonçalves, N.S., Fantini, M.C.A., Martins, T.S., 2017. SBA-15:TiO_2 nanocomposites: II. Direct and post-synthesis using acetylacetone. Microporous Mesoporous Mater. 239, 235–243.

Kanakaraju, D., Kockler, J., Motti, C.A., Glass, B.D., Oelgemöller, M., 2015. Titanium dioxide/zeolite integrated photocatalytic adsorbents for the degradation of amoxicillin. Appl. Catal. B. 166-167, 45–55.

Li, Y., Li, S.G., Wang, J., Li, Y., Ma, C.H., Zhang, L., 2014. Preparation and solarlight photocatalytic activity of TiO_2 composites: TiO_2/kaolin, TiO_2/diatomite, and TiO_2/zeolite. Russ. J. Phys. Chem. A. 88 (13), 2471–2475.

Liu, C., Li, Y., Xu, P., Li, M., Zeng, M., 2015. Controlled synthesis of ordered mesoporous TiO_2-supported on activated carbon and pore-pore synergistic photocatalytic performance. Mater. Chem. Phys. 149-150, 69–76.

Lofrano, G., Libralato, G., Adinolfi, R., Siciliano, A., Iannece, P., Guida, M., et al., 2016. Photocatalytic degradation of the antibiotic chloramphenicol and effluent toxicity effects. Ecotox. Environ. Saf. 123, 65–71.

Mahmoodia, N.M., Aramib, M., 2009. Numerical finite volume modeling of dye decolorization using immobilized titania nanophotocatalysis. Chem. Eng. J. 146 (2), 189–193.

Mahmoun, H.R., El-Molla, S.A., Saif, M., 2013. Improvement of physicochemical properties of Fe_2O_3/MgO nanomaterials by hydrothermal treatment for dye removal from industrial wastewater. Powder Technol. 249, 225–233.

Malakootian, M., Mansoorian, H.J., Yari, A., 2014. Removal of reactive dyes from aqueous solutions by a non-conventional and low cost agricultural waste: adsorption on ash of aloe vera plant. Iran. J. Health, Saf. Environ. 1 (3), 117–125.

Nagarjuna, R., Roy, S., Ganesan, R., 2015. Polymerizable solegel precursor mediated synthesis of TiO_2 supported zeolite-4A and its photodegradation of methylene blue. Microporous Mesoporous Mater. 211, 1–8.

Omri, A., Lambert, S.D., Geens, J., Bennour, F., Benzina, M., 2014. Synthesis, surface characterization and photocatalytic activity of TiO_2 supported on almond shell activated carbon. J. Mater. Sci. Technol. 30 (9), 894–902.

Ozkal, C.B., Koruyucu, A., Meriç, S., 2016. Heterogenous photocatalytic degradation, mineralization and detoxification of ampicillin under varying pH and incident photon flux conditions. Desal. Wat. Treat. 57 (39), 18391–18397.

Pires, C.A., dos Santos, A.C.C., Jordão, E., 2015. Oxidation of phenol in aqueous solution with copper oxide catalysts supported on γ-Al_2O_3, pillared clay and TiO_2: comparison of the performance and costs associated with each catalyst. Braz. J. Chem. Eng. 32 (4), 837–848.

Singh, P., Vishnu, M.C., Sharma, K.K., Singh, R., Madhav, S., Tiwary, D., et al., 2016. Comparative study of dye degradation using TiO_2-activated carbon nanocomposites as catalysts in photocatalytic, sonocatalytic, and photosonocatalytic reactor. Desalin. Water Treat. 57 (43), 20552−20564.

Wang, C., Li, Y., Shi, H., Huang, J., 2015. Preparation and characterization of natural zeolite supported nano TiO_2 photocatalysts by a modified electrostatic self-assembly method. Surf. Interface Anal. 47, 142−147.

Yener, H.B., Yılmaz, M., Deliismail, Ö., Özkan, S.F., Helvacı, S.S., 2017. Clinoptilolite supported rutile TiO_2 composites: synthesis, characterization, and photocatalytic activity on the degradation of terephthalic acid. Sep. Purif. Technol. 173, 17−26.

Zabihi-Mobarakeh, H., Nezamzadeh-Ejhieh, A., 2015. Application of supported TiO_2 onto Iranian clinoptilolite nanoparticles in the photodegradation of mixture of aniline and 2, 4-dinitroaniline aqueous solution. J. Ind. Eng. Chem. 26, 315−321.

Zhang, W., Ma, Z., Li, K., Yang, L., Li, H., He, H., 2016. Sol-gel synthesis of nano-sized TiO_2 supported on HZSM-5. Curr. Nanosci. 12, 514−519.

Zhang, Z., Xu, Y., Ma, X., Li, F., Liu, D., Chen, Z., et al., 2012. Microwave degradation of methyl orange dye in aqueous solution in the presence of nano-TiO_2-supported activated carbon (supported-TiO_2/AC/MW). J. Hazard. Mater. 209-210, 271−277.

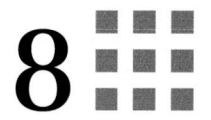

8

Application of Nanomaterials for the Removal of Heavy Metal From Wastewater

Fahmida Parvin[1], Sharmin Yousuf Rikta[2], Shafi M. Tareq[1]

[1]DEPARTMENT OF ENVIRONMENTAL SCIENCES, JAHANGIRNAGAR UNIVERSITY, DHAKA, BANGLADESH [2]DEPARTMENT OF ENVIRONMENTAL SCIENCE, BANGLADESH UNIVERSITY OF PROFESSIONALS, DHAKA, BANGLADESH

8.1 Introduction

Heavy metal pollution in water is thought to be one of the serious environmental problems all over the world (Bhuiyan et al., 2015; Reza and Singh, 2010; Wong et al., 2002). In the big cities of developing countries, this issue is being highly menaced mainly due to the unplanned discharge from different industries. Heavy metals can enter into the environment through both anthropogenic and natural sources (Akoto et al., 2008; Wong et al., 2002). Mining activities, discharge of untreated industrial wastewater, excessive use of heavy metal-rich pesticides and fertilizers in agricultural activities, dry and wet deposition from particulate matter are the main anthropogenic sources of heavy metals (He et al., 2013; Juang et al., 2009; Nouri et al., 2008). As for the natural source, chemical weathering of heavy metal enrich rocks and minerals makes them available to aquatic environments and sediment (Reza and Singh, 2010). However, the natural source of heavy metal is related to the mineralogy of the specific region (Karbassi et al., 2008). Heavy metals likely need special consideration because of their carcinogenic effect on human and detrimental impacts on the environment (Anawar et al., 2001; Godt et al., 2006; He et al., 2013). Compared with the traditional organic pollutant, heavy metals are nonbiodegradable (Farkas et al., 2007) and, consequently, accumulate in various environmental compartments (i.e., water and soil). Hence, there is a great chance of bioaccumulation and transfer of them to the ecological pyramid through food chain (Anawar et al., 2001; Reza and Singh, 2010).

Removal of heavy metals is a challenging issue, which is highly required to prevent/reduce their toxic effects on human body and disruption of ecological balance. Recently, nanotechnology is being integrated with most of the conventional techniques for the removal of heavy metals from water system to improve the removal efficiency. Nanotechnology is a rapidly growing sector in current developed world and nanoparticles (NPs)/nanomaterials

Nanotechnology in Water and Wastewater Treatment. DOI: https://doi.org/10.1016/B978-0-12-813902-8.00008-3

are the substances, which are used to produce nanoenabled engineered products in the field of technologies. NPs are defined as nanomaterials with a size between 1 and 100 nm on at least one dimension (ISO, 2008), having unique physicochemical properties differing from their bulk forms due to their greater surface area to volume ratio (Fig. 8-1). The larger surface area to volume ratio of these materials makes them more capable to react with pollutants/contaminants (Chen et al., 2017; Chowdhury and Yanful, 2010; Di et al., 2006; Feng et al., 2010; Hossein Beyki et al., 2017; Singh et al., 2017). They can penetrate deeper into the pollutants that enhances their reactivity and ultimately removal efficiency, which is generally not plausible in traditional wastewater treatment process (Bystrzejewski et al., 2009). Recently, nanostructured adsorbent, nanocatalysts as well as polymer nanocomposites are being integrated with conventional techniques (i.e., ion exchange, reverse osmosis, membrane technology) and thought to be very effective to remove heavy metals as well as organic pollutants from wastewater in comparison to the conventional techniques (Lou et al., 2017; Qi et al., 2015; Sheela and Nayaka, 2012; Visa, 2016; Xu et al., 2012). To remove heavy metals from wastewater, numerous nanomaterials have been developed, which are cost effective, have high removal efficiencies and easy to regenerate. They have their own unique properties and functions for the purification of wastewater (Gupta et al., 2011).

In this chapter, efficiency of different of nanomaterials as adsorbent for the removal of heavy metals (i.e., arsenic, lead, cadmium, and chromium) from wastewater have been discussed. The synthesis of the nanomaterials and heavy metal adsorption process has also been described here.

FIGURE 8-1 Scanning electron photomicrographs of (A) CuO nanoparticle in the range of 75−100 nm (Singh et al., 2017), (B) Fe_3O_4-SO_3H magnetic nanoparticle (Chen et al., 2017), (C) magnetite−maghemite nanoparticles (20−40 nm) (Chowdhury and Yanful, 2010), (D) γ-alumina nanocomposite (80 nm) (Hossein Beyki et al., 2017).

8.2 Synthesis and Adsorption Process of Nanomaterials

Nanomaterials are generally used to overcome the limitations of conventional adsorbents in wastewater treatment through their very high specific surface area, surface chemistry and related active adsorption sites (Qu et al., 2013). They can adsorb toxic substances by ion exchange, ion precipitation, and adsorption. When the nanoadsorbent is added into the water containing heavy metals, heavy metals diffuse on external surface of the nanoadsorbents due to diffusion potential characterized by the concentrations of heavy metals and available external surface area on the adsorbent. After diffusion on the external surface of the adsorbent, heavy metal is diffused on the available pores of the adsorbent. The exposed active sites of the nanoadsorbent are occupied by the metals during the adsorption process by means of physisorption or chemisorption. However, in polymer-based adsorbents, the adsorption behavior is governed by complexation and electrostatic attraction (Lata and Samadder, 2016). Nanoadsorbents can be categorized in various categories, like, carbon-based, metal-based, and polymeric nanoadsorbents, which are usually applied for removal of heavy metals from the wastewater (Anjum et al., 2016). Many researchers have given effort to incorporate nanoparticles in membrane technology to improve its efficiency for the effective removal of heavy metal from aqueous solution. The synthesis of different types of nanoadsorbent and the mechanisms of heavy metal adsorption are discussed here.

8.2.1 Carbon-Based Nanoadsorbents

Carbon nanotube (CNT) is a well-known carbon-based nanoadsorbent used for the removal of heavy metal from water in its oxidized state due to fast kinetics (Li et al., 2003). CNTs can absorb metal ions from water by different mechanisms, like physical adsorption, chemical interaction, electrostatic attraction and sorption−precipitation (Ihsanullah et al., 2016). The adsorption capacity of CNTs can be enhanced by surface modification by means of acid treatment and grafting functional groups. By acid modification, different functional groups have been introduced in the surface of CNT, which increased the absorption capacity of CNT through electrostatic interactions (Fig. 8-2) (Rao et al., 2007). The metal ions are sorbed on the CNTs surface and consequently release the H^+ from the surface, which leads to the lowering of pH of water (Ihsanullah et al., 2016).

The surface functional groups (e.g., carboxyl, hydroxyl, and phenol) of CNTs are the major adsorption sites for metal ions, mainly through electrostatic attraction and chemical bonding (Bhanjana et al., 2017; Rao et al., 2007). These surface oxidation of CNTs enhance the metal adsorption and substantially exceed the adsorption capacities of CNTs over activated carbons (Bystrzejewski et al., 2009). Among the different grafting techniques for introducing functional groups on the surface of CNTs, plasma technique requires less energy and the process is reported to be environmental friendly (Ihsanullah et al., 2016). Regeneration of most of the carbon-based and polymeric nanoadsorbents can be performed by changing solution pH. By lowering the pH of the solution CNTs can be regenerated where metal recovery is above 90% (Peng et al., 2005; Qu et al., 2013).

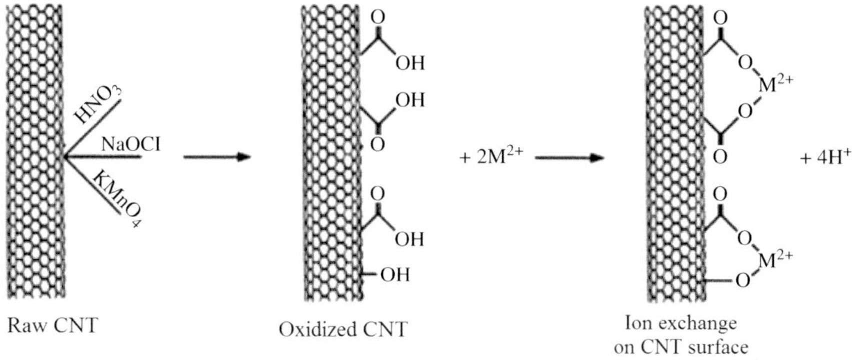

FIGURE 8-2 Schematic diagram showing the acid modification of CNTs, followed by divalent heavy metals adsorption by ion exchange on the surface of modified CNTs. Source: *Reprinted from Rao, G., Lu, C., Su, F., 2007. Sorption of divalent metal ions from aqueous solution by carbon nanotubes: a review. Sep. Purif. Technol. 58, 224–231. doi:10.1016/j.seppur.2006.12.006, with permission from Elsevier.*

Another carbon-based nanomaterial is graphene oxide (GO). Graphene, a thin two-dimensional carbon-based nanomaterial, has interesting physio-chemical properties. Recently, heavy metal adsorption on graphene-based materials has been demonstrated with high adsorption properties (Gopalakrishnan et al., 2015; Mukherjee et al., 2016). The GO nanoparticles can be synthesized by the modified Hummers method in which graphite powder is oxidized in the aqueous medium to graphite oxide and then exfoliated to graphene oxide. The preparation of GO nanosheets from graphite using the modified Hummer's method introduces many oxygen-containing functional groups such as hydroxyl and carboxyl groups, on the surfaces of GO nanosheets. These functional groups are essential for the high-sorption of heavy metal ions. In spite of having high adsorption efficiency, GO has some limitations, such as the possibility of leaching of GO nanoparticles, and high cost of synthesis. However, this problem can be overcome by impregnating GO on mixed matrix membrane (Gopalakrishnan et al., 2015).

8.2.2 Metal Oxide-Based Nanoadsorbents

Uses of metal-based nanoparticles are popular technology for heavy metals removal because of low cost, high adsorption capacity, as well as easy regeneration. Iron oxide (Fe_3O_4), titanium dioxide (TiO_2), nickel oxide (NiO), zinc oxide (ZnO), cupric oxide (CuO), cerium oxide (CeO_2) and aluminum oxide (Al_2O_3) nanomaterials are the well-known metal-based nanoadsorbents for the removal of heavy metals (Qu et al., 2013; Recillas et al., 2011; Sheela and Nayaka, 2012; Tabesh et al., 2018). Adsorption capacity of the metal oxide nanoparticles increased with decreasing the particle size of the metal oxide nanoparticles (Mayo et al., 2007). Metal hydroxide nanoparticles can be impregnated onto the skeleton of activated carbon or other porous materials to achieve simultaneously removal of metal and organic cocontaminants (Bée et al., 2011; Hristovski et al., 2009). These metal-based nanoparticles can also be regenerated by changing the pH of the solution and they are effective enough after several regenerations (Qu et al., 2013; Xu et al., 2012).

Fe_3O_4 nanoadsorbents are effective and economical adsorbents for rapid removal and recovery of metal ions from wastewater effluents (Recillas et al., 2011). Again, iron oxide nanomaterials can be modified by certain chemical reagents and incorporated with bio-based matrix to prepare adsorbent and catalyst materials, which exhibits better performance than that of materials without magnetic particle (Deliyanni et al., 2006; Zhou et al., 2018). In addition to Fe_3O_4 nanoparticles, zero-valent iron based nanoadsorbents have attracted much attention for its use in wastewater treatment. There are many methods for the synthesis of zero-valent iron nanoparticles. These particles can be prepared in aqueous solution through the reduction of ferrous (Fe^{2+}) or ferric (Fe^{3+}) ion with sodium borohydrate ($NaBH_4$) (Mwamulima et al., 2018), the method is known as liquid phase reduction. Zero-valent iron nanoparticles can also be produced from organic solvent decomposition of iron pentacarbo-nyl (Karlsson et al., 2005), hydrogen reduction of Fe_3O_4 (Sun et al., 2006) and chemical wet synthesis mediated by amino acid (Siskova et al., 2013).

CuO, NiO, and ZnO are found to be effective for the removal of heavy metals (Mahdavi et al., 2012; Singh et al., 2017). CuO can be synthesized from different precursors, such as copper nitrate ($Cu(NO_3)_2$) and copper chloride ($CuCl_2$) using precipitation method (Phiwdang et al., 2013). CuO nanoparticle can also be synthesized from natural sources, using *Tamarindus indica* pulp extract, a rich source of natural polyphenolic compounds like catechin, which act as reducing agents (Fig. 8-3). When *Tamarindus indica* pulp extract is mixed with the metal salt solution, the Cu^{2+} reduced to Cu by forming complex with the hydroxyl groups of the catechin. Thus formed metallic copper atoms react with the available atmospheric oxygen to form most stable oxide, that is, CuO. Further, the formed CuO mole-cules come together to nucleate followed by further growth which results the formation of nanoparticles (Singh et al., 2017).

8.2.3 Bimetal Oxide Magnetic Nanomaterials

Bimetal oxide magnetic nanomaterials (i.e., Fe-Mn, Fe-Cu, Fe-Ce, Ce-Ti, Mn-Al, GO−MnFe, etc.) are found to show higher adsorption capacity for heavy metal than the single metal oxide magnetic nanomaterials. Metal composite containing two or more metals shows syner-gistic effect of higher adsorption capacity than individual metal oxide (Babaee et al., 2018; Li et al., 2010; Parsons et al., 2009; Zhang et al., 2007, 2005; Zhang et al., 2010). Parsons et al. (2009) found that among Fe_3O_4, Mn_3O_4, and $MnFe_2O_4$ nanoadsorbent, bimetal oxide showed 20−50 times higher arsenic adsorption capacity. Paramagnetic nature of these magnetic nanomaterials make its heavy metal separation easy from solution with magnetic field (Zhang et al., 2010). Bimetal oxide magnetic nanomaterials adsorb heavy metal from waste-water by sites binding, magnetic selective adsorption, electrostatic interaction, and modified ligands combination (Lata and Samadder, 2016). Usually the hydroxyl groups on the adsor-bent surface involves in the sorption of heavy metal from water. Bimetal oxide nanoadsor-bent can be synthesized by hydrolysis and subsequent precipitation of the both metal oxides

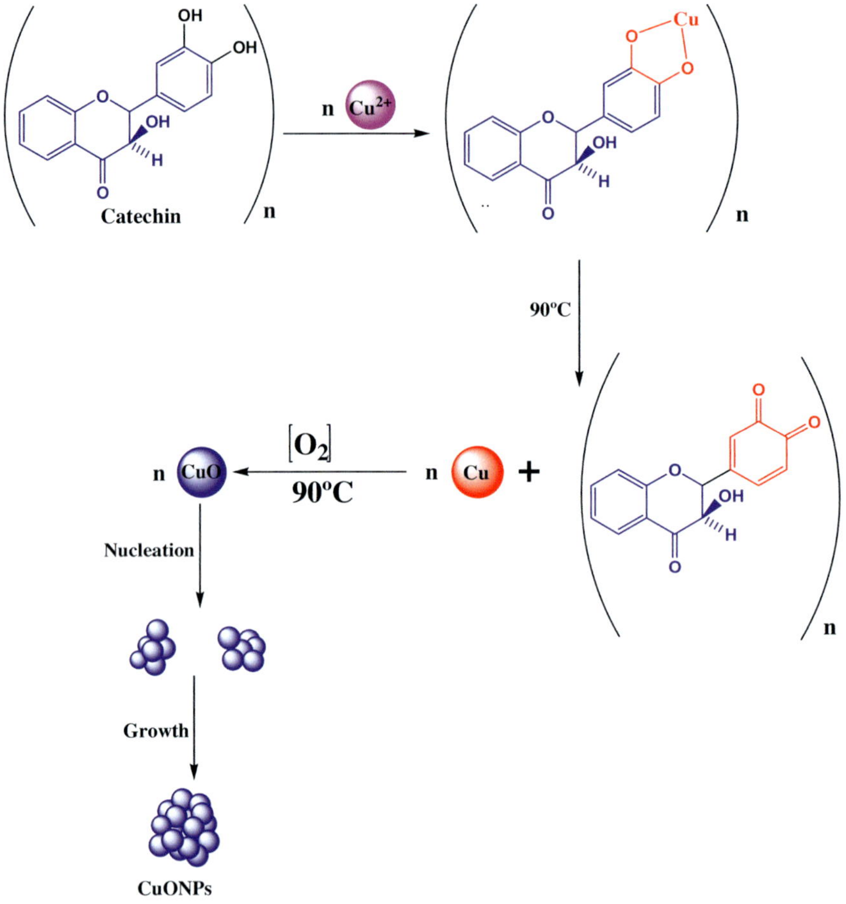

FIGURE 8-3 Schematic representation of the formation of CuO nanoparticle by polyphenolic compound "Catechin" present in the *Tamarindous indica* pulp extract. Source: *Reprinted from Singh, D.K., Verma, D.K., Singh, Y., Hasan, S.H., 2017. Preparation of CuO nanoparticles using Tamarindus indica pulp extract for removal of As(III): optimization of adsorption process by ANN-GA. J. Environ. Chem. Eng. 5, 1302−1318. doi:10.1016/j.jece. 2017.01.046, with permission from Elsevier.*

together in presence of supporting material, followed by magnetic separation, drying, grinding and sieving of nanomaterial (Kumar et al., 2014; Li et al., 2010).

8.2.4 Polymer-Based Nanocomposites

Different nanomaterials can effectively remove heavy metal from water, but their agglomeration restricts their use. However, the problem of agglomeration can be minimized by converting nanomaterials to polymer-based nanocomposites (Pandey et al., 2015; Shi et al., 2011). In **polymer-based** nanocomposite, nanoparticles are impregenated on polymer skeleton (Bée et al., 2011; Geng et al., 2009; Kenawy et al., 2018; Qi et al., 2015; Wang et al., 2014).

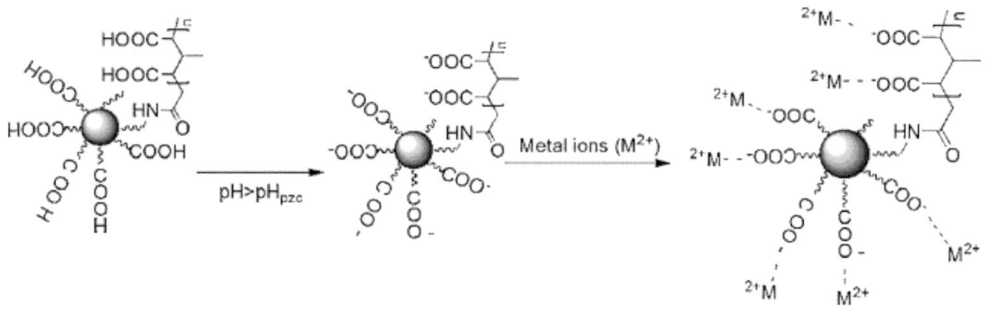

FIGURE 8-4 Possible mechanism for adsorption of metal ions by Fe_3O_4 magnetic nanoparticles modified with 3-aminopropyltriethoxysilane and copolymers of acrylic acid and crotonic acid. Source: *Reprinted from Ge, F., Li, M.-M., Ye, H., Zhao, B.-X., 2012. Effective removal of heavy metal ions Cd2 +, Zn2 +, Pb2 +, Cu2 + from aqueous solution by polymer-modified magnetic nanoparticles. J. Hazard. Mater. 211−212, 366−372. doi:10.1016/j. jhazmat.2011.12.013, with permission from Elsevier.*

The metal ions mainly interact with the polymer-based nanoadsorbent by chelation between the metal ions and the carboxylate anion (Ge et al., 2012; Shokati Poursani et al., 2017). This process is shown in Fig. 8-4. Polymer-based nanocomposite can be synthesized by encapsulating the metal oxide in polymeric resin by acid catalyzed polymerization process (Kenawy et al., 2018; Pandey et al., 2015). One example of polymer-based nanocomposite is resin supported nanoscale zero-valent iron (R-nZVI). Here, resin cannot adsorb the metals. Resin only plays a role as a carrier for zero-valent iron nanoparticles, which mainly adsorb the heavy metal from water. Metal ion reacts with zero-valent iron nanoparticles and form precipitation (Fu et al., 2013).

8.3 Heavy Metal Removal Using Different Nanoadsorbents

While removal of the heavy metal from water using the nanoadsorbents, the anions (i.e., phosphate, sulfate) and dissolved humic substances, which are ubiquitous in the aquatic environment can considerably affect the adsorption process owing to competition for binding with the active site of adsorbing material (Qu et al., 2013; Wang et al., 2010). The existence of anions, like phosphate, not only competes with heavy metal in water for surface sites, but also results in releasing sorbed metals. However, the interference can be minimized by optimizing the initial concentrations of the solutes (heavy metals, anions) as well as pH of the solution (Chowdhury and Yanful, 2010). Further, with the consideration of excellent removal efficiency of the nanomaterials, we also need to think about the production cost of those nanomaterials. The production cost of maghemite, hematite (Fe_2O_3), magnetite (Fe_3O_4) and the mixture of maghemite-magnetite is reported to be very much lower than iron hydroxide and zero-valent iron (Fe^0) (Chowdhury and Yanful, 2010). Apart from these factors, there are several other factors that we should take into consideration for selecting nanoadsorbent for the removal of heavy metal from aquous solution, like adsorption potential, high surface area, physical and chemical stability, holding materials, separation process of nanoadsorbents from aqueous solution, field applicability, environmental impact and risk assessment (Lata and Samadder, 2016). Arsenic, lead, cadmium and chromium are considered as the priority heavy metals because of their high toxicity, prevalent existence and persistence in the environment (He et al., 2013). In this section, we discuss the previous and current studies on the application of nanomaterials for the removal of those environmentally important heavy metals from water.

8.3.1 Removal of Arsenic

Arsenic contaminations in water have been considered as serious health problems. Long-term drinking water containing arsenic has been reported to cause various cancer as well as skin lesions, gangrene in leg, skin, lung, bladder, and liver (Anawar et al., 2001). The World Health Organization (WHO) has amended the maximum permissible limit for arsenic concentration in drinking water from 50 to 10 μg/L, and the US Environmental Protection Agency (EPA) has adopted an arsenic maximum contaminant level of 10 μg/L. Reduction in human exposure to arsenic can be done by reducing arsenic concentration in water by filtering. In the last decade, many efforts (Table 8-1) have been given in developing nano-based adsorbents to use as a filter membrane for the removal of arsenic from wastewater, as well as from drinking water.

In natural water, arsenic predominantly remain as arsenite (As(III)) and arsenate (As(V)) forms (Mehta et al., 2015). However, the presence of monomethylarsonic acid (MMA) and dimethylarsinic acid (DMA) in natural waters has also been reported (Anderson and Bruland, 1991). Removal of arsenic from contaminated water depends on pH, contact time, initial concentration of arsenic or chromium, PO_4^{3-} concentration in water, and adsorbent concentration (Qu et al., 2013). As(III) is more toxic and more difficult to remove from water than As(V). Removal of As(III) using simple treatment is difficult and hence an oxidation step is always necessary to achieve higher removal. Fe$-$Mn binary oxide nanomaterial is found to completely oxidize As(III) to As(V) and is effective for both As(V) and As(III) removal, particularly the As(III). Moreover, the adsorption of As(III) by Fe$-$Mn binary oxide nanomaterial is reported to be two times higher than by $MnFe_2O_4$ and $CoFe_2O_4$ (Zhang et al., 2007; Zhang et al., 2010), as the binary oxide magnetic nanomaterials possess increased number of the surface hydroxyl groups. This adsorption is controlled dominantly by intraparticle diffusion. The applicability of an adsorbent lies on its regeneration capacity also. Zhang et al. (2010) found that about 80%$-$90% of Fe-Mn binary oxide nanomaterial can be regenerated using 0.1 M NaOH. However this process has the limitation of long contact time (24 hours). Similar to Fe-Mn binary oxide nanomaterial, GO-$MnFe_2O_4$ nanohybrid and hybrid nanocrystalline surfactant-modified akaganeite [β-FeO(OH)] also shows high removal efficiency of toxic As(V) and As(III) from aqueous solution. The advantage of using GO-$MnFe_2O_4$ nanohybrid and β-FeO(OH) over Fe$-$Mn binary oxide nanomaterial is the short contact time (2 hours) (Deliyanni et al., 2006; Kumar et al., 2014). As(III) reacts with hydroxyls of the material and during this sorption, electron transfer from adsorbate to substrate occurred (Deliyanni et al., 2006).

Phosphate is the greatest competitor with arsenic in water for adsorptive sites on the adsorbent. In natural groundwater containing more than 5 mg/L phosphate and 1.13 mg/L

Table 8-1 Removal of Arsenic From Water by Different Nanoadsorbents

Adsorbent	Adsorbate	Removal of Adsorbate (mg)/g of Adsorbent	Favorable Conditions	References
Hybrid surfactant-Akaganeite (β-FeO(OH)) nanoparticle	As(II)	100–120 mg/g	pH 7.5, temp. 25°C, CT 2 h	Deliyanni et al. (2006)
Nano zero-valent iron (Fe_2O_3) with a BET surface area 37.2 m²/g, AD 1 g/L	As(II)	23.8 mg/g	pH 6.5, temp. 25°C, CT 12 h	Khodabakhshi et al. (2011)
GO-$MnFe_2O_4$ magnetic nanohybrid	As(III) and As(V)	146 mg/g for As(III), 207 mg/g for As(V)	pH 4.0–6.5, room temp	Kumar et al. (2014)
Mixed magnetite-maghemite nanoparticle	As(III) and As(V)	3.69 mg/g for As(III), 3.71 mg/g for As(V)	pH 2.0	Chowdhury and Yanful (2010)
CeO_2-CNTs (ceria supported on carbon nanotubes)	As(V)	78–81 mg/g	pH of natural water, CT 24 h presence of >10 mg/L Ca^{2+} and Mg^{2+}	Peng et al. (2005)
Cupric oxide (CuO) nanoparticles	As(III)	1.15 mg/g	pH 7.0, temp. 25°C	Singh et al. (2017)
Reduced graphene oxide and starch supported Fe-Mn binary oxide ($FeMnO_x$)	As(III) and As(V)	78.74 mg/g for As(III), 55.56 mg/g for As(V)	pH 7.0 ± 0.1, temp. 25°C, 150 rpm, CT 24 h	Lou et al. (2017)
Ferrimagnetic cobalt ferritenanoparticles	As(III)	25 mg/g	pH 2.0, temp. 22°C, 250 rpm	Martinez-Vargas et al. (2017)
γ-Fe_2O_3 (6 nm) encapsulated in macroporous silica foams (pore size 100 nm)	As(V) and As(III)	248 mg/g for As(III) and 320 mg/g for As(V)	24 h shaking in 200 rpm at 25°C temperature	Yang et al. (2014)
Magnetic nanoparticles impregnated chitosan beads	As(III) and As(V)	35.3 mg/g for As(III) and 35.7 mg/g for As(V)	pH 6.8, CT 24 h	Wang et al. (2014)
Fe–Mn binary oxide impregnated chitosan bead (1.6–1.8 mm in diameter)	As(V) and As(III)	39.1 mg/g for As(III) and 54.2 mg/g for As(V)	pH 7.0, shaking at 180 rpm for 36 h at 25 ± 1°C temp.	Qi et al. (2015)
$MnFe_2O_4$, with a surface area of 138 m²/g	As(III) and As(V)	90–94 mg/g	pH 3.0 for As (V), CT 2 h	Zhang et al. (2010)
$CoFe_2O_4$, with a surface area of 101 m²/g	As(III) and As(V)	100 mg/g for As(III) and 70 mg/g for As(V)	pH 3.0 for As (IV), CT 2 h	Zhang et al. (2010)
Fe–Mn binary oxide material with a surface area of 265 m²/g, particle size 26 μm	As(III) and As(V)	132.6 mg/g for As(III) and 70 mg/g for As(V)	pH 5.0, CT 24 h	Zhang et al. (2007)
Magnetite nanoparticle + 3 mg/L of Zn^{2+}	As(III) and As(V)	0.99 mg/g for As(V) and 0.95 mg/g for As(III)	Neutral to slightly alkaline pH	Yang et al. (2010)
Fe/Cu nanoparticles (13.17 nm in diameter)	As(III) and As(V)	19.68 mg/g for As(III) and 21.32 mg/g for As(V)	pH 7.0	Babaee et al. (2018)
TiO_2 in anatase form with a BET surface area of 329 m²/g	Monomethylarsonic acid (MMA) and dimethylarsinic acid (DMA)	100% removal for MMA at pH 7.5, 65% removal of DMA at pH 5.5	pH 6.0	Jing et al. (2005)
Polymer-supported (styrene-divinylbenzene anion exanger) hydrated iron(III) oxide nanoparticles	Arsenic	More than 90% of the influent arsenic removed after 10,000 bed volume	pH 7.3, superficial liquid velocity 0.60 m/h, Empty bed CT 3.9 min	Cumbal and Sengupta (2005)

CT: contact time, temp: temperature.

of arsenic, less than 60% arsenic uptake is possible using magnetite−maghemite nanoparticles (Chowdhury and Yanful, 2010). However, the adsorption capacity of this adsorbent is lower compared to other nanoadsorbents. Nanoscale hydrated iron (III) oxide (HFO) particles exhibit high sorption affinity toward both arsenates and arsenites within a short time (4 minutes). The polymeric sorbent, where cation and anion exchangers act as a host material for dispersing HFO nanoparticles within the polymer phase, can effectively remove arsenic (Cumbal and Sengupta, 2005).

The presence of Ca^{2+} and Mg^{2+} in natural water is found to enhance As(V) adsorption capacity of nanomaterial (Peng et al., 2005). CeO_2-CNTs exhibit high As(V) adsorption capacity in presence of Ca^{2+} and Mg^{2+} at a concentration of 10 mg/L, due to the ternary surface complex reaction among solid surface, cations, and As(V) anions. The cations (Ca^{2+}, Mg^{2+}) exchange with the surface groups first and then As(V) anions is combined to the liquid−solid interface, and the following surface complex forms. Again, the As(V)-loaded CeO_2-CNTs can be efficiently regenerated 94% by NaOH solution (Peng et al., 2005).

Compared to different types of nanoadsorbent, composites of iron oxide encapsulated in macroporous silica (Fe_XMOSF) showed excellent As(V) and As(III) absorption capacity, which is 4−6 times higher than those of nano-sized iron oxides and other nanoadsorbents (Table 8-1). In addition, this composite shows excellent arsenic absorption capacity in case of real wastewater sample (Yang et al., 2014). Yang et al. (2014) argued the applicability of Fe_XMOSF composites in arsenic removal from real wastewater without any pretreatment.

As for the removal of organic arsenic (MMA and DMA) from water, nanocrystaline TiO_2 is found to be effective at normal pH condition (Jing et al., 2005).

8.3.2 Removal of Lead and Cadmium

Lead (Pb(II)) and cadmium (Cd(II)) are the toxic nonessential heavy metals present in water and soil from different sources. Pb(II) poisoning in humans causes severe damage to the kidney, nervous system, reproductive system, brain and causes death (World Health Organization, 2016). Chronic exposure to elevated level of Cd(II) is known to cause osteoporosis and tubular kidney diseases (Godt et al., 2006; Johri et al., 2010). The World Health Organization guidelines suggest an allowable limit of Cd(II) and Pb(II)in water is 0.003 and 0.05 mg/L, respectively. Because of the toxic effects of these two metals, many scientists are giving efforts to develop different nanoadsorbents for the removal of Cd(II) and Pb(II) (Table 8-2).

Many studies examined the effectiveness of metal oxide-based nanoadsorbent (NiO, ZnO, Al_2O_3, CeO_2, Fe_3O_4, and TiO_2) for the removal of Cd(II) and Pb(II) from water (Gupta and Nayak, 2012; Recillas et al., 2011; Sheela et al., 2012; Sheela and Nayaka, 2012; Tabesh et al., 2018). Among the different type of metal-based nanoadsorbents, NiO and ZnO possess highest adsoption capacity toward Cd(II) and Pb(II) from water at slightly acidic condition (Sheela et al., 2012; Sheela and Nayaka, 2012). Sheela and Nayaka (2012) found that, NiO nanoparticles have higher affinity towards Pb(II) ions than that of Cd(II) ions. NiO-nanoparticles prepared by the organic solvent method is more active than that prepared by the precipitation method for removal of Pb(II) and Cd(II) ions from aqueous solutions

Table 8-2 Potential Application of Different Nanoadsorbents for the Removal Lead and Cadmium From Water

Adsorbent	Removal of Cadmium (mg/g) of Adsorbent	Removal of Lead (mg/g) of Adsorbent	Favorable Conditions	References
Zeolite materials obtained from fly as {SiO_2 (3.67%); Na, K Alumino silicates (84.24%); Fe_2O_3 (4.24%)}	26 mg/g	88 mg/g	pH 6.0−7.5, CT 90 min, absorbent dose 6 g/L	Visa (2016)
Nickel oxide nanoparticles adsorbent (surface area 128.330 m^2/g) prepared by organic solvent method	–	Removal efficiency more than 68% for initial conc of 100 mg/L	pH of solution 5.8, temp., 298K, CT 2 h	Mahmoud et al. (2015)
Fe_3O_4 sulfonated magnetic nanoparticle	80 mg/g	108.93 mg/g	Temp. 25°C	Chen et al. (2017)
γ-Al_2O_3 nanoparticles	78 mg/g	217 mg/g	pH 5.0, CT 20 min for Pb(II) and 30 min for Cd(II), temp. 25°C	Tabesh et al. (2018)
Highly mesoporous silica (containing nanospheres) anchored with 2,5-dimercapto-1,3,4-thiadiazole	–	67.20 mg/g	pH 4.0, shaking for 15 min	Shahat et al. (2018)
Magnetic Fe_3O_4 nanomaterial fabricated with the coating of 3-aminopropyltriethoxysilane on the surface of activated cyclosorusinterruptus	–	133.3 mg/g	pH 4.0−7.0, shaking for 12 h at 303K temperature	Zhou et al. (2018)
Zero-valent iron fixed on bentonite-fly ash	25 mg/g	21 mg/g	pH 4.0 ± 0.5, shaking with a speed of 120 rpm at 30°C	Mwamulima et al. (2018)
Kaolin supported nanoscale zero-valent iron	–	440.5 mg/g^{-1} with initial Pb concentration of 500 mg/L	pH 5.0−6.0, CT 1 h, shaking at 250 rpm at 30°C	Zhang et al. (2010)
Fe_3O_4 magnetic nanoparticles modified with 3-aminopropyltriethoxysilane and copolymers of acrylic acid and crotonic acid	29 mg/g	166.1 mg/g	pH 5.5, temp. 298 K, CT 45 min	Ge et al. (2012)
CeO_2, Fe_3O_4, and TiO_2 nanoparticles	–	189 mg/g for CeO_2, 83 mg/g for Fe_3O_4 and 159 mg/g for TiO_2, with initial Pb concentration of 200 mg/L	Room temperature, pH 7.0, continuously stirring at 150 rpm	Recillas et al. (2011)
Humic acid coated Fe_3O_4 magnetic nanoparticle, with a surface area 64 m^2/g	50 mg/g	92 mg/g	Room temperature (20°C), pH 3, CT 15 min	Liu et al. (2008)
Maghemite nanoparticle (γ-Fe_2O_3) encapsuled incalcium alginate beads	–	100 mg/g	pH 4.0, CT 100 min, temp. 30°C	Bée et al. (2011)
NiO nanoparticles	625 mg/g, with initial concentration of 600 mg/L	909 mg/g, with initial concentration of 600 mg/L	pH 6.0, temp. 303K, CT 2 h	Sheela and Nayaka (2012)
ZnO nanoparticles	387 mg/g	–	pH 5.5, temp. 303K, 200 rpm	Sheela et al. (2012)

(Continued)

Table 8-2 (Continued)

Adsorbent	Removal of Cadmium (mg/g) of Adsorbent	Removal of Lead (mg/g) of Adsorbent	Favorable Conditions	References
Amino-functionalized $Fe_3O_4@SiO_2$ magnetic nanomaterial	—	76 mg/g	pH 7.0, temp. 25°C	Wang et al. (2010)
Chitosan beads impregnated with nano-γ-Al_2O_3	31.67 mg/g	183 mg/g	Fixed bed column system	Shokati Poursani et al. (2017)
Polylysine—resorcinol coated alumina nanotube	220 mg/g	—	pH 7.0, CT 11 min	Hossein Beyki et al. (2017)
Thioglycolic acid modified mesoporous silica nanoparticles	91.3 mg/g with an initial concentration of 100 μg/L	—	pH 6.0, shaking at fixed 250 rpm speed for 30 min at room temperature	Kenawy et al. (2018)
Orange peel—Fe_2O_3 nanoparticles	71.43 mg/g, with an initial concentration of 16 mg/L	—	Shaking at 200 rpm, pH 7.0, room temperature	Gupta and Nayak (2012)
Magnetic hydroxyapatite nanoparticles, with a surface area of 142.5 $m^2\ g^{-1}$	1.964 mmol/g	—	Room temp. ($25 \pm 1°C$), pH 5.0, CT 24 h	Feng et al. (2010)
Multiwalled carbon nanotubes	181.8 mg/g, with an initial concentration of 100 mg/L	—	Sonication of solution for 90 s, centrifugation at 10,000 rpm for 20 min, pH 7.0	Bhanjana et al. (2017)
10% Al_2O_3-impregnated carbon nanotube (CNT-Al_2O_3) membrane	54 mg/g	—	pH 7.0, CT 2 h, transmembrane pressure difference of 15 psi	Ihsanullah et al. (2016)
Carbon-encapsulated magnetic nanoparticles	1.77 mg/g	—	pH 9.0, CT 4 h	Bystrzejewski et al. (2009)
GO-$MnFe_2O_4$magnetic nanohybrid		673, with an initial concentration of 100 mg/L	pH 5.0, room temp	Kumar et al. (2014)

CT: contact time, *temp*: temperature.

(Mahmoud et al., 2015; Sheela and Nayaka, 2012). γ-Al_2O_3 nanoparticles, synthesized by the modified Pechini type sol-gel method can adsorb Pb(II) and Cd(II) within a very short contact time compared to other metal oxide-based nanoadsorbents and can be reused 3 times. Similar to NiO and Fe_3O_4, γ-alumina nanoparticles also preferentially adsorbs Pb(II) ions (217 mg/g) over Cd(II) ions (78 mg/g) onto the gamma-alumina surface due to the high electronegativity and more hydrolysis of lead ions (Ge et al., 2012; Sheela and Nayaka, 2012; Tabesh et al., 2018). However, polylysine−resorcinol wrapped γ-alumina nanotube showed excellent Cd(II) adsorption capacity (220 mg/g) compared to γ-alumina nanoparticles, synthesized by the modified Pechini (Hossein Beyki et al., 2017).

Further, among CeO_2, Fe_3O_4, and TiO_2 nanoparticles, CeO_2 showed higer Pb(II) removal efficiency than Fe_3O_4 and TiO_2. However, CeO_2 presents a high phytotoxicity, whereas TiO_2 and Fe_3O_4 nanoparticles do not exhibit any toxicity (Recillas et al., 2011). Modification and fabrication of Fe_3O_4 nanoparticles by polymers, sulfonated group, amino group increases the adsorption capacity, even in the presence of other interfering cations, humic acid or alkali/earth metal ions in water. The most promising fact of this kind of adsorbent is its reusability (10 cycles) (Chen et al., 2017; Ge et al., 2012; Wang et al., 2010). When Pb(II) remain in high concentration in water, nanoscale zero-valent iron supported on kaolin is reported to be very much effective for removal (Zhang et al., 2010). Adsorption of Pb(II) using highly mesoporous silica (containing nanospheres) anchored with 2,5-dimercapto-1,3,4-thiadiazoleis not so high compare to other adsorbent (Zhang et al., 2010). However, the promising fact is that the adsorbent is applicable to adsorb Pb(II) from wastewater and biological samples such as blood and viscera. In addition, the reusability in multiple cycles without significant loss is implying the cost−effective materials to be used in practically (Shahat et al., 2018).

As for carbon-based nanoadsorbent, GO-$MnFe_2O_4$ magnetic nanohybrids is reported to be very effective for the removal (673 mg/g) of Pb(II) from water (Kumar et al., 2014). In addition, another advantage of this nanoadsorbent is its reusability (5 times). Multiwall CNT also exhibits higher Cd(II) adsorption efficacy compared to CNT and is recommended to be effective as adsorbent material for removal of cadmium as part of waste-water treatment (Bhanjana et al., 2017).

8.3.3 Removal of Chromium

Among the heavy metal ions, chromium holds a distinct position due to its high toxic nature to biological systems. Numerous industrial activities, including iron and steel manufacturing, tanning, chromium plating, and other anthropogenic sources released hexavalent and trivalent chromium (Cr(VI) and Cr(III)) into the environment. Cr(VI) is one of the most toxic and carcinogenic water contaminants. Compare to Cr(VI), Cr(III) is less toxic and is be a nutrient for humans at low concentrations. However, the uptake of higher dosages and long-term exposure to Cr(III) can cause harmful health effects, like allergic skin reactions and cancer (He et al., 2013; Wilburg et al., 2000). Hence, it becomes an obligation for the industries to reduce the chromium concentration in their effluents to an acceptable level before discharging into municipal sewers. Many researchers synthesized different nanomaterials and examined the adsorption efficiency for Cr(VI) and Cr(lll) (Table 8-3), where

Table 8-3 Removal of Chromium From Water Using Different Types of Nanoadsorbents

Adsorbent	Adsorbate	Removal of Adsorbate (mg/g) of Adsorbent	Favorable Conditions	References
Resin supported nanoscale zero-valent iron	Cr(VI) and Cr(III)	Removal efficiency was 84.4%, with initial concentration of Cr is 20 mg/L	pH 5.0, CT 20 min, resin concentration 20.0 g/L; nZVI load 41.3 mg/g	Fu et al. (2013)
Magnetite-maghemite nanoparticle	Cr(VI)	2.4 mg/g	pH 2.0	Chowdhury and Yanful (2010)
Magnetite derivatives nanoparticles	Cr(VI)	12.5 mg/g with an initial concentration of 20 mg/L	pH 2.0, room temp. (-20°C), CT 1–160 min	Tahar et al. (2018)
Maghemite (γ-Fe_2O_3) nanoparticles	Cr(VI)	19.2 mg/g	pH 2.0–3.0, room temp. (25°C), CT 15 min	Hu et al. (2005a)
Surface-modified jacobsite ($MnFe_2O_4$) nanoparticles	Cr(VI)	31.5 mg/g	pH 2.0, temp. 22.5°C, CT 5 min	Hu et al. (2005b)
Multiwall carbon nanotubes	Cr(III)	Removal efficiency 82% at pH 5.0 and 88% at pH 6.0	pH 5.0–6.0, CT 60 min, column bed depth of 1 cm and diameter of 0.3 mm, temp. 25°C, agitation speed 150 rpm	Gupta et al. (2011)
Bentonite-supported nanoscale zero-valent iron nanoparticle	Cr(VI)	7.3 mg/g from an actual electroplating wastewater, where Cr(VI) initial concentration was 73 mg/L	pH 2.0, CT 4 h, temp 25°C and mixing at 250 rpm,	Shi et al. (2011)
Chitosan-Fe^0 nanoparticles	Cr(VI)	60.2 mg/g when the initial concentration is 70 mg/L	Shaking at 220 rpm, pH 6.0, temp. 20°C	Geng et al. (2009)
Ceria nanoparticles (CeO_2) supported on aligned carbon nanotubes	Cr(VI)	35.3 mg/g	pH 7.0, room temp. 25 °C, shaking for 24 h	Di et al. (2006)
Cerium oxide (CeO_2) nanoparticles stabilized with hexamethylenetetramine	Cr(VI)	121.95 mg/g with initial Cr(VI) concentration 80 mg/L	pH 7.0, continual stirring at 150 rpm at room temperature for 3 h	Recillas et al. (2010)
Chitosan beads impregnated with nano-γ-Al_2O_3	Cr(VI)	158 mg/g	Fixed bed column system	Shokati Poursani et al. (2017)
GO impregnated mixed matrix membrane	Cr(VI)	154 mg/g	pH 3.5	Mukherjee et al. (2016)

CT: contact time, *temp*: temperature.

various factors influence the adsorption of chromium, for example, pH, temperature, initial concentration, and adsorbent dosage, etc.

Among different type of nanoadsorbents, GO impregnated on mixed matrix membrane, polymer supported CeO_2 and γ-Al_2O_3 nanomembranes shows higher adsorption capacity for Cr(Vl) (Mukherjee et al., 2016; Recillas et al., 2010; Shokati Poursani et al., 2017). GO impregnated on mixed matrix membrane shows preferentially higher affinity toward Cr(VI) among different type of heavy metals at acidic condition. Compared to this, polymer supported CeO_2 shows high Cr(VI) removal efficiency at normal pH of water. Among these different types of nanoadsorbent chitosan beads impregnated on γ-Al_2O_3 nanomembranes shows highest adsorption Cr (VI) capacity (158 mg/g of adsorbent) in fixed bed column system.

Among the different type of iron nanomaterials, maghemite (γ-Fe_2O_3) nanoparticles is found to be more effective for the removal of Cr(VI) species from water than magnetite (Fe_3O_4) derivatives nanoparticles (Hu et al., 2005a; Tahar et al., 2018). Maghemite (γ-Fe_2O_3) possess selective adsorption for Cr(VI) from wastewater and for this reason, the competition of Cr(VI) species with common coexisting ions in water such as Na^+, Ca^{2+}, Mg^{2+}, Cu^{2+}, Ni^{2+}, NO_3^-, and Cl^- for binding with maghemite (γ-Fe_2O_3) nanoparticles is ignorable (Hu et al., 2005a). However, manganese-iron oxide ($MnFe_2O_4$) nanoparticles shows the highest Cr(VI) adsorption capacity by electrostatic attraction within shortest contact time (Hu et al., 2005b). Hu et al. (2007) compared the performance of different metal-iron oxide nanomembranes for the removal of Cr(VI) and the adsorption capacities followed the order: $MnFe_2O_4 > MgFe_2O_4 > ZnFe_2O_4 > CuFe_2O_4 > NiFe_2O_4 > CoFe_2O_4$. Contact time required for all types of ferrite particles is reported to be relatively short (5−60 minutes). In spite of having high adsorption capacity, using iron nanomaterial has one limitation for the removal of metal is forming aggregation. However, this problem can be reduced by introducing bentonite as a support material for the nanoparticle (Shi et al., 2011).

8.4 Conclusions

The present study critically reviewed the past and recent available studies on the potential application of nanomaterials for the removal of heavy metal from water. Nanomaterials are attractive for the removal of heavy metal from water because of their some key physicochemical properties, like larger surface areas than bulk particles; ability for functionalization using different chemical groups to enhance their affinity to a given compound; and reusability for several times. Removal of different heavy metal from water by using an adsorbent depends on the preferential affinity of those heavy metals toward the adsorbent. From different studies, it can be concluded that, iron oxide encapsulated in macroporous silica and binary metal oxides nanoparticle show relatively higher arsenic absorption capacity. Lead shows preferential affinity toward NiO and GO-$MnFe_2O_4$ nanohybrid. For cadmium removal from water, NiO, ZiO, multiwall CNTs are highly effective. Bimetal oxides nanoparticle and polymer supported CeO_2 shows higher affinity for chromium. However, the efficacy of adsorbents for the adsorption of heavy metals depend on the various experimental conditions such as the solution pH, contact time, initial metal concentration, adsorbent's dosage, and temperature of the system.

References

Akoto, O., Bruce, T.N., Darko, G., 2008. Heavy metals pollution profiles in streams serving the Owabi reservoir. Afr. J. Environ. Sci. Technol. 2, 354−359.

Anawar, H.M., Akai, J., Mostofa, K.M.G., Safiullah, S., Tareq, S.M., 2001. Arsenic poisoning in groundwater: health risk and geochemical sources in Bangladesh. Environ. Int. 27, 597−604. Available from: https://doi.org/10.1016/S0160-4120(01)00116-7.

Anderson, L.C.D., Bruland, K.W., 1991. Biogeochemistry of arsenic in natural waters: the importance of methylated species. Environ. Sci. Technol. 25, 420−427. Available from: https://doi.org/10.1021/es00015a007.

Anjum, M., Miandad, R., Waqas, M., Gehany, F., Barakat, M.A., 2016. Remediation of wastewater using various nano-materials. Arab. J. Chem. Available from: https://doi.org/10.1016/j.arabjc.2016.10.004.

Babaee, Y., Mulligan, C.N., Rahaman, M.S., 2018. Removal of arsenic (III) and arsenic (V) from aqueous solutions through adsorption by Fe/Cu nanoparticles. J. Chem. Technol. Biotechnol. 93, 63−71. Available from: https://doi.org/10.1002/jctb.5320.

Bée, A., Talbot, D., Abramson, S., Dupuis, V., 2011. Magnetic alginate beads for Pb(II) ions removal from wastewater. J. Colloid Interface Sci. 362, 486−492. Available from: https://doi.org/10.1016/j.jcis.2011.06.036.

Bhanjana, G., Dilbaghi, N., Kim, K.-H., Kumar, S., 2017. Carbon nanotubes as sorbent material for removal of cadmium. J. Mol. Liq. 242, 966−970. Available from: https://doi.org/10.1016/j.molliq.2017.07.072.

Bhuiyan, M.A., mi, H., Dampare, S.B., Islam, M.A., Suzuki, S., 2015. Source apportionment and pollution evaluation of heavy metals in water and sediments of Buriganga River, Bangladesh, using multivariate analysis and pollution evaluation indices. Environ. Monit. Assess. 187, 4075. Available from: https://doi.org/10.1007/s10661-014-4075-0.

Bystrzejewski, M., Pyrzyńska, K., Huczko, A., Lange, H., 2009. Carbon-encapsulated magnetic nanoparticles as separable and mobile sorbents of heavy metal ions from aqueous solutions. Carbon N. Y. Available from: https://doi.org/10.1016/j.carbon.2009.01.007.

Chen, K., He, J., Li, Y., Cai, X., Zhang, K., Liu, T., et al., 2017. Removal of cadmium and lead ions from water by sulfonated magnetic nanoparticle adsorbents. J. Colloid Interface Sci. 494, 307−316. Available from: https://doi.org/10.1016/j.jcis.2017.01.082.

Chowdhury, S.R., Yanful, E.K., 2010. Arsenic and chromium removal by mixed magnetite−maghemite nanoparticles and the effect of phosphate on removal. J. Environ. Manage. 91, 2238−2247. Available from: https://doi.org/10.1016/j.jenvman.2010.06.003.

Cumbal, L., Sengupta, A.K., 2005. Arsenic removal using polymer-supported hydrated iron(III) oxide nanoparticles: role of Donnan membrane effect. Environ. Sci. Technol. 39, 6508−6515. Available from: https://doi.org/10.1021/es050175e.

Deliyanni, E.A., Nalbandian, L., Matis, K.A., 2006. Adsorptive removal of arsenites by a nanocrystalline hybrid surfactant−akaganeite sorbent. J. Colloid Interface Sci. 302, 458−466. Available from: https://doi.org/10.1016/j.jcis.2006.07.007.

Di, Z.C., Ding, J., Peng, X.J., Li, Y.H., Luan, Z.K., Liang, J., 2006. Chromium adsorption by aligned carbon nanotubes supported ceria nanoparticles. Chemosphere 62, 861−865. Available from: https://doi.org/10.1016/j.chemosphere.2004.06.044.

Farkas, A., Erratico, C., Viganò, L., 2007. Assessment of the environmental significance of heavy metal pollution in surficial sediments of the River Po. Chemosphere 68, 761−768. Available from: https://doi.org/10.1016/j.chemosphere.2006.12.099.

Feng, Y., Gong, J.-L., Zeng, G.-M., Niu, Q.-Y., Zhang, H.-Y., Niu, C.-G., et al., 2010. Adsorption of Cd (II) and Zn (II) from aqueous solutions using magnetic hydroxyapatite nanoparticles as adsorbents. Chem. Eng. J. 162, 487−494. Available from: https://doi.org/10.1016/j.cej.2010.05.049.

Fu, F., Ma, J., Xie, L., Tang, B., Han, W., Lin, S., 2013. Chromium removal using resin supported nanoscale zero-valent iron. J. Environ. Manage. 128, 822–827. Available from: https://doi.org/10.1016/j.jenvman.2013.06.044.

Ge, F., Li, M.-M., Ye, H., Zhao, B.-X., 2012. Effective removal of heavy metal ions Cd2 + , Zn2 + , Pb2 + , Cu2 + from aqueous solution by polymer-modified magnetic nanoparticles. J. Hazard. Mater. 211–212, 366–372. Available from: https://doi.org/10.1016/j.jhazmat.2011.12.013.

Geng, B., Jin, Z., Li, T., Qi, X., 2009. Kinetics of hexavalent chromium removal from water by chitosan-FeO nanoparticles. Chemosphere 75, 825–830. Available from: https://doi.org/10.1016/j.chemosphere.2009.01.009.

Godt, J., Scheidig, F., Grosse-Siestrup, C., Esche, V., Brandenburg, P., Reich, A., et al., 2006. The toxicity of cadmium and resulting hazards for human health. J. Occup. Med. Toxicol. 1, 22. Available from: https://doi.org/10.1186/1745-6673-1-22.

Gopalakrishnan, A., Krishnan, R., Thangavel, S., Venugopal, G., Kim, S.J., 2015. Removal of heavy metal ions from pharma-effluents using graphene-oxide nanosorbents and study of their adsorption kinetics. J. Ind. Eng. Chem. 30, 14–19. Available from: https://doi.org/10.1016/j.jiec.2015.06.005.

Gupta, V.K., Nayak, A., 2012. Cadmium removal and recovery from aqueous solutions by novel adsorbents prepared from orange peel and Fe_2O_3 nanoparticles. Chem. Eng. J. 180, 81–90. Available from: https://doi.org/10.1016/j.cej.2011.11.006.

Gupta, V.K., Agarwal, S., Saleh, T.A., 2011. Chromium removal by combining the magnetic properties of iron oxide with adsorption properties of carbon nanotubes. Water Res. 45, 2207–2212. Available from: https://doi.org/10.1016/j.watres.2011.01.012.

He, B., Yun, Z.J., Shi, J.B., Jiang, G.B., 2013. Research progress of heavy metal pollution in China: sources, analytical methods, status, and toxicity. Chinese Sci. Bull. Available from: https://doi.org/10.1007/s11434-012-5541-0.

Hossein Beyki, M., Ghasemi, M.H., Jamali, A., Shemirani, F., 2017. A novel polylysine–resorcinol base γ-alumina nanotube hybrid material for effective adsorption/preconcentration of cadmium from various matrices. J. Ind. Eng. Chem. 46, 165–174. Available from: https://doi.org/10.1016/j.jiec.2016.10.027.

Hristovski, K.D., Westerhoff, P.K., Möller, T., Sylvester, P., 2009. Effect of synthesis conditions on nano-iron (hydr)oxide impregnated granulated activated carbon. Chem. Eng. J. 146, 237–243. Available from: https://doi.org/10.1016/j.cej.2008.05.040.

Hu, J., Chen, G., Lo, I.M.C., 2005a. Removal and recovery of Cr(VI) from wastewater by maghemite nanoparticles. Water Res. 39, 4528–4536. Available from: https://doi.org/10.1016/j.watres.2005.05.051.

Hu, J., Lo, I.M.C., Chen, G., 2005b. Fast removal and recovery of Cr(VI) using surface-modified jacobsite ($MnFe_2O_4$) nanoparticles. Langmuir 21, 11173–11179. Available from: https://doi.org/10.1021/la051076h.

Hu, J., Lo, I., Chen, G., 2007. Comparative study of various magnetic nanoparticles for Cr(VI) removal. Sep. Purif. Technol. 56, 249–256. Available from: https://doi.org/10.1016/j.seppur.2007.02.009.

ISO, N, 2008. Terminology and Definitions for Nano-objects — Nanoparticle, Nanofibre and Nanoplate. International Organisation for Standardisation (ISO), Genève, Switzerland.

Ihsanullah, Abbas, A., Al-Amer, A.M., Laoui, T., Al-Marri, M.J., Nasser, M.S., Khraisheh, M., et al., 2016. Heavy metal removal from aqueous solution by advanced carbon nanotubes: critical review of adsorption applications. Sep. Purif. Technol. Available from: https://doi.org/10.1016/j.seppur.2015.11.039.

Jing, C., Meng, X., Liu, S., Baidas, S., Patraju, R., Christodoulatos, C., et al., 2005. Surface complexation of organic arsenic on nanocrystalline titanium oxide. J. Colloid Interface Sci. 290, 14–21. Available from: https://doi.org/10.1016/j.jcis.2005.04.019.

Johri, N., Jacquillet, G., Unwin, R., 2010. Heavy metal poisoning: the effects of cadmium on the kidney. BioMetals 23, 783–792. Available from: https://doi.org/10.1007/s10534-010-9328-y.

Juang, D.F., Lee, C.H., Hsueh, S.C., 2009. Chlorinated volatile organic compounds found near the water surface of heavily polluted rivers. Int. J. Environ. Sci. Technol. 6, 545−556. Available from: https://doi.org/10.1007/BF03326094.

Karbassi, A.R., Monavari, S.M., Nabi Bidhendi, G.R., Nouri, J., Nematpour, K., 2008. Metal pollution assessment of sediment and water in the Shur River. Environ. Monit. Assess. 147, 107−116. Available from: https://doi.org/10.1007/s10661-007-0102-8.

Karlsson, M.N.A., Deppert, K., Wacaser, B.A., Karlsson, L.S., Malm, J.O., 2005. Size-controlled nanoparticles by thermal cracking of iron pentacarbonyl. Appl. Phys. A Mater. Sci. Process. 80, 1579−1583. Available from: https://doi.org/10.1007/s00339-004-2987-1.

Kenawy, I.M.M., Abou El-Reash, Y.G., Hassanien, M.M., Alnagar, N.R., Mortada, W.I., 2018. Use of microwave irradiation for modification of mesoporous silica nanoparticles by thioglycolic acid for removal of cadmium and mercury. Microporous Mesoporous Mater. 258, 217−227. Available from: https://doi.org/10.1016/j.micromeso.2017.09.021.

Kumar, S., Nair, R.R., Pillai, P.B., Gupta, S.N., Iyengar, M.A.R., Sood, A.K., 2014. Graphene oxide-MnFe$_2$O$_4$ magnetic nanohybrids for efficient removal of lead and arsenic from water. ACS Appl. Mater. Interfaces 6, 17426−17436. Available from: https://doi.org/10.1021/am504826q.

Lata, S., Samadder, S.R., 2016. Removal of arsenic from water using nano adsorbents and challenges: a review. J. Environ. Manage 166, 387−406. Available from: https://doi.org/10.1016/j.jenvman.2015.10.039.

Li, Y.-H., Ding, J., Luan, Z., Di, Z., Zhu, Y., Xu, C., et al., 2003. Competitive adsorption of Pb2 + , Cu2 + and Cd2 + ions from aqueous solutions by multiwalled carbon nanotubes. Carbon N. Y 41, 2787−2792. Available from: https://doi.org/10.1016/S0008-6223(03)00392-0.

Li, Z., Deng, S., Yu, G., Huang, J., Lim, V.C., 2010. As(V) and As(III) removal from water by a Ce−Ti oxide adsorbent: behavior and mechanism. Chem. Eng. J. 161, 106−113. Available from: https://doi.org/10.1016/j.cej.2010.04.039.

Liu, J., Zhao, Z., Jiang, G., 2008. Coating Fe$_3$O$_4$ magnetic nanoparticles with humic acid for high efficient removal of heavy metals in water. Environ. Sci. Technol. 42, 6949−6954. Available from: https://doi.org/10.1021/es800924c.

Lou, Z., Cao, Z., Xu, J., Zhou, X., Zhu, J., Liu, X., et al., 2017. Enhanced removal of As(III)/(V) from water by simultaneously supported and stabilized Fe-Mn binary oxide nanohybrids. Chem. Eng. J. 322, 710−721. Available from: https://doi.org/10.1016/j.cej.2017.04.079.

Mahdavi, S., Jalali, M., Afkhami, A., 2012. Removal of heavy metals from aqueous solutions using Fe$_3$O$_4$, ZnO, and CuO nanoparticles. J. Nanoparticle Res. 14, 846. Available from: https://doi.org/10.1007/s11051-012-0846-0.

Mahmoud, A.M., Ibrahim, F.A., Shaban, S.A., Youssef, N.A., 2015. Adsorption of heavy metal ion from aqueous solution by nickel oxide nano catalyst prepared by different methods. Egypt. J. Pet. 24, 27−35. Available from: https://doi.org/10.1016/j.ejpe.2015.02.003.

Martinez-Vargas, S., Martínez, A.I., Hernández-Beteta, E.E., Mijangos-Ricardez, O.F., Vázquez-Hipólito, V., Patiño-Carachure, C., 2017. Arsenic adsorption on cobalt and manganese ferrite nanoparticles. J. Mater. Sci. 52, 6205−6215. Available from: https://doi.org/10.1007/s10853-017-0852-9.

Mayo, J.T., Yavuz, C., Yean, S., Cong, L., Shiple, H., Yu, W., et al., 2007. The effect of nanocrystalline magnetite size on arsenic removal. Sci. Technol. Adv. Mater. 8, 71−75. Available from: https://doi.org/10.1016/j.stam.2006.10.005.

Mehta, D., Mazumdar, S., Singh, S.K., 2015. Magnetic adsorbents for the treatment of water/wastewater-a review. J. Water Process Eng. 7, 244−265. Available from: https://doi.org/10.1016/j.jwpe.2015.07.001.

Mukherjee, R., Bhunia, P., De, S., 2016. Impact of graphene oxide on removal of heavy metals using mixed matrix membrane. Chem. Eng. J. 292, 284−297. Available from: https://doi.org/10.1016/j.cej.2016.02.015.

Mwamulima, T., Zhang, X., Wang, Y., Song, S., Peng, C., 2018. Novel approach to control adsorbent aggregation: iron fixed bentonite-fly ash for Lead (Pb) and Cadmium (Cd) removal from aqueous media. Front. Environ. Sci. Eng. 12. Available from: https://doi.org/10.1007/s11783-017-0979-6.

Nouri, J., Mahvi, A.H., Jahed, G.R., Babaei, A.A., 2008. Regional distribution pattern of groundwater heavy metals resulting from agricultural activities. Environ. Geol. 55, 1337−1343. Available from: https://doi.org/10.1007/s00254-007-1081-3.

Pandey, N., Shukla, S.K., Singh, N.B., 2015. Zinc oxide-urea formaldehyde nanocomposite film as low-cost adsorbent for removal of Cu(II) from aqueous solution. Adv. Mater. Lett. 6, 172−178. Available from: https://doi.org/10.5185/amlett.2014.5604.

Parsons, J.G., Lopez, M.L., Peralta-Videa, J.R., Gardea-Torresdey, J.L., 2009. Determination of arsenic(III) and arsenic(V) binding to microwave assisted hydrothermal synthetically prepared Fe_3O_4, Mn_3O_4, and $MnFe_2O_4$ nanoadsorbents. Microchem. J. 91, 100−106. Available from: https://doi.org/10.1016/j.microc.2008.08.012.

Peng, X., Luan, Z., Ding, J., Di, Z., Li, Y., Tian, B., 2005. Ceria nanoparticles supported on carbon nanotubes for the removal of arsenate from water. Mater. Lett. 59, 399−403. Available from: https://doi.org/10.1016/j.matlet.2004.05.090.

Phiwdang, K., Suphankij, S., Mekprasart, W., Pecharapa, W., 2013. Synthesis of CuO nanoparticles by precipitation method using different precursors. Energy Procedia 34, 740−745. Available from: https://doi.org/10.1016/j.egypro.2013.06.808.

Qi, J., Zhang, G., Li, H., 2015. Efficient removal of arsenic from water using a granular adsorbent: Fe−Mn binary oxide impregnated chitosan bead. Bioresour. Technol. 193, 243−249. Available from: https://doi.org/10.1016/j.biortech.2015.06.102.

Qu, X., Alvarez, P.J.J., Li, Q., 2013. Applications of nanotechnology in water and wastewater treatment. Water Res. 47, 3931−3946. Available from: https://doi.org/10.1016/j.watres.2012.09.058.

Rao, G., Lu, C., Su, F., 2007. Sorption of divalent metal ions from aqueous solution by carbon nanotubes: a review. Sep. Purif. Technol. 58, 224−231. Available from: https://doi.org/10.1016/j.seppur.2006.12.006.

Recillas, S., Colón, J., Casals, E., González, E., Puntes, V., Sánchez, A., et al., 2010. Chromium VI adsorption on cerium oxide nanoparticles and morphology changes during the process. J. Hazard. Mater. 184, 425−431. Available from: https://doi.org/10.1016/j.jhazmat.2010.08.052.

Recillas, S., García, A., González, E., Casals, E., Puntes, V., Sánchez, A., et al., 2011. Use of CeO_2, TiO_2 and Fe_3O_4 nanoparticles for the removal of lead from water. Desalination 277, 213−220. Available from: https://doi.org/10.1016/j.desal.2011.04.036.

Reza, R., Singh, G., 2010. Heavy metal contamination and its indexing approach for river water. Int. J. Environ. Sci. Technol. 7, 785−792. Available from: https://doi.org/10.1007/BF03326187.

Shahat, A., Hassan, H.M.A., Azzazy, H.M.E., El-Sharkawy, E.A., Abdou, H.M., Awual, M.R., 2018. Novel hierarchical composite adsorbent for selective lead(II) ions capturing from wastewater samples. Chem. Eng. J. 332, 377−386. Available from: https://doi.org/10.1016/j.cej.2017.09.040.

Sheela, T., Nayaka, Y.A., 2012. Kinetics and thermodynamics of cadmium and lead ions adsorption on NiO nanoparticles. Chem. Eng. J. 191, 123−131. Available from: https://doi.org/10.1016/j.cej.2012.02.080.

Sheela, T., Nayaka, Y.A., Viswanatha, R., Basavanna, S., Venkatesha, T.G., 2012. Kinetics and thermodynamics studies on the adsorption of Zn(II), Cd(II) and Hg(II) from aqueous solution using zinc oxide nanoparticles. Powder Technol. 217, 163−170. Available from: https://doi.org/10.1016/j.powtec.2011.10.023.

Shi, L., Zhang, X., Chen, Z., 2011. Removal of chromium (VI) from wastewater using bentonite-supported nanoscale zero-valent iron. Water Res. 45, 886−892. Available from: https://doi.org/10.1016/j.watres.2010.09.025.

Shokati Poursani, A., Nilchi, A., Hassani, A., Tabibian, S., Asad Amraji, L., 2017. Synthesis of nano-γ-Al_2O_3/ chitosan beads (AlCBs) and continuous heavy metals removal from liquid solution. Int. J. Environ. Sci. Technol. 14, 1459−1468. Available from: https://doi.org/10.1007/s13762-017-1357-4.

Singh, D.K., Verma, D.K., Singh, Y., Hasan, S.H., 2017. Preparation of CuO nanoparticles using *Tamarindus indica* pulp extract for removal of As(III): optimization of adsorption process by ANN-GA. J. Environ. Chem. Eng. 5, 1302−1318. Available from: https://doi.org/10.1016/j.jece.2017.01.046.

Siskova, K.M., Straska, J., Krizek, M., Tucek, J., Machala, L., Zboril, R., 2013. Formation of zero-valent iron nanoparticles mediated by amino acids. Procedia Environ. Sci. 18, 809−817. Available from: https://doi.org/10.1016/j.proenv.2013.04.109.

Sun, Y.-P., Li, X., Cao, J., Zhang, W., Wang, H.P., 2006. Characterization of zero-valent iron nanoparticles. Adv. Colloid Interface Sci. 120, 47−56. Available from: https://doi.org/10.1016/j.cis.2006.03.001.

Tabesh, S., Davar, F., Loghman-Estarki, M.R., 2018. Preparation of γ-Al$_2$O$_3$ nanoparticles using modified sol-gel method and its use for the adsorption of lead and cadmium ions. J. Alloys Compd. 730, 441−449. Available from: https://doi.org/10.1016/j.jallcom.2017.09.246.

Tahar, L.B., Oueslati, M.H., Abualreish, M.J.A., 2018. Synthesis of magnetite derivatives nanoparticles and their application for the removal of chromium (VI) from aqueous solutions. J. Colloid Interface Sci. 512. Available from: https://doi.org/10.1016/j.jcis.2017.10.044.

Visa, M., 2016. Synthesis and characterization of new zeolite materials obtained from fly ash for heavy metals removal in advanced wastewater treatment. Powder Technol. 294, 338−347. Available from: https://doi.org/10.1016/j.powtec.2016.02.019.

Wang, J., Zheng, S., Shao, Y., Liu, J., Xu, Z., Zhu, D., 2010. Amino-functionalized Fe$_3$O$_4$@SiO$_2$ core−shell magnetic nanomaterial as a novel adsorbent for aqueous heavy metals removal. J. Colloid Interface Sci. 349, 293−299. Available from: https://doi.org/10.1016/j.jcis.2010.05.010.

Wang, J., Xu, W., Chen, L., Huang, X., Liu, J., 2014. Preparation and evaluation of magnetic nanoparticles impregnated chitosan beads for arsenic removal from water. Chem. Eng. J. 251, 25−34. Available from: https://doi.org/10.1016/j.cej.2014.04.061.

Wilburg, S., Ingerman, L., Citra, M., Osier, M., Wohlers, D., 2000. Toxicological profile for chromium. Public Health 421.

Wong, S.C., Li, X.D., Zhang, G., Qi, S.H., Min, Y.S., 2002. Heavy metals in agricultural soils of the Pearl River Delta, South China. Environ. Pollut. 119, 33−44. Available from: https://doi.org/10.1016/S0269-7491(01)00325-6.

World Health Organization, 2016. Lead poisoning and health [WWW Document]. Fact Sheet. URL http://www.who.int/mediacentre/factsheets/fs379/en/#.V6ORzdZF7hU.email.

Xu, P., Zeng, G.M., Huang, D.L., Feng, C.L., Hu, S., Zhao, M.H., et al., 2012. Use of iron oxide nanomaterials in wastewater treatment: a review. Sci. Total Environ. 424, 1−10. Available from: https://doi.org/10.1016/j.scitotenv.2012.02.023.

Yang, W., Kan, A.T., Chen, W., Tomson, M.B., 2010. pH-dependent effect of zinc on arsenic adsorption to magnetite nanoparticles. Water Res. 44, 5693−5701.

Yang, J., Zhang, H., Yu, M., Emmanuelawati, I., Zou, J., Yuan, Z., et al., 2014. High-content, well-dispersed γ-Fe$_2$O$_3$ nanoparticles encapsulated in macroporous silica with superior arsenic removal performance. Adv. Funct. Mater. 24, 1354−1363. Available from: https://doi.org/10.1002/adfm.201302561.

Zhang, G., Qu, J., Liu, H., Liu, R., Wu, R., 2007. Preparation and evaluation of a novel Fe−Mn binary oxide adsorbent for effective arsenite removal. Water Res. 41, 1921−1928. Available from: https://doi.org/10.1016/j.watres.2007.02.009.

Zhang, S., Niu, H., Cai, Y., Zhao, X., Shi, Y., 2010. Arsenite and arsenate adsorption on coprecipitated bimetal oxide magnetic nanomaterials: MnFe$_2$O$_4$ and CoFe$_2$O$_4$. Chem. Eng. J. 158, 599−607. Available from: https://doi.org/10.1016/j.cej.2010.02.013.

Zhang, X., Lin, S., Lu, X.Q., Chen, Z.L., 2010. Removal of Pb(II) from water using synthesized kaolin supported nanoscale zero-valent iron. Chem. Eng. J. 163, 243−248. Available from: https://doi.org/10.1016/j.cej.2010.07.056.

Zhang, Y., Yang, M., Dou, X.M., He, H., Wang, D.S., 2005. Arsenate adsorption on an Fe-Ce bimetal oxide adsorbent: role of surface properties. Environ. Sci. Technol. 39, 7246−7253. Available from: https://doi.org/10.1021/es050775d.

Zhou, J., Liu, Y., Zhou, X., Ren, J., Zhong, C., 2018. Magnetic multi-porous bio-adsorbent modified with amino siloxane for fast removal of Pb(II) from aqueous solution. Appl. Surf. Sci. 427, 976−985. Available from: https://doi.org/10.1016/j.apsusc.2017.08.110.

9

Application of Nanoparticles for Disinfection and Microbial Control of Water and Wastewater

Sharmin Yousuf Rikta

DEPARTMENT OF ENVIRONMENTAL SCIENCE, BANGLADESH UNIVERSITY OF PROFESSIONALS, DHAKA, BANGLADESH

9.1 Introduction

In most of the developing countries, surface water is heavily polluted by untreated/partially treated industrial wastewater containing various pollutants including microbial agents. Poorly developed sanitary latrines and inappropriately designed domestic wastes discharge channels, which find their ends to river/cannel systems aggravate this situation. Inadequately constructed sanitary landfill sites at improper locations around the cities could be another potential source of pollutants (e.g., pathogens, heavy metals, etc.) of aquatic environment during monsoon if it is flooded and finds connection with neighboring surface water channels (Rikta et al., 2018). Along with the pollution scenario, water demand is rapidly rising because of the increasing population, ever growing urbanization, industrialization, agriculture, and domestic usages. This entire dilemma has placed a tremendous stress on global water resource. As a consequence, about half of the world's population will be living in water-stressed areas by 2025 (WHO, 2017).

Safe and clean water is important for drinking, domestic use, food production and other industrial purposes to ensure good public health. Drinking water is not only related with ingestion but also used for cooking, showering, washing, cleaning, and many other purposes. These usages increase the exposure routes of pollutants and pathogens in many folds (Wang et al., 2007). In food production, growing of food requires huge amounts of water through irrigation where certain quality should be maintained depending on crops. During food processing, water quality must be insured as the poor quality water used for food processing can directly affect human health.

In most of the developing and industrialized countries, various hazardous and toxic pollutants are entering water systems (Shannon et al., 2008). These contaminants include heavy metals, micropollutants, recalcitrant organic compounds, emerging contaminants, endocrine disrupting chemicals and many others. But current rising water demand, safe water shortage and environmental concern make the urgency to rejuvenate this contaminated water. Though much concentration has been given to the water quality in many countries, still microbial pollution in water and wastewater is one of the major concerns due to the high mortality rate by waterborne diseases. At least 2 billion people around the world use drinking water sources polluted by feces (WHO, 2017). Every year, around 2 million deaths are caused by waterborne diarrheal diseases, where the most sufferers are children under the age of 5 (WHO, 2017; Li et al., 2008). On one hand, microbial pathogens in water spread waterborne diseases but, on the other hand, they create potable water crisis. Contamination of water by pathogen makes the valuable water sources unusable and creates artificial resource crisis by increasing demand and decreasing supply. About 844 million people around the world lack even a basic drinking water service (WHO, 2017). Hence, water disinfection is a vital step for safe water supply where the problems prevail.

Effective microbial decontamination is a prerequisite to prevent waterborne disease and control its transmission (Abbaszadegan et al., 1997). Disinfection is the process which is applied to inactivate the microbial pathogens (e.g., bacteria, virus, and protozoa) in water by using particular disinfectant. Conventional disinfection methods can be categorized into two different types: chemical disinfection and physical disinfection processes (Kraft, 2008). In chemical disinfection, chemical disinfectant such as chlorine, chlorine dioxide are used to inactivate pathogens whereas in physical disinfection process microbial inactivation is done by applying radiation, ultrasound, high temperature, membrane filtration and other physical means (Kraft, 2008). Conventional disinfection methods have many limitations but fortunately recent research have found out several advanced processes to overcome these limitations by applying nanoparticles. These processes are quite promising because of their simplicity, economic reasonability, and disinfection efficiency.

In this chapter, an overview on the major applications of nanoparticles for the disinfection of water and wastewater has been provided. Conventional methods for water disinfection and microbial control with their limitations in this field have been critically addressed. General mechanisms for microbial disinfection have been discussed. Any harmful impact of nanoparticles on public health and environment as well as the possible interferences with different disinfection processes are beyond the scope of this chapter.

9.2 Limitations of Conventional Methods for Water and Wastewater Disinfection

Demand for controlling infectious waterborne pathogens requires effective water and wastewater treatment technologies. Along with various physicochemical water treatment technologies (e.g., sedimentation, coagualation, fluccolation) many conventional disinfection treatment technologies are applied to enhance the removal efficiency of microbial pathogens (Shannon et al., 2008). Generation of toxic disinfection byproducts (DBPs), cost of equipment and chemicals as well as resistance of microbial pathogen against inactivation are some major drawbacks of current disinfection processes. Table 9-1 represents currently used disinfection methods along with their limitation.

Table 9-1 Currently Used Water and Wastewater Disinfection Technologies and Their Limitations

Disinfection Method	Microbial Component Removed	Limitations of Use	References
Chlorination	*Escherichia coli, Proteus*	Presence of *Klebsiella pneumoniae*	Elmi et al. (2014)
UV radiation	*Klebsiella pneumoniae, Escherichia coli*	Presence of *Staphylococcus epidermidis*	Elmi et al. (2014)
UV radiation	*Escherichia coli*, Total coliform	Biofouling, reactivation/regrowth of bacteria	Haakon et al. (2014), Lee et al. (2015)
Chlorination	*Escherichia coli*	Production of trihalomethanes	Gomez-Lopez et al. (2015)
Combination of ozonation and chlorination	*Escherichia coli, Proteus*	Generation of halonitromethanes	Song et al. (2010), Elmi et al. (2014)
Electrochemical disinfection	*Pseudomonas aeruginosa*	Formation of undesirable inorganic disinfection by-products (e.g., chlorate and perchlorate)	Rajab et al. (2015)
Ozonation	*Escherichia coli*, Enterococci	Presence of *Pseudomonas aeruginosa*, formation of aldehydes, and carboxylic acids	Liu et al. (2015), Alexander et al. (2016)
H_2O_2 disinfection	Fecal Coliform	Long contact time, high concentration required	Casani et al. (2005), Drogui et al. (2001)
Electrochemical disinfection	*Cryptosporidium parvum* oocysts, *Escherichia coli*, *Clostridium perfringens*	Generation of trihalomethanes, inhibitory effect of $H_2PO_4^{2-}$, HCO_3^-, and CO_3^-	Kerwick et al. (2005)

The most common traditional method for disinfection of water and wastewater is the use of free chlorine (Kerwick et al., 2005). Chlorination gained wide acceptance in disinfection because of its germicidal properties but the formation of harmful DBPs like trihalomethanes, chlorite, haloacetic acids, chloropicrin, and chlorophenols (Scholz and Martin, 1997; Simpson, 2008) makes the method lagging behind. Generation of DBPs depends on the dose of chlorine, contact time, pH of solution/water/wastewater, temperature, natural organic matter content and presence of bromide ion (Sadiq and Rodriguez, 2004; Amy et al., 1987). Consumption of these toxic DBPs can cause carcinogenic effects on human health including enhanced risk of colon and bladder cancer (Cantor et al., 1985; Simpson, 2008). On the other hand, many microbial pathogens such as *Giardia, Cryptosporidium* have developed resistance against conventional chemical disinfectants including chlorine treatment (Li et al., 2008; Betancourt and Rose, 2004). In biofilm, *Mycobacterium avium* and another waterborne pathogen *Cryptosporidium parvum* are resistant to free chlorine inactivation (Shannon et al., 2008). Thus, higher concentration/dosage is required to inactivate the pathogens resulting more residual chlorine left in the water, which creates objectionable chorine smell and unpleasant taste in drinking water.

In ultraviolet (UV) radiation treatment, microbial pathogens are photochemically inactivated by the light of visible to ultraviolet region. Ozonation is popular for its germicidal impacts on pathogens in disinfection process. Many pathogens including *Escherichia coli, Pseudomonas aeruginosa, Proteus* can be inactivated by the action of UV radiation and ozone, whereas some organisms (e.g., *Staphylococcus epidermidis*) show viability against their actions (Elmi et al., 2014). A number of species of adenovirus show higher resistance to UV radiation inactivation. Where UV and chlorination disinfection processes face difficulties in controlling viruses, ozonation provide a good efficiency (Shannon et al., 2008). Despite of this advantage, generation of DBPs such as bromate and brominated DBPs during the ozonation process limits its application for disinfection (Richardson et al., 1999). Formation of bromate and brominated DBPs during this process depends on the presence of ammonia ion and bromide concentration (von-Guntenetal., 1995; Sadiq and Rodriguez, 2004). On the other hand, these methods require expensive equipment related to higher cost and process complexity (Dalrymple et al., 2010).

Filtration process only removes large pathogens whereas advanced filtration is much more reliable but requires costly membrane. There is a chance of biofouling and deposition of particle cake which lead to high membrane regeneration costs (Zhou and Smith, 2002). In addition, high energy consumption is related to maintain high cross flow velocity (Zhou and Smith, 2002). During the process of electrochemical disinfection, harmful DBPs are produced (Rajab et al., 2015) and the efficiency of inactivation depends on electrolyte composition, electrode materials, cell configuration, current intensity, flow rate and other factors (Kerwick et al., 2005). Application of hydrogen peroxide (H_2O_2) for disinfection is less effective than ozone treatment or chlorination as its oxygen potential is lower. Though it does not produce toxic DBPs, relatively high concentration and longer contact time requirements make the applicability of this process limited (Kraft, 2008; Casani et al., 2005).

9.3 Wastewater Disinfection Mechanism of Nanoparticles

Disinfection of microorganisms has been studied by many researchers (Foster et al., 2011; Qi et al., 2004; Gazit, 2007) but there are still debates regarding the mechanism which actually results in inactivation of microbes. Most of the researchers agreed that cell membrane destruction is one of the vital processes of microbial inactivation. Inactivation or killing mechanism involves destruction of cell wall and cytoplasmic membrane by the action of produced reactive oxygen species (ROS) (e.g., hydrogen peroxide, hydroxyl radical). These ROS perform the lysis of the cell followed by complete mineralization of the microorganisms which ultimately results in death or inactivation (Foster et al., 2011). A systematic destruction of microbial cell through targeting the microbes/pathogens extracellular and intracellular target sites has been discussed by Dalrymple et al. (2010). A generalized mechanism of microbial disinfection by nanoparticles has been presented in Fig. 9-1.

In case of extracellular target sites, cell wall and cell membrane are most important which mainly consist of peptidoglycan layer, lipopolysaccharide layer, and phospholipid bilayer. Peptidoglycan maintains the shape of the cell and internal pressure by controlling rigidity. It is susceptible to the attack of free radicals from ROS generation as this porous

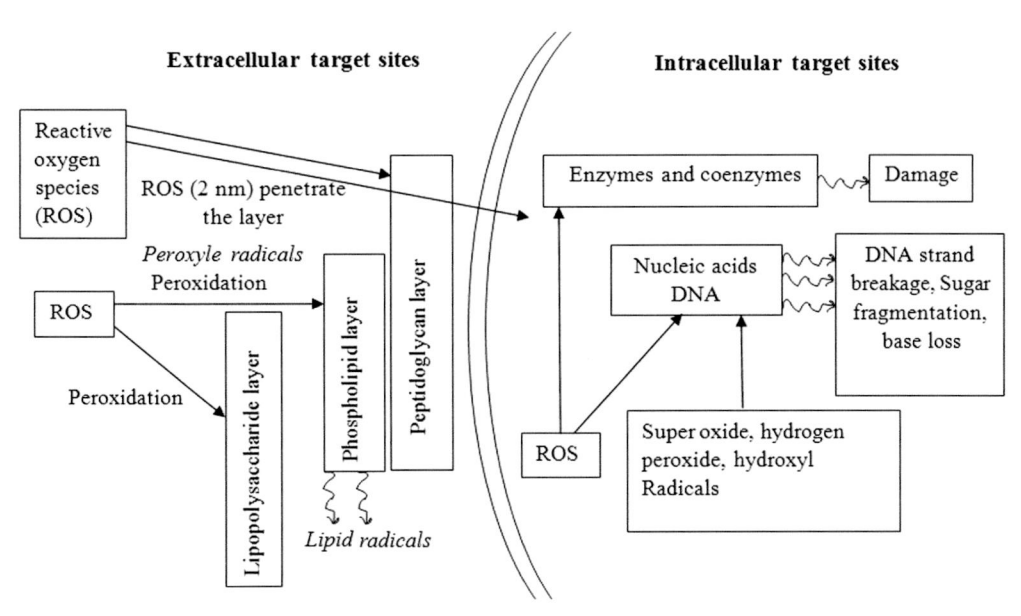

FIGURE 9-1 Generalized mechanism of microbial disinfection by nanoparticles.

material allows the oxidative species of 2 nm size to pass through (Lu et al., 2003; Demchick and Koch, 1996). Though this penetration mainly depends on the thickness of the peptido-glycan layer, it also depends on the reactivity of the species to the external environment (Dalrymple et al., 2010). Lipopolysaccharide is the most important target for ROS attack (Dalrymple et al., 2010; Halliwell and Gutteridge, 1989). Lipopolysaccharide layer and phospholipid bilayer both are susceptible to peroxidation due to the presence of fatty acids in their structure. Phospholipid bilayer primarily consists of alternating layers of lipid which can be attacked by ROS (Dalrymple et al., 2010). Fatty acids in lipid enhance the process because the interaction of radicals with unsaturated fatty acids in the presence of oxygen produce peroxyle radicals which can initiate chain reaction through forming lipid radicals by reacting with nearby lipid molecules and ultimately leading to the oxidation of biomole-cules (Dalrymple et al., 2010; Pryor, 1976; Halliwell and Gutteridge, 1989). The resulting effects are alteration of cell permeability and destruction of cell membrane (Dalrymple et al., 2010) which eventually inactivate the microorganisms by inhibiting membrane medi-ated respiration and releasing cytoplasm from the cell (Maness et al., 1999; Cheng et al., 2007; Saito et al., 1992).

For intracellular target sites, nucleic acids, enzymes, and coenzymes are important for the inactivation of microorganisms. Many researchers have found that superoxide is effective in inactivating some enzymes (Blum and Fridovich, 1985; Kono and Fridovich, 1982). Oxidation of coenzymes also play essential role in cell inactivation (Matsunaga et al., 1985). Attack to the DNA is a vital process of inactivation. DNA can be attacked by ROS as it is susceptible to oxidative stress. Attack of ROS to DNA sugar results in the breakage of DNA strand, fragmentation of sugar and loss of base. Generation of superoxide accelerate the destruction of nucleic acid (Dalrymple et al., 2010; Gogniat and Dukan, 2007). Superoxide and hydrogen peroxide can create reactive hydroxyl radicals that can react in intracellular target sites and damage biomolecules.

9.4 Application of Nanoparticles for Water and Wastewater Disinfection

Rapid growth of nanotechnology has drawn the attraction to use nanoparticles in various environmental fields (Li et al., 2008). Limitations of currently used conventional water and wastewater disinfection technologies make the scopes to experiment and utilize nanoparticles in this sector. Nanoparticles are unique in their physical and chemical properties and act as excellent catalysts and adsorbents because of their stability, reactivity and huge specific surface area (Li et al., 2008; Rikta and Tareq, 2017; Rikta, 2016). Nanoparticles (NPs) are expected not to produce toxic DBPs like other chemical disinfectants (Li et al., 2008). Various NPs have shown better efficiency in disinfection process (Table 9-2) when they are applied properly, which increases the acceptance of these materials in water disinfection process. To be applied in water and wastewater disinfection, NPs are broadly categorized into three main categories: carbon-based engineered NPs, metal and metal oxides NPs and naturally occurring antimicrobial substance (modified to NPs) (Li et al., 2008).

9.4.1 Carbon-Based Engineered NPs for Water and Wastewater Disinfection

Carbon nanotubes (CNTs), fullerenes and fullerenes derivatives are well-known engineered NPs for their disinfection properties against microbial pathogens. Both single walled carbon nanotubes (SWCNTs) and multi walled carbon nanotubes (MWCNTs) show effectiveness due to their unique structures and antimicrobial mechanisms. Inhibition mechanism of CNTs toward bacteria is either by oxidative stress or physical contact which destruct the integrity of cell membrane (Narayan et al., 2005; Kang et al., 2007). Thus, physical contact between target microorganisms and CNTs is important to inactivate the pathogens (Kang et al., 2007). Functionalized CNTs are generally used in disinfection as non-functionalized CNTs in suspension do not provide good contact with microbes (Li et al., 2008). Generally, carbon-based NPs show greatest ability to inactivate bacteria, improved capacity for the adsorption of bacterial and viral spores because of their larger surface area (Smith and Rodrigues, 2015). CNTs provide high adsorption capacity because of their high aspect ratio and fibrous structures which allow remarkably larger surface for spores attachment (Upadhyayula et al., 2009a; Smith and Rodrigues, 2015). Fig. 9-2 shows the scanning electron microscope (SEM) and transmission electron microscope (TEM) image of CNT and AU-Ag nanoparticles, respectively.

CNTs show great inactivation efficiency for bacteria, viruses, and protozoa. *Escherichia coli*, *Salmonella*, *Streptococcus mutans*, *Micrococcus lysodeikticus* bacteria are reported to be

Table 9-2   Application of Nanoparticles for the Disinfection and Microbial Control of Water

Nanoparticles	Microbial Component Removed	Removal Efficiency	Experimental Conditions for Removal	Inhibition Mechanism	References
Graphene oxide nanosheets impregnated with silver nanoparticles	*Escherichia coli*, *Staphylococcus aureus*	100% and 87.6% removal efficiency for *Escherichia coli* and *Staphylococcus aureus*, respectively	Incubation at 37°C for 4 h, shaking in rotary shaker	Cell death or inhibited growth due to damage of proteins and genetic materials	Bao et al. (2011)
Silver nanoparticles integrated in macroporous methacrylic acid copolymer beads	*Bacillus subtilis* ATCC 6051, *Staphylococcus aureus* ATCC 25923, *Escherichia coli* ATCC 8739, *Pseudomonas aeruginosa* ATCC 9027	100% removal efficiency except for *Bacillus subtilis* (99.9%) with an initial bacterial concentration of $10-300 \times 10^6$ CFU/mL using 100 mg nanoparticles	Shaking at 37°C for 6 h	Rupture of bacterial cell wall	Gangadharan et al. (2010)
Silver nanoparticles-alginate composite beads	*Escherichia coli* K12	Over 5-log disinfection efficiency with 1 min hydraulic retention time	Composite beads were packed in a chromatography column (1.6 cm diameter, 16 cm length, volume 32.2 mL), porosity of packing 0.38 ± 0.02, pore volume of 12.1 mL	Bacterial toxicity due to ROS production	Lin et al. (2013)
Silver nanoparticle coated polyurethane foam	*Escherichia coli* ATCC 25922, *Escherichia coli* MTCC 1302	100% removal efficiency	Flow rate 0.5 L/min, foam thickness 6 mm	–	Jain and Pradeep (2005)
Paper (cellulose fibers) embedded silver nanoparticles	*Escherichia coli* ATCC 11229, *Enterococcus faecalis* ATCC 47077	Over log 6 and log 3 reduction for *Escherichia coli* and *Enterococcus faecalis*, respectively	Bacterial suspension passed through Ag nanoparticle paper sheet without suction	Penetration of cell membrane by Ag nanoparticles	Dankovich and Gray (2011)
TiO$_2$/Ag nanoparticle photocatalytic system	*Candida glabrata*	Removal efficiency 94%	Presence of fluorescent UV lamps (5 mW/cm^2), contact time 30 min	Destruction of cell wall and cell membrane, alteration of nucleic acid, protein, inhibition of metabolic activities	Prikrylova et al. (2017)
Carbon nanoparticles	*Escherichia coli* bacteria ATCC 25922	6-log reduction at a dose of 4 mg/50 mL	pH 7.9, Contact time 150 min, exposed to solar radiation, suspension aerated for 10 min, turbidity 15 NTU	Damaging cell membrane	Maddigpu et al. (2018)

(Continued)

Table 9-2 (Continued)

Nanoparticles	Microbial Component Removed	Removal Efficiency	Experimental Conditions for Removal	Inhibition Mechanism	References
Ag nanoparticles loaded kaolin clay	*Salmonella spp.*, *Escherichia coli*	9 % for *Salmonella Spp.*, 80% for *Shigella flexneri*, *Klebsiella Pneumonae*, *Escherichia coli* and 70% for *Klebsiella aerogenes*	Contact time 2 h, 0.1 mg/L dose	Damaging of bacterial cell	Hassouna et al. (2017)
Ag nanoparticles loaded chitosan cryogels	*Escherichia coli* and *Bacillus subtilis*	3-log reduction with a silver content 7.5 mg/g	Sterilization of lab wares before test, gentle shaking of the suspension by hand during the swelling of cryogels, contact time 5 min	Penetration/destruction of cell wall by silver nanoparticles	Fan et al. (2018)
Zinc phosphide nanowires	*Escherichia coli*	>4 log reduction	Contact time 5 min, visible light exposure, anaerobic condition, temperature 25°C	Generation of hydroxyl radical which damaged bacterial cell	Vance et al. (2018)
Vanadium Tetrasulfide (VS_4)- carbon powder nanocomposite	*Escherichia coli*	9.7 log reduction	30 min contact time, dose 0.1 g/L, solution pH 7, presence of visible light	Membrane damage, degradation of DNA	Zhang et al. (2018)
Silver nanoparticle (45 nm) containing polysulfone membranes	*Escherichia coli*	90% removal efficiency	Membrane sterilization at 121°C for 15 min and incubation for 12 h at 37°C in orbital shaker at 150 rpm	Inactivation of bacterial cell by silver ion	Andrade et al. (2015)
Magnetic Fe_2O_3-AgBr photocatalyst	*Escherichia coli* K-12	Inactivation rate 18.25/h	Sterilization of bacteria and photocatalyst using 0.9% NaCl solution, light intensity 100 mW/cm^2, pH 7.67, temperature 37°C, dose 50 mg/L	Inactivation of bacterial cell by photo generated proton and the oxidation of H_2O_2 generation from the conduction band of Fe_2O_3	Ng et al. (2016)
Silver-lysozyme nanoparticles (based on Montmorillonite clay)	*Escherichia coli* and *Staphylococcus aureus*	96.78 ± 3.09 % and 90.07 ± 6.19% removal efficiency for *Escherichia coli* and *Staphylococcus aureus*, respectively	Suspension was incubated for 40 min with a shaking speed of 150 rpm, dose 160 µg/mL	Damage of bacterial cell by the generation of reactive oxygen species	Jiang et al. (2016)
Gold Nanoparticles (15.6 ± 3.4 nm)	*Escherichia coli*	Zone of inhibition 16 mm	Dose 400 µg/mL, nanoparticle loaded discs 3 mm in diameter	Decreased viability of bacterial cell through inhibitory effect of nanoparticle	Mata et al. (2016)
TiO_2–Fe_2O_3 nanocomposite	*Escherichia coli*	99.28% removal efficiency	Contact time 120 min, exposure to natural sunlight	Generation of ROS, destruction of cell membrane	Sharma et al. (2018)

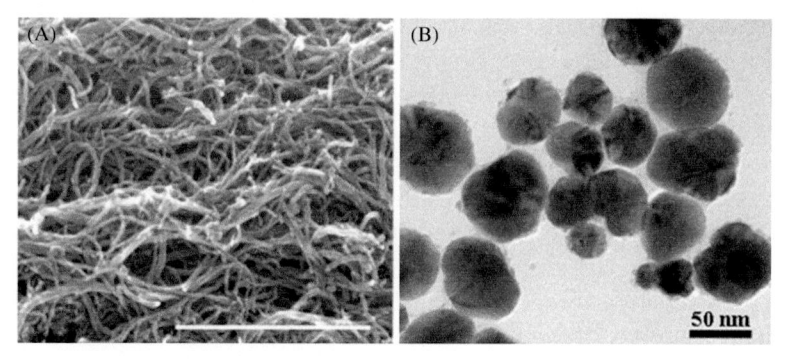

FIGURE 9-2 (A) CNT morphology in SEM image (Yoon et al., 2013) and (B) TEM image of AU-Ag nanoparticles (average size 55 ± 10 nm) (Kahraman et al., 2009). Source: *Images are used with proper permission.*

inactivated/killed by CNTs (Srivastava et al., 2004; Kang et al., 2007; Nepal et al., 2008; Arias and Yang, 2009; Akasaka and Watari, 2009). Polio virus, bacteriophage MS2, and *Tetrahymena pyriformis* protozoa species are susceptible to CNTs microbial inactivation (Srivastava et al., 2004; Brady-Estevez et al., 2008; Zhu et al., 2006). About 27 to 37 times more *Bacillus subtilis* spores can be adsorbed by SWNTs in aqueous solution than with powdered activated carbon (Upadhyayula et al., 2009b). Vertically aligned CNT membrane exhibited better biofouling resistant in wastewater treatment while *Pseudomonas aeruginosa* PAO1 GFP used as a test bacterial strain and this resistance is the result of bacterial inactivation by antimicrobial properties of CNTs (e.g., physical damage of microorganisms by oxidative stress) (Baek et al., 2014). A polyacrylonitrile hollowfiber membrane coated with Ag-MWNT composite was developed by Gunawan et al. (2011). Application of this Ag-MWNT composite layer on hollowfiber membrane for disinfection purpose efficiently hindered biofilm formation on membrane surface and prevented the growth of bacteria (*Escherichia coli*) in filtration module (Gunawan et al., 2011). More than 80% microbial cell inactivation was found by Ahmed et al. (2012) by applying poly-N-vinylcarbazole-SWNT nanocomposite. Therefore, application of CNTs assumed to be very much effective in preventing biofouling formation and microbial growth (Li et al., 2008) in water and wastewater disinfection processes.

Fullerenes (e.g., C_{60}, C_{70}) and fullerenes derivatives are also known for their disinfection capacities (Lyon et al., 2005; Spesia et al., 2007). Cytotoxic effects of C_{60} in eukaryotic are mainly for the toxicity exerted by generated ROS and in prokaryotic by direct oxidation of the microbial cell (Sayes et al., 2005; Fang et al., 2007; Markovic et al., 2007; Lyon and Alvarez, 2008). C_{60} are reported to show antimicrobial activity against *Bacillus subtilis* and *Escherichia coli* by oxidative stress (Lyon and Alvarez, 2008; Lyon et al., 2008).

Graphene and graphene oxide (GO) nanoparticles offer great opportunities in water disinfection process in their composite forms with various metals and metal oxides (Sreeprasad et al., 2011). GO nanosheets are generally obtained through ultrasonication of graphite oxides which are layered carbon material and hydrophilic in nature (Bao et al., 2011). Reduced GO has antimicrobial property that prevent bacterial growth hindering generation of biofilm on filter surface (Hu et al., 2010). GO nanosheets fabricated with Ag nanoparticles show strong antimicrobial activities against *Staphylococcus aureus* and *Escherichia coli* bacteria. Integration of Ag nanoparticles in GO nanosheets increases the antimicrobial efficiency of GO by 25% and 45% for *Staphylococcus aureus* and *Escherichia coli*, respectively (Bao et al., 2011). Gollavelli et al. (2013) reported that, 100% removal efficiency of *Escherichia coli* was found with the application of smart magnetic graphene derived from GO and ferrocene precursors. Inactivation of over 80% Gram positive and Gram negative bacteria was reported by using GO-SWNT nanocomposite (Mejias Carpio et al., 2012). Production of toxic ROS due to the interaction between GO and wastewater could be the possible mechanism to destroy/inactivate microorganisms (Ahmed and Rodrigues, 2013) through the damage of cell proteins and genetic materials (Bao et al, 2011). Sharp edges of CNTs, graphene and GO play important role in microbial disinfection through physical damage of the microorganisms (Smith and Rodrigues, 2015). Also, several studies suggested that, inhibition of nutrients diffusion into microbial cells due to the wrapping of cells by graphene sheets has significant contribution in microbial growth suppression (Liu et al., 2011; Chen et al., 2014; Mejias Carpio et al., 2012).

9.4.2 Metal and Metal Oxides NPs for Water and Wastewater Disinfection

Several metals and metal oxides show antimicrobial ability in water disinfection. Among various metals silver (Ag) ions and Ag compounds are very common for their antimicrobial activity, strong germicidal impacts and they are being used in wide range of fields for disinfection purposes (Li et al., 2008; Bosetti et al., 2002). Antimicrobial activities Ag^+ and Ag compounds were observed in *Pseudomonas aeruginosa, Legionella pneumophila, Clostridium perfringens* and *Escherichia coli* bacteria, MS2 phage and *Haemophilus influenza* phage cell (Xu et al., 2004; Kim et al., 2008; Park et al., 2017; Gogoi et al., 2006; Rahn et al., 1973) as well as in many other microbial pathogens by the action of generated ROS and penetration of microbial cell (Matsumura et al., 2003; Li et al., 2008). Ag^+ shows affinity to sulfhydryl groups and binding to this group impair the respiration of bacteria (Thurman et al., 1989). Effective antimicrobial activity of AgNP against *Escherichia coli* and *Staphylococcus aureus* was reported by Jung et al. (2008). Greater antimicrobial effect against *Pseudomonas aeruginosa* PAO1 was shown by MWNT when coated with AgNP (Kim et al., 2012).

Titanium dioxide (TiO_2) is semiconductor photocatalyst which is commonly used in disinfection purpose. TiO_2 is very much effective in inactivating bacteria, viruses, and protozoa. Generation of ROS in disinfection process facilitates the killing of pathogens through the damage of cell membrane and cell wall (Kikuchi et al., 1997). Gram negative bacteria are more sensitive to TiO_2 inactivation and bacterial killing dose varies from 100 and 1000 ppm based on light intensity (wavelength) and TiO_2 particle size (Wei et al., 1994). Inactivation of *Escherichia coli, Bacillus subtilis* bacteria and hepatitis B virus, MS2 bacteriophage, Herpes simplex virus and poliovirus 1 has been reported by many researchers (Reddy et al., 2007; Cho et al., 2005; Zan et al., 2007; Hajkova et al., 2007; Watts et al., 1995). Zinc oxide (ZnO) NP also shows strong antimicrobial activities against wide range of bacterial community

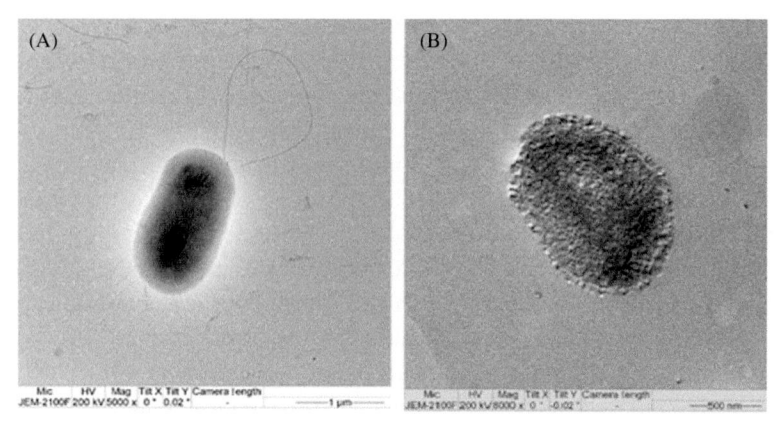

FIGURE 9-3 Morphology of *Escherichia coli* cell obtained by TEM before (A) and after (B) adding carbon supported Vanadium Tetrasulfide nanocomposites (Zhang et al., 2018). Source: *Image is used with proper permission.*

including *Escherichia coli, Lactobacillus helveticus* (Liu and Yang, 2003; Karunakaran et al., 2011). Though the disinfection mechanism of ZnO is still under investigation, photocatalytic H_2O_2 generation is presumed to be the most accepted one (Sawai, 2003). $TiO_2-Fe_2O_3$ nanocomposite is reported as effective microbial disinfectant, which can remove *Escherichia coli* through photocatalytic activity. Photogeneration of strong oxidizing agents ($\cdot OH$ and O_2^-) damages bacterial cells by destructing polyunsaturated phospholipids of the cells (Sharma et al., 2018). Inactivation of microbial cells by metal and metal oxide nanoparticles is mainly by physical damage of the cells (Fig. 9-3).

9.4.3 Naturally Occurring Antimicrobial Substance for Water and Wastewater Disinfection

Certain peptides and chitin which occur naturally are well-known for their antimicrobial activity. Recent advancement makes these substances more efficient by transforming them into NPs through engineering (Qi et al., 2004). They are modified with respective size and morphology to serve the specific purposes. The use of these NPs is more feasible for developing countries due to their low price, availability and simple disinfection process (Li et al., 2008). Chitosan is a naturally occurring antimicrobial substance which can be obtained from the chitin of arthropod shells (Qi et al., 2004; Li et al., 2008). Chitosan modified into NP is popular for the disinfection of water and wastewater. It exhibits higher effectiveness for the inactivation of viruses and fungus than bacteria, and Gram positive bacteria are more vulnerable to its antibacterial activity (Rabea et al., 2003; Don et al., 2005). Inactivation of microbial pathogen was preformed through the binding of charged particle of chitosan with oppositely charged particles of microbial cell wall which increase the permeability of the cell and eventually lead to leakage as well as rupture cytoplasmic constituents (Qi et al., 2004). *Escherichia coli, Candida albicans, Staphylococcus aureus,* bacteriophage and many other pathogens are prone to the chitosan antimicrobial ability (Bonnett et al., 2006; Chirkov, 2002; Qin et al., 2006). According to Hang et al. (2010), addition of $AgNO_3$ into chitosan-polyvinyl alcohol blend solution for microbial control (against *Escherichia coli*) exhibit greater inactivation efficiency.

Integration of current methods with nanoparticles is promising in achieving greater disinfection efficiency, zero/lower DBPs generation and destruction of resistant pathogens. Unique combination of conventional technologies to nanoparticles based on disinfection demand can bring best results. As example, to increase the efficiency of UV treatment, photosynthetic NPs can be combined which is more efficient to inactivate/kill resistant pathogens (e.g., *Cryptosporidium*) (Li et al., 2008). In addition, these types of combinations reduce/ remove other contaminants (i.e., natural organic matter) which actually suppress the regrowth of microbial pathogens.

9.5 Conclusion

In this chapter, a critical review on the application of NPs for water and wastewater treatment has been provided. All the discussions reveal that integration of NPs in disinfection process is an endowed idea to control microbial contaminations in water. The unique properties of NPs offer wide range of disinfection opportunities. Increased microbial pathogen inactivation capacities, relatively low pricing and increased reactivity to resistant infectious waterborne pathogens make them popular in water disinfection sector. Combination of antimicrobial NPs with conventional water disinfection technologies can greatly enhance the disinfection process. Though it is clear that, NPs is promising for water disinfection, their use, handling, and disposal must be carried out carefully as they are also considered as emerging pollutants. Besides this, it should be kept in mind that, outbreak of disinfection resistant microbial strain in future is one of the great challenges that emphasize the appropriate utilization of nanoparticles for disinfection purpose in forthcoming time.

References

Abbaszadegan, M., Hasan, M.N., Gerba, C.P., Roessler, P.F., Wilson, B.R., Kuennen, R., et al., 1997. The disinfection efficacy of a point-of-use water treatment system against bacterial, viral and protozoan waterborne pathogens. Water Res. 31 (3), 574−582.

Ahmed, F., Rodrigues, D.F., 2013. Investigation of acute effects of graphene oxide on wastewater microbial community: a case study. J. Hazard. Mater. 256−257, 33−39.

Ahmed, F., Santos, C.M., Vergara, R.A.M.V., Tria, M.C.R., Advincula, R., Rodrigues, D.F., 2012. Antimicrobial applications of electroactive PVK-SWNT nanocomposites. Environ. Sci. Technol. 46 (3), 1804−1810.

Akasaka, T., Watari, F., 2009. Capture of bacteria by flexible carbon nanotubes. Acta Biomater. 5, 607−612.

Alexander, J., Knopp, G., Dötsch, A., Wieland, A., Schwartz, T., 2016. Ozone treatment of conditioned wastewater selects antibiotic resistance genes, opportunistic bacteria, and induce strong population shifts. Sci. Total Environ. 559, 103−112.

Amy, G.L., Chadik, P.A., Chowdhury, Z.K., 1987. Developing models for predicting trihalomethane formation potential kinetics. J. Am. Water Works Assoc. 79 (7), 89−96.

Andrade, P.F., de Faria, A.F., Oliveira, S.R., Arruda, M.A.Z., 2015. Gonçalves, M.D.C., Improved antibacterial activity of nanofiltration polysulfone membranes modified with silver nanoparticles. Water Res. 81, 333–342. Available from: https://doi.org/10.1016/j.watres.2015.05.006.

Arias, L.R., Yang, L., 2009. Inactivation of bacterial pathogens by carbon nanotubes in suspensions. Langmuir 25, 3003–3012.

Baek, Y., Kim, C., Seo, D.K., Kim, T., Lee, J.S., Kim, Y.H., et al., 2014. High performance and antifouling vertically aligned carbon nanotubemembrane for water purification. J.Memb. Sci. 460, 171–177.

Bao, Q., Zhang, D., Qi, P., 2011. Synthesis and characterization of silver nanoparticle and graphene oxide nanosheet composites as a bactericidal agent for water disinfection. J. Colloid Interface Sci. 360 (2), 463–470.

Betancourt, W.Q., Rose, J.B., 2004. Drinking water treatment processes for removal of *Cryptosporidium* and *Giardia*. Vet Parasitol. 126 (1-2), 219–234.

Blum, J., Fridovich, I., 1985. Inactivation of glutathione peroxidase by superoxide radical. Arch. Biochem. Biophys. 240, 500–508.

Bonnett, R., Krysteva, M.A., Lalov, I.G., Artarsky, S.V., 2006. Water disinfection using photosensitizers immobilized on chitosan. Water Res. 40 (6), 1269–1275.

Bosetti, M., Masse, A., Tobin, E., Cannas, M., 2002. Silver coated materials for external fixation devices: in vitro biocompatibility and genotoxicity. Biomaterials 23 (3), 887–892.

Brady-Estevez, A.S., Kang, S., Elimelech, M., 2008. A single walled carbon nanotube filter for removal of viral and bacterial pathogens. Small 4, 481–484.

Cantor, K.P., Hoover, R., Hartge, P., Mason, T.J., Silverman, D.T., Levin, L.I., 1985. Drinking water source and risk of bladdercancer: a case-control study. In: Jolley, R.L., Bull, R.J., Davis, W.P., Katz, S., Roberts, M.H., Jacobs, V.A. (Eds.), Water Chlorination: Chemistry, Environmental Impact and Health Effects, vol. 5. Lewis Publishers, Chelsea.

Casani, S., Rouhany, M., Knøchel, S., 2005. A discussion paper on challenges and limitations to water reuse and hygiene in the food industry. Water Res 39, 1134–1146.

Chen, J., Peng, H., Wang, X., Shao, F., Yuan, Z., Han, H., 2014. Grapheneoxide exhibits broad-spectrum antimicrobial activityagainst bacterial phytopathogens and fungal conidia by intertwining and membrane perturbation. Nanoscale 6 (3), 1879–1889.

Cheng, Y.W., Chan, R.C.Y., Wong, P.K., 2007. Disinfection of *Legionella pneumophila* by photocatalytic oxidation. Water Res. 41 (4), 842–852.

Chirkov, S.N., 2002. The antiviral activity of chitosan (review). Appl. Biochem. Microbiol. 38 (1), 1–8.

Cho, M., Chung, H., Choi, W., Yoon, J., 2005. Different inactivation behavior of MS-2 phage and *Escherichia coli* in TiO_2 photocatalytic disinfection. Appl. Environ. Microbiol. 71 (1), 270–275.

Dalrymple, O.K., Stefanakos, E., Trotz, M.A., Goswami, D.Y., 2010. A review of the mechanisms and modeling of photocatalytic disinfection. Appl. Catal., B 98, 27–38.

Dankovich, T.A., Gray, D.G., 2011. Bactericidal paper impregnated with silver nanoparticles for point-of-use water treatment. Environ. Sci. Technol. 45, 1992–1998.

Demchick, P., Koch, A.L., 1996. The permeability of the wall fabric of *Escherichia coli* and *Bacillus subtilis*. J. Bacteriol. 178, 768–773.

Don, T.M., Chen, C.C., Lee, C.K., Cheng, W.Y., Cheng, L.P., 2005. Preparation and antibacterial test of chitosan/PAA/PEGDA bilayer composite membranes. J. Biomater. Sci. Polym. Ed. 16 (12), 1503–1519.

Drogui, P., Elmaleh, S., Rumeau, M., Bernard, C., Rambaud, A., 2001. Hydrogen peroxide production by water electrolysis: application to disinfection. J. Appl. Electrochem. 31, 877–882.

Elmi, F., Alinezhad, H., Moulana, Z., Salehian, F., Tavakkoli, S.M., Asgharpour, F., et al., 2014. The use of antibacterial activity of ZnO nanoparticles in the treatment of municipal wastewater. Water Sci. Technol. 70 (5), 763–770.

Fan, M., Gong, L., Huang, Y., Wang, D., Gong, Z., 2018. Facile preparation of silver nanoparticle decorated chitosan cryogels for point-of-use water disinfection. Sci. Total Environ. 613−614, 1317−1323.

Fang, J., Lyon, D.Y., Wiesner, M.R., Dong, J., Alvarez, P.J.J., 2007. Effect of a fullerene water suspension on bacterial phospholipids and membrane phase behavior. Environ. Sci. Technol. 41 (7), 2636−2642.

Foster, H.A., Ditta, I.B., Varghese, S., Steele, A., 2011. Photocatalytic disinfection using titanium dioxide: spectrum and mechanism of antimicrobial activity. Appl. Microbiol. Biotechnol. 90, 1847−1868. Available from: https://doi.org/10.1007/s00253-011-3213-7.

Gangadharan, D., Harshvardan, K., Gnanasekar, G., Dixit, D., Popat, K.M., Anand, P.S., 2010. Polymeric microspheres containing silver nanoparticles as a bactericidal agent for water disinfection. Water Res. 44 (18), 5481−5487.

Gazit, E., 2007. Self-assembled peptide nanostructures: the design of molecular building blocks and their technological utilization. Chem. Soc. Rev. 36 (8), 1263−1269.

Gogniat, G., Dukan, S., 2007. TiO_2 photocatalysis causes DNA damage via Fenton reaction-generated hydroxyl radicals during the recovery period. Appl. Environ. Microbiol. 73, 7740−7743.

Gogoi, S.K., Gopinath, P., Paul, A., Ramesh, A., Ghosh, S.S., Chattopadhyay, A., 2006. Green fluorescent proteinexpressing *Escherichia coli* as a model system for investigating the antimicrobial activities of silver nanoparticles. Langmuir 22 (22), 9322−9328.

Gollavelli, G., Chang, C.C., Ling, Y.C., 2013. Facile synthesis of smart magnetic graphene for safe drinking water: heavy metal removal and disinfection control. ACS Sustain. Chem. Eng. 1 (5), 462−472. Available from: https://doi.org/10.1021/sc300112z.

Gomez-Lopez, V.M., Gil, M.I., Pupunat, L., Allende, A., 2015. Cross-contamination of *Escherichia coli* O157: H7 is inhibited by electrolyzed water combined with salt under dynamic conditions of increasing organic matter. Food Microbiol. 46, 471−478. Available from: https://doi.org/10.1016/j.fm.2014.08.024.

Gunawan, P., Guan, C., Song, X., Zhang, Q., Leong, S.S.J., Tang, C., et al., 2011. Hollow fiber membrane decorated with Ag/MWNTs: toward effective water disinfection and biofouling control. ACS Nano 5 (12), 10033−10040.

Haaken, D., Dittmar, T., Schmalz, V., Worch, E., 2014. Disinfection of biologically treated wastewater and prevention of biofouling by UV/electrolysis hybrid technology: influence factors and limits for domestic wastewater reuse. Water Res. 52, 20−28.

Hajkova, P., Spatenka, P., Horsky, J., Horska, I., Kolouch, A., 2007. Photocatalytic effect of TiO_2 films on viruses and bacteria. Plasma Process. Polym. 4, S397−S401.

Halliwell, B., Gutteridge, J., 1989. Free Radicals in Biology and Medicine, second ed. Clarendon Press, Oxford.

Hang, A.T., Tae, A., Park, J.S., 2010. Non-woven mats of poly(vinyl alcohol)/chitosan blends containing silver nanoparticles: fabrication and characterization. Carbohydr.Polym. 82, 472−479.

Hassouna, M.E.M., Bably, M.A.E., Mohammed, A.N., Nasser, M.A.G., 2017. Assessment of carbon nanotubes and silver nanoparticles loaded clays as adsorbents for removal of bacterial contaminants from water sources. J Water Health 15 (1), 133−144. Available from: https://doi.org/10.2166/wh.2016.304.

Hu, W., Peng, C., Luo, W., Lv, M., Li, X., Li, D., et al., 2010. Graphene-based antibacterial paper. ACS Nano 4, 4317−4323.

Jain, P., Pradeep, T., 2005. Potential of silver nanoparticle-coated polyurethane foam as an antibacterial water filter. Biotechnol. Bioeng. 90 (1), 59−63. Available from: https://doi.org/10.1002/bit.20368.

Jiang, J., Zhang, C., Zeng, G.M., Gong, J.L., Chang, Y.N., Song, B., et al., 2016. The disinfection performance and mechanisms of Ag/lysozyme nanoparticle ssupported with Montmorillonite clay. J. Hazard. Mater. 317, 416−429. Available from: https://doi.org/10.1016/j.jhazmat.2016.05.089.

Jung, W.K., Koo, H.C., Kim, K.W., Shin, S., Kim, S.H., Park, Y.H., 2008. Antibacterial activity andmechanism of action of the silver ion in *Staphylococcus aureus* and *Escherichia coli*. Appl. Environ. Microbiol. 74 (7), 2171−2178.

Kahraman, M., Aydın, Ö., Çulha, M., 2009. Oligonucleotide-mediated Au−Ag core−shell nanoparticles. Plasmonics 4, 293−301. Available from: https://doi.org/10.1007/s11468-009-9105-3.

Kang, S., Pinault, M., Pfefferle, L.D., Elimelech, M., 2007. Single walled carbon nanotubes exhibit strong antimicrobial activity. Langmuir 23, 8670−8673.

Karunakaran, C., Abiramasundari, G., Gomathisankar, P., Manikandan, G., Anandi, V., 2011. Preparation and characterization of ZnO−TiO$_2$ nanocomposite for photocatalytic disinfection of bacteria and detoxification of cyanide under visible light. Mater. Res. Bull. 46 (10), 1586−1592.

Kerwick, M.I., Reddy, S.M., Chamberlain, A.H.L., Holt, D.M., 2005. Electrochemical disinfection, an environmentaly acceptable method of drinking water disinfection? Electrochim. Acta 50, 5270−5277.

Kikuchi, Y., Sunada, K., Iyoda, T., Hashimoto, K., Fujishima, A., 1997. Photocatalytic bactericidal effect of TiO$_2$ thin films: dynamic view of the active oxygen species responsible for the effect. J. Photochem. Photobiol. A. Chem. 106, 51−56.

Kim, E.S., Hwang, G., Gamal El-Din, M., Liu, Y., 2012. Development of nanosilver and multi-walled carbon nanotubes thin-film nanocomposite membrane for enhanced water treatment. J. Membr. Sci. 394, 37−48.

Kim, J.Y., Lee, C., Cho, M., Yoon, J., 2008. Enhanced inactivation of E. coli and MS-2 phage by silver ions combined with UV-A and visible light irradiation. Water Res. 42 (1−2), 356−362.

Kono, Y., Fridovich, I., 1982. Superoxide radical inhibits catalase. J. Biol. Chem. 257, 5751−5754.

Kraft, A., 2008. Electrochemical water disinfection: a short review. Platinum Metals Rev. 52 (3), 177−185. Available from: https://doi.org/10.1595/147106708X329273.

Lee, O.M., Kim, H.Y., Park, W., Kim, T.H., Yu, S., 2015. A comparative study of disinfection efficiency and regrowth control of microorganism in secondary wastewater effluent using UV, ozone, and ionizing irradiation process. J. Hazard. Mater. 295, 201−208.

Li, Q., Mahendra, S., Lyon, D.Y., Brunet, L., Liga, M.V., Li, D., et al., 2008. Antimicrobial nanomaterials for water disinfection and microbial control: potential applications and implications. Water Res. 42, 4591−4602.

Lin, S., Huang, R., Cheng, Y., Liu, J., Lau, B.L.T., Wiesner, M.R., 2013. Silver nanoparticle-alginate composite beads for point-of-use drinking water disinfection. Water Res. 47 (12), 3959−3965.

Liu, C., Tang, X., Kim, J., Korshin, G.V., 2015. Formation of aldehydes and carboxylic acids in ozonated surface waterand wastewater: a clear relationship with fluorescence changes. Chemosphere 125, 182−190.

Liu, H.L., Yang, T.C.K., 2003. Photocatalytic inactivation of *Escherichia coli* and *Lactobacillus helveticus* by ZnO and TiO$_2$ activated with ultraviolet light. Process Biochem. 39, 475−481.

Liu, S., Zeng, T.H., Hofmann, M., Burcombe, E., Wei, J., Jiang, R., et al., 2011. Antibacterial activity of graphite, graphite oxide, graphene oxide, and reduced graphene oxide: membraneand oxidative stress. ACS Nano 5 (9), 6971−6980.

Lu, Z.X., Zhou, L., Zhang, Z.L., Shi, W.L., Xie, Z.X., Xie, H.Y., et al., 2003. Cell damage induced by photocatalysis of TiO$_2$ thin films. Langmuir 19, 8765−8768.

Lyon, D.Y., Alvarez, P.J.J., 2008. Fullerene water suspension (nC$_{60}$) exerts antibacterial effects via ROS-independent protein oxidation. Environ. Sci. Technol. 42 (21), 8127−8132. Available from: https://doi.org/10.1021/es801869m.

Lyon, D.Y., Fortner, J.D., Sayes, C.M., Colvin, V.L., Hughes, J.B., 2005. Bacterial cell association and antimicrobial activity of a C$_{60}$ water suspension. Environ. Toxicol. Chem. 24 (11), 2757−2762.

Lyon, D.Y., Brown, D.A., Alvarez, P.J.J., 2008. Implications and potential applications of bactericidalfullerene water suspensions: effect of nC60 concentration, exposure conditions and shelf life. Water Sci. Technol. 57 (10), 1533−1538.

Maddigpu, P.R., Sawant, B., Wanjari, S., Goel, M.D., Vione, D., Dhodapkar, R.S., et al., 2018. Carbon nanoparticles for solar disinfection of water. J. Hazard. Mater. 343, 157−165.

Maness, P.C., Smolinski, S., Blake, D.M., Huang, Z., Wolfrum, E.J., Jacoby, W.A., 1999. Bactericidal activity of photocatalytic TiO_2 reaction: toward an understanding of its killing mechanism. Appl. Environ. Microbiol. 65, 4094−4098.

Markovic, Z., Todorovic-Markovic, B., Kleut, D., Nikolic, N., Vranjes-Djuric, S., Misirkic, M., et al., 2007. The mechanism of cell-damaging reactive oxygen generation by colloidal fullerenes. Biomaterials 28 (36), 5437−5448.

Mata, R., Bhaskaran, A., Sadras, S.R., 2016. Green-synthesized gold nanoparticles from *Plumeria alba* flower extract to augment catalytic degradation of organic dyes and inhibit bacterial growth. Particuology 24, 78−86. Available from: https://doi.org/10.1016/j.partic.2014.12.014.

Matsumura, Y., Yoshikata, K., Kunisaki, S., Tsuchido, T., 2003. Mode of bactericidal action of silver zeolite and its comparison with that of silver nitrate. App. Environ. Microbiol. 69 (7), 4278−4281.

Matsunaga, T., Tomoda, R., Nakajima, T., Wake, H., 1985. Photoelectrochemical sterilization of microbial cells by semiconductor powders. FEMS Microbiol. Lett. 29, 211−214.

Mejias Carpio, I.E., Santos, C.M., Wei, X., Rodrigues, D.F., 2012. Toxicity ofa polymer-graphene oxide composite against bacterial planktonic cells, biofilms, and mammalian cells. Nanoscale 4 (15), 4746−4756.

Narayan, R.J., Berry, C.J., Brigmon, R.L., 2005. Structural and biological properties of carbon nanotube composite films. Mater. Sci. Eng. B 123, 123−129.

Nepal, D., Balasubramanian, S., Simonian, A.L., Davis, V.A., 2008. String antimicrobial coatings: single walled carbon nanotubes armored with biopolymers. Nano Lett. 8, 1896−1902.

Ng, T.W., Zhang, L., Liu, J., Huang, G., Wang, W., Wong, P.K., 2016. Visible-light-driven photocatalytic inactivation of *Escherichia coli* by magnetic Fe_2O_3-AgBr. Water Res. 90, 111−118.

Park, S.J., Park, H.H., Ko, Y.S., Lee, S.J., Le, S., Woo, K., et al., 2017. Disinfection of various bacterial pathogens using novel silver nanoparticle-decorated magnetic hybrid colloids. Sci. Total Environ. 609, 289−296.

Prikrylova, K., Polievkova, E., Drbohlavova, J., Vesela, M., Hubalek, J., 2017. Nanostructured titania decorated with silver nanoparticles for photocatalytic water disinfection. Monatsh. Chem. 148, 1913−1919. Available from: https://doi.org/10.1007/s00706-017-2046-1.

Pryor, W.A. (Ed.), 1976. Free Radicals in Biology. Academic Press, Inc, New York.

Qi, L., Xu, Z., Jiang, X., Hu, C., Zou, X., 2004. Preparation and antibacterial activity of chitosan nanoparticles. Carbohydr. Res. 339 (16), 2693−2700.

Qin, C., Li, H., Xiao, Q., Liu, Y., Zhu, J., Du, Y., 2006. Water-solubility of chitosan and its antimicrobial activity. Carbohydr. Polym. 63 (3), 367−374.

Rabea, E.I., Badawy, M.E., Stevens, C.V., Smagghe, G., Steurbaut, W., 2003. Chitosan as antimicrobial agent: applications and mode of action. Biomacromolecules 4 (6), 1457−1465.

Rahn, R.O., Setlow, J.K., Landry, L.C., 1973. Ultraviolet irradiation of nucleic acids complexed with heavy atoms. 3. Influence of Ag^+ and Hg^{2+} on the sensitivity of phage and of transforming DNA to ultraviolet radiation. Photochem. Photobiol. 18 (1), 39−41.

Rajab, M., Heim, C., Letzel, T., Drewes, J.E., Helmreich, B., 2015. Electrochemical disinfection using boron-doped diamond electrode − the synergetic effects of in situ ozone and free chlorine generation. Chemosphere 121, 47−53.

Reddy, M.P., Venugopal, A., Subrahmanyam, M., 2007. Hydroxyapatite-supported Ag−TiO2 as *Escherichia coli* disinfection photocatalyst. Water Res. 41, 379−386.

Richardson, S.D., Thruston Jr., A.D., Caughran, T.V., Chen, P.H., Collette, T.W., Floyd, T.L., et al., 1999. Identification of new ozone disinfection by-products in drinking water. Environ. Sci. Technol. 33, 3368−3377.

Rikta, S.Y., 2016. Engineered nanoparticles (ENPs) in the aquatic environment: a review. Int. Res. J. Environment. Sci. 5 (3), 75−79.

Rikta, S.Y., Tareq, S.M., 2017. Impacts of engineered nanoparticles (ENPs) on aquatic organisms and public health: a review. JDWP 6, 3−8.

Rikta, S.Y., Tareq, S.M., Uddin, M.K., 2018. Toxic metals (Ni^{2+}, Pb^{2+}, Hg^{2+}) binding affinity of dissolved organic matter (DOM) derived from different ages municipal landfill leachate. Appl. Water Sci. 8 (5). Available from: https://doi.org/10.1007/s13201-018-0642-9.

Sadiq, R., Rodriguez, M.J., 2004. Disinfection by-products (DBPs) in drinking water and predictivemodels for their occurrence: a review. Sci. Total Environ. 321, 21−46.

Saito, T., Iwase, T., Horie, J., Morioka, T., 1992. Mode of photocatalytic bactericidal action of powdered semiconductor TiO_2 on mutant streptococci. J. Photochem. Photobiol. B Biol. 14, 369−379.

Sawai, J., 2003. Quantitative evaluation of antibacterial activities of metallic oxide powders (ZnO, MgO and CaO) by conductimetric assay. J. Microbiol. Methods 54 (2), 177−182.

Sayes, C.M., Gobin, A.M., Ausman, K.D., Mendez, J., West, J.L., Colvin, V.L., 2005. Nano-C_{60} cytotoxicity is due to lipid peroxidation. Biomaterials 26 (36), 7587−7595.

Scholz, M., Martin, R., 1997. Ecological equilibrium on biologicalactive carbon. Water Res. 31 (12), 2959−2968.

Shannon, M.A., Bohn, P.W., Elimelech, M., Georgiadis, J.G., Mariñas, B.J., Mayes, A.M., 2008. Science and technology for waterpurification in the coming decades. Nature 452 (7185), 301−310. Available from: https://doi.org/10.1038/nature06599.

Sharma, B., Boruah, P.K., Yadav, A., Das, M.R., 2018. TiO_2−Fe_2O_3 nanocomposite heterojunction for superior charge separation and the photocatalytic inactivation of pathogenic bacteria in water under direct sunlight irradiation. J. Environ. Chem. Eng. 6, 134−145.

Simpson, D.R., 2008. Biofilm processes in biologically active carbon waterpurification. Water Res. 42, 2839−2848.

Smith, S.C., Rodrigues, D.F., 2015. Carbon-based nanomaterials for removal ofchemical and biological contaminants from water: a review of mechanisms and applications. Carbon 91, 122−143.

Song, H., Addison, J.W., Hu, j, Karanfil, T., 2010. Halonitromethanes formation in wastewater treatment plant effluents. Chemosphere 79, 174−179.

Spesia, M.B., Milanesio, M.E., Durantini, E.N., 2007. Synthesis, properties and photodynamic inactivation of *Escherichia coli* by novel cationic fullerene C(60) derivatives. Eur. J. Med. Chem. 43 (4), 853−861.

Sreeprasad, T.S., Maliyekkal, S.M., Lisha, K.P., Pradeep, T., 2011. Reduced graphene oxide-metal/metal oxide composites: facile synthesisand application in water purification. J. Hazard. Mater. 186 (1), 921−931.

Srivastava, A., Srivastava, O.N., Talapatra, S., Vajtai, R., Ajayan, P.M., 2004. Carbon nanotube filters. Nat. Mater. 3, 610−614.

Thurman, R.B., Gerba, C.P., Bitton, G., 1989. The molecular mechanisms of copper and silver ion disinfection of bacteria and viruses. Crit. Rev. Environ. Cont. 18 (4), 295−315. Available from: https://doi.org/10.1080/10643388909388351.

Upadhyayula, V.K.K., Deng, S., Mitchell, M.C., Smith, G.B., 2009a. Application of carbon nanotube technology for removal of contaminants in drinking water: a review. Sci. Total Environ. 408 (1), 1−13.

Upadhyayula, V.K.K., Deng, S., Smith, G.B., Mitchell, M.C., 2009b. Adsorption of *Bacillus subtilis* on single-walled carbonnanotube aggregates, activated carbon and NanoCeram™. Water Res. 43 (1), 148−156.

Vance, C.C., Vaddiraju, S., Karthikeyan, R., 2018. Water disinfection using zinc phosphide nanowires under visible light conditions. J. Environ. Chem. Eng. 6, 568−573.

WHO (World Health Organization), 2017. Drinking-water, Fact Sheet. Available at http://www.who.int/media-centre/factsheets/fs391/en/, (accessed 23.01.17.).

Wang, W., Ye, B., Yang, L., Li, Y., Wang, Y., 2007. Risk assessment on disinfection by-products of drinking water of differentwater sources and disinfection processes. Environ. Int. 33, 219−225.

Watts, R.J., Kong, S., Orr, M.P., Miller, G.C., Henry, B.E., 1995. Photocatalytic inactivation of coliform bacteria and viruses in secondary wastewater effluent. Water Res. 29, 95−100.

Wei, C., Lin, W.Y., Zainal, Z., Williams, N.E., Zhu, K., Kruzic, A.P., et al., 1994. Bactericidal activity of TiO_2 photocatalyst in aqueous media: toward a solar-assisted water disinfection system. Environ. Sci. Technol. 28 (5), 934−938.

Xu, X.H., Brownlow, W.J., Kyriacou, S.V., Wan, Q., Viola, J.J., 2004. Real-time probing of membrane transport in living microbial cells using single nanoparticle optics and living cell imaging. Biochemistry 43 (32), 10400−10413.

von-Gunten, U., Hoigne, J., Bruchet, A., 1995. Bromate formation during ozonation of bromide-containing waters. Water Supp. 13, 45−50.

Yoon, I., Hamaguchi, K., Borzenets, I.V., Finkelstein, G., Mooney, R., Donald, B.R.V., 2013. Intracellular neural recording with pure carbon nanotube probes. PLoS One 8 (6), e65715. Available from: https://doi.org/10.1371/journal.pone.0065715.

Zan, L., Fa, W., Peng, T.P., Gong, Z.K., 2007. Photocatalysis effect of nanometer TiO_2 and TiO_2-coated ceramic plate on Hepatitis B virus. J. Photochem. Photobiol. B. Biol. 86 (2), 165−169.

Zhang, B., Zoub, S., Caia, R., Lia, M., He, Z., 2018. Highly-efficient photocatalytic disinfection of *Escherichia coli* under visible light using carbon supported Vanadium Tetrasulfide nanocomposites. Appl.Catal. B 224, 383−393.

Zhou, H., Smith, D.W., 2002. Advanced technologies in water and wastewatertreatment. J. Environ. Eng. Sci. 1, 247−264. Available from: https://doi.org/10.1139/S02-020.

Zhu, Y., Ran, T., Li, Y., Guo, J., Li, W., 2006. Dependence of cytotoxicity of multi walled carbon nanotubes on the culture medium. Nanotechnology 17, 4668−4674.

10

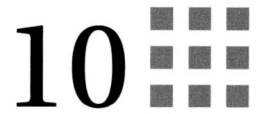

Forward Osmosis (FO) for Removal of Heavy Metals

Norfadhilatuladha Abdullah, Muhammad Hanis Tajuddin, Norhaniza Yusof

ADVANCED MEMBRANE TECHNOLOGY RESEARCH CENTRE (AMTEC), SCHOOL OF CHEMICAL AND ENERGY ENGINEERING, FACULTY OF ENGINEERING, UNIVERSITI TEKNOLOGI MALAYSIA, JOHOR BAHRU, MALAYSIA

10.1 Introduction

Nowadays, water contamination has become a global issue to the worldwide community. This issue has been aggravated due to the rapid rate of industrialization that release huge amount of toxic wastages into water resources. One of the commonly released contaminants found in water is heavy metals (Abdullah et al., 2016). The term of heavy metals is somehow imprecise; however, it is mostly refers to any metals with atomic atom more than 20 and has density more than 56 g/L (Sardar et al., 2013). In general, heavy metals are found in water as a result from anthropogenic and natural activities. Various anthropogenic sources come for domestic, industrial, agricultural and mining. On the other hand, natural activities are linked to the weathering of soil, rocks as well as from volcanic eruptions. Excessive disposal or exposure of heavy metals leads to the entry of these mobile ions into plants, animals and human tissues. The small size of heavy metals could easily penetrate into biological system and consequently bind and interfere to the vital cellular components (Tchounwou et al., 2012; Parmar and Thakur, 2013). Table 10-1 elucidates the some information of arsenic, cadmium, chromium, lead, and mercury in which these are the common hazardous metals found in water sources.

Referring to Table 10-1, it is obvious that the exposure of these heavy metals has acute effects. Due to the deteriorating water quality caused by heavy metals contamination, many strategies have been employed to find a feasible and robust technique to remove heavy metals from water. Membrane technology on the other hand, has been profound to effectively remove various kinds of water contaminants including dyes, oils, salts, heavy metals, inorganic contaminants, and organic contaminants. Due to its versatility of membranes tailoring, membrane technology is used in various applications in food industry, gas separation, desalination, oil refinery and water/wastewater treatment. There are five different membrane processes; namely microfiltration (MF), ultrafiltration (UF), nanofiltration (NF), reverse osmosis (RO) and forward osmosis (FO) (Favre, 2010). In fact, many studies have been

Nanotechnology in Water and Wastewater Treatment. DOI: https://doi.org/10.1016/B978-0-12-813902-8.00010-1

Table 10-1 Sources of Release and Potential Risk of Exposure of Heavy Metals Towards Human (Tchounwou et al., 2012)

Type of Metals	Forms of Existence in Water	Occurrence/Sources	Potential Risk of Exposure
Arsenic	Trivalent arsenic, pentavalent arsenic, monomethylarsonic acid, dimethylarsinic acid, trimethylarsenine oxide	Volcanic eruption, soil erosion, agricultural activities, veterinary medicinal applications	Cardiovascular and peripheral vascular diseases, developmental anomalities, neurological disorder, cancers, hematologic disorder
Cadmium		Manufacturing of alloys, stabilizers, pigments and batteries, mining, smelting	Abdominal pain, nausea, vomiting, gastrointestinal tract erosion, pulmonary injury, prostatic proliferative lesions
Chromium	Chromium (II), chromium (III), ferrochromite, Chromium (VI)	Metal processing, tannery facilities, chromate production, metallurgical industries, stainless steel welding, pigment production, leather tanning	Respiratory problems, cardiovascular disease, hematological, hepatic, renal effects, neurological disorder
Lead	Lead oxide, divalent lead, lead chromate	Fossil fuel burning, mining, agricultural activities, domestic activities, metal products piping, amunitions, X-rays, paint, and ceramic products	Blood level poisoning, lower IQ, impaired neurobehavioral developments, impaired growth, reproductive problems, abortions, brain damage, kidney, and gastrointestinal problems, damage to nervous system
Mercury	Mercury (I), mercurous (II), methylmercury	Electrical industries, dentistry, production of caustic soda, nuclear reactors, solvent, pharmaceutical industries	Gastrointestinal damage, neurotoxicity, nephrotoxicity, kidney problems

conducted for all of these membrane processes upon removing heavy metals from water, in exemption of MF as no significant rejection could be obtained due to the large pore size of MF (Abdullah et al., 2016). In recent years, FO has attracted a very great interest among researchers for heavy metals removal as it requires a very low/nonenergy, low fouling rate, satisfactory pure water permeation, and high solute rejection (Cui et al., 2014). In addition, few studies conducted on heavy metals by FO process showed a very great potential of this membrane processes to treat water laden by heavy metals. Thus, in this book chapter, focus will be given to the development of FO membranes for removing heavy metals from water.

10.2 Forward Osmosis Technology

10.2.1 Concept and Principle of Forward Osmosis

FO is generally derived from osmosis process. In principle, osmosis is a phenomenon of solute movement passing a semipermeable membrane by means of diffusion from lower solute concentration to higher solute concentration (Cath et al., 2006). There are two basic factors that act as the driving forces for the movement of solute from one side to another. These are, (1) osmotic pressure, and (2) osmotic gradient. In water treatment, there are three types of osmosis processes which are FO, pressure-retarded osmosis (PRO), and RO (Cath et al., 2006).

Fig. 10-1 shows the difference of water movement between FO and RO. FO differs to RO where FO uses natural osmotic pressure as the driving force between feed solution (FS) and draw solution (DS) streams to induce mass transport of water from lower solute concentration to higher solute concentration (Yang et al., 2015). On the other hand, RO requires high hydraulic pressure to force water from high solute concentration to low solute concentration. In comparison to UF, NF and RO, FO has been proven to be flexible and efficient in water/wastewater treatment where no energy output is needed. In addition, DS such as salts and sugars are generally harmless to the environment. Another advantage of FO is lower fouling rate due to the absence of hydraulic pressure that accelerates the attachment of foulants on membrane surface (Choi et al., 2017).

The basic laboratory setup of FO process is illustrated in Fig. 10-2. There are two operational modes can be conducted in FO system; (1) FO mode where the thin selective layer facing the FS and (2) PRO mode where the thin selective layer facing the DS (Cath et al., 2006). Both orientations have different impacts on the water flux as well as the internal concentration polarization (CP) which will be discussed in the next section. Based on Fig. 10-2, peristaltic pump will initiate the simultaneous movement of solutes from where the FS will be in contact to the active layer while the DS will be in contact with the support layer. As DS has higher osmotic pressure, fresh water will be withdrawn via active layer and eventually passing the support layer. At the end of the process, the DS will be diluted as a result of incoming water from the FS. Visual observation of decrease in weight of the FS and increase water level in the DS will be seen.

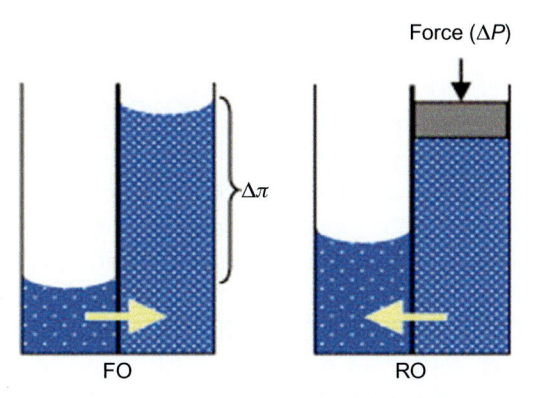

FIGURE 10-1 Principle of RO and FO process (Cath et al., 2006). Source: *Reprinted with written permission from Elsevier.*

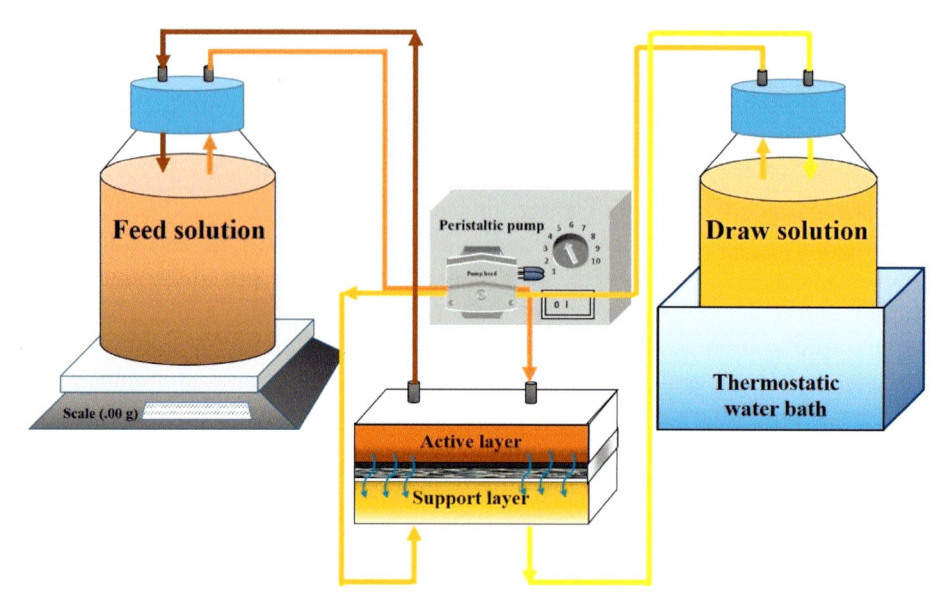

FIGURE 10-2 Basic setup of FO system (Wu et al., 2016). Source: *Reprinted with written permission from Elsevier.*

10.2.2 Development of Forward Osmosis Membrane

10.2.2.1 Concentration Polarization

The design of good FO process should meet these three major requirements: (1) high water flux; (2) high solute retention; (3) low CP and fouling effects (Chun et al., 2017). Other features of FO membranes include high mechanical and chemical stability. The state-of-art for FO membrane is the thin film composite (TFC) FO membranes comprised of polyamide (PA) dense layer with thickness of few hundred nanometers which is supported by porous substrate (Shaffer et al., 2015). This is the active layer that responsible for salt rejection. As compared to the commercial FO without TFC, FO membranes with TFC have been shown to elucidate higher water flux with better solute rejection. Nevertheless, there are few limitations faced by TFC membranes. These include low water flux, RSD and high internal concentration polarization (ICP) (Shaffer et al., 2015). In general, in comparison to RO membranes which the water transport occurs at only on selective layer of membrane towards the substrate, the FS and DS will be simultaneously present on each side of membrane in FO process (Chun et al., 2017). This condition leads to concentration polarization (CP) and RSD. Theoretically, RSD is defined as the movement of dissolved draw solute from DS to FS or vice versa. The rate of RSD occurring during the process is majorly controlled by CP. CP is a phenomenon of cake layer deposition on (external CP) or within (internal CP) FO membranes (Yang et al., 2015; Shaffer et al., 2015). It has been discussed that ECP does not has major effect on the flux of FO membranes; instead, ICP has been profound to greatly occur in FO membranes and greatly reduce water flux. ICP basically refers to the occurrence of cake layer within the substrate layer as a result of solute unable to penetrate into the dense selective layer (Mccutcheon and Elimelech, 2006).

Depending on the orientation of the membrane, there are two types of ICP, which are concentrative ICP or dilutive ICP, as shown in Fig. 10-3. The concentrative ICP is illustrated in Fig. 10-3A under PRO mode when the active layer is facing the DS. Without any consideration of ICP, the water flux can be measured by using the equation proposed by Lee et al. (1981) and Loeb (1976) (Eq. (10.1)):

$$J_w = \frac{1}{S} \ln \frac{\pi_{D,b}}{\pi_{F,b}} \tag{10.1}$$

where π_{HI} and π_{LOW} are the osmotic pressure of DS (D_b) and FS (F_b) and S is the constant showing of solute diffusion resistance within the membrane. The equation of osmotic pressure $(<)$ used is in accordance to Van't Hoff's law (Eq. (10.2)):

$$\pi = iMRT \tag{10.2}$$

where π is the osmotic pressure (atm), i is the amount of molecules per mole of solute, M is the molarity of one solution, T is temperature (kelvin) and R is the gas constant. On the other hand, the value of S is generally influenced by three factors which are thickness, tortuosity, and porosity of membrane and is governed by Eq. (10.3):

$$S = \frac{t\tau}{\varepsilon D_s} \tag{10.3}$$

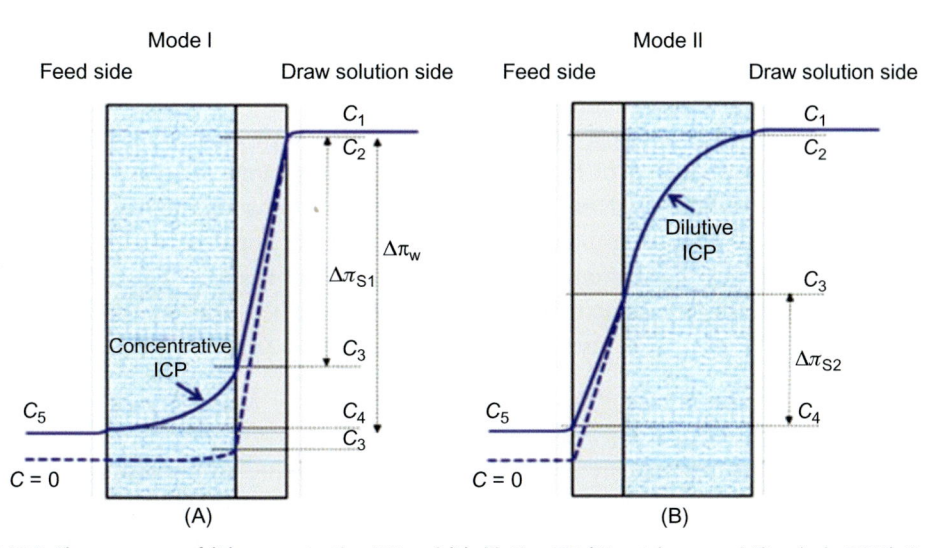

FIGURE 10-3 Phenomenon of (A) concentrative ICP and (B) dilutive ICP (Mccutcheon and Elimelech, 2006). *Source: Reprinted with written permission from Elsevier.*

where t is membrane thickness, τ is the membrane tortuosity ε is membrane porosity and D_s is the coefficient for solute diffusion. To quantify the severity of concentrative ICP, the below equation is developed by combining Eq. (10.1) within Eq. (10.3) to yield Eq. (10.4):

$$S = \left(\frac{1}{J_w}\right)\ln\left(\frac{B + A\pi_{D,b} - J_w}{B + A\pi_{F,b}}\right) \tag{10.4}$$

By performing salt rejection experiment using RO or NF system, the value of solute permeability coefficient, B, can be determined by using Eq. (10.5):

$$B = \frac{(1 - R)A(\Delta P - \Delta\pi)}{R} \tag{10.5}$$

Where R is the rejection for salt used as DS. The value of B is negligible if the membranes attain a very high rejection at high water flux. The bigger the value of S, it therefore could be assumed that the occurrence of concentrative ICP is severe.

On the other hand, the dilutive ICP is when FO mode is run is illustrated by Fig. 10-3B. Dilutive ICP occurs as a result of formation of boundary layer at the permeate side. For this case, the DS is diluted by the permeate water within the porous support and the dilutive ICP can be measured by using Eq. (10.6) (Mccutcheon and Elimelech, 2006):

$$S = \left(\frac{1}{J_w}\right)\ln\left(\frac{B + A\pi_{D,b}}{B + J_w + A\pi_{F,b}}\right) \tag{10.6}$$

Depending on which mode is being used in FO system, Mccutcheon and Elimelech (2006) strongly suggested that it is crucial to measure the flux in governed to the ICP so that accurate results could be obtained.

Apart of finding the optimized operational conditions, the rate of ICP could be minimized by improving the properties of membrane. There are two ways in which the modification can be achieved; either by physical modification or chemical modification. Physical modification simply refers to addition of hydrophilic nanomaterials into the TFC-FO membranes while chemical modification involves the chemical reactions between the hydrophilic materials on the existing TFC-FO membranes. In this regard, the two strategies generally; (1) improving the properties of substrate layer of TFC-FO membranes which will increase the hydrophilicity, porosity, pore structures and hence reducing the ICP; or (2) improving the properties of thin active/selective layer (Zirehpour et al., 2017; Xu et al., 2017).

10.2.2.2 Modification of Polyamide Selective Layer

Due to the inherent hydrophobicity of PA layer, low permeability and membrane fouling are observed. Modification of selective layer of has been widely conducted by incorporating various types of nanomaterials in the active layer. Some of these nanomaterials include titanium oxide, carbon nanotubes, zeolite, silica, graphene oxide, or surfactants (Lau et al., 2012). They are dispersed into either organic phase or aqueous phase during the interfacial polymerization.

Wu et al. (2017) coating pore forming agent polyvinylpyrolidone (PVP) on the GO sheet as modifier of the PA active layer. In their experiment, PVP/GO nanoparticle was dispersed into MPD aqueous phase. Coating PVP on the GO nanosheet revealed to not only improve the dispersion and dis-aggregation of the nanomaterials within MPD solution, but also improving the hydrophilicity and lowering the reverse flux. It was found out that the reverse flux was significantly reduced to 1.25 g/m^{2}h and the water flux was 33.2 L/m^{2}h when the 1 M of NaCl was used under AL-FS mode. On the other hand, Zirehpour et al. (2017) incorporated silver−based nano-sized metal organic frameworks (MOFs) into the organic solution of TMC during the preparation of the selective thin layer. In general, MOFs is a porous nanomaterial comprised of metal ions surrounded by organic groups. It has been suggested that MOFs exhibits a very good affinity towards polymer in comparison to the inorganic materials. In their experiment, it is demonstrated that TFC incorporated MOFs displayed more hydrophilic surface assigned by lower contact angle measurement and better water flux which was 39 L/m^{2}h under AL-FS mode when 1 M of NaCl was used as the DS. In fact, the antifouling property of the TFC membrane was also enhanced due to the presence of Ag which possesses antibacterial characteristic. Shakeri et al. (2017) on the other hand, incorporated chitosan, a natural polymer containing amine groups as additive into the organic solution during preparation of the PA layer. In the study, negatively charged sulfonated polyethersulfone (SPES) was used; the sulfonic functional groups in SPES allows strong interaction with positive amine groups from chitosan to result to excellent attachment of PA layer with the substrate. Results indicated that higher concentration of chitosan led to a dense yet thin PA layer with improved hydrophilicity. Their study showed an increment of water flux as compared to the control membrane where water flux achieved was 16.6 L/m^{2}h under AL-FS mode when 2 M of NaCl was used.

Besides of incorporation of nanomaterial into FO membrane, surface grating or chemical modification is often used to modify the properties of TFC-FO membranes via crosslinking certain functional groups on the surface of selective layer. For example, Shen et al. (2017) grafting self-catalyzed tripodal amine-tris (2-aminoethyl)amine (TAEA) on the PA layer. These tertiary amines comprised of a small molecular size and three primary amine groups where it functions as crosslinker, amine monomer, or catalyst that can accelerate the reaction between active layer and the unreacted acyl chloride groups of the substrate's surface. Under AL-FS mode, better water permeability and low RSD were observed which are 74 L/m^{2}h and 2.63 g/m^{2}h, respectively. Faria et al. (2017) on the other hand, crosslinking silver-GO nanosheet on the PA active layer for improving the properties of TFC-FO membranes as well as improving the antifouling properties of the membrane. The sequence of TFC modification is shown in Fig. 10-4. Initially, the PA layer comprised of free carboxyl groups is treated with ethylene diamine (ED) and NHS to form amino terminated membrane surface (Fig. 10-4A). At the same time, when immersed in a buffered solution containing EDC and NHS, the carbonyl groups on GO layer is activated and converted to intermediate esters (Fig. 10-4B). The latter is later linked to PA layer by forming amide bond with the amine group of the PA layer (Fig. 10-4C). Apart of improved water flux which is 33.8 L/m^{2}h, the biofouling feature of the membrane has been enhanced where the static microbial assay

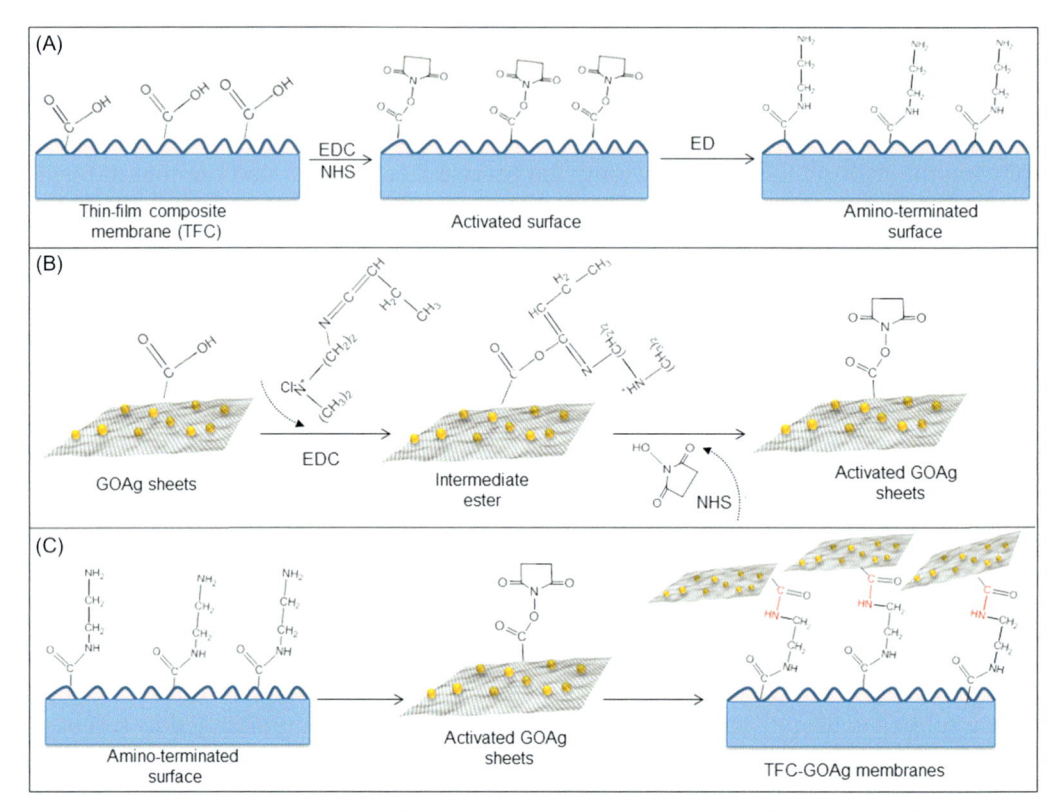

FIGURE 10-4 Schematic diagram of crosslinking process of Ag-GO nanosheet on PA layer (Faria et al., 2017). Source: *Reprinted with written permission from Elsevier.*

showed that the membrane inhibited the growth of biofilm by 30%. Another work of surface grating on TFC was conducted by Xiong et .al. (2017) using MPD combined with diamine monomer N-[3-(trimethoxysilyl) propyl] ethylenediamine (NPED) in order to fabricate TFC selective layer (Xiong et al., 2016). As NPED consists of diamine and siloxane groups, the latter will be hydrolyzed into hydrophilic functional groups (Si-OH) while the diamine groups will be reacting with the organic solution. Under AL-FS mode by using 0.5 M NaCl as DS, enhanced water flux was achieved, from 9.67 L/m^{2}h by using pristine membrane to 16.67 L/m^{2}h when NPED was added. Despite better antifouling properties was seen, slight reduction of salt rejection was also observed in which further actions are needed to tackle the limitation.

10.2.2.3 Modification of Substrates

The modification of substrate of TFC-FO membranes was initially started in 2012 when Ma et al. conducted a study of various loading of zeolite into PSf substrate. Their study showed that increasing zeolite content in PSf caused an 50% increment of water flux than pristine

TFC membrane which is of 32 (AL-DS mode) and 15 L/m²h (AL-FS mode) when 1 M of NaCl was used. The improved water flux is attributed to the nature of zeolite which is porous and thicker thin selective layer. Amini and coworkers studied the effect of functionalized MWCNTs and SiO_2 nanoparticles on the performance of TFC-FO (Amini et al., 2013). Incorporation of low loading of SiO_2 (0.1%) led to 36 (Al-DS) and 23 L/m²h (AL-FS) when tested with 2 M NaCl. In contrast, addition of 0.1% MWCNTs yielded an excellent water flux which was 96 (AL-DS) and 37 L/m²h (AL-FS), respectively. Nevertheless, high RSD was also observed. Morales et al. (2016) had also incorporating carbon based nanomaterials which are MWCNTs, functionalized MWCNTs (fMWCNTs), graphene oxide (GO) and carbon-TiO_2 composites into PSf support. Their studies demonstrated that membrane incorporated with fMWCNTs and GO yielded higher water flux and lower RSD as compared to PSf loaded with MWCNTs and pristine membrane. When combining fMWCNTs with TiO_2 as composite nanofillers into PSf substrate, highest flux obtained was 20.3 L/m²h with low RSD of 0.66 g/L. Initially, when fMWCNTs is added into PSf, more porous and larger finger like structures will be induced as a result of improved membrane hydrophilic surface. Further addition of TiO_2 led to elongated finger like structures, enabling the membrane to become more hydrophilic without sacrificing the low RSD. Emadzadeh et al. (2014) embedded commercial TiO_2 into the microporous substrate. Their experimental results showed that addition of 0.5 wt% TiO_2 resulting to 57 (AL-DS) and 30 L/m²h (AL-FS) water flux which is approximately 90% and 71.5% higher than the commercial FO membranes. However, it should be noted that increasing the loading of additives into substrate might be disadvantageous as reverse flux solute diffusion and ICP phenomenon will be aggravated. This could be due to the excessive loading of nanoparticles cause agglomeration on the surface of substrate.

Besides on the incorporation of nanomaterials into the substrate, it can also be seen that recent trend of published papers in 2017 and 2018 showed that instead of incorporating nanofiller into microporous UF or MF membrane, the use of nanofiber-based membrane as the substrate seems to be promising (Park et al., 2018). The utilization of nanofiber membranes in comparison to common porous UF/MF substrate is beneficial to reduce the ICP as NFM has lower tortuosity (higher *S* factor) due to interconnected voids. Some studies even suggested that nanofibrous substrate demonstrated about 2−5 times higher than the commercial membranes in which most studies blending hydrophilic/hydrophobic polymers as NF membranes (Shokrollahzadeh and Tajik, 2018). The very recent study of Park et al. (2018) had used PVDF nanofiber membranes coated by hydrophilic polyvinyl alcohol (PVA) as substrate for PA layer. In comparison to the uncoated PVDF NF, the PVA-coated NF showed higher pure water flux which is 24.8 L/m²h and lower reverse flux. This could be devoted to the improved hydrophilicity of the substrate layer that allows better solution diffusivity and hence lowering the ICP value. On the other hand, Shokrollahzadeh and Tajik (2018) had using electrospun PSf/PAN as the porous substrate for TFC membrane. Not only the water flux of the membranes had been reported to be higher that control TFC membrane which was 38.3 L/m²h under FO mode, the reverse flux had also retained to be low which was 10.1 g/m²h. Pan et al. (2017) fabricated pure PAN NF membranes by electrospinning process. Prior of interfacial polymerization, the PAN NF was initially laminated by using

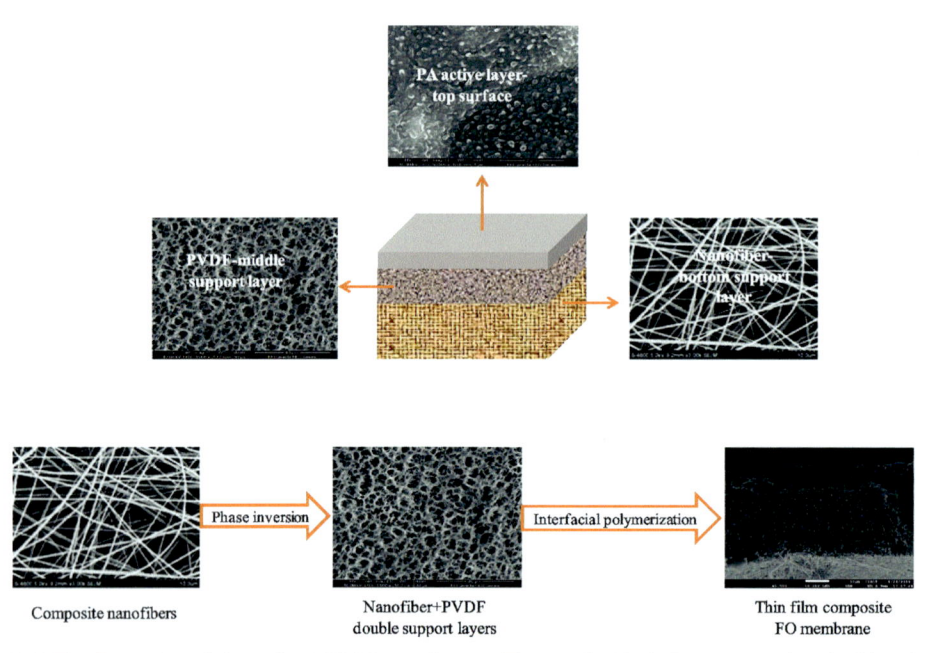

FIGURE 10-5 The formation of three tiers TFC-FO membranes (Tian et al., 2017). Source: *Reprinted with written permission from Elsevier.*

paper laminator. Under FO mode, the water flux obtained was high which is 57 L/m^{2}h when 2 M of NaCl was used, apart of an improved in term of mechanical stability. Tian et al. (2017) had fabricated a three tier FO membranes as shown in Fig. 10-5. From the figure, the membrane comprised of hydrophilic/hydrophobic polyvinyl alcohol/polyethyleneimine (PVA/PEI) electrospun nanofibers, microporous PVDF as middle support layer and PA active layer. The authors claimed that the presence of NF facilitates the mitigation of ICP which is severely seen on common porous substrate due to its interconnected pores that has low tortuosity. Nevertheless, the structure causes a very poor adhesion of the PA layer on the NF. Thus in their study, the NF acts as a sub-layer that is localized beneath porous PVDF substrate layer. It has been found out that the S value decreased when the thickness of double substrate is decreasing, even on par than the commercial FO membranes under AL-FS and AL-DS mode, the water flux obtained by the membrane were 30.6 and 22.4 L/m^{2}h respectively.

Other modification of substrate includes the use of layer by layer membrane assembly, as the one conducted by Pardeshi et al. (2017). In their study a composite polyvinyl chloride (PVC) and layered sodium hydroxide (LSHs) were used as substrate. In their case, LSHs is utilized as this material is positively charged materials as well as highly crystalline and functionalized. Increasing the LdH by 2% resulted to enhancement in term of pore size, porosity, and connectivity between sublayers. No further improvement of membrane can be seen when the amount of LdH is added higher than 2% due to formation of disturbed finger like structures. Under FO process, the water flux achieved were 50.9 (AL-DS) and 37.4 L/m^{2}h (AL-FS) when 1 M of NaCl was tested. Salehi et al. (2017) improved the hydrophilicity of TFC membranes by using carboxylic polyether sulfone as substrate. In the study, polyethersulfone was synthesized via acetylation process followed by oxidation treatment. It has been showed that using CPES as substrate led to an improvement in term of hydrophilicity due to the presence of polar groups on the surface of the membrane as well as its porous and interconnected morphological structures. As compared to the pristine membrane, the water flux of the CPES membrane under AL-FS and AL-DS mode increased to be 12.1 and 30.2 L/m^{2}h from 10.5 and 19.1 L/m^{2}h, respectively.

10.3 Draw Solutions

One of the key to the excellent performances in FO process is the suitable DS. Many research have been devoted to study the suitable DS in FO process (Zirehpour et al., 2017; Salehi et al., 2017; Akther et al., 2015). Generally, good characteristics of DS should have high water flux, minimal reverse draw solute, no toxicity, reasonably low cost, and easy to recover. In 2010, Achilli et al. proposed a protocol as illustrated in Fig. 10-6 in selection of appropriate DS (Luo et al., 2014). An ideal DS not only helps in efficiency of FO process, but also needs to have low toxicity and cost-effective (Ge et al., 2013). Besides that, draw solutes also should be easily and inexpensively separated from treated water. Therefore, an ideal DS should have the following criteria (Achilli et al., 2010):

1. The ability to generate high osmotic pressure. The osmotic pressure of DS needs to be higher that the FS to ensure positive permeate flux as the osmotic pressure difference is a driving force between DS and FS.
2. The reverse draw solute must be in minimal amount. The RSD probably occurs due to significant difference of concentration between DS and FS. Moreover, majority of FO membrane are not hypothetically semipermeable, thus, the draw solute most likely diffuse from DS tank into the FS tank (Achilli et al., 2010; Zhao et al., 2012).
3. Easy regeneration for diluted DS after FO process. Normally, FO process is integrated with another process for example RO, NF, UF and membrane distillation. On the contrary, these processes require external hydraulic pressure or heating thus consuming extra energy. Hence, easy regeneration or recovery of DS is highly desirable to minimize energy consumption and overall operation costs.
4. Small molecular weight and low viscosity in aqueous solution is favorable as a DS agent in FO process. As above-mentioned the ICP is unavoidable in FO process partially contributed by low diffusion coefficient of draw solute. The diffusion coefficient of draw solute is inversely proportional to molecular weight and solution viscosity therefore a larger molecular weight of draw solute and high viscosity have low diffusion coefficient. Hence, small molecular weight of draw solute with low viscosity is preferable in FO process.

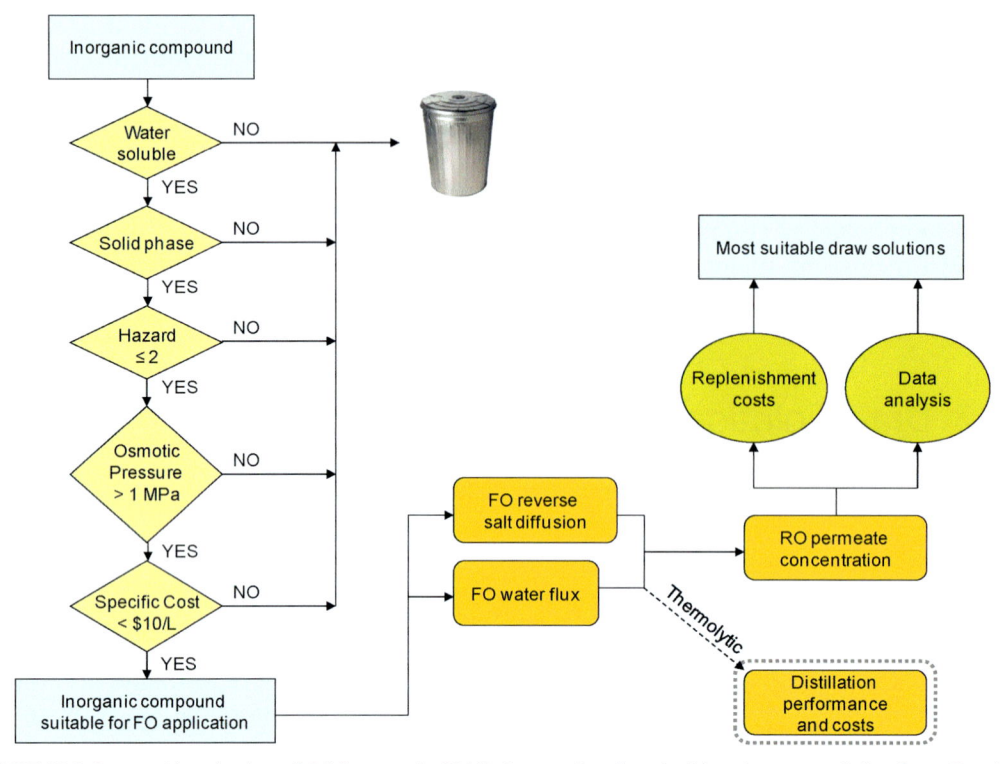

FIGURE 10-6 Protocol in selection of DS (Luo et al., 2014). Source: *Reprinted with written permission from Elsevier.*

10.3.1 Recovery Methods of Data Solution

One the major research challenge that hindering the commercialization of FO application is the efficient recovery process of DS. However, several DS such as sugar based materials (glucose and fructose) do not require recovery process specifically in food industry application. Wang et al. (2017). As the FO process does not produce pure water directly, the secondary step for recovery of DS is necessary. To address the aforementioned issue, recovery of DS is one of the important processes so that an energy-saving and affordable cost FO system can be attained. Depending on which DS has been used, majority of researchers agreed that RO, NF, and UF can be used for regeneration of DS (Luo et al., 2014) (Fig. 10-7).

10.3.1.1 Reverse Osmosis

As FO process is highly effective in the removal of wide range of contaminants such a trace organic compounds, it also helps to improve the lifespan of RO membrane through the reduction cleaning frequency. Implementation of RO as a practical approach for recovery of DS can be ascribed to high water recovery rate and high salt rejection rate. Bamaga et al. proposed a hybrid FO-RO system by using seawater and RO brine as a DS to investigate

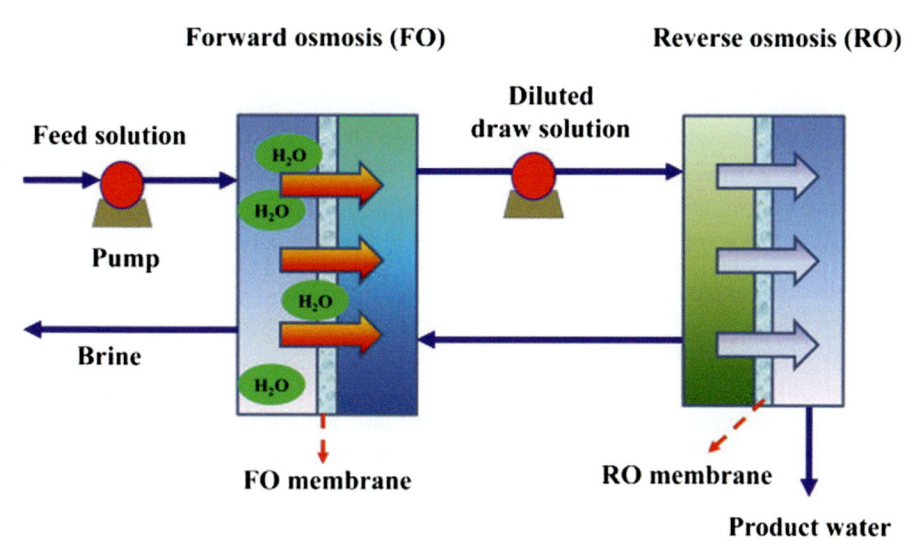

FIGURE 10-7 Schematic diagram of integrated system FO-RO process (Luo et al., 2014). Source: *Reprinted with written permission from Elsevier.*

osmotic energy recovery (Zhao et al., 2016). The experimented results showed that the water flux of FO membrane is lower and it was suggested that the FO process apply as for pretreatment prior to RO to decrease the fouling problem. Based on reported research by Cath et al. (2010), the integration of FO/RO for purification of impaired water seawater desalination (Cath et al., 2010). Firstly, the impaired water was treated with FO process with seawater as DS then the dilute DS subsequently subjected to RO units to produce clean water. The hybrid process could achieve high quality of water and economically save up to 63%. Even though RO process can have suffered with severe fouling problem (cited), the high salt rejection rate and low-molecular weight cut-off is an intrinsic property especially when it comes to filtering monovalent salts. In spite of that, FO process integrated with RO was infeasible due to this hybrid system actually consume more energy than a standalone RO process (Chun et al., 2017; Akther et al., 2015). Moreover, RO recovery process also consume more energy probably caused by their high operating pressure (Luo et al., 2014).

10.3.1.2 Nanofiltration Process

Instead of RO process used in recovery of DS, NF also coupled with FO process. NF have the properties lies between UF and RO with high multivalent salt rejection, low operating pressure and high-water flux. Works by Zhao and coworkers, proposed the FO-NF process for brackish desalination process by using Na_2SO_4 and $MgSO_4$ as DS (Zhao et al., 2012). Based on their research works, the hybrid FO-NF process offered more advantages over standalone RO such as lower operating pressure (10 bar against 30 bar), high water flux and less membrane fouling. Another reported research by Zhao conducted experiment based by using metal-EDTA salts and recovery process based on four different membranes (Zhao et al., 2016). As the metal

EDTA salts have larger molecular weight they can easily separate from water with pressure driven membrane that have a bigger pore size than RO. According to their experimental results, EDTA-ZnNa2 and EDTA-MgNa2 have solute retention more than 96% when using DL NF membrane and 98% when using TS-80 membrane. The enhanced rejection is highly influenced by the larger molecular weights possess by EDTA-ZnNa2 and EDTA-MgNa2 which smaller than any of 4 membrane that being tested with TS-80 being the smallest. However, the implementation of NF as a recovery are restricted to divalent salts and large organic molecules.

10.3.1.3 Low Pressure Process

UF process has attracted significant attention in drinking water production due to their low energy cost and high-water flux compared with NF and RO. Despite having such interesting properties, the separation of ionic species frequently hindered. In 2012, Ge et al. utilized sodium polyacrylate (PAA-Na) as a DS and regeneration process through UF process (Ge et al., 2012). Basically, larger DS can decrease the RSD and leads to higher rejection. Based on their experimental data, the results displayed higher rejection up to 99% with no PSA leakage during water recovery process. Chung and coresearchers, employed magnetic nanoparticles (MNPs) coated with polyacrylic acid (PAA) for desalination process and used UF as recovery of DS (Ling and Chung, 2011). The results showed PAA-NPs can be recycled five times with UF system without losing their osmolality and increment in their sizes. Although UF recovery process requires lower energy compared with RO and NF, they are not feasible to recover DS with lower molecular weight frequently hindered due huge pore sizes of UF process. Moreover, a larger DS have a higher tendency to increase the ICP in the membrane support layer probably caused by smaller diffusion coefficients thus resulting in declining of water flux in FO process. Representative publications related to DS and its recovery process are presented in Table 10-2.

10.4 Forward Osmosis Membranes for Heavy Metals Removal

So far, UF, NF, RO, and FO have been effectively removed wide range of metals under optimized condition. However, each of the membrane processes possesses their own limitations as tabulated in Table 10-3. For example, sole UF membrane cannot eliminate metallic ions as the membrane attained large pore which allow the ions to be easily passing along the pores (Al Mamun et al., 2015). Due to that, prior of UF process, complexing polymers or surfactants will be added into the FS so that a large complex (complexation enhanced UF) or micelles (micellar enhanced UF) will be formed and retained as retentate (Bessbousse et al., 2008; Bade and Lee, 2011). Nevertheless, both methods require detailed technique and proper selection of ligands or surfactant as they could also resulted to secondary pollutant as surfactant is toxic (Bade and Lee, 2011). Another method for UF membranes is by fabricating

Table 10-2 Summary of Different DS in FO Process and Its Regeneration Techniques

Draw Solution	Method of Recovery	Advantages	Disadvantages	References
$CuSO_4$	Precipitation method	No energy required for recovery	Cannot desalinate seawater Additional chemical required	Alnaizy et al. (2013)
NaCl	RO MD	High osmotic pressure Low cost Low viscosity	High reverse solute diffusion	Johnson et al. (2017) Achilli et al. (2010)
$CaCl_2$	Not available	High water flux	Scaling problem	Shu et al. (2016)
$MgCl_2$	Not available	High water flux High rejection for heavy metal up to 97% Low reverse solute diffusion	High ICP in AL-FS orientation	Liu et al. (2017)
$FeSO_4$	Precipitation method	High solubility in water Abundantly available in the form of coquimbite Pure product of water can be recovered without extra energy	Slowly dissolve in water Aqueous solution of FeSO4 is acidic	Qasim et al. (2017)
NaCl, $LaCl_3$, $MgCl_2$	NF	High osmotic pressure	High reverse solute flux	Kim et al. (2018)
EDTA-ZnNa2 EDTA-MgNa2	NF	High salt retention up to 99% Low energy consumption	Expensive materials	Zhao et al. (2016)
PAA-Na	NF	High water flux Low solute diffusion	Expensive materials	
MNPs -PAA	UF	High recovery process Low operating pressure	Low salt rejection especially for low MWs DS	Ling and Chung (2011)
Organic Salts	NF	Biodegradable	Low water flux	Bowden et al. (2012)
Sugar based	Not available	Large molecule Low operating pressure	Low osmotic pressure	Hamdan et al. (2015)

adsorptive UF mixed matrix membrane in which large amount of adsorbents is added into porous UF membranes that act as a host media for the adsorbents. Despite many studies have been conducted about the adsorptive UF MMMs, membrane leaching as consequence of too high amount of adsorbent added and high fouling tendency have become its major issues (Abdullah et al., 2016, 2018). For the case of NF and RO, various studies have reported the efficiency of NF and RO for heavy metal removal by means of steric size exclusion and Donnan exclusion (Barakat, 2011). Nevertheless, both membrane processes require high

Table 10-3   General Comparison of Membrane Processes for Heavy Metal Removal

Description	UF	NF	RO	FO
Type of membrane	Porous UF membranes aided with ligand (complexation enhanced UF) or micelle (micellar enhanced UF)	TFC-NF membranes	TFC-RO membranes	TFC-FO membranes
Requirement of pressure	Low pressure (0−5 bar)	Intermediate pressure (Up 10 bar)	High pressure (up to 50 bar)	Natural osmotic pressure
Rejection of metals	Low concentration of metal	Up to 100−1000 ppm	Up to 100−1000 ppm	Up to 2000 ppm
Mechanism of rejection	Size-exclusion mechanism, adsorption	Size-exclusion, Donnan-exclusion mechanism	Size exclusion mechanism	Osmosis diffusivity
Disadvantages	Large pore size, fouling, generation of secondary pollutants	High energy requirement, fouling	High energy requirement, fouling	Selection of proper DS potential of concentration polarization

energy operation and has high fouling tendency, which resulted in reduced productivity and extra operational cost.

On the other hand, FO has drawn attention of many scientific communities as a potential candidate for heavy metals separation. Looking at the properties of FO membranes which is nearly similar to NF and RO (comprised of substrate and selective thin film layer except does not utilized hydraulic pressure), FO can be potentially for this particular application.

The removal of metals by FO membranes is shown in Table 10-4. Based on the table, the study of heavy metals removal was first demonstrated 2012 by Jin et al. whom studied the removal of boron and arsenic by studying the influence of membrane orientation towards metals rejection. Their study revealed that 58% and 10% rejection of boron and 90% and 50% rejection of arsenic were obtained under AL-FS and AL-DS mode. As compared to AL-FS orientation, the FO process by AL-DS mode revealed lower rejection and the gap of rejection is significant between both modes. One of the major reasons is due to the occurrence of ICP where there is solutes build up within porous substrate layer. Consequently, higher solute concentration gradient that led to lower metal rejection was observed. This study also focused on the comparison performance of metal rejection by FO and RO. Their results demonstrated that the rejection of As and Br by FO overpassed the performance of RO with higher membrane flux was seen on FO process. As both FO and RO membrane are TFN membrane, the nonwoven structure present underneath the FO membrane substrate allows expansion of membrane pores to allow water molecules to pass through.

Table 10-4 Removal of Heavy Metals by FO Membranes

Type of Membrane	Membrane Orientation	Contact Angle (°)	Type of Metal/ Feed Solution	Concentration of FS (mg/L)	Draw Solution	Concentration of DS (M)	Water Flux (L/ m²h)	Rejection (%)	References
Commercial membrane	AL-FS	NS	Boron	10	NaCl	5	7	58	Jin et al. (2012)
	AL-DS							10	
	AL-FS		As	10	NaCl	1		90	
	AL-DS							50	
Commercial membrane	AL-FS	45	Pb	10	NaCl	1	2.4	99.9	Butler et al. (2013)
			As	10	NaCl	1		99.5	
			Cr	10	NaCl	1		99.5	
PA/polyimide	AL-FS	NS	Pb	2000	Na-Co-CA	1	11	99.6	Cui et al. (2014)
			Cd	2000	Na-Co-CA	1		99.6	
			Cu	2000	Na-Co-CA	1		99.6	
			Hg	2000	Na-Co-CA	1		99.6	
Commercial membrane	AL-FS	67.15	As	150	Glucose	NS	2.5	90	Mondal et al. (2014)
Commercial membrane	AL-FS	NS	Hg	0.02	NaCl	1	8	98.2	Wu et al. (2016)
Commercial membrane	AL-FS	76	Ni	100	MgCl$_2$	1	5.8	99.9	Zhao et al. (2016)
	AL-DS						98.98	99.65	
CTA-TFC	AL-FS	45	Ni	100	NaCl	2	145	98.77	
	AL-DS							98.15	
Commercial membrane	AL-FS	52.15	Co	20	NaCl	1	14.8	98.6	Liu et al. (2017)
	DS-FS			20		1	21	93.2	
CTA/polyester	AL-FS	70.7	Co	20	NaCl	1	23.4	99.4	
	DS-FS			20		1	15.5	95	
CTA/polyester with nonwoven support	AL-FS	79.1	Co	20	NaCl	1	15	98.04	
	DS-FS			20		1	6	87.6	

(Continued)

Table 10-4 (Continued)

Type of Membrane	Membrane Orientation	Contact Angle (°)	Type of Metal/ Feed Solution	Concentration of FS (mg/L)	Draw Solution	Concentration of DS (M)	Water Flux (L/ m^2h)	Rejection (%)	References
L-b-L PEI/Sodium alginate on PDA functionalized PVDF substrate	AL-FS	NS	Ni	5	$MgCl_2$	1	15	99.8	Liu et al. (2017)
	AL-DS						25	97.75	
	AL-FS		Cu	5	$MgCl_2$	1	14	99.9	
	AL-DS						24	98.7	
	AL-FS		Pb	5	$MgCl_2$	1	15	99.7	
	AL-DS						23	96	
	AL-FS		Zn	5	$MgCl_2$	1	16	99.7	
	AL-DS						27	98.5	
	AL-FS		Cd	5	$MgCl_2$	1	16	99.6	
	AL-DS						26	96.2	
Silica thin film inorganic on standard steel mesh	AL-FS	NS	Cd	200	NaCl	2	69	94	You et al. (2017)
			Pb	200	NaCl	2		93	
			Cu	200	NaCl	2		92.5	
			Zn	200	NaCl	2		93	
Commercial membrane	AL-FS	36	Co	2.33	NaCl	1	12	99	Vital et al. (2018)
			Cu	615	NaCl	1		98	
			Zn	68.5	NaCl	1		99.9	
BSA embedded PA on glass nanofiber as substrate	AL-FS	50	Cu	2	NaCl	2	50	99.5	Zhao and Liu (2018)
			Pb	2	NaCl	2	47	98.9	
			Cd	2	NaCl	2	45	98.5	

NS, not stated.

Butler et al. (2013) evaluated FO system from Hydration Technology Innovation (HTI) in which the membrane module is in spiral wound form contained membrane comprised of cellulose triacetate with nonwoven mesh structures. The viability of the system for potential of point-of-use-water-treatment (PoUWT) technology (technology used during water crisis) by studying the removal of Pb(II) under AL-FS mode was studied. In order to be classified as PoUWT technology, there are few criteria need to be complied which are production capacity, contaminant removal/deactivation, process economics, operations and maintenance and materials availability. In their study, evaluation was made in term of inorganic ions removal in which comparison was conducted between FO and RO. Similar performance was obtained for rejection of Pb(II) was found between FO and RO which is more than 98% when 1 M of NaCl was used. Despite of its technical cost, the capabilities of HTI's system in term of contaminant removal surpassed existing PoUWT technologies as potable water can be directly produced from heavily contaminated water. Improvement to the current system to minimize the capital cost could benefited FO as a promising PoUWT technology.

Cui et al. (2014) later conducted a much thorough study on metal removal by FO process. In their study, they prepared TFC on a macrovoid-free polyimide support (commercial Matrimid substrate) via interfacial polymerization. Using novel hydroacid complex $Na_4[Co(C_6H_4O_7)_2]\bullet2H_2O$ (referred to as Na−Co−CA) as the DS, the study had successfully removed Pb, Cr, As, Cd, and Cu and other heavy metals with rejection of 99.5% and the concentration of heavy metal tested at 2000 mg/L with water flux of 16.5 L/m^{2}h. When more concentrated draw solute (1.5 M) and FS (5000 mg/L) was introduced, high rejection of metal (99.5%) was maintained. The selection of Na-CO-CA as DS was made due to the cobaltic complex performs better than NaCl. In addition to that, Na-Co-CA has been found to have lower RSD, higher water flux, lower production of total dissolved solids in the FS, lower production cost. As compared to the previous studies, they comparison study of performance was conducted between FO and NF had also been conducted and the results are demonstrated in Table 10-5. Since that FO do not require any hydraulic pressure, thus, the mechanism of metal rejection is solution-diffusion mechanism. Therefore, the heavy metals ions that attains larger hydrated radii will be retained from diffused across the membrane.

Mondal et.al (2014) studied for the removal of As groundwater with the presence of by using commercial FO membrane. They found out that using glucose as DS caused rejection about 90% of As from FS. Nevertheless, the rejection of As is found to be reduced by about 10% when Ca^{2+} and Mg^{2+} (represent hardness in water), sulfate, phosphate, and silicate present. This phenomenon could be explained due to the competition between these ions for being rejected by FO membrane. In contrast to Jin et al. (2012) whom stated that the presence of humic acid as fouling agent is likely to reduce the performance of FO process, the result of humic acid presence in this study is in contradict to the former study. The result could be ascribed due to two reasons: (1) humic acid which caused gel-like fouling layer contributes to higher negative charge on membrane's surface that simultaneously improved As rejection via electrostatic interaction, and (2) there are strong repulsion between the hydroxyl and carbonyl group in humic acid layer improved the rejection of As ions. Nevertheless, despite improved As rejection was seen, the water flux of the membrane is

Table 10-5 Comparison of Heavy Metal Removal Efficiency in FO and NF Processes (Cui et al., 2014)

	Solute Rejection		
	Nanofiltration (%)		Forward Osmosis (%)
	Commercial NF Membrane	Lab Fabricated TFC Membrane	Lab Fabricated TFC Membrane
$Na_2Cr_2O_7$	8.93 ± 1.34	81.50 ± 3.60	99.87 ± 0.18
Na_2HAsO_4	60.69 ± 1.04	93.86 ± 0.49	99.74 ± 0.04
$Pb(NO_3)_2$	24.72 ± 3.36	57.46 ± 4.06	99.37 ± 0.43
$CdCl_2$	12.37 ± 2.19	75.89 ± 1.05	99.9 ± 0
$CuSO_4$	83.21 ± 3.14	96.10 ± 1.88	99.78 ± 0.03
$Hg(NO_3)_2$	28.95 ± 3.94	51.34 ± 3.05	99.77 ± 0.16

significantly dropped attributed to the fouling that hindered the diffusion of water molecules across the membrane.

Meanwhile, Wu et al. (2016) attempted to remove the trace amount of mercury from wastewater by using commercial membrane. At all tested pH, mercury removal was found to be consistent, approximately 92%. In addition, increasing the pH also resulted to an increased water flux. Nevertheless, the lowest RSD was observed at neutral pH. The behavior of the membrane can be explained by few factors such as altered membrane hydrophobicity induced by pH solution which indirectly affects the permeability of Hg ions. Other than that, the functional groups present on surface of membrane may also contribute to electrostatic repulsion which opening membrane pores and allowing water molecules to pass through. This study also give a comprehensive comparison between the feasibility of univalent (NaCl) and multivalent ($MgCl_2$) as DS in the FO process. When tested using $MgCl_2$ and NaCl, $MgCl_2$ had shown that it is more effective than NaCl where 98% rejection was achieved. As NaCl possess smaller hydrated radii than $MgCl_2$, it could have diffused through FO membrane, lead to severe reverse flux. $MgCl_2$ on the other hand, enhance the Donnan effect of FO membrane as well as the osmotic gradient. However, one of the major limitation of $MgCl_2$ as DS solution is it may lead to the formation of complex in the matrix membrane.

On the other hand, Zhao et al. (2016) investigated the removal of Ni(II) from high salinity wastewater. The removal was carried out by comparing the performance of commercial cellulose triacetate (CTA) FO membrane and polyamide-based TFC FO membrane. The commercial CTA FO membrane attained asymmetrical membrane embedded with non-woven polyester mesh while the TFC FO membranes attained of the presence of selective layer on top of substrate. By comparing both CTA-FO and commercial membrane, the rejection of Ni was found to not be significantly different, ranging from 98%–99%, except the water flux obtained for CTA-FO is slightly lower that the PA-based TFC FO.

Liu et al. (2017) on the other hand prepared layer-by-layer (L-b-L) membrane assemblies of PEI and sodium alginate (SA) bilayers on polydopamine (PDA) functionalized- PVDF support for the removal of Cu(II), Ni(II), Pb(II), Zn(II), and Cd(II). The use of L-b-L is possible as heavy metals usually exist in divalent forms, and thus possess large hydrated radii which can be rejected by this kind of membrane. Their study has successfully rejected the five metals with the rejection rate of more than 95.9% and 99.3% with water flux of 23.5 and 14 L/m^{2}h under PRO and FO modes when 1.0 M of MgCl$_2$ was used. The mechanism of removal of the membrane is illustrated by Fig. 10-4. Despite the rejection of metals is satisfactory, it should be noted that the L-b-L membranes might be unable to remove monovalent ions due to their smaller hydrated radii. The mechanism of removal of this membrane is illustrated in Fig. 10-8. Referring to the figure, the removal of heavy metals is happened via two possible mechanisms. First, the decreased diffusivity caused metal ions to be easily rejection via solution-diffusion mechanism in which water will be diffused to the DS while the metal solution will be concentrated. Second, the prepared membranes possessed amine groups on its membrane's surface that attributed to the positive charge of the membrane. As a result, electrostatic repulsive forces are generated to repulse the incoming metal ions.

Another study conducted by You et al. (2017) investigated the rejection of Cu^{2+}, Cd^{2+}, Pb^{2+}, and Zn^{2+} by using nanoporous thin film inorganic prepared by tetraethylorthosilicate-driven sol gel process. During FO process in which NaCl served as the DS, all of the metals have been successfully rejected up to 94% at the FS of 200 mg/L. The membrane also yielded higher water flux which was 69.0 L/m^2.

Two of the latest studies are from Vital et al. (2018) and Zhao and Liu (2018). Vital et.al investigating the feasibility of commercial FO membranes for the removal of Co(II), Cu(II) and Zn(II) from acid mine drainage. Their study showed that the FO process successfully removed the previously mentioned heavy metals with rejection rate of 99.0%, 98.0% and 99.9%, respectively by using 1.0 M NaCl as the DS. On the other hand, Zhao and Liu (2018)

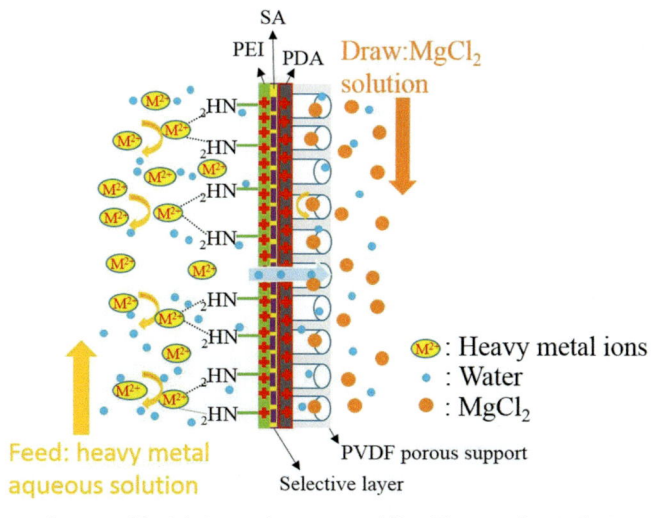

FIGURE 10-8 Mechanisms of removal by L-b-L membrane assemblies (Zhao et al., 2012). Source: *Reprinted with written permission from Elsevier.*

conducted an interesting study in which they prepared a novel TFC-FO membranes comprised of glass nanofiber as substrate layer and PA-embedded BSA as the active layer. Results as displayed in Fig. 10-9 indicated that increasing the amount of BSA resulted to granular protrusion on the surface (Fig. 10-9A). When comparing to PVDF and PES, it can be clearly seen that the water flux of the new TFC-FO membranes is superior due to smaller turtousity factor of the nanofiber substrate (Fig. 10-9B). During FO process, the nanofiber provides sufficient channels for the solution while decreasing the mass transfer resistance. On the other hand, the BSA molecules present on the PA layer creates hydrophilic channel for water molecules to easily diffused via dissolution-diffusion process while the metal ions is retained in the FS tank (Fig. 10-9C).

10.5 Issue and Challenges of Heavy Metal Removal by Forward Osmosis Process

Application of FO membranes offers tremendous advantage in the field of heavy metals removal. Recent advancement makes it realistic to support that FO might be one of the most

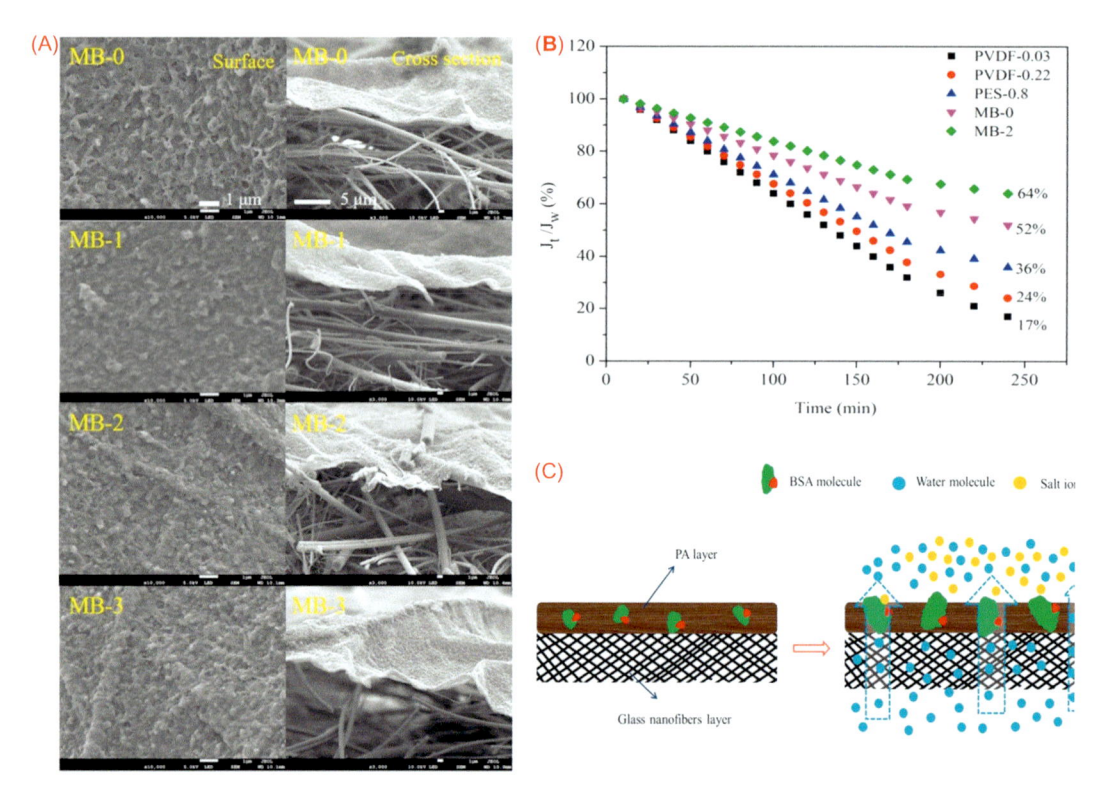

FIGURE 10-9 (A) Morphological structures of membrane; (B) water flux obtained as compared to commercial membranes; (C) mechanism of metal removal by membrane (Goh and Ismail, 2017). Source: *Reprinted with written permission from Elsevier.*

reliable technologies to lower the energy consumption as compared to pressure driven membrane processes such as RO and NF. However, several critical issues still remains as major constraints for FO to be as a feasible method for heavy metals removal since FO technology still is in its infancy stage and the applications explored just at a laboratory scale. One of the most critical challenges which may hinder their further development is associated to the regeneration step needed in FO system which requires additional energy, which in some cases may push the total energy costs above that of alternatives, such as RO or membrane distillation (MD) (Johnson et al., 2017).

Another critical challenge is that, if the FO process is to compete with other more established membrane processes in term of heavy metals removal, it requires higher flux rates (Goh and Ismail, 2017). In order to achieve this, critical selection of DS capable of producing high osmotic pressures to drive the membrane flux is needed. Selection of an appropriate DS is the key to efficient and cost effective operation of FO. An effective DS must be abundant, be capable of generating a high osmotic potential, have low RSD, be nontoxic and noncorrosive, inexpensive and easy to regenerate (Johnson et al., 2017). The modification on thin film active layer and substrates also was found to improve the hydrophilicity and lowering the reverse flux and thus increasing the water flux (Wu et al., 2017).

In addition to that, ineffective membranes and reverse solute leakage also among the challenges hindering the growth of FO-wastewater applications (Lutchmiah et al., 2014). The use of hybrid systems, that is, FO-MD, FO-RO, FO-NF and/or integrated with seawater desalination system can enhance the overall performance of FO and thus lead it to be more feasible for wastewater recovery. However, as compared to the economic benefit, the overall energy balances for integrated systems are still lacking. A better understanding of these concepts will further promote the use of this technology to allow it to be comparable to the existing techniques for heavy metals removal. Nevertheless, since waste streams contain high concentrations of heavy metals, proper concentrated disposal must be deliberated well. Energy-efficient pretreatment of the FS is an option. Higher quality water is in demand, due to the imposition of new and ever-changing water quality standards. Therefore interest in FO technology is growing as a potential, cost-competitive and reliable alternative, and can stand as a versatile option, comparable to the existing techniques.

10.6 Conclusion

This chapter described the heavy metals removal by means of FO process. Development of FO membrane relies on the rate of ICP, where modification on either the substrate or PA layer is necessary to maintain an efficient performance of membrane in FO process. Different type of DS and recovery system of DS is also play major role in defining FO as an excellent integrated system. In term of heavy metals removal, literature findings showed that FO exhibits very promising results, comparable to existing techniques. However, further improvements and studies are needed to enhance the overall performance of FO process.

References

Abdullah, N., Yusof, N., Jaafar, J., Ismail, A.F., Hasbullah, H., Othman, F.E., et al., 2016. Preparation of polyacrylonitrile (PAN)/manganese oxide based activated carbon nanofiber (ACNFs) for adsorption of cadmium (II) from aqueous solution, IOP Conf Ser. Environ. Sci. 36, 1−6.

Abdullah, N., Gohari, R.J., Yusof, N., Ismail, A.F., Juhana, J., Lau, W.J., et al., 2016. Polysulfone/hydrous ferric oxide ultrafiltration mixed matrix membrane: preparation, characterization and its adsorptive removal of lead (II) from aqueous solution. Chem. Eng. J. 289, 28−37.

Abdullah, N., Yusof, N., Gohari, R.J., Ismail, A.F., Jaafar, J., Lau, W.J., et al., 2018. Characterizations of polysulfone/ferrihydrite mixed matrix membranes for water/wastewater treatment. Water Environ. Res. 90, 64−73. Available from: https://doi.org/10.2175/106143017X15054988926541.

Achilli, A., Cath, T.Y., Childress, A.E., 2010. Selection of inorganic-based draw solutions for forward osmosis applications. J. Memb. Sci. 364, 233−241. Available from: https://doi.org/10.1016/j.memsci.2010.08.010.

Akther, N., Sodiq, A., Giwa, A., Daer, S., Arafat, H.A., Hasan, S.W., 2015. Recent advancements in forward osmosis desalination: a review. Chem. Eng. J. 281, 502−522. Available from: https://doi.org/10.1016/j.cej.2015.05.080.

Al Mamun, A., Ahmed, Y.M., AlKhatib, M.F.R., Jameel, A.T., AlSaadi, M.A.H.A.R., 2015. Lead sorption by carbon nanofibers grown on powdered activated carbon—kinetics and equilibrium. Nano. 10, 1550017. Available from: https://doi.org/10.1142/S1793292015500174.

Alnaizy, R., Aidan, A., Qasim, M., 2013. Copper sulfate as draw solute in forward osmosis desalination. J. Environ. Chem. Eng. 1, 424−430. Available from: https://doi.org/10.1016/j.jece.2013.06.005.

Amini, M., Jahanshahi, M., Rahimpour, A., 2013. Synthesis of novel thin film nanocomposite (TFN) forward osmosis membranes using functionalized multi-walled carbon nanotubes. J. Memb. Sci. 435, 233−241. Available from: https://doi.org/10.1016/j.memsci.2013.01.041.

Bade, R., Lee, S.H., 2011. A review of studies on micellar enhanced ultrafiltration for heavy metals removal from wastewater. J. Water Sustain. 1, 85−102.

Barakat, M., 2011. New trends in removing heavy metals from industrial wastewater. Arab. J. Chem. 4, 361−377. Available from: https://doi.org/10.1016/j.arabjc.2010.07.019.

Bessbousse, H., Rhlalou, T., Verchère, J.F., Lebrun, L., 2008. Removal of heavy metal ions from aqueous solutions by filtration with a novel complexing membrane containing poly(ethyleneimine) in a poly(vinyl alcohol) matrix. J. Memb. Sci. 307, 249−259. Available from: https://doi.org/10.1016/j.memsci.2007.09.027.

Bowden, K.S., Achilli, A., Childress, A.E., 2012. Organic ionic salt draw solutions for osmotic membrane bioreactors. Bioresour. Technol. 122, 207−216. Available from: https://doi.org/10.1016/j.biortech.2012.06.026.

Butler, E., Silva, A., Horton, K., Rom, Z., Chwatko, M., Havasov, A., et al., 2013. Point of use water treatment with forward osmosis for emergency relief. Desalination 312, 23−30. Available from: https://doi.org/10.1016/j.desal.2012.12.013.

Cath, T.Y., Childress, A.E., Elimelech, M., 2006. Forward osmosis: principles, applications, and recent developments. J. Memb. Sci. 281, 70−87. Available from: https://doi.org/10.1016/j.memsci.2006.05.048.

Cath, T.Y., Hancock, N.T., Lundin, C.D., Hoppe-Jones, C., Drewes, J.E., 2010. A multi-barrier osmotic dilution process for simultaneous desalination and purification of impaired water. J. Memb. Sci. 362, 417−426. Available from: https://doi.org/10.1016/j.memsci.2010.06.056.

Choi, H., Son, M., Choi, H., 2017. Integrating seawater desalination and wastewater reclamation forward osmosis process using thin-film composite mixed matrix membrane with functionalized carbon nanotube blended polyethersulfone support layer. Chemosphere. 185, 1181−1188. Available from: https://doi.org/10.1016/J.CHEMOSPHERE.2017.06.136.

Chun, Y., Mulcahy, D., Zou, L., Kim, I.S., 2017. A. short review of membrane fouling in forward osmosis processes. Membranes (Basel) 7. Available from: https://doi.org/10.3390/membranes7020030.

Cui, Y., Ge, Q., Liu, X., Chung, T., 2014. Novel forward osmosis process to effectively remove heavy metal ions. J. Memb. Sci. 467, 188−194. Available from: https://doi.org/10.1016/j.memsci.2014.05.034.

Cui, Y., Ge, Q., Liu, X.Y., Chung, T.S., 2014. Novel forward osmosis process to effectively remove heavy metal ions. J. Memb. Sci. 467, 188−194. Available from: https://doi.org/10.1016/j.memsci.2014.05.034.

Emadzadeh, D., Lau, W.J., Matsuura, T., Rahbari-Sisakht, M., Ismail, A.F., 2014. A novel thin film composite forward osmosis membrane prepared from PSf-TiO_2 nanocomposite substrate for water desalination. Chem. Eng. J. 237, 70−80. Available from: https://doi.org/10.1016/j.cej.2013.09.081.

Faria, A.F., Liu, C., Xie, M., Perreault, F., Nghiem, L.D., Ma, J., et al., 2017. Thin-film composite forward osmosis membranes functionalized with graphene oxide−silver nanocomposites for biofouling control. J. Memb. Sci. 525, 146−156. Available from: https://doi.org/10.1016/j.memsci.2016.10.040.

Favre, E., 2010. Comprehensive Membrane Science and Engineering, 2010. doi:10.1016/B978-0-08-093250-7.00020-7.

Ge, Q., Su, J., Amy, G.L., Chung, T.S., 2012. Exploration of polyelectrolytes as draw solutes in forward osmosis processes. Water Res. 46, 1318−1326. Available from: https://doi.org/10.1016/j.watres.2011.12.043.

Ge, Q., Ling, M., Chung, T.S., 2013. Draw solutions for forward osmosis processes: developments, challenges, and prospects for the future. J. Memb. Sci. 442, 225−237. Available from: https://doi.org/10.1016/j.memsci.2013.03.046.

Goh, P.S., Ismail, A.F., 2017. A review on inorganic membranes for desalination and wastewater treatment. Desalination. 0−1. Available from: https://doi.org/10.1016/j.desal.2017.07.023.

Hamdan, M., Sharif, A.O., Derwish, G., Al-Aibi, S., Altaee, A., 2015. Draw solutions for forward osmosis process: osmotic pressure of binary and ternary aqueous solutions of magnesium chloride, sodium chloride, sucrose and maltose. J. Food Eng. 155, 10−15. Available from: https://doi.org/10.1016/j.jfoodeng.2015.01.010.

Jin, X., She, Q., Ang, X., Tang, C.Y., 2012. Removal of boron and arsenic by forward osmosis membrane : influence of membrane orientation and organic fouling. J. Memb. Sci. 389, 182−187. Available from: https://doi.org/10.1016/j.memsci.2011.10.028.

Johnson, D.J., Suwaileh, W.A., Mohammed, A.W., Hilal, N., 2017. Osmotic's potential: an overview of draw solutes for forward osmosis. Desalination. 0−1. Available from: https://doi.org/10.1016/j.desal.2017.09.017.

Kim, D.I., Choi, J., Hong, S., 2018. Evaluation on suitability of osmotic dewatering through forward osmosis (FO) for xylose concentration. Sep. Purif. Technol. 191, 225−232. Available from: https://doi.org/10.1016/j.seppur.2017.09.036.

Lau, W.J., Ismail, A.F., Misdan, N., Kassim, M.A., 2012. A recent progress in thin film composite membrane: a review. Desalination. 287, 190−199. Available from: https://doi.org/10.1016/j.desal.2011.04.004.

Lee, K.L., Baker, R.W., Lonsdale, H.K., 1981. Membranes for power generation by pressure-retarded osmosis. J. Memb. Sci. 8, 141−171. Available from: https://doi.org/10.1016/S0376-7388(00)82088-8.

Ling, M.M., Chung, T.S., 2011. Desalination process using super hydrophilic nanoparticles via forward osmosis integrated with ultrafiltration regeneration. Desalination. 278, 194−202. Available from: https://doi.org/10.1016/j.desal.2011.05.019.

Liu, C., Lei, X., Wang, L., Jia, J., Liang, X., Zhao, X., et al., 2017. Investigation on the removal performances of heavy metal ions with the layer-by-layer assembled forward osmosis membranes. Chem. Eng. J. 327, 60−70. Available from: https://doi.org/10.1016/j.cej.2017.06.070.

Liu, X., Wu, J., Liu, C., Wang, J., 2017. Removal of cobalt ions from aqueous solution by forward osmosis. Sep. Purif. Technol. 177, 8−20. Available from: https://doi.org/10.1016/j.seppur.2016.12.025.

Loeb, S., 1976. Production of energy from concentrated brines by pressure-retarded osmosis. J. Memb. Sci. 1, 49−63. Available from: https://doi.org/10.1016/S0376-7388(00)82257-7.

Luo, H., Wang, Q., Zhang, T.C., Tao, T., Zhou, A., Chen, L., et al., 2014. A review on the recovery methods of draw solutes in forward osmosis. J. Water Process Eng. 4, 212−223. Available from: https://doi.org/10.1016/j.jwpe.2014.10.006.

Lutchmiah, K., Verliefde, A.R.D., Roest, K., Rietveld, L.C., Cornelissen, E.R., 2014. Forward osmosis for application in wastewater treatment: a review. Water Res. 58, 179−197. Available from: https://doi.org/10.1016/j.watres.2014.03.045.

Ma, N., Wei, J., Liao, R., Tang, C.Y., 2012. Zeolite-polyamide thin film nanocomposite membranes: Towards enhanced performance for forward osmosis. J. Memb. Sci. 405−406, 149−157. Available from: https://doi.org/10.1016/j.memsci.2012.03.002.

Mccutcheon, J.R., Elimelech, M., 2006. Influence of concentrative and dilutive internal concentration polarization on flux behavior in forward osmosis. J. Memb. Sci. 284, 237−247. Available from: https://doi.org/10.1016/j.memsci.2006.07.049.

Mondal, P., Thi, A., Tran, K., Van Der Bruggen, B., 2014. Removal of As (V) from simulated groundwater using forward osmosis : effect of competing and coexisting solutes. Desalination. 348, 33−38. Available from: https://doi.org/10.1016/j.desal.2014.06.001.

Morales-Torres, S., Esteves, C.M.P., Figueiredo, J.L., Silva, A.M.T., 2016. Thin-film composite forward osmosis membranes based on polysulfone supports blended with nanostructured carbon materials. J Memb Sci 520, 326−336. Available from: https://doi.org/10.1016/j.memsci.2016.07.009.

Pan, Y.H., Zhao, Q.Y., Gu, L., Wu, Q.Y., 2017. Thin film nanocomposite membranes based on imologite nanotubes blended substrates for forward osmosis desalination. Desalination. 421, 160−168. Available from: https://doi.org/10.1016/j.desal.2017.04.019.

Pardeshi, P.M., Mungray, A.K., Mungray, A.A., 2017. Polyvinyl chloride and layered double hydroxide composite as a novel substrate material for the forward osmosis membrane. Desalination. 421, 149−159. Available from: https://doi.org/10.1016/j.desal.2017.01.041.

Park, M.J., Gonzales, R.R., Abdel-Wahab, A., Phuntsho, S., Shon, H.K., 2018. Hydrophilic polyvinyl alcohol coating on hydrophobic electrospun nanofiber membrane for high performance thin film composite forward osmosis membrane. Desalination. 426, 50−59. Available from: https://doi.org/10.1016/J.DESAL.2017.10.042.

Parmar, M., Thakur, L.S., 2013. Review Article Heavy Metal Cu, Ni And Zn : Toxicity, Health Hazards and Their Removal Techniques by Low Cost Adsorbents : A Short Overview 1 M. Tech Research Scholar, Department of Chemical Engineering, Ujjain Engineering College, Ujjain, pp. 143−157.

Qasim, M., Mohammed, F., Aidan, A., Darwish, N.A., 2017. Forward osmosis desalination using ferric sulfate draw solute. Desalination. 423, 12−20. Available from: https://doi.org/10.1016/j.desal.2017.08.019.

Salehi, H., Shakeri, A., Rastgar, M., 2017. Carboxylic polyethersulfone: a novel pH-responsive modifier in support layer of forward osmosis membrane. J. Memb. Sci. Available from: https://doi.org/10.1016/j.memsci.2017.10.044.

Sardar, S.A. Kamran, Ali, Shafaqat, Samra Hameed, Zal, Fatima, M.B.S. Samar, Saima Aslam Bharwana, H.M. T., 2013. Heavy metals contamination and what are the impacts on living organisms. Greener J. Environ. Manag. Public Saf. 2, 172−179.

Shaffer, D.L., Werber, J.R., Jaramillo, H., Lin, S., Elimelech, M., 2015. Forward osmosis: where are we now? Desalination. 356, 271−284. Available from: https://doi.org/10.1016/j.desal.2014.10.031.

Shakeri, A., Salehi, H., Rastgar, M., 2017. Chitosan-based thin active layer membrane for forward osmosis desalination. Carbohydr. Polym. 174, 658−668. Available from: https://doi.org/10.1016/j.carbpol.2017.06.104.

Shen, L., Zuo, J., Wang, Y., 2017. Tris(2-aminoethyl)amine in-situ modified thin-film composite membranes for forward osmosis applications. J. Memb. Sci. 537, 186−201. Available from: https://doi.org/10.1016/j.memsci.2017.05.035.

Shokrollahzadeh, S., Tajik, S., 2018. Fabrication of thin film composite forward osmosis membrane using electrospun polysulfone/polyacrylonitrile blend nanofibers as porous substrate. Desalination 425, 68−76. Available from: https://doi.org/10.1016/j.desal.2017.10.017.

Shu, L., Obagbemi, I.J., Liyanaarachchi, S., Navaratna, D., Parthasarathy, R., Ben Aim, R., et al., 2016. Why does pH increase with CaCl$_2$ as draw solution during forward osmosis filtration. Process Saf. Environ. Prot. 104, 465−471. Available from: https://doi.org/10.1016/j.psep.2016.06.007.

Tchounwou, P.B., Yedjou, C.G., Patlolla, A.K., Sutton, D.J., 2012. Heavy Metals Toxicity and the Environment. NCBI 101, 1−30. Available from: https://doi.org/10.1007/978-3-7643-8340-4.

Tian, E., Wang, X., Zhao, Y., Ren, Y., 2017. Middle support layer formation and structure in relation to performance of three-tier thin film composite forward osmosis membrane. Desalination 421, 190−201. Available from: https://doi.org/10.1016/j.desal.2017.02.014.

Vital, B., Bartacek, J., Ortega-bravo, J.C., Jeison, D., 2018. Treatment of acid mine drainage by forward osmosis : heavy metal rejection and reverse fl ux of draw solution constituents. Chem. Eng. J. 332, 85−91. Available from: https://doi.org/10.1016/j.cej.2017.09.034.

Wang, Y.N., Goh, K., Li, X., Setiawan, L., Wang, R., 2017. Membranes and processes for forward osmosis-based desalination: recent advances and future prospects. Desalination. 0−1. Available from: https://doi.org/10.1016/j.desal.2017.10.028.

Wu, C., Mouri, H., Chen, S., Zhang, D., Koga, M., 2016. Journal of water process engineering removal of trace-amount mercury from wastewater by forward osmosis. J. Water Process Eng. 14, 108−116. Available from: https://doi.org/10.1016/j.jwpe.2016.10.010.

Wu, X., Field, R.W., Wu, J.J., Zhang, K., 2017. Polyvinylpyrrolidone modified graphene oxide as a modifier for thin film composite forward osmosis membranes. J. Memb. Sci. 540, 251−260. Available from: https://doi.org/10.1016/j.memsci.2017.06.070.

Xiong, S., Zuo, J., Ma, Y.G., Liu, L., Wu, H., Wang, Y., 2016. Novel thin film composite forward osmosis membrane of enhanced water flux and anti-fouling property with N-[3-(trimethoxysilyl) propyl] ethylenediamine incorporated. J. Memb. Sci. 520, 400−414. Available from: https://doi.org/10.1016/j.memsci.2016.07.034.

Xu, W., Chen, Q., Ge, Q., 2017. Recent advances in forward osmosis (FO) membrane: chemical modifications on membranes for FO processes. Desalination 419, 101−116. Available from: https://doi.org/10.1016/j.desal.2017.06.007.

Yang, Q., Lei, J., Sun, D.D., Chen, D., 2015. Forward osmosis membranes for water reclamation. Sep. Purif. Rev. 45, 93−107. Available from: https://doi.org/10.1080/15422119.2014.973506.

You, S., Lu, J., Tang, C.Y., Wang, X., 2017. Rejection of heavy metals in acidic wastewater by a novel thin-film inorganic forward osmosis membrane. Chem. Eng. J. 320, 532−538. Available from: https://doi.org/10.1016/J.CEJ.2017.03.064.

Zhao, P., Gao, B., Yue, Q., Liu, S., Shon, H.K., 2016. The performance of forward osmosis in treating high-salinity wastewater containing heavy metal Ni2 + . Chem. Eng. J. 288, 569−576. Available from: https://doi.org/10.1016/j.cej.2015.12.038.

Zhao, S., Zou, L., Tang, C.Y., Mulcahy, D., 2012. Recent developments in forward osmosis: opportunities and challenges. J. Memb. Sci. 396, 1−21. Available from: https://doi.org/10.1016/j.memsci.2011.12.023.

Zhao, S., Zou, L., Mulcahy, D., 2012. Brackish water desalination by a hybrid forward osmosis-nanofiltration system using divalent draw solute. Desalination 284, 175−181. Available from: https://doi.org/10.1016/j.desal.2011.08.053.

Zhao, X., Liu, C., 2018. Efficient removal of heavy metal ions based on the optimized dissolution-diffusion-flow forward osmosis process. Chem. Eng. J. 334, 1128−1134. Available from: https://doi.org/10.1016/j.cej.2017.11.063.

Zhao, Y., Ren, Y., Wang, X., Xiao, P., Tian, E., Wang, X., et al., 2016. An initial study of EDTA complex based draw solutes in forward osmosis process. Desalination 378, 28−36. Available from: https://doi.org/10.1016/j.desal.2015.09.006.

Zirehpour, A., Rahimpour, A., Ulbricht, M., 2017. Nano-sized metal organic framework to improve the structural properties and desalination performance of thin film composite forward osmosis membrane. J. Memb. Sci. 531, 59−67. Available from: https://doi.org/10.1016/J.MEMSCI.2017.02.049.

11

Nanoparticles Enhanced Coagulation of Biologically Digested Leachate

A.Y. Zahrim, I. Azreen, S.S. Jie, C. Yoiying, J. Felijia, H. Hasmilah, C. Gloriana, I. Khairunis

FACULTY OF ENGINEERING, UNIVERSITI MALAYSIA SABAH, JALAN UMS, KOTA KINABALU, SABAH, MALAYSIA

11.1 Introduction

Landfilling is one of the most applied method for disposal of municipal solid waste (MSW) and sanitary landfilling technology are extensively recognized and used due to its financial feasibility (Abu Amr et al., 2017; Comstock et al., 2010). However, leachate produced from the landfills becomes a major environmental problem accompanying this disposal method (Zainol et al., 2012). Leachate contains high concentrations of biodegradable and nonbiodegradable substances which are hazardous and toxic. If not properly treated and disposed, it may become a potential source for groundwater and surface water contamination.

Landfill leachate is among the most challenging effluents to treat owing to its challenging composition, highly variable characteristics and strength (Gálvez et al., 2009; Tugtas et al., 2013). Various physical/chemical and biological treatment processes are being applied to treat leachates. Both physical/chemical and biological treatment processes have their own advantages and disadvantages, therefore typically these processes are combined for a more effective leachate treatment (Kargi and Yunus Pamukoglu, 2003; Renou et al., 2008; Tugtas et al., 2013). Selection of a suitable leachate treatment process is dependent on contaminants that need to be removed from the leachates. Coagulation is one of the most commonly used treatment for leachate (Dolar et al., 2016; Ishak et al., 2017; Kamaruddin et al., 2017; Shu et al., 2016; Zamri et al., 2016). During coagulation, coagulant is added that will destabilize particles and form an insoluble end product which is able to remove undesirable substances found in leachate through ionic mechanism (Kamaruddin et al., 2015). This review paper will discuss on coagulation method in treating biologically digested leachate, along with types of coagulant available and advancement of coagulation using nano particles.

In their review paper, Kamaruddin et al. (2017) discussed briefly on mechanism of coagulants and flocculant aids in landfill leachate treatment. They classified the coagulants into three types which are metallic salts (or hydrolyzing metallic salts), polymeric metallic salts (or prehydrolyzed/prepolymerized metallic salts), and polymers. Different types of flocculant aids applied in water and wastewater treatment were also compared. The review however, did not discuss in detail for each of the coagulant and flocculant aids except for polyaluminum chloride (PAC) and none is mentioned on application of nano particles as coagulant. Rui et al. (2012) also reviewed on leachate treatment by coagulation flocculation using different type of coagulants and polymer. However, the paper only discussed on synthetic polymers, aluminum sulfate ($Al_2(SO_4)_3$), ferric chloride ($FeCl_3$), and PAC. Torretta et al. (2017) discussed on novel and conventional technologies for landfill leachate treatment, covering biological and physical treatment. The review looked at probable integrated systems of various leachate treatment as revised by other researchers, and shared latest development on the techniques. Under coagulation flocculation, they briefly mentioned about effectiveness of $FeCl_3$ as coagulant in leachate treatment, as well as some other types of coagulant.

A review on coagulation flocculation for color removal was done by Verma et al. (2012) but focusing on textile wastewater. Prehydrolyzed coagulants such as PAC, polyferrous sulfate (PFS), polyferric chloride (PFC), and polyaluminum ferric chloride (PAFC) provides efficient color removal at low temperature, generating less sludge volume even though at low dosage and affectivity at wider wastewater's pH range. Comparing with other hydrolyzing metallic salts, ferrous sulfate was considered as a better coagulant. The review also discussed on application of natural coagulant as potential coagulants as well as flocculant aids in treating textile wastewater especially as pretreatment that will not impede the subsequent biological treatment.

Advancement of various flocculants for industrial wastewater treatment, together with flocculation performance and mechanism were reviewed by Lee et al. (2014) extensively. Coagulation flocculation and direct flocculation methods were compared, and it was concluded that the type of wastewater to be treated affects the selection between these two methods. In coagulation flocculation, the cationic inorganic metal salts are applied first as common coagulants to the wastewater to be treated, followed by the long chains nonionic or anionic polymers which acts as flocculants. Whereas, in direct flocculation, normally a cationic polymer with moderate charge density and high molecular weight is used. The polymer will neutralize the negative charges of the colloidal particles and concurrently forming flocs by bridging the combined destabilized particles. Direct flocculation is frequently applied to organic-based wastewater such as textile, paper and pulp, and food effluents which are concentrated with suspended and colloidal solids. Potential application of conventional flocculants, bioflocculants, and grafted flocculants in wastewater treatment had been reviewed in the paper however, none is mentioned on application of nano materials as flocculant.

By looking at review by others as mentioned above, the review is mainly on raw leachate or general wastewater, and lack of discussion on natural coagulant and application of nano particles in coagulation. Hoping to cover the gap of review papers by others, this paper will provide a more comprehensive review on coagulation process applied for biologically digested leachate, incorporating various types of coagulants used in leachate treatment including metal coagulant, polymer (organic, inorganic, and natural) coagulant, as well as application of nano particles in coagulation.

11.2 Leachate Generation and Characteristics

Municipal solid waste is one of key environment pollution contributor that generates leachate. Leachate production is mainly caused by percolation of rain through refuse dumped in a landfill. Composition of landfill leachate includes mostly of dissolved organic matter, heavy metals, ammoniacal nitrogen, phosphate, sulfide, phenol, salinity, alkalinity, acidity, hardness, inorganic salts, solids, and other hazardous substances (Renou et al., 2008; Wang et al., 2002; Zainol et al., 2012). It contains the waste composition and had been through the biochemical processes themselves. Since leachate is produced from the decomposition of waste inside the waste cell, the leachate characteristic will change from time to time. In Malaysia, recent study presented that more than 26,500 tons of solid wastes are being dumped at 166 operating landfills every day generating around 1.1 kg/day of solid waste. This becomes contributing factor for generation of landfill leachate that has been emblematic dilemmas of the landfilling method (Kamaruddin et al., 2017).

Landfill leachate characteristic knowledge is necessary to figure out the constantly changing performance level in leachate treatment either by physical, biological or physical/chemical process (Kamaruddin et al., 2015). McBean et al. (1995), presented that the variation in composition of leachate may be affected by several factors which include age of refuse, operation period of landfill, climate, hydrological condition, solid waste composition, moisture content, temperature, pH value, biological and chemical activities and degree of stabilization.

According to Renou et al. (2008), landfill leachate characteristic is usually defined by the basic parameter which are chemical oxygen demand (COD), biochemical oxygen demand (BOD_5), nitrogen-ammonium, total Kjeldahl nitrogen, BOD_5/COD (biodegradability), color, pH, and conductivity. Oloibiri et al. (2017) pointed out that apart from pH, COD, and conductivity, dissolved organic matter (DOM) is also important to be measured in the assessment of landfill leachate treatment.

Classification of landfill leachate according to composition changes is as represented in Table 11-1. Young landfill leachate (age < 5 years) is generally characterized by high BOD and COD concentrations, high ratio of BOD/COD, moderately high ammonium nitrogen and

Table 11-1   Classification of Landfill Leachate According to the Composition Changes (Foo and Hameed, 2009; Peng, 2017)

Type of Leachate	Young	Intermediate	Stabilized
Age (years)	<5	5–10	>10
COD (mg/L)	>10,000	4000–10,000	<4000
pH	<6.5	6.5–7.5	>7.5
Ammonia nitrogen (mg/L)	<400	–	>400
BOD_5/COD	0.5–1.0	0.1–0.5	<0.1
Heavy metals (mg/L)	Low to medium	Low	Low

a pH value as low as 4. Stabilized or old leachate (age > 10 years) is presented by a high refractory compounds which are not easily degradable, moderately high strength of COD, high strength of ammonia nitrogen, and a low BOD/COD ratio of less than 0.1 (Foo and Hameed, 2009; Peng, 2017). Quite differently, Alvarez-Vazquez et al. (2004) classify young leachate as less than 1 year old with BOD/COD ratio ranging from 0.5 to 1.0, and old leachate as more than 5 years old with BOD/COD ratio ranging from 0 to 0.3.

11.2.1 Biologically Digested Leachate

Biological digestion is a sequence of biological processes in which microorganisms decompose biodegradable material. In this paper, biologically digested leachate refers to leachate that has undergone pretreatment with biological process, as well as stabilized leachate. Characterization of an old (mature) or stabilized (less biodegradable) leachates are low in COD (<3000 mg/L), BOD/COD ratio (<0.1), and heavy metals (<2 mg/L), but high in pH (>7.5) and NH_4-N (>400 mg/L) (Ghafari et al., 2010). This agrees with findings by Singh Yadav and Dikshit (2017) and Bashir et al. (2009), where very less BOD_5/COD ratio of sample leachate was found which indicates less biodegradable organics, relatively low COD, and high concentration of heavy metals suggesting it was old/stabilized leachate. With the increase of landfill age, the microorganisms break down the organic materials into CH_4 and CO_2. As CO_2 is reduced with hydrogen, the pH rises and the organic compounds become less biodegradable. The leachate then turns to stabilized leachate (Liu, 2013; Mårtensson et al., 1999). Like stabilized leachate, biologically treated leachate also contains less biodegradable materials. Fig. 11-1 is a sample of biologically treated leachate taken from a local MSW landfill.

Biological treatment methods can be divided into aerobic and anaerobic based on availability of dissolved oxygen. In aerobic condition, organic pollutants are mostly converted into CO_2 and sludge by using atmospheric O_2 transmitted to the wastewater. In anaerobic treatment, organic matter is transformed into biogas, mainly consisting of CO_2 and CH_4 and in a minor part into biological sludge in the absence of O_2 (Abbas et al., 2009). Fig. 11-2 is a

FIGURE 11-1 Sample of biologically treated leachate.

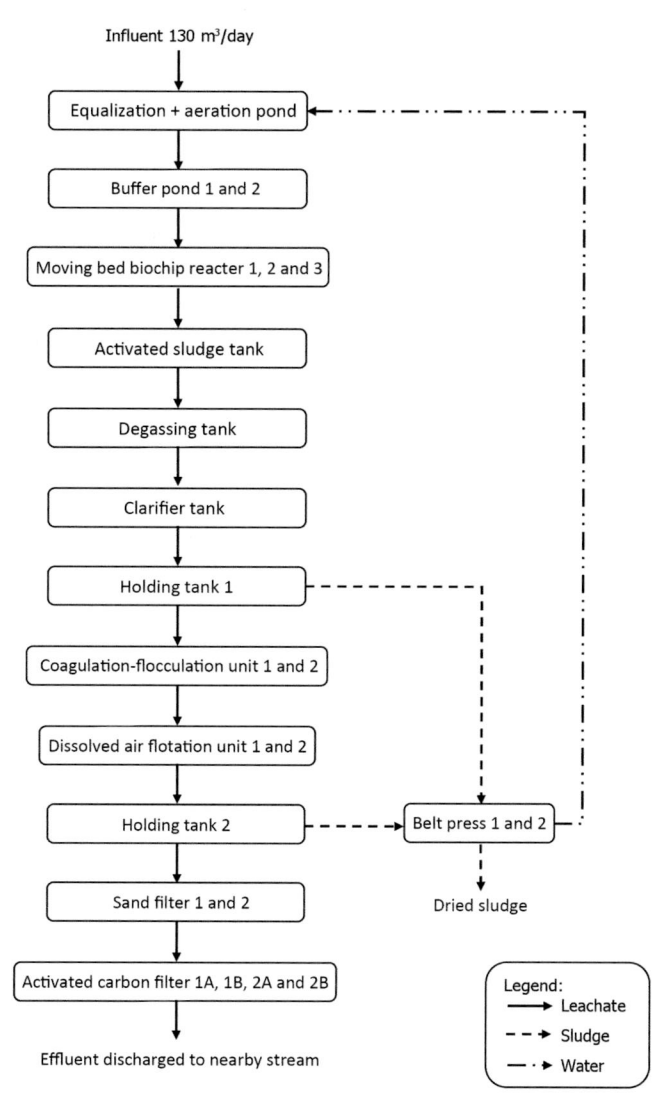

FIGURE 11-2 Process flow diagram of leachate treatment plant.

typical process flow diagram of a leachate treatment plant showing biological treatment as part of the treatment process.

Biological processes were considered to be most effective in treatment of young leachates with high BOD/COD ratio (>0.5) for removal of organic and nitrogenous matter (Abbas et al., 2009). Usually, a combination of aerobic/anaerobic system has been used to treat leachate more efficiently for the removal of biodegradable compounds. Table 11-2 shows characteristics of biologically digested leachate taken from various landfills. However, as the age of landfill increase, it is difficult to treat the stabilized leachate which is categorized as

Table 11-2 Characteristic of Biologically Digested Landfill Leachate From Various Landfill Site

Landfill Site	Color (Pt.Co)	pH	COD (mg/L	BOD (mg/L)	TS (mg/L)	TSS (mg/L)	Ammonia Nitrogen (NH_3-N)(mg/L)	BOD_5/ COD	Conductivity (mS/cm)	References
Alachua County Southwest Landfill, Florida (stabilized)	3280	8.01	2076	57.8	--	--	--	0.028	--	(Comstock et al., 2010)
Union County New River Regional Landfill, Florid (stabilized)	12700	8.12	4843	435	--	--	--	0.090	--	
Brazil landfill (biologically treated)	6983	7.9	1500	--	11770	477.5	17.9	--	--	(Do Carmo Nascimento et al., 2016)
Alor Pongsu Landfill, Malaysia (anaerobic stabilized)	12475	8.29	3125	274	--	--	1674	--	10.788	(Abu Amr et al., 2016)
La zoreda landfill, Spain (biologically treated)	0.622[a]	6.8	1143	13.5	--	--	403	--	11.430	(Oulego et al., 2015)
Dennis Town Landfill, Malaysia (anaerobic, partially stabilized)	2570	7.51	551.13	150.6	--	80.67	292	--	--	(Aziz and Ramli, 2014)
Municipal solid Landfill, Wuhan (biologically treated)	NS	8.42	625	NS	NS	78	28	--	--	(Ye et al., 2016)
Jeram Sanitary Landfill, Malaysia (biologically treated)	10200	7.98	5600	--	--	--	600	--	14.84	(Jumaah et al., 2016)
Pulau Burung Landfill, Malaysia (Semi-aerobic, stabilized)	3334	8.2	935	83	6271	1437	483	0.09	--	(Kamaruddin et al., 2015)
Kulim Landfill, Malaysia (Anaerobic)	1936	7.76	1892	326	4041	6336	300	0.205	--	

-- not measured.

[a]measured as color number (CN).

very tough organic with very low biodegradability due to the major presence of nonbiode-gradable organics (Abu Amr et al., 2017; Guo, 2016). Main portion of stabilized or biologically treated leachate also consists of recalcitrant organic molecules that are not easily removed using biological treatment (Abbas et al., 2009).

A characterization study on biologically treated leachate collected from Jeram Sanitary Landfill, Malaysia shows a very high color values which indicates that the biological treatment system alone would not be effective in reducing the color (Jumaah et al., 2016). According to Torretta et al. (2017), biological treatment is not quite efficient for removal of certain substances such as halogenated bio-refractory organic compounds (AOX) and metals. Biologically treated leachate also has remaining amount of COD due to the complexity of the leachate (Ye et al., 2016). Therefore, conventional biological treatment methods are no longer sufficient and efficient to be utilized for stabilized landfill leachate treatment, hence physical/chemical treatments are recommended as a refining step for biologically digested leachate and also to remove refractory substances from stabilized leachate (Kamaruddin et al., 2015; Torretta et al., 2017). Coagulation is considered as a more effective technology in removal of nonbiodegradable organic material and anthropogenic organic chemicals during the pretreatment or posttreatment of stabilized leachate as the process is cost-effective, easy to implement (Comstock et al., 2010) and capable of reducing pollutants contents for the succeeding biological process (Guo, 2016; Zheng et al., 2009).

11.3 Coagulation

Coagulation flocculation has been widely used in landfill leachate treatment to reduce total suspended solids, color, and organic content such as dissolved organic matter and recalcitrant compounds in order to improve treatment's efficiency and to prevent fouling of the subsequent treatment of leachate (Comstock et al., 2010). It can also improve the biodegradability of high concentration refractory pollutants and saline compound in leachate membrane concentrate (Long et al., 2017).

Stephenson and Stuetz (2009) stated that coagulation is the first stage that needs to be done in wastewater treatment process. It is a process of neutralizing the charge on the surface of particles to let the particles to grow into a larger size that are easier to settle by means of various types of coagulants (Sillanpää et al., 2018; Stephenson and Stuetz, 2009). Coagulation also helps to reduce the factors that cause the stability of colored effluent through formation of flocs (Zahrim et al., 2013) as well as decrease color, turbidity, and remove pathogens and suspended solids (Volk et al., 2000).

The advantages of using coagulation flocculation are that it is a controllable, economically feasible, excellent color removal and simple physical-chemical treatment of stabilized and old landfill leachate (Amokrane et al., 1997; Verma et al., 2012). Its behavior can be effected by a lot of parameter including the coagulant dosages and type, as well as the operating condition (Aguilar et al., 2005; Chakraborti et al., 2003; Sillanpää et al., 2018). Selecting the most suitable coagulant type is crucial since this is a major factor influencing the operating conditions of coagulation, which then determine the effectiveness of the treatment (Ghafari et al., 2010; Kumar and Bishnoi, 2017; Zhu et al., 2016).

The typical coagulant and flocculant aids include natural, manufactured organic and inorganic polymer and metal salts. Generally, metal coagulant can be classified into two which are iron, and aluminum based. Examples of iron based coagulants were ferric chloride, ferric sulfate, ferrous sulfate and ferric chloride sulfate, while the examples of aluminum based coagulants are aluminum sulfate, sodium aluminate, aluminum chloride, and aluminum chlorohydrate (Bratby, 2016).

Metal coagulants are widely utilized as conventional coagulants due to their advantages such that they are readily available and relatively low cost. Besides that, they are capable to form multicharged polynuclear complexes in a solution which improved adsorption characteristics thus making them effective coagulants (Bratby, 2016). In leachate treatment, metal coagulants can remove the COD, BOD, suspended solid, turbidity and color of the leachate (Amor et al., 2015; Dolar et al., 2016; Kamaruddin et al., 2015).

The examples of inorganic polymer coagulants are polyferric sulfate (PFS), polyaluminum chloride (PAC), and polyferric chloride (PFC). The coagulants are prepared by combining other additive compounds to inorganic coagulants. By doing so, their coagulation flocculation efficiency is greater than the individual coagulant (Wei et al., 2016).

However, regardless of the great performance and low-cost of inorganic and synthetic coagulants, their drawbacks include generation of excessive volumes of sludge, requirement of pH and alkalinity adjustments, and their increasing concentration of residuals (aluminum) in treated water can be associated to neurotoxic and carcinogenic effects, as well as neurodegenerative diseases such as Alzheimer's (Camacho et al., 2017; Teh and Wu, 2014). Thus, coagulation is usually applied as a pretreatment or posttreatment in an integrated water and wastewater treatment system so that the drawbacks can be solved, and a better treatment of leachate can be achieved. As a substitute for an environmentally friendly coagulant, natural coagulants derived from animals or plants are being utilized (Shak and Wu, 2015). Gelatin galactomannans, starches, cellulose derivatives, microbial polysaccharides, chitosan, glues, alginate, Nirmali seeds, Moringa oleifera, tannin, and cactus are examples of natural coagulants that have been used in wastewater treatments (Kumar et al., 2017). Chitosan is a natural high molecular polymer that formed by deacetylation of chitin and is an attractive alternative because of its abundance (Rizzo et al., 2008).

Hybrid materials or composites have emerged as another type of coagulant with remarkable potential in wastewater treatment. Hybrid materials offer better coagulation performance than conventional inorganic-based coagulants, plus hybrid materials are cheaper compared to organic-based flocculants (Wang et al., 2006). Review done by Lee et al. (2012) discussed on the preparation, classification and usage of hybrid materials in coagulation flocculation process. Hybrid materials could possibly be as combination of inorganic-organic, organic-organic, inorganic-inorganic, inorganic-biopolymer, organic-natural polymer, as well as inorganic-natural polymer. These combinations could be classified under functionally-hybridized materials, chemically-bound-hybridized materials or structurally-hybridized materials (composites). Preparation methods of hybrid materials include copolymerization, physical blending at ambient or elevated temperature, chemical grafting/crosslinking and hydroxylation−prepolymerization. Application of hybrid materials could lessen operation time, since wastewater treatment can be performed by adding one chemical in one tank instead of two separate tanks in the conventional coagulation flocculation system (Lee et al., 2010). PAC-FeCl$_3$, PFS-PAM, PFC-PDMDAAC, PFM-PDMDAAC, PAC-chitosan are some example of hybrid materials used as coagulant.

Since it is important and necessary to design an advanced wastewater treatment that can offer great treatment efficiency and require less capital cost, latest advanced processes in nano-material sciences and nanotechnology have been considered as promising (Anjum et al., 2016; Zhang et al., 2016a,b). MFPAC and carbon nanotubes are examples of coagulants derived from nano material which has been applied to treat landfill leachates (Joseph et al., 2013; Liu et al., 2017).

11.3.1 Coagulation Mechanism

Colloidal particles carry charges on their surface, which lead to the stabilization of the suspension. Hence in coagulation process, the suspended particles are removes by combining small particles into large particles (Li et al., 2006). By adding some chemicals, the surface property of such colloidal particles can be changed or dissolved material can be precipitated so as to facilitate the separation of solids by gravity or filtration. There are certain type of mechanisms to destabilize natural water in coagulation which are double layer compression, adsorption and charge neutralization, entrapment of particles in precipitate and absorption, sweep coagulation and bridging between particles (Davis, 2011; Freitas et al., 2018; Li et al., 2006).

The sufficient amount of counter ions presents on the diffuse layer in a colloidal dispersion, resulting in an electrical double layer. It is used to balance the electrical charge on the particle. During the compression of electrical double layer of water suspension, colloidal particles are more bonded together due to the effect of van der Waals attraction forces and Brownian motion. Besides, colloidal particles can be destabilized via adsorption and charge neutralization mechanism using hydrolyzed metal salt or cationic polymers as positively charged coagulants (Davis, 2011).

To ensure efficient coagulation, proper agitation is necessary to produce intense mixing that will provide uniform dissemination of the coagulant into the solution by contacting the particles in suspension prior to reaction completion. Coagulation is usually followed by flocculation. In flocculation, slow stirring is provided to evade breaking the already formed flocs (Freitas et al., 2018). In laboratory, coagulation flocculation is performed using Jar Test equipment as shown in Fig. 11-3.

11.3.2 Coagulation of Biologically Digested Leachate

Depending on characteristic of wastewater to be treated, coagulation can be carried out just by using coagulant process alone, without the need for flocculation process. The coagulant commonly used for leachate treatment include conventional metal coagulant and polymer

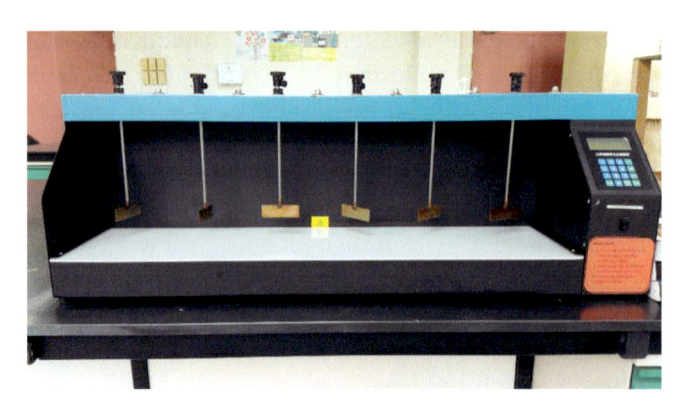

FIGURE 11-3 Jar Test equipment.

coagulant which comprise of organic, inorganic as well as natural substances. A more advanced type of coagulant to treat leachate is hybrid materials which is combination of coagulants that could improve the performance offered by single coagulant.

Ferric chloride ($FeCl_3$) is the most popular metal coagulant to be used in leachate treatment based on previous studies (Amor et al., 2015; Kumar and Bishnoi, 2017; Oloibiri et al., 2017; Singh et al., 2012; Zhang et al., 2016a,b). Amor et al. (2015) reported that at operating conditions of pH 5 and a dose of 2 g/L, $FeCl_3$ could accomplish removals of 63% for COD and 71% for total polyphenols for the pretreatment of stabilized leachate. Whereas in another study, at pH 3 and a dose of 300 mg/L of $FeCl_3$, up to 17.8% of COD, 81.9% of total carbon, 0.3% of Hg, 27.5% of Pb, 36.1% of NH^{4+}, and 7.7% of As can be removed from stabilized leachate (Vedrenne et al., 2012).

Aluminum sulfate ($Al_2(SO_4)_3$), also known as alum is the most common aluminum-based metal coagulant implemented in leachate treatment. The optimal pH and dosage were found to be 7.0 and 5.0 g/L respectively during the post treatment of biologically treated by SBR using alum resulting in removal of 36.29% COD, 0.74% NH_3-N, 43.18% TSS and 56.76%. color (Yong, Bashir, Ng, Sethupathi, & Lim, 2018).

Gandhimathi et al. (2013) performed coagulation process using alum and $FeCl_3$ on fresh and stabilized leachate. 59% and 35% COD removal efficiency were obtained at optimum dose of 0.7 and 0.6 g/L $FeCl_3$ for stabilized and fresh leachate respectively. The removal efficiency becomes maximum of 75% and 55% at 0.6 and 0.8 g/L alum for stabilized and fresh leachate respectively. This showed that compared to $FeCl_3$, alum is more efficient to remove COD.

Performance of lime ($Ca(OH)_2$), $FeCl_3$, and alum are being compared to treat the stabilized landfill leachate of the Mediouna municipal landfill of Casablanca city (Morocco). 66.25% of COD, 98% of turbidity, 80% of color with a low sludge volume generation (30%) were obtained using lime at optimum coagulant dosages of 52.5 g $Ca(OH)_2$/L. At optimum coagulant dose of 12 g $FeCl_3$/L, 62.5% of COD, 92.5% of turbidity and 82% of color are removed with less sludge volume generation (21%). Meanwhile, alum only allow for 11%

COD and 6% of turbidity removal at 22.5 g $Al_2(SO_4)_3$/L but with a slightly lower sludge volume generation (11%). $FeCl_3$ could eliminate most of the detected metal elements with a high affinity toward Cd, Cu, Ni, Cr, and Zn, while alum is very suitable for Ni, Pb, Cr elimination with 95%, 94%, and 93% removal efficiency, respectively (Zainol et al., 2017).

Apart from metal coagulant, polyferric sulfate (PFS) and PAC are also commonly used to treat leachate. Li et al. (2010) conducted coagulation study on stabilized leachate using alum, $FeCl_3$, PAC, and PFS for the removal of SS, turbidity and COD. The optimum working pH was 5.5−6.0, which suggested that main mechanism of coagulation flocculation process is charge neutralization. The optimum dosages were 0.3 g Fe^{3+}/L for PFS, 0.6 g Fe^{3+}/L for $FeCl_3$, and 0.6 gAl^{3+}/L for $Al_2(SO_4)_3$ and PAC, respectively. PFS showed the highest removal efficiency of 70% for COD, 93% for SS, 97% for turbidity and 74% for toxicity reduction as well as producing the least sludge volume. Meanwhile, V. Oloibiri et al. (2015) reported that $FeCl_3$ offers better organic matter removal as compared to PAC for treatment of biologically stabilized leachate as shown by higher COD, turbidity and α_{254} removal plus better settling characteristics of the produced sludge.

The performance of alum and PAC in treating partially stabilized leachate was shown in the study by Ghafari et al. (2010) by comparing removal of turbidity, COD, TSS and color. At optimum coagulant dose and pH of 9.4 g/L and 7 for alum and 1.9 g/L and 7.5 for PAC respectively, physical parameters of leachate achieved almost complete removals using PAC but lower removal using alum. However, results showed that alum was more efficient than PAC for COD removal with removal efficiency of 84.50% for alum and 56.76% for PAC.

In their study, Liu et al. (2012) applied RSM to optimize the performance of coagulation flocculation on stabilized leachate using PFS, $FeCl_3$ and ferric sulfate ($Fe_2(SO_4)_3$). Good removals of humic acid (HA) and COD suggested that coagulation flocculation when applied as pretreatment could increase the biodegradability of leachate, remove recalcitrant compounds effectively, and reduce pollutant capacity for the following biological process. From result, PFS and $FeCl_3$ can improve the biodegradability of leachate better than $Fe_2(SO_4)_3$, and PFS is suggested for landfill leachate treatment in a large scale.

Less studies were performed on application of hybrid materials on biologically digested leachate. However Li et al. (2015) conducted a study on raw landfill leachate using polyferric-magnesium polydimethyldiallylammonium chloride (PFM-PDMDAAC), a novel inorganic−organic composite coagulant which was prepared using $MgSO_4$, $FeSO_4$ and PDMDAAC as raw materials. PFM-PDMDAAC shows better coagulation performance than PFM + PDMDAAC at the same dosage, suggesting that the optimized composite coagulant was more efficient than the simple sequential addition of the same reagents. XRD and FTIR spectroscopy revealed that the PFM-PDMDAAC was not a simple mechanical mixing of PFM and PDMDAAC. A composite system with inorganic−organic complex interpenetration networks was formed as the PDMDAAC dispersedly interweaved into the PFM, which is why PFM-PDMDAAC provides better coagulation efficiency than the single coagulants. The new PFM-PDMDAAC reagent can be added in just one step in the coagulation process, thus overcoming the limitation of the PFM + PDMDAAC where two reagent addition systems are needed. Nevertheless, as compared to PFM, the PFM + PDMDAAC still showed better coagulation performance. As conclusion, superior COD removal can be achieved with composite PFM-PDMDAAC prepared using a PDMDAAC/PFM weight ratio at 0.07 and Mg/Fe molar ratio of 1/3 over single PFM and PDMDAAC coagulants.

Natural coagulant has also been applied widely as coagulant to treat leachate. Feasibility of chitosan as primary coagulant in the post treatment of landfill leachate has been tested by Do Carmo Nascimento et al. (2016). In this research, chitosan performance has been studied and compared with alum performance in terms of color and turbidity removal as well as acute toxicity by using the duckweed Lemna minor and the guppy fish Poecilia reticulata as test organisms. The optimum process variables were found to be 960 mg/L of chitosan at pH 8.5 and 1,610 mg/L of alum at pH 9.5. Under these conditions, the maximum efficiency removal of true color was 80% and the turbidity removal reached 91% using chitosan. Alternatively, alum coagulant reached 87% true color removal efficiency and 81% turbidity removal, indicating that chitosan as a primary coagulant performed as well as alum in the removal of recalcitrant organic matter. However, the results of the acute toxicity tests showed that organisms were sensitive to biologically treated leachate after coagulation flocculation using chitosan and results suggested that chitosan has a toxicity level which is not negligible. Therefore, it was concluded that chitosan is not suitable to be the primary coagulant in the posttreatment of landfill leachate.

Aziz and Ramli (2014) has studied coagulation flocculation of anaerobic landfill leachate using $FeCl_3$, Aloe Vera (AV) and Chitosan (CS) by evaluating the performance of the three coagulants to remove suspended solid (SS) and color. AV and CS also were tested to know the suitability as flocculant aids. When used alone, at dosage of 1100 mg/L, $FeCl_3$ has removal rate of 91.3% and 93.5% for color and SS. Whereas, 66.7% of color was removed by AV at dosage 10 000 mg/L and no SS was removed. Meanwhile, CS reduced SS up to 18.8 % of turbidity at 5 mg/L and 37.74% of color at 20 mg/L.

Aziz and Sobri (2015) reported that color, SS, turbidity and COD of semi-aerobic landfill leachate can be reduced by 13.1%, 27.9%, 0.0% and 8.2% respectively by solely dosing 7000 mg/L of native sago trunk starch (NSTS) coagulant. Similarly, the associated reductions were 15.1%, 29.5%, 0.0% and 28.0 % for a 6000 mg/L dosage of solely commercial sago starch (CSS). The researchers concluded that the removal efficiencies were not effective using either NSTS or CSS as sole coagulant.

Another study presented that starch flour extracted from oil palm trunk waste (OPTS) gave reduction of 45.5% and 26% of COD and suspended solid removal respectively from partially stabilized semiaerobic landfill leachate (Zamri et al., 2016). They stated that OPTS acts as better coagulant as compared to PAC in terms of COD removal but not effective for SS removal. The researchers then suggested that OPTS can be applied as flocculant aid to PAC for a 12%−16% coagulation improvement.

Mohd Zin et al. (2016) also investigated on the performance of tapioca starch (TS) in coagulation of real partially stabilized leachate. They describe that at optimum values of pH 4 and 2.5 g/L, 12%, 54.7% and 13.2% of SS, color, and ammonia removals can be achieved. At higher doses, the linkage produced bigger floc but weak floc strength. Based on the result obtained, the researchers stated that tapioca starch may not be the best coagulant as compared to chemical coagulants, but still able to remove suspended solids and color of partially stabilized leachate. They then suggested that coagulation flocculation performance might be improved using TS as flocculant aid together with chemical coagulant.

Vijayaraghavan et al. (2011) reported that there is cationic protein present in the Moringa oleifera seed extract which consists dimeric proteins with molecular weight in the range of 6.5−14 kDa, in which these properties enable M. oleoifera to acts as main coagulant component in coagulation. Oliveira et al. (2016) suggested an optimum dosage of 37500 mg/L Moringa oleifera Lam as a natural polymer at pH of 7.07 able to reduce 78.66%, 79.74%, 43%, 87.91%, and 61% for COD, BOD, SS, color, and turbidity of raw leachate respectively. Another study presented by Sivakumar (2013) stated that the Moringa oleifera seeds extraction is able to perform 84.5% and 82.6% on COD and total dissolved solid (TDS) reduction. The dosage was only 100 mg/L to treat municipal solid waste leachate. Table 11-3 shows performance of coagulation treatment of leachate using natural coagulants.

11.3.3 Coagulation With Flocculant Aids

In some coagulation processes, flocculant aids are added to supplement the orthokinetic flocculation process and improve floc characteristics such as settleability, compressibility, filterability, permeability, shear strength, density and floc size (Bratby, 2016). Coagulation is usually proceeded with the flocculation of unstable particles into bulky flocs. As the particle size increase, they can then settle more easily (Zhang et al., 2013).

Long et al. (2017) reported that the highest TOC, COD and chromaticity removal of biologically treated leachate, 81%, 82% and 97% respectively were achieved using 5000 mg/L $FeCl_3$ and 70 mg/L polyacrylamide (PAM) at pH 4. The biologically treated leachate is obtained from leachate membrane bioreactor (MBR) concentrate which has undergone nanofiltration. The researchers first compare the coagulation performance with various coagulants such as $FeCl_2$, $FeSO_4$, and PAC at varying pH and dosage with PAM as flocculant aids. According to them, there is no significant increase in TOC removal with the rose of $FeCl_3$ dosage. They suggested that iron(III) was more suitable than ferrous and aluminum coagulants due to its reactivity, large specific surface areas, and low solubility of iron(III) hydroxides. However, the research team did not further their research to determine the optimum dosage of $FeCl_2$, $FeSO_4$, and PAC but on ferric coagulant anion. Coagulation performance was analyzed with Fe(III) added as $FeCl_3$, $Fe_2(SO_4)_3$, $Fe(NO_3)_3$ at varies concentration (5−43 mmol/L). The team then finally reported the minimal initial Fe(III) concentration was 26 mmol/L for all examined coagulants which contributed 78% TOC removal. In another study on the coagulation flocculation process applied for the raw landfill leachate treatment, 80% removal of COD was attained at optimum conditions of pH 3.36, coagulant dosage of 0.87 g/L $FeCl_3$, flocculant aid dosage of 26 mg/L cationic PAM, mixing speed in the flocculation step of 48 rpm, and setting time of 30 min (Smaoui et al., 2016).

The decolorization of aeration lagoon biotreated leachate can be obtained by implementing dual coagulant of alum and barley with a 98% reduction (Shaylinda Mohd Zin and Azraff Zulkapli, 2017). By applying this method, the researcher described that barley, which acts as

Table 11-3 Performance of Coagulation Treatment of Leachate Using Natural Coagulants

Source of Leachate	Natural Coagulant	pH	Temperature (K)	Dosage (mg/L)	Settling time (min)	Rapid (rpm)	Slow (rpm)	COD	BOD	SS	Color	Turbidity	References
Saravan Dumpsite, Iran	Alum : *Ocimum basilicum L.*	7	—	1:1 (500—5000)	30	120, 5 min	40,	64.4	—	—	77.8	58.8	(Rasool et al., 2016)
Pulau Burung Landfill Site (PBLS), Byram Forest Reserve, Penang	Native sago trunk starch (NSTS)	4	—	7000	120	80, 2 min	30, 30 min	1.7	—	27.9	13.1	0.0	(Aziz and Sobri, 2015)
	Commercial sago starch (CSS)	4	—	6000	120	80, 2 min	30, 30 min	28	—	29.5	15.1	—	
Matang Landfill Site (MLS), Taiping, Perak	Tapioca Starch (TS)	4	—	2500	60	100, 4 min	40, 25 min	—	—	12	54.7	—	(Mohd Zin et al., 2016)
Muribeca CTR-Candeias Landfill	Moringa Oleifera Lam extract (LBMS)	7.07	—	37500	—	—	—	78.66	79.74	43	87.91	61	(Oliveira et al., 2016)
	Abelmoschus esculentus (L.) Moench (LBQ)	7.89	—	37500	—	—	—	78	71.92	52	72	34	
	Combination of Moringa & Okra (LBEMQ)	7.07	—	37000 + 20000	—	—	—	52	86	8	78	46	
Thanjavur Municipality Dumping Yard (Blending of MSW and water)	Moringa Oleifera seeds	7	318	100	60	100	20	84.5	—	82.6 (TDS)	—	—	(Sivakumar, 2013)

Semi-aerobic landfill leachate, Pulau Burung Landfill Site, Penang	Starch flour extracted from oil palm trunk waste	7	298	500	45	120, 3 min	20, 15 min	45.5	—	26	—	—	(Zamri et al., 2016)
Air Hitam Sanitary Landfill at Puchong	Moringa oleifera seed extract	—	299	175	60	100, 8 min	40, 20 min	—	—	9.9	6.1	2.5	(Ameen et al., 2011)
Air Hitam Sanitary Landfill at Puchong	Moringa oleifera stock solution	8.51	325	175	60	100, 8 min	30, 20 mn	0	—	19.7	NS	3.6	(Muyibi et al., 2002)

potential flocculant aid produce better strength of floc, increase the weight of floc and minimizes 33% usage of alum (Zin et al., 2014). Besides, possible cost of treatment and toxicity effect of the chemical coagulant are indicated by the reduction dose of alum. Similar outcome demonstrated by Aziz and Sobri (2015) where native sago trunk starch (NSTS) and commercial sago starch (CSS) acts as effective flocculant aid which able to reduce 35.5% dosage of PAC in treating semiaerobic landfill leachate. Application of flocculant aids gave similar removal efficient of 99.2%, 94.7%, and 98.9% on suspended solid, color, and turbidity respectively when compared to PAC as sole coagulant.

Yang et al. (2009) reported that the risk associated with chemical flocculants can be lessen by applying composite of bioflocculant and chemical flocculant in a coagulation flocculation process, since chemical flocculants' dose was reduced to minimum. In this study, composite of alkaline-thermal-treated sludge (as a bioflocculant) and PFS was studied in stabilized leachate pretreatment. Using RSM optimization, at bioflocculant dose of 19.3 mg/L, PFS dose of 12.7 g/L, agitation speed of 202 rpm and pH 6.9, this composite could remove HA, turbidity, color and COD at 79.3%, 85.1%, 64.9%, and 72.6%, respectively. Application of coagulation flocculation process as a pretreatment could effectively remove recalcitrant compounds from landfill leachate as indicated by significant removal of COD and HA (Guo, 2016).

As known, PAC and $FeCl_3$ are the most commonly used coagulants however, $FeCl_3$ coagulation needs pH adjustment, whereas PAC does not. Flocculation effect of cationic PAM is better than that of the anionic and neutral PAM, thus cationic PAM was used in study by Zhang et al. (2013) for treatment of anaerobic−oxic biologically treated leachate. After the treatment at optimal conditions of PAC dosage of 350 mg/L, cationic PAM dosage of 8.0 mg/L, and 0.4−0.6 mm ceramsite media in the filtration process, the residual NH^+_4-N, TN, Mn and As in the leachate was able to meet the maximum permissible values for landfill leachate discharge in China but not for COD and total phosphorus.

Zainol et al. (2017) has conducted research on the coagulation-flocculation performances for stabilized leachate treatment using Diplazium esculentum (a type of fern plant) Stock Solution (DEaqs) and *D. esculentum* powder (DEpowder) as flocculant aid to PAC as main coagulant. Fig. 11-4 shows image of *Diplazium esculentum*. The treatment is performed for

FIGURE 11-4 Diplazium esculentum.

two difference source of leachate, which resulting in different optimum dosage of DEaqs and PAC for their highest COD, suspended solid, color, and turbidity removal.

For stabilized leachate from Kulim Landfill Site, dosage of 500 mg/L PAC and 150 mg/L DEaqs resulted in 34%, 88%, 76%, and 87% reduction of COD, suspended solid, color and turbidity respectively; while for stabilized leachate from Kuala Sepetang landfill site, dosage of 1000 mg/L PAC and 500 mg/L DEaqs result in 55%, 88%, 76%, and 76% reduction of COD, suspended soild, color, and turbidity respectively. For both leachates, DEaqs was reported to be more effective flocculant aid as compare to DEpowder. In this dual coagulant system, the inorganic coagulant encourages the destabilization of colloids by charge neutralization, while the polyelectrolyte or polymer can perform as a bridging element that speeds the formation of more easily settling flocs, thus enhancing the performance of the coagulation flocculation process (Zainol et al., 2017).

Study on coagulation-flocculation of anaerobic landfill leachate by Hamidi Abdul Aziz and Siti Fatihah Ramli (2014) found out that removal rate was at the highest, 98.64% of color and 100% of SS, when 25 mg/L of chitosan is added as flocculant aid in coagulation-flocculation process which contains 1100 mg/L of $FeCl_3$ at pH of 6. 97.2% SS and 97.3% color reduction was achieved when 4000 mg/L of aloe vera was used as flocculant aid at the same conditions of process. It is concluded that chitosan is superior to aloe vera as flocculant aid due to its highly positive charges.

Table 11-4 shows the performance of coagulation treatment of leachate using coagulants with flocculant aid. It can be observed that by using flocculant aid, the dosage of primary coagulant which is mostly metal and synthetic polymer which has disadvantages of adding to sludge volume and posing health effect can be reduced. Nevertheless, metal and synthetic polymer are cost effective as compared to natural coagulant aids and bioflocculant. Therefore, it is recommended that more study should be done to look for the most viable combination of coagulant with flocculant aids or hybrid materials that is more effective, safe and affordable.

11.4 Nanoparticles in Water and Wastewater Treatment

Nanomaterials have the size at a nanoscale of 1−100 nm which are the smallest structures that humans have developed (Chaturvedi et al., 2012; Theron et al., 2008). More precisely, they are nanoparticles that have structure components with one dimension at least less than 100 nm (Amin et al., 2014). From the past few decades, nanotechnology has gained widespread consideration in water and wastewater treatment (Santhosh et al., 2016; Thines et al., 2017; Yang et al., 2013) and is considered one of the largest engineering innovations since the Industrial Revolution (Wang et al., 2013). Ipso facto, different nanomaterials have been produced and utilized for the removal of aquatic pollutants (Gautam and Chattopadhyaya, 2016). And with regards to water and wastewater treatment, the development of nanomaterials presents opportunities to come out with grounded and practical solutions to overcome global water pollution. Though, limited collective knowledge is available in this context.

Table 11-4 Performance of Coagulation Treatment of Leachate Using Coagulants With Flocculant Aid

Source of Leachate	Coagulant and Flocculant Aid	pH	Temperature (K)	Dosage (mg/L)	Settling Time (min)	Rapid (rpm)	Slow (rpm)	COD	BOD	SS	Color	Turbidity	References
					Condition					**Removal Efficiencies (%)**			
Pulau Burung Landfill Site (PBLS), Byram Forest Reserve, Penang	PAC + Native sago trunk starch (NSTS)	4	—	2000 (PACl), 6000 (NSTS)	120	80, 2 min	30, 30 min	—	—	99.2	94.7	98.9	(Aziz and Sobri, 2015)
	PAC + Commercial sago starch (CSS)	4	—	2000 (PACl), 5000 (CSS)	120	80, 2 min	30, 30 min	—	—	99.2	94.7	98.9	
Simpang Renggam landfill site (SRL), Johor.	Alum + Barley (coagulant aid)	6	—	3000 + 800	30	200, 4 min	30, 15 min	—	—		98	—	(Shaylinda Mohd Zin and Azraff Zulkapli, 2017)
Kulim Landfill Site (KLS)	Diplazium esculentum Stock Solution (DEaqs), Polyaluminum chloride (PAC), D. esculentum powder (DEpowder)	6	—	500 PACl + 150 DEaqs	30	150, 3 min	20, 10 min	34	—	88	76	87	(Zainol et al., 2017)
Kuala Sepetang Landfill Site (KSLS)	(DEaqs), (PAC), (DEpowder)	6	—	1000 PACl + 500 DEaqs	30	200, 3 min	20, 20 min	55	—	88	76	76	

Nanomaterials are categorized into four main classes; nanoadsorbents, nanocatalysts, nano-membranes, and finally the integration of nanotechnology with biological processes (Anjum et al., 2016).

In the field of wastewater treatment application, various efficient, environmentally friendly and cost-effective nanomaterials have been developed for potential water remediation of different water sources such as industrial effluents, surface water, ground water, and drinking water because of their unique functionalities (Brumfiel, 2003; Theron et al., 2008; Tyagi et al., 2017). Nanotechnology is one of the most advanced processes for water and wastewater treatment. Generally, some of the benefits of using nanomaterials are high capacity, fast kinetics, fast dissolution, high dispersion ability, great finite-size effect, tunable pore size and surface chemistry, high activity for (photo)catalysis, superparamagnetism for particle separation, strong antibacterial activity for disinfection and biofouling control, strong sorption to achieve charge stabilization, specific affinity towards certain contaminants, and among other merits. One of the major advantageous attributes of nanomaterials is high "aspect ratio," which is the ratio of surface area to volume. The same could be described using the BET (Brunauer-Emmett-Teller) theory. The smaller the particle size, the greater the aspect ratio. This and greater surface free energy enhance the reactivity of the nanoparticles with the surrounding molecules (Hassan et al., 2016). Not only that, nanomaterials usually have higher density of active sites per unit mass because of this high specific surface area characterization (Zhang et al., 2016a,b). Besides, they are microporous and have great numbers of activated functionalized sites. All in all, nanoparticles can penetrate deeper and treat water or wastewater which is generally not possible by using traditional technologies (Prachi et al., 2013).

In the current scenario, nanotechnology has been widely regarded as one the of most ideal optional solutions for advanced wastewater treatment to fulfill the desperate need to improve quality of water, remove chemical and biological pollutants so that it can be applied in local industries. There is no doubt of efficiency of utilization of nanomaterials in wastewater remediation. Despite it all, there are some downsides that should be considered. For instance, nanoparticles might be released into the environment during preparation and treatment processes in which they can accumulate and cause serious risks. And as aforementioned, nanotechnology is rarely adopted to mass processes and because of that, they are not cost-competitive as compared with conventional materials. Some effects that come with synthesizing nanoparticles through conventional methods are critical conditions of temperature and pressure cost of chemicals, long refluxing time of reaction, toxicity of the byproducts, etc. (Hassan et al., 2016). Moreover, nanoparticles face shortcomings in regards to large-scale application or mass transport in water treatment. Firstly, there is aggregation (nanomaterials are unstable in nature due to van der Waals forces and other interactions) in fluidized, fixed bed or any other flow through systems, causing major activity loss and pressure drop (Lofrano et al., 2016). Secondly, there is a problem of challenging separation (difficult to recycle or reuse) of most of the exhausted nanoparticles (except for magnetic nanoparticles) from the treated water for reuse. This would be undesirable in terms of economical considerations (Al-Hamadani et al., 2015; Qu et al., 2012). Thirdly, there is also leakage into contact water and potential adverse effect caused to environment and human well-being (Zhang et al., 2016a,b). Additionally, although energy conservation leading to lower cost is possible due to their small sizes, overall usage cost of it still needs to be compared with other techniques in the industries (Crane and Scott, 2012). With this, there is a limited understanding on the behavior and fate of the nanomaterials at the end of it all and how it affects specifically the aquatic environment and human health that could become the bottlenecks of the application of nanotechnology in water and wastewater treatment (Dale et al., 2015; Varma, 2012).

Thus, nanocomposites are developed as they have the potential to minimize the release of the nanomaterials whilst maintaining their high reactivity. Nanocomposite could be defined as a multiphase material which at least one dimension of the constituent phases is <100 nm (Tesh and Scott, 2014). They are produced with combination of nanoparticles with different supporting materials (organic or inorganic). Thus, they can exhibit advantages of both the hosts and the impregnated functional nanoparticles. Case in point, hosts can enhance dispersion and stability of the nanoparticles and also facilitates the diffusion of contaminants in hosts, further improving the interfacial interaction. Basically, the novel combination of these substances has been realized to create a solid synergistic outcome that can greatly enhance the efficiency of water remediation. Some of them are also recyclable, cost-effective and work well with existing infrastructure (Lofrano et al., 2016; Yin and Deng, 2015) as nanotechnology-based multifunctional and highly efficient processes do not rely on large infrastructure or centralized system (Amin et al., 2014). They are also flexible in system size and configuration (Qu et al., 2012). On a different note, synthesizing nanoparticles through green techniques also has gained recently major importance and it became one of the most adequate and suggested methods. The benefits of this are simple, good stability of the nanoparticles, inexpensive, nontoxic by-products, less time consuming and large-scale synthesis (Awwad et al., 2013; Moritz and Geszke-Moritz, 2013).

Hence, future usage of this technology will only focus on efficient processes where only small amount of nanomaterials is needed as nanomaterials are relatively expensive as well. This is because the production of nanoparticles often needs the use of an aggressive chemical reducing agent and a capping agent and may additionally involve a volatile organic solvent (Varma, 2012). But long term reusability of nanomaterials could enhance their cost-effectiveness (Qu et al., 2012). Most importantly, more work is necessary to develop cost effective ways to generate nanomaterials and test their efficiency at large scale for successful field application (Anjum et al., 2016). The application of nanomaterials in lab scale has already demonstrated positive outcomes. The main challenge now would be the application of this lab scale concept in the field as the readiness for commercialization of nanomaterials differs greatly from each other (Thines et al., 2017).

11.4.1 Coagulation Using Nanoparticles

There is no doubt that coagulation is one of the most widely known physiochemical means utilized for leachate treatment (Daud et al., 2012; Maleki et al., 2009; Tatsi et al., 2003). One of the reasons is that it is a somewhat uncomplicated approach that could be assigned in the treatment of old landfill leachates (Amor et al., 2015). However, the choosing of appropriate coagulants is important to get better coagulation performance (Yaser and Pogaku, 2017). On the other hand, for performance optimization, choosing of the right coagulant, selection of the optimal experimental settings, evaluation of pH aftermath and studies on appropriate reagents dosage are important (Amor et al., 2012; Maleki et al., 2009).

The process mainly involves nanoparticles react with suspended and colloidal particles in water or wastewater to enhance them to bind together, thus leading to their removal in the subsequent treatment process (Hassan et al., 2016). The mechanism of aggregation consists of the combination of charge neutralization, entrapment, adsorption and/or complexion with the coagulant ions to form insoluble precipitate. Therefore, the nanoparticles and colloids can be removed. Coagulation process forms an integral part of the conventional water/wastewater treatment and has been employed to decrease turbidity, color and to remove pathogens (Verma et al., 2012). More often than not, nanomaterials are used as nanoabsorbents for adsorption and there are only limited studies which are dedicated to using nanoparticles as coagulants. Thus, this review paper will focus on the practicality and application of a number of nanoparticles in water remediation through mainly coagulation process.

Novel nano-size metallic calcium/iron dispersed reagent nano-Ca/CaO and nano-Fe/Ca/CaO was synthesized and used as coagulant/catalyst in hybrid zero valent iron (ZVI)/H_2O_2 oxidation process to treat MSW tipping hall leachate and MSW landfill leachate (MSWLL). The coagulation with nano- Fe/Ca/CaO was optimized to achieve the best starting conditions for the Fenton process by generating a low COD and high Fe concentration. In this study, nano-Fe/Ca/CaO was first use as coagulant and then, the nonprecipitated Fe remaining in dissolution was used as catalyst in the oxidation process. During nano-coagulation process, the colloidal particles electrostatic surface potential is neutralized and destabilized particles stick together into larger flocs which then aggregate with suspended polluting matter. In the zero valent iron (ZVI)/H_2O_2 oxidation process, the nano-iron from the excess of coagulant (nFe/Ca/CaO) ferrous ion reacts with hydrogen peroxide, producing a hydroxyl radical (.OH). As result, respective removal of 64.0%, 56.0%, and 20.7% for color, COD, and TSS were obtained with 2.5 g/L of nano-Ca/CaO, while 67.5%, 60.2%, and 37.7% removal for color, COD, and TSS were attained respectively, with only 1.0 g/L of nano-Fe/Ca/CaO, in MSWLL. Successfully, 91%−99% removal efficiency of heavy metal was attained after treatment with nano-Fe/Ca/CaO in both leachate samples. The performance was further enhanced after applying hybrid coupled coagulation-Fenton process where 95% color, 96% COD, and 66% TSS removal were achieved using a 1.0 g/L of nano-Fe/Ca/CaO and 20 mM H_2O_2 doses. After this treatment, the color, COD, TSS and heavy metals were significantly decreased. A hybrid coupled zero valent iron (ZVI)/H_2O_2 oxidation process with novel nano-sized metallic calcium/iron dispersed reagent appears to be a suitable treatment of leachate (Lee et al., 2017).

Composite coagulants are introduced to maximize coagulation efficiencies (Moussas and Zouboulis, 2009; Wei et al., 2009). Instead of adding coagulants separately which requires two reagent addition system (Ovenden and Xiao, 2002), the composite coagulant is added to water in one go. However, there is a limited amount of research done on the mechanism of magnetic coagulation and the characterizations of the formed flocs (Zhang et al., 2012). Nanomagnetite is one of the most commonly used nanoparticles in water coagulation. A study has shown that nano-Fe_3O_4 particles can improve floc density and settling performance by wrapping up Fe_3O_4, thus enhancing the removal rate of *Microcystis aeruginosa* (Sun et al., 2016). There is also downsizing in the use of chemical products, including sludge volume and low cost of materials for the use of this material (Lakshmanan et al., 2013; Okoli et al., 2012). One of the most used composite coagulant types for coagulation is nanomagnetic composite coagulant. Mostly in the researches, the nanomagnetite is combined with an inorganic polymer such as PAC, PFS, PFC, and PFAC. Since Fe_3O_4 is magnetic, it can combine with the flocculation body of the coagulant to form a more compact magnetic flocculation with increasing dosage. Then, there will be an increase in attraction between magnetic particles, thus enhancing aggregation that leads to an efficient coagulation process (Kushida et al., 2013).

After a novel ferromagnetic nanoparticle combined with PAC has been produced, it has shown that the addition of Fe_3O_4 increases COD removal and speeds up the sedimentation rate of particles during coagulation (Fosso-Kankeu et al., 2016; Santos et al., 2016; Zhang et al., 2015; Zhang et al., 2012). It also results in better performance in both turbidity and dissolved organic carbon (DOC) removal when compared with using PAC alone. For the further study of the novel ferromagnetic nanoparticle combined with PAC, MFPAC is prepared to treat mature landfill leachate (Liu et al., 2017). Fig. 11-5 shows SEM micrograph of MFPAC. An experiment has shown that the coagulation and flocculation performance of MFPAC is better than that using PAC alone in COD, color, humic, and fulvic acid removal. The results obtained are similar in the forms of COD and color removal in another study as well (Liu et al., 2017). The mass ratio of PAC and Fe_3O_4 is 3:1. X-ray diffraction pattern analysis indicates that the chemical composition of MFPAC does not change in a major way. This means

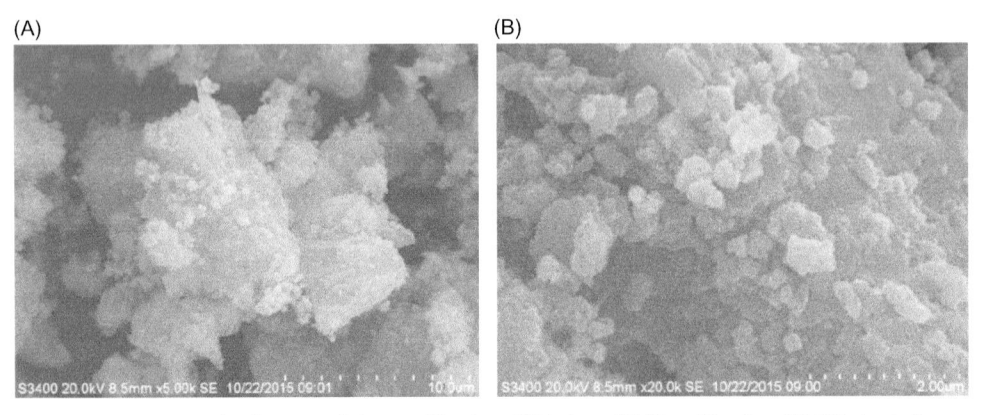

FIGURE 11-5 SEM micrograph of MFPAC: (A) Magnification: 5000 times (B) Magnification: 20,000 times. Source: *From Liu, Z., Duan, X., Zhan, P., Liu, R., Nie, F., 2017. Coagulation performance and microstructural morphology of a novel magnetic composite coagulant for pre-treating landfill leachate. Int. J. Environ. Sci. Technol., 14 (11), 2507–2518. doi: 10.1007/s13762-017-1338-7.*

that the two materials have combined together comprehensively to exhibit chemical characteristics of both of them. The magnetic components may improve coagulation behaviors and, side note, involve in magnetic separation after the coagulation process (Lakshmanan and Kuttuva Rajarao, 2014). Scanning electron microscope spectral also shows that the composite coagulant is similar to those of typical porous-type materials, giving a larger specific surface, smaller average pore diameter and more pore volume of a structure. There is also a midlevel passage pore connecting the narrow cracks. This composite can be used to remove *Microcystis aeruginosa* more efficiently in removal rate at a wider pH range as well compared to using PAC alone (Zhang et al., 2015). Though the mass ratio of the composite coagulant here is 4:1 (PAC to nano-Fe_3O_4). Nanomagnetite combine with PAC to form clusters and generate flocs with algae afterwards. The more nanomagnetite it is, the bigger the size of the floc will be. The nanomagnetic composite coagulant provides nucleation sites for larger flocs to integrate with *M. aeruginosa*. Increased floc density also improves the removal of the algae. MFPAC was also used by Liu et al. (2017) to study the pretreatment of landfill leachate using the combined coagulation process and aged-refuse adsorption process. The results showed COD and color removals of 62.6% and 66.5%, respectively indicating that MFPAC offered a better coagulation efficiency than PAC.

MFPFS coagulation combined with sulfate radical oxidation was applied to treat biologically pretreated leachate using a novel magnetic composite coagulant of MFPFS, prepared from Fe_3O_4 nanoparticles and PFS. Better color and COD removals were realized using the MFPFS magnetic coagulant as compared to when using PFS alone. The removals of color and COD reached up to 80% and 60%, respectively at working conditions of 1.2 g/L MFPFS dosage, 1:3 Fe_3O_4/PFS mass ratios and 50 minutes sedimentation time, without pH adjustment, (Liu et al., 2016).

Another example of a nanomagnetic composite coagulant is the combination of Fe_3O_4 and PFC (Jiang et al., 2010) to remove algae and it has shown promising coagulation performance in higher removal values and slighter pH dependence. Nanomagnetite can be combined with PFS which has a great effect in removing arsenic in large area, unexpected arsenic leakage situation and in ground water arsenic treatment (Li et al., 2010). Controllable conditions are mixing time and dosages. Because of surface charge interaction, mesh structure of iron oxide will cause precipitation on the magnetic particles.

Besides, nano-Fe_3O_4 can be composited with an organic flocculant (chitosan or polyacrylamide) to produce an inorganic-organic composite flocculant that can treat wastewater (Hong et al., 2007; Liu et al., 2008). For chitosan, it can be used to treat algae containing water as well (Liu et al., 2014). Though, few resources are available on nano-Fe_3O_4 as a flocculant aid to remove algae in water remediation. Another example of nano-Fe_3O_4 combined with an organic component is the gum karaya hydrogel nanocomposite flocculant which has been utilized to remove metal ions from mine effluents (Fosso-Kankeu et al., 2016). It is made from gum karaya-grafted poly (acrylamide-co-acrylic acid) incorporated onto iron oxide magnetic nanoparticles (nano-Fe_3O_4) hydrogel nanocomposite. Not only that, *Moringa oleifera* Lam, a natural coagulant has been functionalized with magnetic iron oxide nanoparticles to produce flake (Santos et al., 2016). This is a great alternative in terms of being cost effective and environmentally friendly compared to the use of synthetic compound in the water treatment process.

There are also other nanoparticles that could be used as coagulants. Nanosized manganese dioxide ($nMnO_2$) is used as an enhanced coagulant to remove Tl in simulated water and surface waters (Huangfu et al., 2017) by improving flocculation performance for high and low turbidity water. Enhanced coagulation is carried out with a lower cost and extra convenience without complicated structures. Another example is the use of graphene oxide (GO) as a novel flocculant to remove contaminants with different surface charges from water, including two particulate ones (kaolin and hematite) and two soluble ones (humic acid) and cationic light yellow 7GL dye (7GL). PAC and original graphite are tested for comparison (Yang et al., 2013). GO is observed to perform better for positively charged contaminants (hematite and 7GL) via patching effect for hematite suspension and charge neutralization effect for 7GL solution. For negatively charged contaminants (kaolin and HA), it is done through sweeping flocculation effect under acidic and neutral conditions.

One of the examples of application of green chemistry in this field is green synthesized ZnO nanoparticles (from black tea solid waste and Zn acetate dehydrate) are used to remove pharmaceutical active compounds (PACs) — Ibuprofen, Ephedrine, and Propranolol from urine via chemical coagulation for safe reuse (agriculture purposes) without any environmental threat (Hassan et al., 2016). A semipilot plant is designed and operated to treat the urine through a mixing tank for coagulation using the ZnO nanoparticles. Fig. 11-6 shows SEM image of green synthesized ZnO nanoparticles at different magnification. The synthetic polymers used in the industries are mostly made from oil-based, nonrenewable raw materials. There has been interest in replacing them with more sustainable natural bio-based alternatives. Anionic nanofibrillated dicarboxylated cellulose (DCC) bioflocculants with variable charge densities are used to treat municipal waste water and are taken to compare with a commercial coagulant and a synthetic polymeric flocculant. The results are high charge density and high nanofibril content of the DCC flocculant is able to flocculate waste water effectively and requires similar dosage as when using the commercial flocculant.

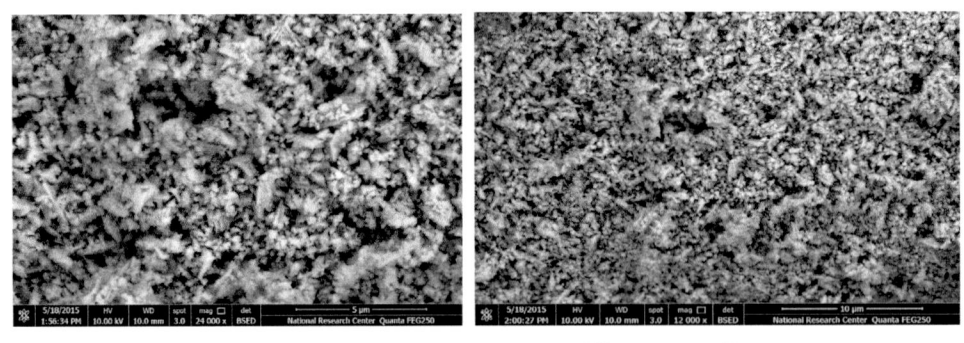

FIGURE 11-6 SEM image of green synthesized ZnO nanoparticles at different magnification. Source: *From Hassan, S.S., Abdel-Shafy, H.I., Mansour, M.S., 2016. Removal of pharmaceutical compounds from urine via chemical coagulation by green synthesized ZnO-nanoparticles followed by microfiltration for safe reuse. Arab. J. Chem.*

Positively charged carbon nanotubes as a type of nanomaterial can be used as heterogeneous coagulants and/or flocculants in the pretreatment of brewery wastewater instead of the traditional $FeCl_3$ through the surface charge neutralization between CNTs and colloidal particles. Though $FeCl_3$ turns out to perform more effectively than the CNTs (Simate et al., 2012) and the two coagulants have no synergetic effect when used together. Single-walled carbon nanotubes (SWCNTs) and multiwalled carbon nanotubes (MNCNTs) are also utilized to remove natural organic matter and organic micropollutants in various water sources and powdered activated carbon is used as a comparison. SWCNTs turn out to outperform MWCNTs in low DOC source water because of its smaller diameter and larger surface area (Joseph et al., 2012). CNTs and specifically SWCNTs are used to remove bisphenol A and 17α-ethinyl estradiol in other applications (Joseph et al., 2013; Joseph et al., 2011). Powdered activated carbon is used for comparison. SWCNTs again outperform MWCNTs. The highest removals are in young leachate for both CNTs. But through it all, powdered activated carbon often outperforms both SWCNTs and MWCNTs, making it doubtful whether the CNTs are suitable to be used as coagulants in the real field when compared to other conventional coagulants. Fig. 11-7 shows SEM image of aligned CNTs.

However, compared with one-dimensional CNTs, 2D graphene based materials with atomic thickness and larger surface area are more interesting (Yang et al., 2012). One significant branch of that is GO, an oxidized form of GO (Chen et al., 2012; Sun et al., 2012). GO consists of various groups that can enhance dispersibility and remove pollutants in water. Preparation of GO, using Hummers' method (Hummers and Offeman, 1958), is significantly easier than the special oxidation processes of CNTs and thus offers the potential of cost effective and large scale production. With concern of CNTs to the human health (Simate et al., 2012), GO is revealed to have better biocompatibility (Chen et al., 2012). It can be biodegradable in the presence of certain enzymes (Kotchey et al., 2011). But if it is ingested,

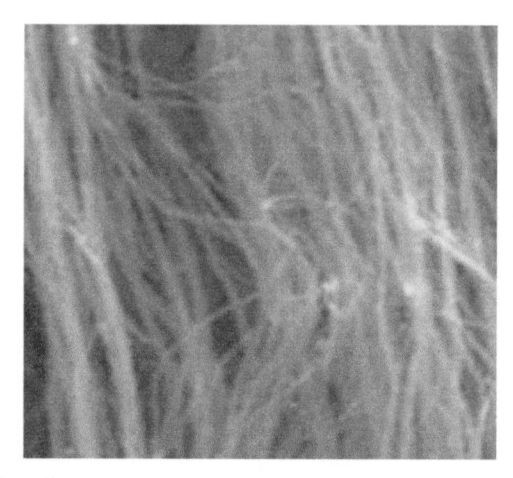

FIGURE 11-7 SEM image of aligned CNTs. Source*: Amin, M., Alazba, A., Manzoor, U., 2014. A review of removal of pollutants from water/wastewater using different types of nanomaterials. Adv. Mater. Sci. Eng. 2014.*

there may still be possible risks to human health as GO can exhibit either nontoxicity or dose-dependent toxicity (Jastrzębska et al., 2012; Sanchez et al., 2011). Moreover, GO is currently relatively expensive because of how it's made (Yang et al., 2013). Table 11-5 summarized the performance of wastewater treated by coagulation using nano materials.

11.5 Conclusion and Future Works

Through this study which has been primarily focusing on coagulation of biologically digested landfill leachate, different coagulants have been touched on, such as metal, polymer (organic synthetic, inorganic, organic-inorganic), natural coagulant (chitosan and plant-based) and nanoparticles. But before that, it is important to understand the generation of leachate in the first place and its characteristics. The landfill leachate treatment has become a vital part of environmental protection. Landfill characteristics of leachate varied by several factors such as age of landfill, operation period of landfill, climate, hydrological condition, moisture content, temperature, pH value, biological, and chemical activities and degree of stabilization. Coagulation process is considered as one process that can improve the performance of treatment of the leachate especially in terms of color depending on the coagulant used and the operating condition. It has ability in separating various kinds of particles from leachate that uses abundant and low cost chemical as compared to other treatment. The application of metal coagulant in leachate treatment has contributed mostly on the removal of COD, color, suspended solid, and turbidity of leachate. There are three major groups of polymer being used in wastewater treatment; organic, inorganic and combination of both. Natural organic is toxic free but have shorter lifespan compared to the synthetic organic polymer. On the other hand, inorganic polymer is formed when additive compound is added to inorganic coagulant in order to enhance its coagulation flocculation efficiency. Since inorganic polymer contains metal, its dosage must be lowered to minimize toxicity. Therefore, organic polymer is added as flocculant aid to improve the coagulation flocculation performance and reduce the toxicity inorganic polymer. Most plant-based coagulants such as okra, barley, and starch act as flocculant aids while Moringa Oleifera is able to act as sole and main coagulant for the coagulation treatment of leachate. Chitosan is also proven to be feasible to treat wastewater. As for nanoparticles, the efficiency of adsorption is scrutinized to analyze the coagulation performance. Nanoadsorbents such as CNTs, GO, nanocellulose, magnetic composite nanoparticles, ZnO, gum karaya hydrogel, nano-Al_{13}, UVA/MgO and nanosized manganese dioxide mainly remove heavy metal ions in contaminant water sources with their high aspect ratios.

11.5.1 Future Perspective

Continuous study must be done on the characteristic of the landfill leachate because the characteristics varies with time, source of leachate and to determine the most effective leachate treatment. It is suggested that further research need to be done by using the organic coagulant instead of metal or chemical coagulant to decrease the bad effect to the

Table 11-5 Performance of Wastewater Treated by Coagulation Using Nano Materials

| Sources | Method | Using | Targeting | Conditions | | | | | | Removal Efficiencies (%) | | | | | | References |
				pH	Dosage (mg/L)	Settling Time (min)	Rapid (rpm)	Slow (rpm)	Zeta Potential (mV)	COD	BOD	DOC	Color	Turbidity	Others	
Synthetic water	Coagulation	MPACl 2.0	HA and Kaolin	4.0	0.08 mmol/L as Al	20	200, 1 min	30, 15 min	−10-0			90		>90		(Zhang et al., 2012)
Culture	Coagulant aid	Nano-Fe$_3$O$_4$ + PAC	M. aeruginosa	11.2	2 as PAC	60	300, 2 min	100, 10 min	−10						173.1 M. Aeruginosa	(Zhang et al., 2015)
Mature landfill leachate	Coagulation	MFPAC		7.5–8.1	1500–1750	30	300, 1 min	70, 15 min		>59			>67			(Liu, Duan, et al., 2017)
Landfill leachate	Coagulation	MFPAC		7.5	1500	30	300, 1 min	70, 15 min		63			67			(Liu, Li, et al., 2017)
FACHB	Coagulation	MPFC	M. aeruginosa	10	3.5 as Fe	60	400, 2 min	50, 10 min	0.13						>95 algae	(Jiang et al., 2010)
Mine effluents	Coagulation	Gum karaya hydrogel nano-composite flocculant	Metal ions	4	100	15	200, 1.5 min	50, 15 min						100	Different types of metal ions	(Fosso-Kankeu et al., 2016)
Surface waters	Coagulation	Nano-Fe$_3$O$_4$ + Moringa oleifera Lam			10	30	100, 3 min	15, 15 min				85		90		(Santos et al., 2016)
Human Urine	Chemical coagulation	Nano-ZnO	PACs	5.1	1500	60	350, 5 min	80, 30 min	−41.3	97.48	96.05			97.96	98.40 Ibuprofen, 91.13 Epheddrine, 98.68 Propranolol	(Hassan et al., 2016)
Surface waters	Enhanced coagulation	nMnO$_2$	Thallium	9.0	60 µM	30	250, 2 min	50, 15 min	−41.96						0.03 µg/L Tl residual	(Huangfu et al., 2017)
Municipal waste water	Coagulation-flocculation	DCC		3.5–7.0	5	30	200, 3 min	40, 15 min	−15	60				80		(Suopajärvi, Liimatainen, Hormi, & Niinimäki, 2013)
Distilled water	Flocculation	GO	Kaolin	3.95	8	40	200, 5 min	50, 15 min	0					8.42 residual	1.92 residual contaminants	(Yang et al., 2013)
Brewery wastewater	Heterogeneous coagulation	CNTs		8.3	60	30	150, 3 min	60, 20 min	−15	33.03				58.82		(Simate et al., 2012)

(Continued)

Table 11-5 (Continued)

Sources	Method	Using	Targeting	Conditions							Removal Efficiencies (%)						References
				pH	Dosage (mg/L)	Settling Time (min)	Rapid (rpm)	Slow (rpm)	Zeta Potential (mV)		COD	BOD	DOC	Color	Turbidity	Others	
Brackish water	Coagulation	SWCNT	NOM	7.5	100	60	250, 3 min	30, 30 min								75	(Joseph et al., 2012)
Seawater		MWCNT		5	80											54	
Young leachate	Coagulation	MCNT	EE2		200	60	250, 3 min	30, 30 min								91.4 EE2	(Joseph et al., 2013)
			BPA													16.1 BPA	
		SWCNT	EE2													>99 EE2	
			BPA													91 BPA	

FACHB, Freshwater Algal Culture Collection of Institute of Hydrobiology.
PACs, pharmaceutical active compounds.
HA, hHumic acid.
NOM, natural organic matter.
EE2-17α-ethinyl estradiol.
BPA, bisphenol A.

environment. The application of metal coagulant also gives negatives drawbacks such as increasing the concentration of aluminum or iron in the leachate that might constitute a potential risk to the quality of the environment itself. Therefore, further studies on the alternative coagulant such as natural polymer to substitute the usage of metal coagulant should be done. Extraction and purification of plant-based coagulant may increase their processing costs and may not be practical as leachate treatment system. Nonetheless, extraction using active coagulating agents is still a noteworthy aspect which may prove useful should they be commercialized or applied in concentrated form for industrial leachate treatment. In short, dosage must be adequate and condition for wastewater to reach maximum removal rate must in between acidic and neutral. Nanoparticles have been known to cause secondary pollutions. Not only that, most of the nanomaterials have not been cost-competitive when compared with conventional materials such as activated carbons. Thus, future applications will focus on efficient processes where only small quantities of nanomaterials are required. Moreover, further work is required on developing a cost effective methods of synthesizing nanomaterials and testing the efficiency at large scale for successful field application.

References

Abbas, A.A., Jingsong, G., Ping, L.Z., Ya, P.Y., Al-Rekabi, W.S., 2009. Review on landwll leachate treatments. J. Appl. Sci. Res. 5 (5), 534−545.

Abu Amr, S.S., Zakaria, S.N.F., Aziz, H.A., 2016. Performance of combined ozone and zirconium tetrachloride in stabilized landfill leachate treatment. J. Mater. Cycles Waste Manage. Available from: https://doi.org/10.1007/s10163-016-0524-x.

Abu Amr, S.S., Zakaria, S.N.F., Aziz, H.A., 2017. Performance of combined ozone and zirconium tetrachloride in stabilized landfill leachate treatment. J. Mater. Cycles Waste Manage. 19 (4), 1384−1390. Available from: https://doi.org/10.1007/s10163-016-0524-x.

Aguilar, M.I., Sáez, J., Lloréns, M., Soler, A., Ortuño, J.F., Meseguer, V., et al., 2005. Improvement of coagulation−flocculation process using anionic polyacrylamide as coagulant aid. Chemosphere 58 (1), 47−56. Available from: https://doi.org/10.1016/j.chemosphere.2004.09.008.

Al-Hamadani, Y.A., Chu, K.H., Son, A., Heo, J., Her, N., Jang, M., et al., 2015. Stabilization and dispersion of carbon nanomaterials in aqueous solutions: a review. Sep. Purif. Technol. 156, 861−874.

Alvarez-Vazquez, H., Jefferson, B., Judd, S.J., 2004. Membrane bioreactors vs conventional biological treatment of landfill leachate: a brief review. J. Chem. Technol. Biotechnol. 79 (10), 1043−1049.

Ameen, E.S.M., Muyibi, S.A., Abdulkarim, M.I., 2011. Microfiltration of pretreated sanitary landfill leachate. Environmentalist 31 (3), 208−215. Available from: https://doi.org/10.1007/s10669-011-9322-0.

Amin, M., Alazba, A., Manzoor, U., 2014. A review of removal of pollutants from water/wastewater using different types of nanomaterials. Adv. Mater. Sci. Eng. 2014.

Amokrane, A., Comel, C., Veron, J., 1997. Landfill leachates pretreatment by coagulation-flocculation. Water Res. 31 (11), 2775−2782. Available from: https://doi.org/10.1016/S0043-1354(97)00147-4.

Amor, C., Lucas, M.S., Pirra, A.J., Peres, J.A., 2012. Treatment of concentrated fruit juice wastewater by the combination of biological and chemical processes. J. Environ. Sci. Health, Part A 47 (12), 1809−1817.

Amor, C., Torres-Socías, E.D., Peres, J.A., Maldonado, M.I., Oller, I., Malato, S., et al., 2015. Mature landfill leachate treatment by coagulation/flocculation combined with Fenton and solar photo-Fenton processes. J. Hazard. Mater. 286 (Supplement C), 261−268. Available from: https://doi.org/10.1016/j.jhazmat.2014.12.036.

Anjum, M., Miandad, R., Waqas, M., Gehany, F., Barakat, M., 2016. Remediation of wastewater using various nano-materials. *Arabian Journal of Chemistry*.

Awwad, A.M., Salem, N.M., Abdeen, A.O., 2013. Green synthesis of silver nanoparticles using carob leaf extract and its antibacterial activity. Int. J. Ind. Chem. 4 (1), 29.

Aziz, H.A. and Ramli, S.F., Coagulation-Flocculation of Anaerobic Landfill Leachate using Ferric Chloride (FeCl3), Aloe Vera (AV) and Chitosan (CS), The Standard International Journal, May 2014, 1−5.

Aziz, H.A., Sobri, N.I.M., 2015. Extraction and application of starch-based coagulants from sago trunk for semi-aerobic landfill leachate treatment. Environ. Sci. Pollut. Res. 22 (21), 16943−16950. Available from: https://doi.org/10.1007/s11356-015-4895-7.

Aziz, H.A., Ramli S.F., 2014. Coagulation-flocculation of anaerobic landfill leachate using ferric chloride (FeCl$_3$), aloe vera (AV) and chitosan (CS). Stand. Int. J. 1−5.

Bashir, M.J.K., Isa, M.H., Kutty, S.R.M., Awang, Z.B., Aziz, H.A., Mohajeri, S., et al., 2009. Landfill leachate treatment by electrochemical oxidation. Waste Manage. 29 (9), 2534−2541. Available from: https://doi.org/10.1016/j.wasman.2009.05.004.

Bratby, J., 2016. Coagulation and Flocculation in Water and Wastewater Treatment. IWA Publishing.

Brumfiel, G., 2003. Nanotechnology: a little knowledge. Nature 424 (6946), 246−248.

Camacho, F.P., Sousa, V.S., Bergamasco, R., Ribau Teixeira, M., 2017. The use of Moringa oleifera as a natural coagulant in surface water treatment. Chem. Eng. J. 313 (Supplement C), 226−237. Available from: https://doi.org/10.1016/j.cej.2016.12.031.

Chakraborti, R.K., Gardner, K.H., Atkinson, J.F., Van Benschoten, J.E., 2003. Changes in fractal dimension during aggregation. Water Res. 37 (4), 873−883. Available from: https://doi.org/10.1016/S0043-1354(02)00379-2.

Chaturvedi, S., Dave, P.N., Shah, N., 2012. Applications of nano-catalyst in new era. J. Saudi Chem. Soc. 16 (3), 307−325.

Chen, D., Feng, H., Li, J., 2012. Graphene oxide: preparation, functionalization, and electrochemical applications. Chem. Rev. 112 (11), 6027−6053.

Chen, L., Hu, P., Zhang, L., Huang, S., Luo, L., Huang, C., 2012. Toxicity of graphene oxide and multi-walled carbon nanotubes against human cells and zebrafish. Sci. China Chem. 55 (10), 2209−2216.

Comstock, S.E.H., Boyer, T.H., Graf, K.C., Townsend, T.G., 2010. Effect of landfill characteristics on leachate organic matter properties and coagulation treatability. Chemosphere 81 (7), 976−983. Available from: https://doi.org/10.1016/j.chemosphere.2010.07.030.

Crane, R., Scott, T., 2012. Nanoscale zero-valent iron: future prospects for an emerging water treatment technology. J. Hazard. Mater. 211, 112−125.

Dale, A.L., Casman, E.A., Lowry, G.V., Lead, J.R., Viparelli, E., Baalousha, M., 2015. Modeling Nanomaterial Environmental Fate in Aquatic Systems. ACS Publications.

Daud, Z., Aziz, A., Abdul Latif, L.M., 2012. Coagulation-flocculation In leachate treatment by using ferric chloride and alum as coagulant.

Davis, M.L., 2011. *Water and Wastewater Engineering: Design Principle and Practice* (International edition, 2011 ed. McGraw-Hill, New York, Americas.

Do Carmo Nascimento, I.O., Guedes, A.R.P., Perelo, L.W., Queiroz, L.M., 2016. Post-treatment of sanitary landfill leachate by coagulation-flocculation using chitosan as primary coagulant. Water Sci. Technol. 74 (1), 246−255. Available from: https://doi.org/10.2166/wst.2016.203.

Dolar, D., Košutić, K., Strmecky, T., 2016. Hybrid processes for treatment of landfill leachate: coagulation/UF/NF-RO and adsorption/UF/NF-RO. Sep. Purif. Technol. 168, 39−46. Available from: https://doi.org/10.1016/j.seppur.2016.05.016.

Foo, K.Y., Hameed, B.H., 2009. An overview of landfill leachate treatment via activated carbon adsorption process. J. Hazard. Mater. 171 (1), 54−60. Available from: https://doi.org/10.1016/j.jhazmat.2009.06.038.

Fosso-Kankeu, E., Mittal, H., Waanders, F., Ntwampe, I., Ray, S.S., 2016. Preparation and characterization of gum karaya hydrogel nanocomposite flocculant for metal ions removal from mine effluents. Int. J. Environ. Sci. Technol. 13 (2), 711−724.

Freitas, T.K.F.S., Almeida, C.A., Manholer, D.D., Geraldino, H.C.L., de Souza, M.T.F., Garcia, J.C., 2018. Review of utilization plant-based coagulants as alternatives to textile wastewater treatment. In: Muthu, S. S. (Ed.), Detox Fashion: Waste Water Treatment. Springer Singapore, Singapore, pp. 27−79.

Gálvez, A., Greenman, J., Ieropoulos, I., 2009. Landfill leachate treatment with microbial fuel cells; scale-up through plurality. Bioresour. Technol. 100 (21), 5085−5091. Available from: https://doi.org/10.1016/j.biortech.2009.05.061.

Gandhimathi, R., Durai, N.J., Nidheesh, P.V., Thanga Ramesh, S., Kanmani, S., 2013. *Use of Combined Coagulation-Adsorption Process as Pretreatment of Landfill Leachate* (Vol. 10).

Gautam, R.K., Chattopadhyaya, M.C., 2016. Nanomaterials for Wastewater Remediation. Butterworth-Heinemann.

Ghafari, S., Aziz, H.A., Bashir, M.J.K., 2010. The use of poly-aluminum chloride and alum for the treatment of partially stabilized leachate: a comparative study. Desalination 257 (1), 110−116. Available from: https://doi.org/10.1016/j.desal.2010.02.037.

Guo, J., 2016. Pretreatment of landfill leachate by using the composite of poly ferric sulfate and bioflocculant MBFR10543. Desalin. Water Treat. 57 (41), 19262−19272. Available from: https://doi.org/10.1080/19443994.2015.1107506.

Hassan, S.S., Abdel-Shafy, H.I., Mansour, M.S., 2016. Removal of pharmaceutical compounds from urine via chemical coagulation by green synthesized ZnO-nanoparticles followed by microfiltration for safe reuse. Arab. J. Chem.

Hong, M.K., Park, B.J., Choi, H.J., 2007. Preparation and physical characterization of polyacrylamide coated magnetite particles. Phys. Stat. Solidi (A) 204 (12), 4182−4185.

Huangfu, X., Ma, C., Ma, J., He, Q., Yang, C., Jiang, J., et al., 2017. Significantly improving trace thallium removal from surface waters during coagulation enhanced by nanosized manganese dioxide. Chemosphere 168, 264−271.

Hummers Jr., W.S., Offeman, R.E., 1958. Preparation of graphitic oxide. J. Am. Chem. Soc. 80 (6), 1339-1339.

Ishak, A.R., Hamid, F.S., Mohamad, S., Tay, K.S., 2017. Removal of organic matter from stabilized landfill leachate using coagulation-flocculation-Fenton coupled with activated charcoal adsorption. Waste Manage. Res. 35 (7), 739−746. Available from: https://doi.org/10.1177/0734242X17707572.

Jastrzębska, A.M., Kurtycz, P., Olszyna, A.R., 2012. Recent advances in graphene family materials toxicity investigations. J. Nanopart. Res. 14 (12), 1320.

Jiang, C., Wang, R., Ma, W., 2010. The effect of magnetic nanoparticles on Microcystis aeruginosa removal by a composite coagulant. Colloids Surf. A 369 (1), 260−267.

Joseph, L., Zaib, Q., Khan, I.A., Berge, N.D., Park, Y.-G., Saleh, N.B., et al., 2011. Removal of bisphenol A and 17α-ethinyl estradiol from landfill leachate using single-walled carbon nanotubes. Water Res. 45 (13), 4056−4068. Available from: https://doi.org/10.1016/j.watres.2011.05.015.

Joseph, L., Flora, J.R.V., Park, Y.-G., Badawy, M., Saleh, H., Yoon, Y., 2012. Removal of natural organic matter from potential drinking water sources by combined coagulation and adsorption using carbon nanomaterials. Sep. Purif. Technol. 95 (Supplement C), 64−72. Available from: https://doi.org/10.1016/j.seppur.2012.04.033.

Joseph, L., Boateng, L.K., Flora, J.R.V., Park, Y.-G., Son, A., Badawy, M., et al., 2013. Removal of bisphenol A and 17α-ethinyl estradiol by combined coagulation and adsorption using carbon nanomaterials and powdered activated carbon. Sep. Purif. Technol. 107 (Supplement C), 37–47. Available from: https://doi.org/10.1016/j.seppur.2013.01.012.

Jumaah, M.A., Othman, M.R., Yusop, M.R., 2016. Characterization of leachate from jeram sanitary landfill-Malaysia. Int. J. ChemTech Res. 9 (8), 571–574.

Kamaruddin, M.A., Yusoff, M.S., Aziz, H.A., Hung, Y.-T., 2015. Sustainable treatment of landfill leachate. Appl. Water Sci. 5 (2), 113–126. Available from: https://doi.org/10.1007/s13201-014-0177-7.

Kamaruddin, M.A., Yusoff, M.S., Rui, L.M., Isa, A.M., Zawawi, M.H., Alrozi, R., 2017. An overview of municipal solid waste management and landfill leachate treatment: Malaysia and Asian perspectives. Environ. Sci. Pollut. Res. Available from: https://doi.org/10.1007/s11356-017-0303-9.

Kamaruddin, M.A., Abdullah, M.M.A., Yusoff, M.S., Alrozi, R., Neculai, O., 2017. Coagulation-flocculation process in landfill leachate treatment: focus on coagulants and coagulants aid. In: IOP Conference Series: Materials Science and Engineering, 209 (1), 012083.

Kargi, F., Yunus Pamukoglu, M., 2003. Simultaneous adsorption and biological treatment of pre-treated landfill leachate by fed-batch operation. Process Biochem. 38 (10), 1413–1420. Available from: https://doi.org/10.1016/S0032-9592(03)00030-X.

Kotchey, G.P., Allen, B.L., Vedala, H., Yanamala, N., Kapralov, A.A., Tyurina, Y.Y., et al., 2011. The enzymatic oxidation of graphene oxide. ACS Nano 5 (3), 2098–2108.

Kumar, S.S., Bishnoi, N.R., 2017. Coagulation of landfill leachate by $FeCl_3$: process optimization using Box–Behnken design (RSM). Appl. Water Sci. 7 (4), 1943–1953. Available from: https://doi.org/10.1007/s13201-015-0372-1.

Kumar, V., Othman, N., Asharuddin, S., 2017. Applications of natural coagulants to treat wastewater – a review. MATEC Web Conf. 103, 06016.

Kushida, M., Koide, T., Osada, I., Imaizumi, Y., Kawasaki, K., Sugawara, T., 2013. Fabrication of Fe_3O_4/SiO_2 core–shell nanoparticle monolayer as catalyst for carbon nanotube growth using Langmuir–Blodgett technique. Thin Solid Films 537, 252–255.

Lakshmanan, R., Kuttuva Rajarao, G., 2014. Effective water content reduction in sewage wastewater sludge using magnetic nanoparticles. Bioresour. Technol. 153, 333–339. Available from: https://doi.org/10.1016/j.biortech.2013.12.003.

Lakshmanan, R., Okoli, C., Boutonnet, M., Järås, S., Rajarao, G.K., 2013. Effect of magnetic iron oxide nanoparticles in surface water treatment: trace minerals and microbes. Bioresour. Technol. 129, 612–615.

Lee, C.S., Robinson, J., Chong, M.F., 2014. A review on application of flocculants in wastewater treatment. Process Saf. Environ. Prot. 92 (6), 489–508. Available from: https://doi.org/10.1016/j.psep.2014.04.010.

Lee, K.E., Teng, T.T., Morad, N., Poh, B.T., Hong, Y.F., 2010. Flocculation of kaolin in water using novel calcium chloride-polyacrylamide ($CaCl_2$-PAM) hybrid polymer. Sep. Purif. Technol. 75 (3), 346–351. Available from: https://doi.org/10.1016/j.seppur.2010.09.003.

Lee, K.E., Morad, N., Teng, T.T., Poh, B.T., 2012. Development, characterization and the application of hybrid materials in coagulation/flocculation of wastewater: a review. Chem. Eng. J. 203, 370–386. Available from: https://doi.org/10.1016/j.cej.2012.06.109.

Lee, S.D., Mallampati, S.R., Lee, B.H., 2017. Hybrid zero valent iron (ZVI)/H2O2 oxidation process for landfill leachate treatment with novel nanosize metallic calcium/iron composite. J. Air Waste Manage. Assoc. 67 (4), 475–487. Available from: https://doi.org/10.1080/10962247.2016.1252449.

Li, S., Lv, Y., Liu, Z., 2015. Preparation of composite coagulant of PFM-PDMDAAC and its coagulation performance in treatment of landfill leachate. J. Water Reuse Desalin. 5 (2), 177–188. Available from: https://doi.org/10.2166/wrd.2015.093.

Li, T., Zhu, Z., Wang, D., Yao, C., Tang, H., 2006. Characterization of floc size, strength and structure under various coagulation mechanisms. Powder Technol. 168 (2), 104–110. Available from: https://doi.org/10.1016/j.powtec.2006.07.003.

Li, W., Hua, T., Zhou, Q., Zhang, S., Li, F., 2010. Treatment of stabilized landfill leachate by the combined process of coagulation/flocculation and powder activated carbon adsorption. Desalination 264 (1), 56−62. Available from: https://doi.org/10.1016/j.desal.2010.07.004.

Li, Y., Wang, J., Zhao, Y., Luan, Z., 2010. Research on magnetic seeding flocculation for arsenic removal by superconducting magnetic separation. Sep. Purif. Technol. 73 (2), 264−270.

Liu, J., Tao, Y., Wu, J., Zhu, Y., Gao, B., Tang, Y., et al., 2014. Effective flocculation of target microalgae with self-flocculating microalgae induced by pH decrease. Bioresour. Technol. 167, 367−375.

Liu, S., 2013. Landfill leachate treatment methods and evaluation of Hedeskoga and Måsalycke landfills.

Liu, X., Hu, Q., Fang, Z., Zhang, X., Zhang, B., 2008. Magnetic chitosan nanocomposites: a useful recyclable tool for heavy metal ion removal. Langmuir 25 (1), 3−8.

Liu, X., Li, X.-M., Yang, Q., Yue, X., Shen, T.-T., Zheng, W., et al., 2012. Landfill leachate pretreatment by coagulation−flocculation process using iron-based coagulants: optimization by response surface methodology. Chem. Eng. J. 200, 39−51. Available from: https://doi.org/10.1016/j.cej.2012.06.012.

Liu, Z., Zhan, P., Nie, F., 2016. Advanced treatment of leachate secondary effluent by a combined process of MFPFS coagulation and sulfate radical oxidation. Polish J. Environ. Stud. 25 (4), 1615−1622. Available from: https://doi.org/10.15244/pjoes/62645.

Liu, Z., Liu, R., Nie, F., 2017. MFPAC coagulation and aged refuse adsorption pretreatment of landfill leachate. Chinese J. Environ. Eng. 11 (1), 393−400. Available from: https://doi.org/10.12030/j.cjee.201509024.

Liu, Z., Duan, X., Zhan, P., Liu, R., Nie, F., 2017. Coagulation performance and microstructural morphology of a novel magnetic composite coagulant for pre-treating landfill leachate. Int. J. Environ. Sci. Technol. 14 (11), 2507−2518. Available from: https://doi.org/10.1007/s13762-017-1338-7.

Liu, Z., Li, X., Xu, L., Liu, R., 2017. A novel combined process of MFPAC coagulation and mineralized refuse adsorption for landfill leachate pretreatment. Desalin. Water Treat. 80, 74−82. Available from: https://doi.org/10.5004/dwt.2017.20961.

Lofrano, G., Carotenuto, M., Libralato, G., Domingos, R.F., Markus, A., Dini, L., et al., 2016. Polymer functionalized nanocomposites for metals removal from water and wastewater: an overview. Water Res. 92, 22−37.

Long, Y., Xu, J., Shen, D., Du, Y., Feng, H., 2017. Effective removal of contaminants in landfill leachate membrane concentrates by coagulation. Chemosphere 167, 512−519. Available from: https://doi.org/10.1016/j.chemosphere.2016.10.016.

Maleki, A., Zazouli, M.A., Izanloo, H., Rezaee, R., 2009. Composting plant leachate treatment by coagulation-flocculation process. Am.-Eurasian J. Agric. Environ. Sci. 5 (5), 638−643.

Mårtensson, A.M., Aulin, C., Wahlberg, O., Ågren, S., 1999. Effect of humic substances on the mobility of toxic metals in a mature landfill. Waste Manage. Res. 17 (4), 296−304. Available from: https://doi.org/10.1034/j.1399-3070.1999.00053.x.

McBean, E.A., Rovers, A.F., Farquhar, J.G., 1995. Solid Waste Landfill Engineering and Design. Prentice Hall, New Jersey.

Mohd Zin, N.S., Aziz, H.A., Tajudin, S.A.A., 2016. Performance of tapioca starch in removing suspended solid, colour and ammonia from real partially stabilized leachate by coagulation-flocculation method. ARPN J. Eng. Appl. Sci. 11 (4), 2543−2546.

Moritz, M., Geszke-Moritz, M., 2013. The newest achievements in synthesis, immobilization and practical applications of antibacterial nanoparticles. Chem. Eng. J. 228, 596−613.

Moussas, P., Zouboulis, A., 2009. A new inorganic−organic composite coagulant, consisting of polyferric sulphate (PFS) and polyacrylamide (PAA). Water Res. 43 (14), 3511−3524.

Muyibi*, S.A., Mohd Noor, M.J., Ahmadun, F.-R., Ameen, E., 2002. Bench scale studies for pretreatment of sanitary landfill leachate with moringa oleifera seeds extract. Int. J. Environ. Stud. 59 (5), 513−535. Available from: https://doi.org/10.1080/00207230212731.

Okoli, C., Boutonnet, M., Järås, S., Rajarao-Kuttuva, G., 2012. Protein-functionalized magnetic iron oxide nanoparticles: time efficient potential-water treatment. J. Nanopart. Res. 14 (10), 1194.

Oliveira, Z.L., Lyra, M.R.C.C., Arruda, A.C.F., Silva, A.M.R.B., Nascimento, J.F., Ferreira, S.R.M., 2016. Efficiency in the treatment of landfill leachate using natural coagulants from the seeds of Moringa oleifera lam and *Abelmoschus esculentus* (L.) moench (Okra). Electron. J. Geotech. Eng. 21 (24), 9735−9752.

Oloibiri, V., Ufomba, I., Chys, M., Audenaert, W.T., Demeestere, K., Van Hulle, S.W., 2015. A comparative study on the efficiency of ozonation and coagulation-flocculation as pretreatment to activated carbon adsorption of biologically stabilized landfill leachate. Waste Manag 43, 335−342. Available from: https://doi.org/10.1016/j.wasman.2015.06.014.

Oloibiri, V., De Coninck, S., Chys, M., Demeestere, K., Van Hulle, S.W.H., 2017. Characterisation of landfill leachate by EEM-PARAFAC-SOM during physical-chemical treatment by coagulation-flocculation, activated carbon adsorption and ion exchange. Chemosphere 186, 873−883. Available from: https://doi.org/10.1016/j.chemosphere.2017.08.035.

Oulego, P., Collado, S., Laca, A., Díaz, M., 2015. Tertiary treatment of biologically pre-treated landfill leachates by non-catalytic wet oxidation. Chem. Eng. J. 273, 647−655.

Ovenden, C., Xiao, H., 2002. Flocculation behaviour and mechanisms of cationic inorganic microparticle/polymer systems. Colloids and Surfaces A: Physicochemical and Engineering Aspects 197 (1), 225−234.

Peng, Y., 2017. Perspectives on technology for landfill leachate treatment. Arab. J. Chem. 10, S2567−S2574. Available from: https://doi.org/10.1016/j.arabjc.2013.09.031.

Prachi, P.G., Madathil, D., Nair, A.B., 2013. Nanotechnology in waste water treatment: a review. Int. J. ChemTech Res. 5 (5), 2303−2308.

Qu, X., Brame, J., Li, Q., Alvarez, P.J., 2012. Nanotechnology for a safe and sustainable water supply: enabling integrated water treatment and reuse. Accounts Chem. Res. 46 (3), 834−843.

Rasool, M.A., Tavakoli, B., Chaibakhsh, N., Pendashteh, A.R., Mirroshandel, A.S., 2016. Use of a plant-based coagulant in coagulation−ozonation combined treatment of leachate from a waste dumping site. Ecol. Eng. 90 (Supplement C), 431−437. Available from: https://doi.org/10.1016/j.ecoleng.2016.01.057.

Renou, S., Givaudan, J.G., Poulain, S., Dirassouyan, F., Moulin, P., 2008. Landfill leachate treatment: review and opportunity. J. Hazard. Mater. 150 (3), 468−493. Available from: https://doi.org/10.1016/j.jhazmat.2007.09.077.

Rizzo, L., Lofrano, G., Grassi, M., Belgiorno, V., 2008. Pre-treatment of olive mill wastewater by chitosan coagulation and advanced oxidation processes. Sep. Purif. Technol. 63 (3), 648−653. Available from: https://doi.org/10.1016/j.seppur.2008.07.003.

Rui, L.M., Daud, Z., Latif, A.A.A., 2012. Treatment of leachate by coagulation-flocculation using different coagulants and polymer: a review. Int. J. Adv. Sci., Eng. Inf. Technol. 2 (2), 114−117.

Sanchez, V.C., Jachak, A., Hurt, R.H., Kane, A.B., 2011. Biological interactions of graphene-family nanomaterials: an interdisciplinary review. Chem. Res. Toxicol. 25 (1), 15−34.

Santhosh, C., Velmurugan, V., Jacob, G., Jeong, S.K., Grace, A.N., Bhatnagar, A., 2016. Role of nanomaterials in water treatment applications: a review. Chem. Eng. J. 306, 1116−1137.

Santos, T.R., Silva, M.F., Nishi, L., Vieira, A.M., Klein, M.R., Andrade, M.B., et al., 2016. Development of a magnetic coagulant based on Moringa oleifera seed extract for water treatment. Environ. Sci. Pollut. Res. 23 (8), 7692−7700.

Shak, K.P.Y., Wu, T.Y., 2015. Optimized use of alum together with unmodified Cassia obtusifolia seed gum as a coagulant aid in treatment of palm oil mill effluent under natural pH of wastewater. Ind. Crops Prod. 76 (Supplement C), 1169−1178. Available from: https://doi.org/10.1016/j.indcrop.2015.07.072.

Shaylinda Mohd Zin, N., Azraff Zulkapli, Z., 2017. Application of dual coagulant (alum + barley) in removing colour from leachate. MATEC Web Conf 103, 06002.

Shu, Z., Lü, Y., Huang, J., Zhang, W., 2016. Treatment of compost leachate by the combination of coagulation and membrane process. Chinese J. Chem. Eng. 24 (10), 1369–1374. Available from: https://doi.org/10.1016/j.cjche.2016.05.022.

Sillanpää, M., Ncibi, M.C., Matilainen, A., Vepsäläinen, M., 2018. Removal of natural organic matter in drinking water treatment by coagulation: a comprehensive review. Chemosphere 190, 54–71. Available from: https://doi.org/10.1016/j.chemosphere.2017.09.113.

Simate, G.S., Iyuke, S.E., Ndlovu, S., Heydenrych, M., 2012. The heterogeneous coagulation and flocculation of brewery wastewater using carbon nanotubes. Water Res. 46 (4), 1185–1197. Available from: https://doi.org/10.1016/j.watres.2011.12.023.

Singh, S.K., Townsend, T.G., Boyer, T.H., 2012. Evaluation of coagulation (FeCl3) and anion exchange (MIEX) for stabilized landfill leachate treatment and high-pressure membrane pretreatment. Sep. Purif. Technol. 96 (Supplement C), 98–106. Available from: https://doi.org/10.1016/j.seppur.2012.05.014.

Singh Yadav, J., Dikshit, A., 2017. Stabilized Old Landfill Leachate Treatment Using Electrocoagulation, vol. 10.

Sivakumar, D., 2013. Adsorption study on municipal solid waste leachate using Moringa oleifera seed. Int. J. Environ. Sci. Technol. 10 (1), 113–124. Available from: https://doi.org/10.1007/s13762-012-0089-8.

Smaoui, Y., Chaabouni, M., Sayadi, S., Bouzid, J., 2016. Coagulation–flocculation process for landfill leachate pretreatment and optimization with response surface methodology. Desalin. Water Treat. 57 (31), 14488–14495. Available from: https://doi.org/10.1080/19443994.2015.1067837.

Stephenson, T., Stuetz, R., 2009. Principle of Water and Wastewater Treatment Processes. IWA Publishing.

Sun, P., Hui, C., Wang, S., Khan, R.A., Zhang, Q., Zhao, Y.-H., 2016. Enhancement of algicidal properties of immobilized *Bacillus methylotrophicus* ZJU by coating with magnetic Fe_3O_4 nanoparticles and wheat bran. J. Hazard. Mater. 301, 65–73.

Sun, Y., Wang, Q., Chen, C., Tan, X., Wang, X., 2012. Interaction between Eu(III) and graphene oxide nanosheets investigated by batch and extended X-ray absorption fine structure spectroscopy and by modeling techniques. Environ. Sci. Technol. 46 (11), 6020–6027. Available from: https://doi.org/10.1021/es300720f.

Suopajärvi, T., Liimatainen, H., Hormi, O., Niinimäki, J., 2013. Coagulation–flocculation treatment of municipal wastewater based on anionized nanocelluloses. Chem. Eng. J. 231, 59–67.

Tatsi, A., Zouboulis, A., Matis, K., Samaras, P., 2003. Coagulation–flocculation pretreatment of sanitary landfill leachates. Chemosphere 53 (7), 737–744.

Teh, C.Y., Wu, T.Y., 2014. The Potential Use of Natural Coagulants and Flocculants in the Treatment of Urban Waters, vol. 39.

Tesh, S.J., Scott, T.B., 2014. Nano-composites for water remediation: a review. Adv. Mater. 26 (35), 6056–6068.

Theron, J., Walker, J., Cloete, T., 2008. Nanotechnology and water treatment: applications and emerging opportunities. Crit. Rev. Microbiol. 34 (1), 43–69.

Thines, R., Mubarak, N., Nizamuddin, S., Sahu, J., Abdullah, E., Ganesan, P., 2017. Application potential of carbon nanomaterials in water and wastewater treatment: a review. Journal of the Taiwan Institute of Chemical Engineers.

Torretta, V., Ferronato, N., Katsoyiannis, I., Tolkou, A., Airoldi, M., 2017. Novel and conventional technologies for landfill leachates treatment: a review. Sustainability 9 (1), 9.

Tugtas, A.E., Cavdar, P., Calli, B., 2013. Bio-electrochemical post-treatment of anaerobically treated landfill leachate. Bioresour. Technol. 128, 266–272. Available from: https://doi.org/10.1016/j.biortech.2012.10.035.

Tyagi, I., Gupta, V., Sadegh, H., Ghoshekandi, R., Makhlouf, A.S.H., 2017. Nanoparticles as adsorbent; a positive approach for removal of noxious metal ions: a review. Sci. Technol. Dev. 34 (3), 195–214.

Varma, R.S., 2012. Greener approach to nanomaterials and their sustainable applications. Curr. Opin. Chem. Eng. 1 (2), 123−128.

Vedrenne, M., Vasquez-Medrano, R., Prato-Garcia, D., Frontana-Uribe, B.A., Ibanez, J.G., 2012. Characterization and detoxification of a mature landfill leachate using a combined coagulation−flocculation/photo Fenton treatment. J. Hazard. Mater. 205 (Supplement C), 208−215. Available from: https://doi.org/10.1016/j.jhazmat.2011.12.060.

Verma, A.K., Dash, R.R., Bhunia, P., 2012. A review on chemical coagulation/flocculation technologies for removal of colour from textile wastewaters. J. Environ. Manage. 93 (1), 154−168. Available from: https://doi.org/10.1016/j.jenvman.2011.09.012.

Vijayaraghavan, G., Sivakumar, T., Adichakkravarthy, V., 2011. Application of Plant Based Coagulants for Waste Water Treatment, vol. 1.

Volk, C., Bell, K., Ibrahim, E., Verges, D., Amy, G., LeChevallier, M., 2000. Impact of enhanced and optimized coagulation on removal of organic matter and its biodegradable fraction in drinking water. Water Res. 34 (12), 3247−3257. Available from: https://doi.org/10.1016/S0043-1354(00)00033-6.

Wang, J., Gerlach, J.D., Savage, N., Cobb, G.P., 2013. Necessity and approach to integrated nanomaterial legislation and governance. Sci. Tot. Environ. 442, 56−62.

Wang, Y., Gao, B.Y., Yue, Q., Wei, J.C., Zhou, Z.W., 2006. Novel Composite Flocculent Ployferric Chloride-Polydimethyldiallylammonium Chloride (PFC-PDMDAAC): Its Characterization and Flocculation efficiency, vol. 1.

Wang, Z.-p, Zhang, Z., Lin, Y.-j, Deng, N.-s, Tao, T., Zhuo, K., 2002. Landfill leachate treatment by a coagulation−photooxidation process. J. Hazard. Mater. 95 (1), 153−159. Available from: https://doi.org/10.1016/S0304-3894(02)00116-4.

Wei, J., Gao, B., Yue, Q., Wang, Y., Li, W., Zhu, X., 2009. Comparison of coagulation behavior and floc structure characteristic of different polyferric-cationic polymer dual-coagulants in humic acid solution. Water Res. 43 (3), 724−732.

Wei, Y., Dong, X., Ding, A., Xie, D., 2016. Characterization and coagulation−flocculation behavior of an inorganic polymer coagulant − poly-ferric-zinc-sulfate. J. Taiwan Inst. Chem. Eng. 58, 351−356. Available from: https://doi.org/10.1016/j.jtice.2015.06.004.

Yang, X., Chen, C., Li, J., Zhao, G., Ren, X., Wang, X., 2012. Graphene oxide-iron oxide and reduced graphene oxide-iron oxide hybrid materials for the removal of organic and inorganic pollutants. RSC Adv. 2 (23), 8821. Available from: https://doi.org/10.1039/c2ra20885g.

Yang, Z., Yan, H., Yang, H., Li, H., Li, A., Cheng, R., 2013. Flocculation performance and mechanism of graphene oxide for removal of various contaminants from water. Water Res. 47 (9), 3037−3046. Available from: https://doi.org/10.1016/j.watres.2013.03.027.

Yang, Z.-H., Huang, J., Zeng, G.-M., Ruan, M., Zhou, C.-S., Li, L., et al., 2009. Optimization of flocculation conditions for kaolin suspension using the composite flocculant of MBFGA1 and PAC by response surface methodology. Bioresour. Technol. 100 (18), 4233−4239. Available from: https://doi.org/10.1016/j.biortech.2008.12.033.

Yaser, A.Z., Pogaku, R., 2017. Recent trends for the removal of colored particles in industrial wastewaters. Environ. Sci. Pollut. Res. Int. 24 (19), 15861−15862. Available from: https://doi.org/10.1007/s11356-017-9555-7.

Ye, Z., Zhang, H., Yang, L., Wu, L., Qian, Y., Geng, J., et al., 2016. Effect of a solar Fered-Fenton system using a recirculation reactor on biologically treated landfill leachate. J. Hazard. Mater. 319, 51−60. Available from: https://doi.org/10.1016/j.jhazmat.2016.01.027.

Yin, J., Deng, B., 2015. Polymer-matrix nanocomposite membranes for water treatment. J. Membr. Sci. 479, 256−275.

Yong, Z.J., Bashir, M.J.K., Ng, C.A., Sethupathi, S., Lim, J.-W., 2018. A sequential treatment of intermediate tropical landfill leachate using a sequencing batch reactor (SBR) and coagulation. J. Environ. Manage. 205 (Supplement C), 244−252. Available from: https://doi.org/10.1016/j.jenvman.2017.09.068.

Zahrim, A.Y., Nurmin, Y., Rosakam, S., 2013. Coagulation/flocculation of anaerobically treated palm oil mill effluent (AnPOME): a review. Developments in Sustainable Chemical and Bioprocess Technology.

Zainol, N.A., Aziz, H.A., Yusoff, M.S., 2012. Characterization of leachate from Kuala Sepetang and Kulim landfills: a comparative study. Energy Environ. Res. 2 (2), 45.

Zainol, N.A., Aziz, H.A., Lutpi, N.A., 2017. Diplazium esculentum leaf extract as coagulant aid in leachate treatment. AIP Conf. Proc. 1835 (1), 020034. Available from: https://doi.org/10.1063/1.4981856.

Zamri, M.F.M.A., Yusoff, M.S., Aziz, H.A., Rui, L.M., 2016. The effectiveness of oil palm trunk waste derived coagulant for landfill leachate treatment. AIP Conf. Proc. 1774 (1), 030017. Available from: https://doi.org/10.1063/1.4965073.

Zhang, B., Jiang, D., Guo, X., He, Y., Ong, C.N., Xu, Y., et al., 2015. Removal of Microcystis aeruginosa using nano-Fe_3O_4 particles as a coagulant aid. Environ. Sci. Pollut. Res. 22 (23), 18731−18740.

Zhang, C., Wang, K., Tan, S., Niu, X., Su, P., 2013. Evaluation and remediation of organics, nutrients and heavy metals in landfill leachate − a case study in Beijing. Chem. Ecol. 29 (8), 668−675. Available from: https://doi.org/10.1080/02757540.2013.841897.

Zhang, M., Xiao, F., Xu, X., Wang, D., 2012. Novel ferromagnetic nanoparticle composited PACls and their coagulation characteristics. Water Res. 46 (1), 127−135.

Zhang, Y., Wu, B., Xu, H., Liu, H., Wang, M., He, Y., et al., 2016b. Nanomaterials-enabled water and wastewater treatment. NanoImpact 3 (Supplement C), 22−39. Available from: https://doi.org/10.1016/j.impact.2016.09.004.

Zhang, Y.-J., Yang, Z.-H., Song, P.-P., Xu, H.-Y., Xu, R., Huang, J., et al., 2016a. Application of TiO_2-organobentonite modified by cetyltrimethylammonium chloride photocatalyst and polyaluminum chloride coagulant for pretreatment of aging landfill leachate. Environ. Sci. Pollut. Res. 23 (18), 18552−18563. Available from: https://doi.org/10.1007/s11356-016-7031-4.

Zheng, Z., Zhang, H., He, P.-J., Shao, L.-M., Chen, Y., Pang, L., 2009. Co-removal of phthalic acid esters with dissolved organic matter from landfill leachate by coagulation and flocculation process. Chemosphere 75 (2), 180−186. Available from: https://doi.org/10.1016/j.chemosphere.2008.12.011.

Zhu, G., Wang, Q., Yin, J., Li, Z., Zhang, P., Ren, B., et al., 2016. Toward a better understanding of coagulation for dissolved organic nitrogen using polymeric zinc-iron-phosphate coagulant. Water Res. 100 (Supplement C), 201−210. Available from: https://doi.org/10.1016/j.watres.2016.05.035.

Zin, N., Aziz, H.A., Adlan, M.N., Ariffin, A., Mohd Suffian, Y., Dahlan, I., 2014. Treatability Study of Partially Stabilized Leachate by Composite Coagulant (Prehydrolyzed Iron and Tapioca Flour).

Further Reading

Chaouki, Z., El Mrabet, I., Khalil, F., Ijjaali, M., Rafqah, S., Anouar, S., et al. (2017). Use of Coagulation-Flocculation Process for the Treatment of the Landfill Leachates of Casablanca City (Morocco), vol. 8.

12

Chitosan Nanocomposite Application in Wastewater Treatments

Muhd Arif Aizat[1,2], Farhana Aziz[3]

[1]ADVANCED MEMBRANE TECHNOLOGY RESEARCH CENTRE (AMTEC), UNIVERSITI TEKNOLOGI MALAYSIA, JOHOR, MALAYSIA [2]FACULTY OF CHEMICAL AND ENERGY ENGINEERING (FCEE), UNIVERSITI TEKNOLOGI MALAYSIA, JOHOR, MALAYSIA [3]ADVANCED MEMBRANE TECHNOLOGY RESEARCH CENTRE (AMTEC), SCHOOL OF CHEMICAL AND ENERGY ENGINEERING, FACULTY OF ENGINEERING, UNIVERSITI TEKNOLOGI MALAYSIA, JOHOR BAHRU, MALAYSIA

12.1 Introduction

Public concerns regarding water pollutants; organic or inorganic, biologically or chemically, has raised awareness among researchers and governments all around the world. Pollutants such as dyes, pesticides, heavy metals, emerging contaminants mainly from pharmaceutical industries and industrial solvents may be introduced to the water streams through multiple ways, either directly or indirectly. Examples of direct introduction is physical dumping of waste such as industrial effluents into the streams and indirect introduction are agricultural leeching of biocides and fertilizers into underground water networks and via rainwater, due to atmospheric gaseous release from factories, open-burning and vehicles. Generally, water-soluble particles and molecules will be distributed and easily solubilize into the water stream and cannot be physically traced. Even though there are rules and regulations set by governmental bodies regarding these matter, lack of enforcement and illegal activities such as logging contributed the highest percentage of water pollution in multiple countries all around the world.

One of the ways to reduce the level of pollutants and subsequently treating the wastewater is by chemical treatments. The usage of chemical compounds such as aluminum sulfate, sodium aluminate, sodium hydroxide and chlorinated metals are favored for decades due to their fast-acting mechanism and low cost (as there are mass produced). These chemicals are used in multiple conventional wastewater treatment process including coagulation, floatation, dewatering, flocculation, gravity settling and membrane separation technology. Despite that, the usage of chemical compounds may raise multiple other problems in the treatment process. The action of chemical compounds towards the pollutants in the wastewater may alter the structure of the pollutants, thus known as chemical intermediates, and subsequently

Nanotechnology in Water and Wastewater Treatment. DOI: https://doi.org/10.1016/B978-0-12-813902-8.00012-5

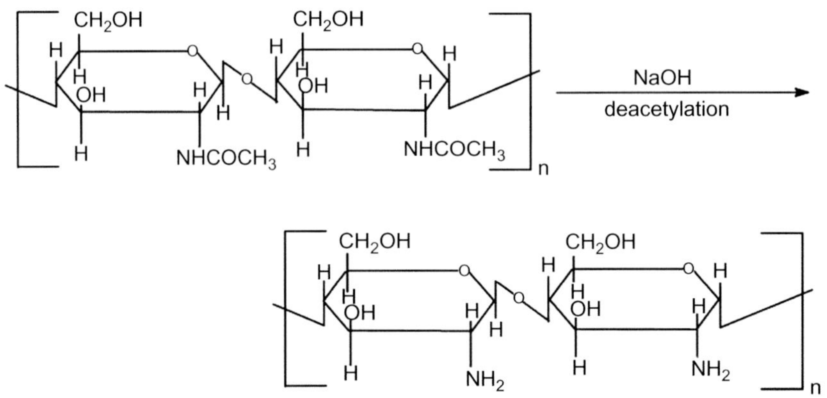

FIGURE 12-1 *N*-deacetylation of chitin to produce chitosan (Ravi Kumar, 2000).

made them much more difficult to be degraded. Other than that, reactions between those chemical compounds and pollutants may form by-products that are as much as harmful, or even more towards aquatic organisms, animals and even human population.

Regarding this problem, researchers, scientists, and engineers all over are opting for safer options in wastewater treatment and management. To overcome these problems with the main objective of finding methods that are cost-efficient, biodegradable, safe to human, animals and aquatic organisms, reduce and eliminate the chances of chemical intermediates' formation while preserving the oxygen demand of the water biologically and chemically, one of the promising discovery is the usage of chitosan in wastewater treatment (Fig. 12-1).

Chitosan is the most important derivative of chitin, which is the second-most abundant natural polymer behind cellulose. Mainly extracted from crustaceans such as shrimp and crabs, chitin can also be found in the exoskeleton of anthropods or in the cell walls of yeasts and fungi as ordered crystalline microfibrils. The extraction of chitin from crustaceans for industrial processing and commercialization are via acid treatment and alkaline extraction to dissolve calcium carbonate and solubilize proteins, respectively. This biopolymer will then be graded in terms of their purity, color and degree of *N*-deacetylation. Partially or fully deacetylate chitin with degree of deacetylation higher of 50% (based on the source of the biopolymer) is needed to produce a weak acid soluble derivative form of chitin, known as chitosan.

The solubility of chitosan took place due to the protonation of the—NH_2 functional group on C-2 position of the repeated polymer unit with pseudo-natural cationic properties. Composed of randomly distributed β-(1→4)-linked D-glucosamine, chitosan is extensively used in researches and experiments around the globe, mainly as a low-cost substitute of conventional chemicals or compounds used multiple industries, ranging from biomedical, pharmaceutical, polymer science and engineering. Chitosan possessed excellent nontoxic, antimicrobial, biocompatible, and biodegradable properties that are proven to be advantageous compared to chitin. Moreover, chitosan can be chemically and structurally modified to increase their porosity, permeability, chelating ability, adsorption rate and solubility.

The absorbency properties of chitosan can be further enhanced with the addition of nano-particles to produce chitosan nanocomposite. The fabrications of chitosan nanocomposites are mainly studied due to the good chelating and absorbent properties of the compound which makes it suitable as support for nanoparticles. This will later be discussed in detail in 4.0.

12.2 Chitosan Properties

Naturally-occurring polysaccharides such as pectin, agarose, carragenans, and cellulose are neutral or acidic state in nature. However, chitin and its main derivative, chitosan are naturally found as highly basic polysaccharides. With this special characteristic, they are known for their ability to form films, chelate metals ions, and formed polyoxysalt (Hench, 1998). Just like cellulose, chitin is naturally occurred as structural polysaccharides for crustaceans and multiple microorganisms, that is, fungi, bacteria and yeast. However, its hydrophobicity and insolubility in water and most organic solvents set it apart from cellulose. On the other hand, chitosan is soluble in weak or diluted acids such as acetic acid and formic acid. Production of chitosan began with the extraction of chitin from crustaceans' waste, including shellfish (shrimp, crab, lobster and krill), squid, and oyster. Typically, the chitin contents from crustaceans ranging from 2% to 12% compared to whole body mass, and it depends greatly on the processing conditions, state of the organisms' nutrition, reproductive stage and their species. Enzymatic reaction and chemical hydrolysis of chitin will then produce chitosan compounds. Commonly, in industrial scale, chemical treatment with sodium hydroxide (NaOH) will be favored. The reaction of chitin with NaOH will deacetylate the compound and causing the formation of amino group (NH_2) to replace the acetyl group in C-2, thus form a new acid-soluble compound known as chitosan. In this chapter, the properties of chitosan will be discussed.

12.2.1 Structure and Characterization

Chitosan is a semicrystalline, heteropolymer in solid state (formed by two or more monomer), compared to homopolymeric cellulose. The main investigations regarding its molecular weights and degree of deacetylation (DA) compared to chitin, preparation of chitosan's derivatives and its solvent and solution properties will be discussed in this subchapter.

12.2.1.1 Degree of N-Deacetylation

The *N*-acetylation of chitin introduced amino group on C-2 of the compound, thus making it soluble in weak acetic and formic aqueous media. Chitosan can be partially or fully deacetylate, based on multiple factors, such as the contact time of NaOH with chitin, concentration of NaOH, pH and temperature. The source of chitin will also affect the degree of deacetylation to produce chitosan. However, there is a fine line to differentiate the acetylation degree of chitin (typically 0.90) and chitosan (degree of acetylation is less than 0.35) (Ravi Kumar, 2000). The degree of deacetylation can also be determined by its ratio of

2-acetamido-2-deoxy-D-glucopyranose to 2-amino-2-deoxy-D-glucopyranose structural units. When the number of 2-amino-2-deoxy-D-glucopyranose units is more than 50 percent, the biopolymer is said to be chitosan and when the number of 2-acetamido-2-deoxy-D-glucopyranose units is higher, the polymer is said to be chitin (Irom et al., 2012). Multiple analytical tools have been used to determine the ratio such as IR spectroscopy, gel permeation chromatography and UV spectrophotometry, pyrolysis gas chromatography, H-NMR spectroscopy, thermal analysis, acid hydrolysis and HPLC, and near-infrared spectroscopy. A study carried out to determine the effect of chitosan additive on the properties of gelatin-based films showed that higher degree of chitosan deacetylation provides better gel strength and elasticity. These results were obtained as chitosan with higher degree of deacetylation provides more reactive groups within one chain to react with gelatin, thus providing a better intermolecular interactions (via hydrogen bonding and hydrophobic interactions) between chitosan-gelatin, leading to film strengthening (Liu et al., 2012).

12.2.1.2 Molecular Weight

Molecular weight of chitosan is very crucial properties to be taken care of when conducting experiments and researches that utilizing this bio-polymer. Based on the study carried out by Chen and Hwa (1996), membranes prepared with high molecular weight chitosan (viscosity average molecular weight, $M_v = 4.10 \times 10^5$) showed better tensile strength, tensile elongation and enthalpy; while the permeability of membrane prepared with low molecular weight chitosan ($M_v = 2.40 \times 10^5$) showed better result. They concluded that as chitosan with higher molecular weight tend to have higher enthalpy, it means that the crystallinity in the membrane and intermolecular interactions are also higher compared to low molecular weight chitosan. Molecular weight distributions of chitosan polymer can be obtained using HPLC. As for the weight-average molecular weight (M_w), the light scattering method is used. Another method that be used for molecular weight determination of chitosan is by viscometry. This simple yet rapid method deals with Mark−Houwink equation as stated below:

$$[\eta] = KM^a$$

Where η = intrinsic viscosity; and constants K and a is a known value of acid and base, respectively.

The first proposed solvent for molecular weight characterization of chitosan are 0.1 M acetic acid/0.2 M NaCl. (K (mL/g) $= 1.81x10^{-3}/a = 0.93$) However, the application of these constant may rise an issue as chitosan is a charged compound, thus the usage of those solvents may promote aggregation and overestimate the molecular weight of chitosan. Due to this matter, Rinaudo (2006) proposed the usage of 0.3 M acetic acid/0.2 M sodium acetate constants (K (mL/g) $= 7.9x10^{-2}/a = 0.796$) as no aggregation problem encountered when the mixture were used. Moreover, the conversion of chitin to chitosan may lower the molecular weight, changes the degree of deacetylation and subsequently change the charge distribution.

12.2.1.3 Solubility

While chitin is known for being insoluble in most organic solvents, chitosan, on the other hand is readily soluble in dilute acidic solutions with pH below than 6.0. The difference occurs as chitosan is considered to be a strong base due to the presence of the primary amino groups with a pKa value of 6.3. This factor alters its charged state and properties, based on pH. At low pH, the amines get protonated and become positively charged, thus making the chitosan a water-soluble cationic polyelectrolyte. However, as the pH increases well above 6, the amines gets deprotonated, thus chitosan loses its charges and become insoluble. The soluble-insoluble transition took place at chitosan's pKa value around pH between 6.0 and 6.5. This concludes that the solubility of chitosan is highly dependent on its deacetylation degree (DD) and method of deacetylation used (Cho et al., 2000).

The role of chitosan's protonation in the presence of acetic acid and hydrochloric acid (Rinaudo et al., 1999) on solubility proved that the pH and the pK of the acid affect the degree of ionization of chitosan. Moreover, these studies also found out that the amount of acid needed to dissolve chitosan depends on the quantity this bio-polymer itself. The proton's concentration must be at least equal to the concentration of amines unit involved. In a nutshell, the solubility of chitosan is a very difficult parameter to control as it is related to acetylation degree, molecular weight and distribution of acetyl group of the chitosan; and the nature, pH and ionic concentration of acid used (Rinaudo, 2006).

On the other hand, recent studies demonstrated that the presence of glycerol-2-phosphate aid the preparation of a water-soluble chitosan at neutral pH (Bhattarai et al., 2005; Chenite et al., 2000, 2001) while the synthesis of carboxymethylchitosan made this bio-polymer soluble in wider range of pH (Mourya et al., 2010; de Abreu and Campana-Filho, 2009; Guo and Gao, 2007).

12.3 Modifications and Derivatives

Modifications of chitosan has been carried out to enhance the polymer, commonly in term of their performance by specifically incorporated desired properties to suit specific needs to the chitosan backbone. The presence of $-NH_2$ and $-OH$ groups in chitosan molecules provide a suitable base for interaction with other monomers, biological molecules, polymers, and nanoparticles. Chitosan can be modified via countless route, but the most common methods used are blending, graft co-polymerization and curing (Shukla et al., 2013). Modifications of chitosan can be subcategorized into two, which are the physical modifications and chemical modifications. Physical modification is simple and easy, yet an effective mode that can be done to modify polymers; while chemical modification often required the usage of chemical compounds to alter the physico-chemical properties of this polymer by introducing new chemical groups to enhance specific properties of chitosan.

12.3.1 Physical Modifications and Its Derivatives

Physical modification is done by blending, or physically mixing at least two polymers to create a new material with different, enhanced physical properties (Payne et al., 1996). Polymer blending is carried out to generate new materials with improvised, or enhanced properties, whether chemically, biologically, structurally, mechanically, or morphologically. Combination of multiple individual polymers and combinatorial methods is required for this procedure to successfully produce new polymers with desired properties (Eidelman and Simon, 2004). Blending also tend to be easier and cheaper to carry out, and one of the less time-consuming technique can be used in polymer synthesis. Another advantage of blending technique is the alteration of starting materials' composition will be resulting materials with a wide range of properties specifically tailored for specific applications. As for the compatibility and miscibility of the resulting materials, the mechanical and thermal properties can be used to determine them (Grizzuti et al., 2000). On the other hand, several methods have been proposed to study the interaction between chitosan and other polymers.

Even though studies regarding the physical modifications of chitosan are decreasing as researchers favored the applications of multiple chemical modifications to tailor the modified chitosan according to specific pollutants, which will be discussed in Section 3.2, several recent researches carried out by means of physical modifications of chitosan will still be discussed as these are equally important as chemical modifications.

In order to synthesize the adsorptive membranes to remove heavy metals ions, including Ca(II), Mg(II), Zn(II) and Cd(II) from aqueous solution, Mahatmanti et al. (2016) has blended rice hull ash (RHA) silica and polyethylene glycol (PEG) into chitosan. The blended membranes were an effective and convenient way to improvize the mechanical and physical properties of pristine chitosan membranes, while increasing the porosity and wettability of the adsorptive membrane itself. The adsorption studies carried out towards metal ions showed that Zn(II) is the best ion to form complex with the fabricated membrane (182 μmol/g), while the Ca(II) and Mg(II) tend to form ionic bonds with chitosan $-NH_2$ groups as these two metal ions are hard acids, according to concept developed by Pearson in 1963.

Modification of chitosan by fabrication of chitosan/cellulose acetate composite nanofiltration membrane was further demonstrated by Ghaee et al. (2015). By immersing the cellulose acetate membrane formed in the beginning of the experiment into the chitosan solution, and later in glutaraldehyde aqueous solution as crosslinking agent, the chitosan/cellulose acetate composite nanofiltration membrane formed was able to provide 81.03% retention towards copper ions with $4.37 \, Lm^2/h$ water flux. The nature of chitosan and cellulose acetate as hydrophilic materials are very suitable to form a structurally stable composite membrane that withstand lengthy experimental conditions under high operating pressure (506.5 atm).

12.3.2 Chemical Modifications and Its Derivatives

The chemical modifications of chitosan mostly carried out in multiple studies with the same objective, to improvize the existing properties of chitosan and these can be done via introduction of various functional groups to the chitosan's structure. Even though there are numerous modifications can be done, this subchapter will only be discussed several chemical modifications of chitosan that found their applications in wastewater treatment and management.

12.3.2.1 Photosensitizer-modified chitosan

One of the most under-studied yet promising chemical modification of chitosan available is the photosensitizer-modified chitosan. Photosensitizer is a chemical compound which able to absorb light energy from certain wavelengths and transfer the energy to the chosen reactants. The utilization of photosensitizers is required when reactions such as photocatalysis but the light sources are not readily available. Currently, most synthesized photosensitive systems are based on synthetic polymers (Ji et al., 2014). However with the advancement of "green technology" and increasing awareness related to the pollution due to the usage of synthetic polymers, the interest in practical applications of natural polymers such as chitosan are gradually increasing since the first successful synthesis by Tanaka et al. (1980).

A study carried out by Nowakowska et al. (2008) revealed that the synthesis of water-soluble photoactive napthyl-substituted chitosans (CHNA) absorbed light from the near UV spectral region. The presence of naphtyl (NP) chromophores via chemical modification of chitosan with 1-napthylacetic acid (NA) can serve as energy and electrons donors to the molecules of suitable acceptors. Other than that, the exclusive attachment of the chromophore to the chitosan chain by the amide bonds confmed by ^{13}C NMR spectra. They concluded that the reactions of fabricated photosensitizers with organic compounds in aqueous solution is useful in the water purification field, as demonstrated by photosensitizing activity with viologen water-soluble compound (SPV)-contained aqueous polymer solutions and oxidation perylene in two different studies with promising results.

On a latter study by Wu et al. (2005) regarding the anthracene chromophore containing chitosan, the absorption of light from UV-vis spectral region electronically excited the polymeric chromophores could lead to the application of environmentally friendly photocatalytic system with visible light to carry out efficient degradation of pollutants. The investigation by using perylene (PE) as energy acceptor was carried out as the thermodynamic feasibility for the singlet-singlet energy transfer. As for the study for chromophore-containing chitosan to participate in the electron-transfer process, methyl viologen dichloride (MV^{2+}) was used and experimental results showed that the electron transfer processes could be used to induce photochemical reactions of organic compounds.

However, more studies need to be carried out regarding this chemical modification as the application of chitosan as photosensitizers is promising and may be relevant for synergistic operation with light-derived catalysts to further enhance the efficiency of organic pollutants' degradation.

12.3.2.2 Carboxymethylation of Chitosan

The second chemical modification of chitosan to be put in context is carboxymethyl chitosan (CMC). CMC is the product of chitosan carboxymetylation, having some of amino and primary hydroxyl sites of the glucosamine units of chitosan substitutes by carboxyl group ($-COOH$). The derivatives obtained via this chemical modification can be further classified into three, which are the N-carboxymethyl chitosan (NCMC), O-carboxymethyl chitosan (OCMC) and N,O-carboxymethyl chitosan (NOCMC). There are numerous studies carried out regarding this modification as the modified chitosan will possessed hydrophilic and water-soluble properties, and these are sought after not only in the membrane technology and wastewater treatment, but also in medical engineering, biomedical, agriculture, tissue engineering field, to name a few (Zinadini et al., 2014; Mourya et al., 2010).

An outstanding fabrication of chemically-modified glutaraldehyde cross-linked, functionalized by poly(aminocarboxymethylation) of chitosan (PCM-Chit) by El-magied et al. (2017) suggested the increased in sorption of erbium (III), a rare earth element is due to the progressive deprotonation of reactive groups, such as R-OH, R-SH, amine and carboxylic acid groups, compared to the glutaraldehyde cross-linked chitosan (GLA-Chit). The fabrication of PCM-Chit took place by the reaction chlorinated GLA-Chit with tetraethylenepentamine (TEPA). The FTIR analysis carried out that the carboxymethylation of amine groups leads to decrease in amine groups' intensity and appearance of signal corresponding to carboxyl group at around $1730\,\mathrm{cm}^{-1}$. This finding is in agreement with study carried out by Tolba et al. (2017).

On the other hand, an earlier study done by Zinadini et al. (2014) regarding the synthesis of novel high-flux antifouling nanofiltration membranes for dye removal utilized O-carboxymethyl chitosan (OCMC) for two main purposes; (1) as a hydrophilic derivative of chitosan with carboxyl group ($-COOH$) replacing the hydroxyl group ($-OH$) on the backbone, which providing hydrophilicity to polyethersulfone (PES) polymer membrane, and (2) provide suitable functional groups to bind onto Fe_3O_4 nanoparticles, which act as an efficient adsorbent and as support for carboxymethylated chitosan. The application of OCMC modified the wetting properties of PES polymer, thus overcame the irreversible fouling problem and subsequently improved the pure water flux of the membranes. Moreover, the addition of OCMC and Fe_3O_4 nanoparticles induced the negative charges on the membrane surface and increased rejection of Direct Red 16 dye.

12.3.2.3 Recent Modifications

In order to enhance the sorption of uranyl/uranium(VI), the representative of toxic and valorizable metal ions, Galhoum et al. (2017) synthesized diethylenetriamine (DETA)-grafted nano-based magnetized chitosan sorbents through combination of hydrothermal treatment and precipitation process. On one hand, the deposition of chitosan active layer onto the magnetic particles and simultaneous conditioning of the layer will generally allowing the production of smaller particles compared to the pure coating process. On the other hand, the DETA-modified chitosan will subsequently increase the sorption sites' density and selectivity towards multiple parameters, such as pH, the composition of solutions etc. The presence of two primary amine groups and a tertiary amine group via DETA grafting caused the significant reduction in protonation of amine groups and carboxylic groups with increasing of pH in chitosan, making it possible for the binding of metal cations (such as UO_2^{2+} and other cationic forms, including $UO_2(OH)^+$, $(UO_2)_2(OH)_2^{2+}$, $(UO_2)_3(OH)_5^+$) in neutral and moderately acidic solutions. With the maximum sorption capacity of uranyl up to 185.2 mg/g and reusability of the sorbents over five to six cycles with low decrease in sorption capacity of nine percent, the synthesis of this chemically-modified magnetized chitosan sorbents appears to be very efficient and promising for fast recovery of uranyl ions.

The efficient sorption of uranyl by chemically-modified magnetic-chitosan was also demonstrated by Hamza et al. (2017). The fabrication of magnetite-chitosan microparticles via hydrothermal co-precipitation method is equivalent to the one prepared by Galhoum et al. (2017), or so-called Massart method (Massart, 1981). This method is advantageous when incorporating nanoparticles' sorbents as (1) reduces the head loss pressure and blocking phenomena when used in fixed-bed columns and (2) overcome the difficulty in the solid/liquid separation at the end of the process in batch systems. It also improves sorption performance, in terms of sorption levels, kinetics and processing (solid/liquid separation). The authors intensified the sorption capability of the magnetic-chitosan particles by grafting of glycine functional groups followed by hydrazinolysis process. These steps were taken as the grafting of glycine to the backbone of chitosan will increase the nitrogen content of the magnetic-chitosan particles by a factor of 2.5, and the subsequent grafting of hydrazide groups will further doubling the nitrogen content of the sorbents. The grafting of these functional groups obviously enhancing the sorption of metal cations as the amine functions increased, especially in higher pH conditions. Progressive deprotonation of these reactive groups reduced the competition effect of protons, thus improves metal binding in pH close to 5.

The composite formation of chitosan with inorganic or synthetic polymers were also studied by Ghiggi et al. (2017) and Wang et al. (2017) in order to synthesize ultrafiltration membrane with antifouling property, and chemically-modified chitosan biosorbents for heavy metal removal in wastewater, respectively. The application of chitosan is important to fabricate a cost-competitive, high adsorption capacity, fast kinetics, favorable reusability, abundant hydrophilicity, promising reactivity, flexibility in batches or flow modes, efficient and reliable platform in wastewater treatment, either in membrane filtration or adsorption process, or whether the removal or organic or inorganic pollutants/matters (Salehi et al., 2016).

The study by Ghiggi et al. (2017) demonstrated a successful fabrication of polyethersulfone (PES)/N-phthaloyl-chitosan (NPhthCs) ultrafiltration membrane in order to reduce the cake layer formation or fouling caused by organic contaminants (in this study, bovine serum albumin was the model pollutant). The reaction of chitosan with phthalic anhydride in dimethylformamide (phthaloylation procedure) caused an increase in permeance and hydrophilicity (contact angle: $56 \pm 5°$) compared to the pristine PES membrane (contact angle: $61 \pm 3°$). The increase in permeance may be related to the increased pore sizes, increased porosity and reduction of membrane thickness. Moreover, the antifouling property possess by the fabricated PES/NPhthCs membrane was proven by the significantly lower concentration polarization and fouling resistances.

On the contrary, the synthesis of a highly monodispersed polyethylenimine-chitosan (PEI-CS) biosorbent was performed in order to create biosorbents with ultra-high

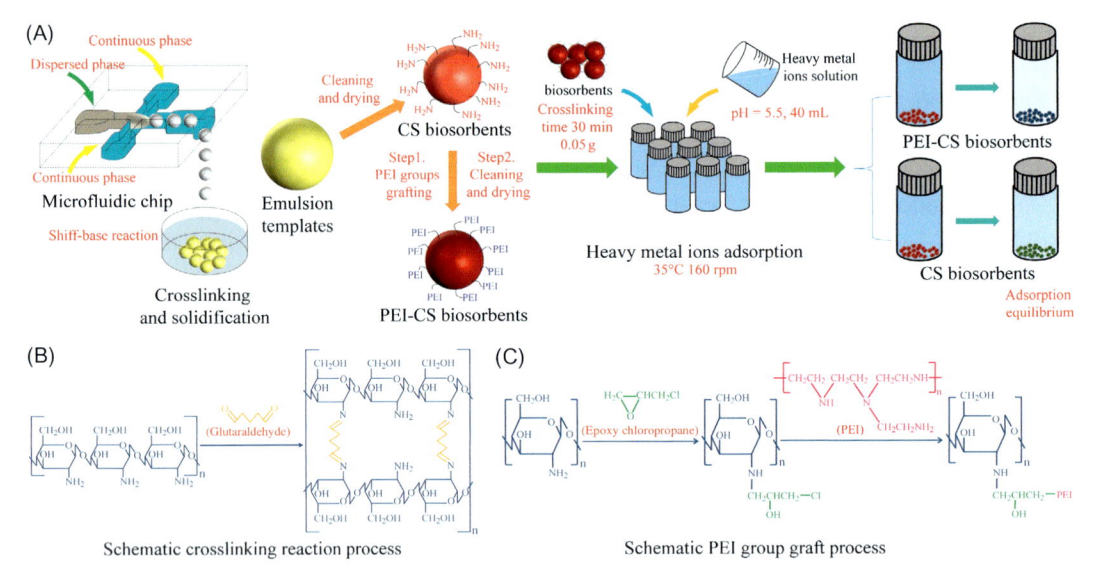

FIGURE 12-2 (A) The overview of the synthesis process of PEI-CS biosorbents and batch adsorption experiments; (B) Schematic crosslinking reaction process; (C) Schematic PEI group graft process (Wang et al., 2017).

performance, mechanical strength and tunable selectivity for heavy metals (copper (Cu) ions were selected as the desired heavy metals in this study) (Wang et al., 2017). Integration of numerous processes including the facile microfluidic emulsion, chemical crosslinking, solvent extraction and chemical modification (Fig. 12-2) were carried out in the synthesis of PEI-CS biosorbent. The chitosan was modified with the addition of *n*-octanol, Span 80 surfactant, glutaraldehyde and epoxy chloropropane in order to successfully fabricate the composite sorbents with PEI. With the final adsorption capacity of 146 mg/g for Cu ions, the selectivity of PEI-CS biosorbent towards Cu ions is well proven when used in aqueous solution containing Cu^{2+}, Na^+, and Al^{3+} ions due to the formation of stable metal chelation of chitosan-Cu^{2+} ions contributed by the good affinity of amine groups towards Cu ions. It is also worth mentioning that the PEI-CS biosorbents showed better mechanical performance, compared to CS-only biosorbents and the reasons rely on multiple factors, including crosslinking degree by reaction, dehydration by *n*-octanol, and being dried twice in the integrated process. Reaching 0.368 MPa after 60 minutes crosslinking time, the improvement in the mechanical strength do not caused any reduction in adsorption capacity, as it was dominated by the amount of grafted PEI groups, and not its crosslink degree.

The studies for adsorption of reactive dyes are as important as heavy metals and organic contaminants found in wastewater due to their high salt content, nonbiodegradability, high water solubility, low degree of fixation on the surfaces and low adsorption ability. Reactive dyes may also threaten aquatic ecosystem, especially due to their toxicity and sunlight transmission reduction. Vakili et al. (2017) studied the adsorption performance of chemically-modified chitosan beads using cationic surfactant and organo-silane, hexadecylamine and

3-aminopropyl triethoxysilane, respectively. With the main objective of increasing the amino groups on chitosan's surface and the surface charge (cationity) of the beads, the adsorption capacity of the chitosan beads rose for 1.48 times, achieving 468.8 mg/g or Reactive Blue 4 (RB4) dye removal, compared to unmodified chitosan beads (317.1 mg/g). Attributed mainly by the positive charges grafted onto the chitosan beads, the protonation of $-NH_2$ groups in pH 4 dye solution increased the electrostatic interaction with $-SO_3-$ groups of RB4 molecules, thus showed the highest uptake value of 99.2%.

Throughout all studies listed within this sub-chapter, a key consistency that can be traced is the bands' patterns observed as FT-IR analysis were carried out. The chemical modifications carried out will affect the broad bands in the region between 3000 and 3600 cm^{-1} of modified chitosan-based materials. The region attributed to the stretching vibration of primary amine and OH stretching groups. The intensity of the band may have varied based on the modification carried out to the chitosan. For example, the presence of APTES and HDA molecules intensified the band due to the accumulation of OH and NH_2 groups on the surface of modified chitosan beads (Vakili et al., 2017), while the decreased in intensity of the band can be observed when the chitosan was subjected to phthaloylation reaction with phthalic anhydride that caused specific reaction towards the amine groups (Ghiggi et al., 2017).

Apart from that, another pattern behavior can be observed between 1550 and 1750 cm^{-1} region. The chemical modification most definitely introduced new peaks, while decreasing the intensity primary amine groups of chitosan within these regions. The introduction of new amide and/or carboxylic group, and the stretching of chitosan $-C=O-$ functional groups suggested the successful modification of the chitosan. For instance, the crosslinking reaction between glutaraldehyde and chitosan introduced a new peak at 1654 cm^{-1}, due to the introduction of $-C=N-$ bond as Schiff-base reaction between $-CHO$ of glutaraldehyde and $-NH_2$ group of chitosan was successful (Wang et al., 2017).

12.4 Chitosan Nanocomposite in Wastewater Treatment

The applications of nanomaterials, including nano-rods, nano-pin, nano-sheet, nano-fiber, and nanoparticles opened up a whole new chapter in wastewater treatment researches. This material is known to work efficiently as they possessed greater Specific Surface Area (SSA) compared to their existence in bulk volume. SSA is the total surface area of a material per unit mass (m^2/kg or m^2/g), thus defined the property of the solids itself. The increased in SSA of the materials will increase the "working site" to remove any specific contaminants found in wastewater. Besides that, the alteration of nanomaterials' electronic properties due to quantum size effect is another momentous aspect that should be highlighted. The reduction of particle size from macro to micro dimensions may not come into play in changing their electronic properties, but as the nanometer range is reached, this effect becomes prominent (Shukla et al., 2013).

One of the most studied nanoparticles in the field of wastewater treatment is titanium dioxide, TiO_2. TiO_2 is chosen as it possesses several good properties such as low-cost, high chemical stability, commercially available, nontoxic, and environmental friendly. However, pristine TiO_2 can only absorbed light in ultraviolet region due to their large band gap of 3.2 eV. Multiple researches were carried out to narrow the band gap, thus allowing the TiO_2 particles to be activated by visible light (Leong et al., 2014).

However, the application of nanoparticles solely may not be good enough. For example, TiO_2 is known to be a very good photocatalytic material and the degradation ability was proven by multiple studies. However, TiO_2 is only effective in degradation of various organic pollutants, but not the nonbiodegradable metal ions, inorganic compounds and heavy metal constituents (Chen and Ray, 2001). It may also build up in concentrations in food chains to toxic levels as they have infinite lifetimes. On the other hand, the common usage of photocatalytic particles suspension method while carrying out studies regarding the photocatalysts possessed few major drawbacks, such as the final recovery step (Ghimici and Nichifor, 2013; Szabó et al., 2013). This step is necessary as the release of nanoparticles in the environment (commonly through streams and underground water) is very dangerous towards human, animals and nature (Reijnders, 2008). Other than that, the risk of poisoning and strenuous regeneration of photocatalysts may hindered the industrial application of this advanced oxidation process (AOP) if these problems are not overcome (Gaya and Abdullah, 2008).

Due to this fact, the immobilization of nanoparticles such as TiO_2 has been carried out as soon as the 1980s and since that, multiple supports and binders that will actually contributed to the photocatalytic process has been investigated (Hashimoto et al., 2005; Gaya and Abdullah, 2008). The utilization of chitosan as the TiO_2 binder is one of the perfect approach as the presence of $-NH_2$ and $-OH$ group on the polymer will serve as the binding and reaction site for those materials via adsorption, thus providing an extensive ways in treating various wastewater pollutants (Liu et al., 2012; Nawi and Sabar, 2012).

12.4.1 Chitosan Nanocomposites for Organic and Emerging Contaminants' Degradation

A study carried out by Hamdi et al. (2015) demonstrated that the photodegradation of aromatic amine organic compound, aniline can be achieved by the preparation of hybrid chitosan-phthalocyanine-TiO_2 (PC/CS-TiO_2) photocatalyst. The successful fabrication of this nanocomposite is due to the efficient stabilization of PC nanoparticles contributed by the adsorption of positively-charged chitosan chains onto the negatively-charged particles, preventing the nanoparticles from aggregates via electrostatic barrier. These ensure the full immobilization of PC nanoparticles within the hybrid nanocomposite and the intimate contact between each particle can be achieved. The PC dye acts as a mediator for transferring electrons from the sensitizer to substrate electron acceptors on TiO_2 surface. The excited electron is then transferred to the conduction band of TiO_2 to form superoxide radicals $O_2\bullet$ and hydroxyl radicals $HO\bullet$ on the TiO_2 surface (Fig. 12-3).

One of the most important parameter that must be observed is the chemical stability of the chitosan nanocomposite formed. This parameter is very crucial as different range of pH will affect the performance of chitosan, especially their adsorbing and chelating ability.

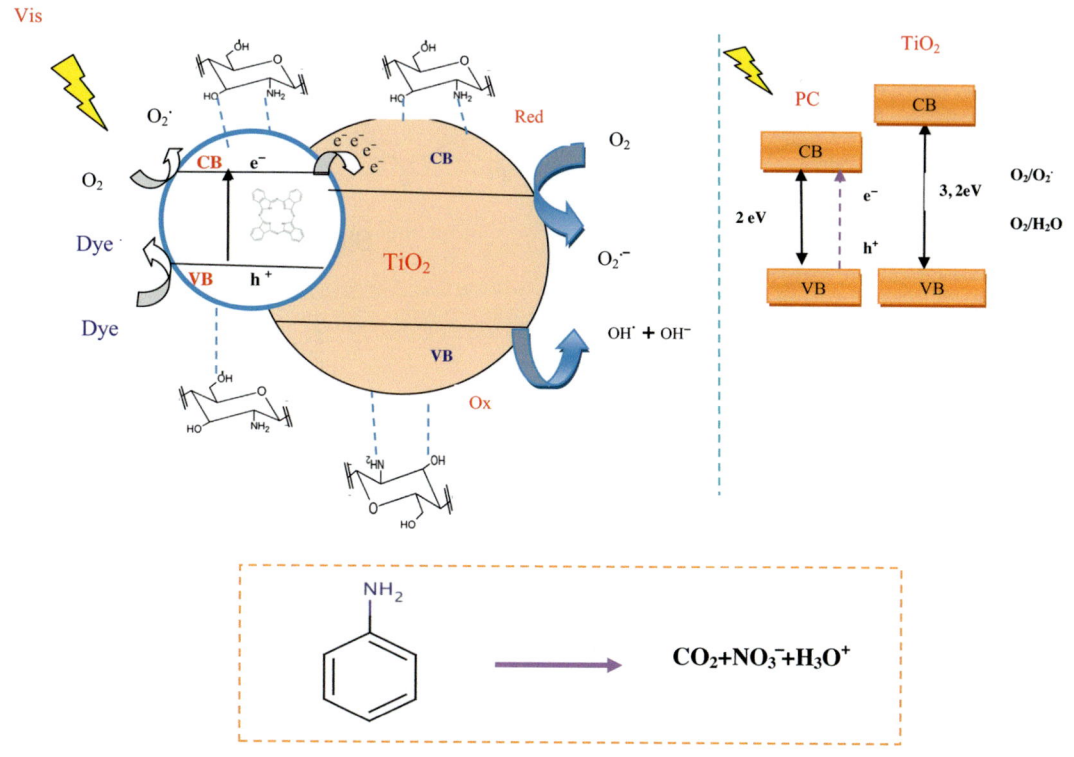

FIGURE 12-3 Photoactivation mechanism of TiO_2 and dye-sensitized under visible light (Hamdi et al., 2015).

The chemical stability can be determined with total carbon and/or nanoparticles leaching in different pH conditions.

A study by Le Cunff et al. (2015) in the removal of herbicide terbuthylazine (TBA) by chitosan-TiO_2 nanocomposite (Fig. 12-4) demonstrated that the titanium leaching was more noticeable at the acidic conditions, as chitosan is much more soluble in lower pH. However, a rather small leaching of Ti (between 16.95 and 170.47 ng/dm) attributed by the solubility of chitosan under irradiation of UV light has been neglected after initial period of time and adsorption equilibrium established as the catalysts remains stable over prolonged reaction time. This observation is confirmed by AAS analysis. At alkaline concentration, the poor solubility of chitosan formed precipitate at pH 9 after the initial increase of dissolved carbon concentration. Due to this, they suggested that the photocatalyst pretreatment should be done by 2 hours conditioning at pH 5 to prevent the effect of homogenous photocatalysis from leached TiO_2 during reactions. As for the removal of TBA under UV light irradiation, this immobilized TiO_2/chitosan nanocomposite exhibited a very high photocatalytic activity, obtaining almost 100 percent degradation of TBA into cyanuric acid after 80 minutes. Nevertheless, the presence of cyanuric acid is nothing to be worried about as cyanuric acid is a nonharmful chemical and can be easily destroyed by microbial degradation in nature,

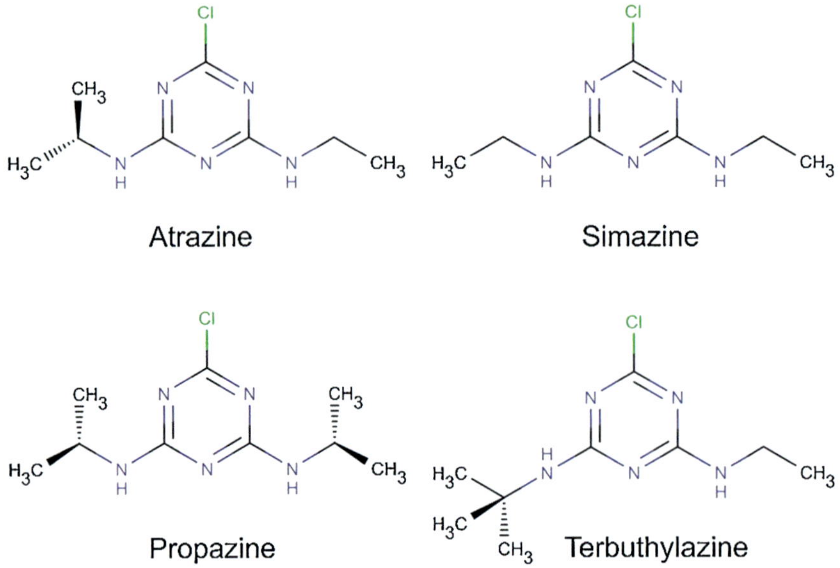

FIGURE 12-4 Schematic representation of terbuthylazine structure (*N-tert*-buthyl-6- cloro-N'-ethyl-1,3,5-triazine-2,4-diamine) and other representatives of *s*-triazine herbicides (Le Cunff et al., 2015).

compared to the formation of other degradation byproducts such as ammeline (2-chloro-4,6-diamino-s-triazine) and ammelide (2-amino-4,6-dihydroxy-s-triazine) through stand-alone photolytic degradation of TBA only.

As demonstrated by Jaiswal et al. (2012), incorporation of copper nanoparticles onto chitosan enhances the adsorption ability for malathion, a broad spectrum organophosphate insecticide commercially used for agricultural, industrial, governmental and residential purposes. The high removal of malathion achieved at pH 2 (81.9% removal) as the pH of the synthetic solution increase from 1 to 2. The high adsorption of malathion by copper-chitosan nanocomposite is mainly contributed by acidic hydrolysis of malathion to dithiophosphate, followed by complexation of copper to form Cu(II) dithiophosphate. However, the increasing pH achieving alkalinity of the synthetic solution decreased the adsorption capability of the nanocomposite, as the competition from hydroxyl ions of chitosan and the solution to bind to the malathion increased. The hydrolysis of malathion under acidic condition is further proven via IR and EDX spectra. To further strengthen their study, the evaluation for the removal of parathion and methyl parathion were carried out under the prevailing conditions. The results obtained shows complete removal of pesticides occurs unto 5 mg/L and as the concentration of pesticides increased to 10 mg/L, 99.9% and 98.9% removal of methyl parathion and parathion were observed, respectively with the dose rate of 1 g/L in 20 mL aqueous solution, similar to the dose used for malathion removal study. No metabolites formed were observed up to 10 mg/L concentration levels. The hydrolysis of parathion and methyl parathion by copper-chitosan nanocomposite into diethyl thiophosphate and dimethyl thiophosphate, respectively with 4-nitrophenol as metabolites for both solutions can only be observed at concentration of 50 mg/L.

12.4.2 Antimicrobial Properties of Chitosan Nanocomposites

Besides their good reputation in degradation of organic and emerging contaminants, studies regarding the antimicrobial properties of chitosan nanocomposites are massively carried out across multiple discipline, including in wastewater treatment and management. The antimicrobial properties of chitosan are arisen from its positively charged properties. The disruption of microbes' normal structure due to reactions between positive charge of chitosan and the negatively-charged cells walls of bacteria and/or protein inside is principally known as the killing mechanism of chitosan. However, this action is limited due to relatively weak positive charge center of amino groups on chitosan backbone. Due to this matter, the strength of positive charge within chitosan must be enhanced (Xiao et al., 2011). Chemically modified chitosan is one of the routes that can be taken in order to achieve this. However, with the rise of nanotechnology, the modification of chitosan can be done by incorporating positively-charged nanoparticles into chitosan matrix and subsequently increase their antimicrobial properties.

Recently, a study carried out by Hajji et al. (2017) demonstrated the antimicrobial properties of nanocomposite consisting of chitosan-poly(vinyl alcohol)-silver nanoparticles against Gram-positive and Gram-negative bacteria via the agar disc diffusion assay. They revealed that the nanocomposite films inhibited the growth of these bacteria and the mechanisms behind the successful inhibition were proposed. The damaging interaction of the poly-cation (protonated amino groups) with the bacteria's negatively-charged surface, leading to membrane depolarization followed by loss of membrane permeability, cell leakage and bactericidal effect.

The advantageous antimicrobial properties possessed by chitosan and their nanocomposites can be further explored for the synthesis of antifouling and antibiofouling nanocomposites. Fouling is one of the problem commonly encountered in membrane-based technology in wastewater treatment, especially in traditional membrane reactor. Membrane fouling could decrease permeate flux and productivity, elevate the energy consumption and cleaning frequency, reduce membrane lifespan, thus leading to higher maintenance cost (Zhang et al., 2016). Biofouling occurs on practically all surface in direct contact with water starting with the adhesion of organic molecules or polysaccharides, followed by bacterial attachment, diatoms and/or microalgae to form a biofilm, and subsequently adhesion of macroorganisms (Al-naamani et al., 2017). Even though the application of this nanotechnology is not deeply explored within the wastewater treatment and management field, the future prospect of it seems promising as the search for biocompatible, biodegradable, low toxicity, ecologically safe, cost effective, and antimicrobial substances is heavily carried out.

A recent study carried out by Natarajan et al. (2017) demonstrated that the incorporation of TiO_2 and silver (Ag) nanoparticles with chitosan greatly enhanced the antifouling mechanism against *Scendesmus* sp. and *Chlorella* sp. algae. The UV light irradiation increased the hydrophilicity of the nanocomposite films and simultaneously increase their toxicity towards both algal species with photocatalytic action of TiO_2 and dissolution of Ag^+ ions from the films, thus decreased the slime formation which cause fouling of the films.

On separate study by Al-naamani et al. (2017), the susceptibility of chitosan nanocomposites for prevention of marine biofouling is also demonstrated by the development of chitosan-zinc nanocomposite coating. The nanocomposite coating showed a significant improvement in wettability by decrease in hydrophilicity, reduced solubility and swelling properties. The chitosan-ZnO nanocomposite also showed significant growth inhibition of marine fouling bacterium *P. nigrifaciens*, fouling diatom *N. incerta* and uncultivated marine microorganisms compared to chitosan-only coatings and uncoated samples. The antimicrobial properties of the nanocomposite lay on the synergistic effect of both particles. Interaction of ZnO nanoparticles with the bacteria causing tremendous damages to its cell wall, proteins, DNA and lipids due to the release of Zn^{2+} ions and generation of reactive oxygen species (ROS) via photocatalysis. On the other hand, the network formation of chitosan and ZnO nanoparticles control the release of Zn^{2+} ions into the environment, thus confirms the stability of the nanocomposite coating. The study was carried with three main strategies: preventing biofouling organisms from attaching to surfaces, reducing adhesion of bio-foulants and killing of bio-foulants. All these strategies are applicable in the area of wastewater treatment in order to develop a robust, multifunctional nanocomposite, especially in membrane technology subdivision.

As demonstrated by Malini et al. (2015), the incorporation of ZnO nanoparticles in the chitosan membrane proven to have bactericidal effect towards both species of bacteria used in this study. Although there was a difference in antibacterial activity of the nanocomposite membrane due to the structural and chemical compositional differences in Gram positive (*Bacillus subtilis*) and Gram negative (*K. planticola*) cell wall nature, the nanocomposites showed better bactericidal activities with the formation of inhibition clear zone around the membrane compared to as-synthesized chitosan membrane as the disk diffusion method study was carried out. Moreover, as the nanocomposite membrane was put to test for their antifouling activities by immersion in collected wastewater, the result is promising as the ZnO-chitosan nanocomposite exhibited relatively clean surface and free from microorganisms' growth, compared to as-synthesized chitosan membrane. The authors suggested that the incorporation of green-synthesized ZnO nanoparticles with macroporous chitosan membrane prevent the initiation and/or formation of biofilms due to the formation of active free radicals on ZnO surfaces. Besides, the strong molecular interaction between them consequently restricted the motion of chitosan and enhanced the mechanical properties of this biopolymer while the stabilization of ZnO nanoparticles within chitosan matrix prevents nano ZnO from further agglomeration.

12.4.3 Chitosan Nanocomposites for Heavy Metal Removals

Studies regarding the removal of heavy metals in wastewater by chitosan nanocomposite are heavily done, especially by adsorption methods, and summarized in Table 12-1. This is because adsorption is considered to be well-established operation for water and wastewater treatment. Commonly, the usage of activated carbon is favored due to their effectiveness, versatility, and good capacity for heavy metals, dyes, and toxic compounds. However, the cheaper alternatives

Table 12-1   Recent Studies Regarding the Adsorption Capacities of Synthesized Chitosan Nanocomposites for Heavy Metals Removal in Wastewater

Chitosan Nanoparticles/Nanocomposite	Metal Ion Studied	Adsorption Capacity	References
Graphene oxide (GO)—poly-ethylenimine (PEI)—chitosan (CS) nanocomposite	Cr(VI)	91.10%	(Perez et al., 2017)
	Cu(II)	78.18%	
Ethylene-diamine-tetra-acetic acid (EDTA)—functionalized magnetic chitosan (CS) graphene oxide (GO) nanocomposites	Pb(II)	206.52 mg/g	(Shahzad et al., 2017)
	Cu(II)	207.26 mg/g	
	As(III)	42.75 mg/g	
Chitosan-magnetite nanocomposite	Cr(VI)	92.33%	(Sureshkumar et al., 2016)
Nano-hydroxyapatite chitosan composites	Cd(II)	122 mg/g	(Salah et al., 2014)
Electro-spun poly ethylene oxide (PEO)/chitosan nanofiber membrane	Ni(II)	175.1 mg/g	(Aliabadi et al., 2013)
	Cu(II)	163.7 mg/g	
	Cd(II)	143.8 mg/g	
	Pb(II)	135.4 mg/g	
Thiourea—modified magnetic ion—imprinted chitosan/TiO$_2$ composite	Cd(II)	256.4 mg/g	(Chen et al., 2012)

are intensely sought after by researchers as the full-scale applications of activated carbon is very costly. One of the prominent material is chitosan, as this biopolymer is practically the second most abundant biopolymer in nature. Multiple mechanisms of removal of heavy metal ions by chitosan and its nanocomposites had been proposed. The electrostatic interaction, ionic exchange, metal chelation and ion-pairs are among the suggested mechanisms.

Shahzad et al. (2017) suggested the adsorption of metal ions (Pb^{2+} and Cu^{2+}) might be due to the strong chemical bonding between the metals ions and oxygenated groups of their fabricated EDTA-magnetic chitosan-graphene oxide nanocomposite, while the adsorption of As^{3+} transitional metallic ions is thought to be due to the presence of chitosan within the nanocomposite as the bonding interactions was established much more weak and slow compared to Pb^{2+} and Cu^{2+} ions that easily formed strong bonding with EDTA's functional groups. Moreover, they suggested that the pseudo-second order kinetic model gave the best fit for the chemical interactions of nanocomposite functional groups and metal ions, and the adsorption process involved chemisorption and to be the rate-limiting step.

On the other hand, the synthesis of response surface methodology (RSM)-optimized chitosan-polyethyleenelmine (PEI)-graphene oxide by Perez et al. (2017) demonstrated the successful removal of anionic Cr(VI) and cationic Cu(II) up to 91.10% and 78.18%, respectively due to two mechanisms, which are the electrostatic attraction between the composite with Cr(VI) ions and the formation of complex structures between the primary amine groups found on the surface of the beads with Cu(II) ions. The multifunctionality and excellent properties of the synthesized nanocomposites (chemical and thermal stability, mesoporous structure, high surface area and abundance of functional groups present on the surface) making it promising to be utilized as adsorbent for wastewater treatment.

12.4.4 Chitosan Nanocomposites for Dyes Removal and Degradation

The applicability of chitosan and its nanocomposites for dyes removal have also been studied, as they are an effective biosorbents due to their intrinsic characteristics of amino and hydroxyl functional groups. Dyes are harmful towards flora and fauna, and known to be mutagenic and carcinogenic towards human. So the need for their removal from wastewater is compulsory (Fan et al., 2013). Even though there are multiple technologies and processes have been proposed for dyes removal in wastewater, adsorption is most widely used due to the ease of operation and cost-effectiveness. However, the common usage of activated carbon in adsorption process possessed two main disadvantages, which are their high cost and need for regeneration (Soltani et al., 2013). Low-cost adsorbents have been studied to replace activated carbon, such as kaolin, clays, and chitosan itself. However, the adsorption capacity of these materials cannot compete with the efficiency of activated carbon. Due to that, the modification of chitosan by addition of functional groups and formation of nanocomposites were heavily studied to further enhanced their adsorption capability, and few recent studies were tabulated in Table 12-2.

A recent study carried out by Nagarpita et al. (2017) showed that the synthesis of chitosan grafted sodium acrylate *co*-acrylamide/nanoclay nanocomposites can successfully remove multiple dyes (crystal violet, napthol green, and sunset yellow) due to their superabsorbent properties. The addition of nanoclay increased the porous interface area of the nanocomposites, while increasing the availability of hydroxyl groups as it was found on the surface of the nanoclay. Moreover, the authors found out that the increased in initial concentration of dyes will enhance the driving force and lowers the mass transfer resistance between the aqueous dye solution and the nanocomposites, subsequently increasing the adsorption efficiency of the nanocomposites towards the dyes used.

Table 12-2 Recent Studies Regarding the Adsorption Capacities of Synthesized Chitosan Nanocomposites for Dyes Removal in Wastewater

Chitosan Nanoparticles/Nanocomposite	Dyes Studied	Adsorption Capacity (mg/g)	pH	References
Chitosan/KSF montmorillonite beads	Remazol blue	310.00	3.0	(Pereira et al., 2017)
Chitosan grafted sodium acrylate-co-acrylamide/ nanoclay superabsorbent	Crystal violet	256.41	4.0	(Nagarpita et al., 2017)
	Sunset yellow	208.33	4.0	
	Naphtol green	221.72	4.0	
Chitosan/SiO$_2$/CNTs	Direct blue 71	61.35	6.8	(Abbasi, 2017)
	Reactive blue 19	97.08	2.0	
Magnetic chitosan—graphene oxide (MCGO)	Methyl orange	398.08	4.0	(Jiang et al., 2016)
Graphene oxide-doped porous chitosan aerogels	Methyl orange	686.89	4.6	(Wang et al., 2015)
	Amido black	573.47	4.0	
Chitosan/bio-silica	Acid red 88	25.84	3.0	(Soltani et al., 2013)
Magnetic β-cyclodextrin—chitosan/graphene oxide	Methylene blue	84.32	11.0	(Fan et al., 2013)

The modification on chitosan structure followed by the addition of foreign material to form a better adsorbent was carried out by Fan et al. (2013). The fabrication of magnetic β-cyclodextrin-chitosan/graphene oxide nano-adsorbent through a facile chemical route demonstrated a good and versatile adsorbent that possessed the high surface area property of graphene oxide, hydrophobicity of β-cyclodextrin, magnetic property of Fe_3O_4, and the abundance of amino and hydroxyl groups of chitosan for excellent adsorption of methylene blue (MB) dyes. Following the pseudo-second order kinetic model and Langmuir isotherm model for adsorption kinetics and isotherms, respectively, the fabricated nanocomposites successfully adsorbed 84.32 mg/g of MB dye onto its surface. Most importantly, the easy and rapid extraction of the nanocomposites by external magnetic field, coupled with high regeneration efficiency were also showed by the nanocomposites by sorption-desorption study.

In a separate study, Soltani et al. (2013) postulated that adsorption of Acid Red 88 (AR88) dye by prepared bio-silica/chitosan nanocomposites occurred via a multistep mechanism (superficial adsorption and subsequent intraparticle or pore diffusion), based on the fitness of experimental data to the intra-particle diffusion model ($R^2 = 0.9756$), although it was not concluded as the sole rate-determining step. Alternatively, the thermodynamic study suggested that adsorption of AR88 dye onto the nanocomposites was simultaneous and exothermic in nature. The incorporation of bio-silica led to grafting of Si-OH groups into chitosan biopolymer, which is desirable for the higher adsorption of AR88 dye that containing negatively-charged sulfonic groups, compared to pristine bio-silica.

12.5 Summary and Conclusions

The application of chitosan and its derivatives in the field of wastewater treatment have been studied by numerous researchers, especially on its adsorption capability and antimicrobial activity. This unique biopolymer can be modified with various functional groups to control their hydrophobicity, cationic and anionic properties. However, up to this day, no specific mechanisms for both activities can be finalized, apart from postulations and hypotheses made. For instance, the adsorption kinetics and isotherms differ from one study to the other reflects the uncertainty of the real adsorption mechanisms of chitosan.

Apart from that, the fabrication of chitosan nanocomposites will be usually based on the fact that chitosan is low-cost biopolymer, environmental-friendly and abundance in availability. However, if the application of chitosan nanocomposites took place in this very moment, is it really that cost-effective? As for now, the fabrication/synthesis of chitosan-based nanocomposites may not be lowered than commercially available adsorbents, coagulants and/or polymeric membrane. Furthermore, the production cost of chitosan is still high and involves the usage of multiple chemicals with high concentration. This may eventually create an enormous impact towards the environment. The more efficient and environmental-friendly way to produce chitosan must be sought after in the near future if chitosan is chosen to be used in wastewater treatment.

Conversely, the usage of chitosan nanocomposites, especially involving the incorporation of nanoparticles must be given additional attention. Incorporation of nanoparticles with chitosan must be followed by leeching studies to show the stability of the composites formed because it may give rise to multiple detrimental effects towards human beings, animals, and plants if they are found in water stream. Moreover, the reusability study must also be prioritized in future researches to prove the efficiency of the fabricated chitosan-based nanocomposites compared to conventional treatments. The study must carry out in periodical manners (days, months, or years) rather than specified cycles strengthen the results.

Lastly, the synthesized chitosan nanocomposites' studies must be demonstrated based on the real-life application. Experiments involving the usage of real industrial effluents, compared to the synthetic wastewater may affect the performance tremendously as pH, temperature, concentration and presence of multiple contaminants cannot be controlled. This experiment will give way to further improvement needed to optimize the materials, and hopefully applied for real world in the near future.

References

Abbasi, M., 2017. Synthesis and characterization of magnetic nanocomposite of chitosan/SiO2/carbon nanotubes and its application for dyes removal. J. Cleaner Prod. 145, 105−113.

Aliabadi, M.M., Irani, M., Ismaeili, J., Piri, H., Parnian, M.J., 2013. Electrospun nanofiber membrane of PEO/Chitosan for the adsorption of nickel, cadmium, lead and copper ions from aqueous solution. Chem. Eng. J. 220, 237−243.

Al-naamani, L., Dobretsov, S., Dutta, J., Burgess, J.G., 2017. Chitosan-zinc oxide nanocomposite coatings for the prevention of marine biofouling. Chemosphere 168, 408−417. Available at: https://doi.org/10.1016/j.chemosphere.2016.10.033.

Bhattarai, N., et al., 2005. PEG-grafted chitosan as an injectable thermosensitive hydrogel for sustained protein release. J. Controlled Release 103 (3), 609−624.

Chen, D., Ray, A.K., 2001. Removal of toxic metal ions from wastewater by semiconductor photocatalysis. Chem. Eng. Sci. 56 (4), 1561−1570.

Chen, R.H., Hwa, H.D., 1996. Effect of molecular weight of chitosan with the same degree of deacetylation on the thermal, mechanical, and permeability properties of the prepared membrane. Carbohydr. Polym. 29 (4), 353−358.

Chen, A., Zeng, G., Chen, G., Hu, X., Yan, M., Guan, S., Shang, C., Lu, L., Zou, Z., Xie, G., 2012. Novel thiourea-modified magnetic ion-imprinted chitosan/TiO2 composite for simultaneous removal of cadmium and 2,4-dichlorophenol. Chem. Eng. J. 191, 85−94.

Chenite, A., et al., 2000. Novel injectable neutral solutions of chitosan form biodegradable gels in situ. Biomaterials 21 (21), 2155−2161. Available at: http://linkinghub.elsevier.com/retrieve/pii/S0142961200001162.

Chenite, a, et al., 2001. Rheological characterisation of thermogelling chitosan/glycerol-phosphate solutions. Carbohydr. Polym. 46, 39−47.

Cho, Y.-W., Jang, J., Park, C.R., Ko, S.-W., 2000. Preparation and solubility in acid and water of partially deacetylated chitins. Biomacromolecules 1 (4), 609−614. Available at: http://pubs.acs.org/doi/abs/10.1021/bm000036j.

de Abreu, F.R., Campana-Filho, S.P., 2009. Characteristics and properties of carboxymethylchitosan. Carbohydr. Polym. 75 (2), 214−221. Available at: https://doi.org/10.1016/j.carbpol.2008.06.009.

Eidelman, N., Simon, C.G.J., 2004. Characterization of combinatorial polymer blend composition gradients by FTIR microspectroscopy. J. Res. Natl. Inst. Standards Technol. 109 (2), 219−231. Available at: http://nvlpubs.nist.gov/nistpubs/jres/109/2/cnt109-2.htm.

El-magied, M.O.A., et al., 2017. Cellulose and chitosan derivatives for enhanced sorption of erbium (III). Colloids Surf. A 529 (March), 580−593.

Fan, L., et al., 2013. Synthesis of magnetic beta-cyclodextrin − chitosan / graphene oxide as nanoadsorbent and its application in dye adsorption and removal. Colloids Surf. B 103, 601−607. Available at: https://doi.org/10.1016/j.colsurfb.2012.11.023.

Galhoum, A.A., et al., 2017. Chemical modifications of chitosan nano-based magnetic particles for enhanced uranyl sorption. Hydrometallurgy 168, 127−134. Available at: https://doi.org/10.1016/j.hydromet.2016.08.011.

Gaya, U.I., Abdullah, A.H., 2008. Heterogeneous photocatalytic degradation of organic contaminants over titanium dioxide: a review of fundamentals, progress and problems. J. Photochem. Photobiol. C 9, 1−12.

Ghaee, A., et al., 2015. Preparation of chitosan/cellulose acetate composite nanofiltration membrane for wastewater treatment. Desalin. Water Treat. July, 1−8. Available at: https://doi.org/10.1080/19443994.2015.1068228.

Ghiggi, F.F., Pollo, L.D., Cardozo, N.S.M., Tessaro, I.C., 2017. Preparation and characterization of polyether-sulfone / N-phthaloyl- chitosan ultra filtration membrane with antifouling property. Eur. Polym. J. 92, 61−70. Available at: https://doi.org/10.1016/j.eurpolymj.2017.04.030.

Ghimici, L., Nichifor, M., 2013. Separation of TiO_2 particles from water and water/methanol mixtures by cationic dextran derivatives. Carbohydr. Polym. 98 (2), 1637−1643. Available at: https://doi.org/10.1016/j.carbpol.2013.07.085.

Grizzuti, N., Buonocore, G., Iorio, G., 2000. Viscous behavior and mixing rules for an immiscible model polymer blend. J. Rheol. 44 (1), 149−164. Available at: http://sor.scitation.org/doi/10.1122/1.551073.

Guo, B.L., Gao, Q.Y., 2007. Preparation and properties of a pH/temperature-responsive carboxymethyl chitosan/poly(N-isopropylacrylamide)semi-IPN hydrogel for oral delivery of drugs. Carbohydr. Res. 342 (16), 2416−2422.

Hajji, S., et al., 2017. Nanocomposite films based on chitosan − poly (vinyl alcohol) and silver nanoparticles with high antibacterial and antioxidant activities. Process Saf. Environ. Prot. 111, 112−121. Available at: https://doi.org/10.1016/j.psep.2017.06.018.

Hamdi, A., Boufi, S., Bouattour, S., 2015. Phthalocyanine/chitosan-TiO_2 photocatalysts: characterization and photocatalytic activity. Appl. Surf. Sci. 339, 128−136.

Hamza, M.F., et al., 2017. Functionalization of magnetic chitosan particles for the sorption of U (VI), Cu (II) and Zn (II)— hydrazide. Materials 10, 539.

Hashimoto, K., Irie, H., Fujishima, A., 2005. TiO_2 photocatalysis: a historical overview and future prospects. Jpn. J. Appl. Phys. 44 (12), 8269−8285.

Hench, L.L., 1998. Biomaterials: a forecast for the future. Biomaterials 19 (16), 1419−1423.

Irom, B.C., Kavitha, K., Rupeshkumar, M., Singh, S.D.J., 2012. Research Journal of Pharmaceutical, Biological and Chemical Sciences Review Article Applications of natural polymer chitosan and chitosan derivatives in drug delivery: a review. Res. J. Pharm., Biol. Chem. Sci. 3 (4), 309−316.

Jaiswal, M., Chauhan, D., Sankararamakrishnan, N., 2012. Copper chitosan nanocomposite: synthesis, characterization, and application in removal of organophosphorous pesticide from agricultural runoff. Environ Sci Pollut Res 19, 2055−2062.

Ji, J., et al., 2014. Chemical modifications of chitosan and its applications. Polym.-Plast. Technol. Eng. 53, 1494−1505.

Jiang, Y., Gong, J.-L., Zeng, G.-M., Ou, X.-M., Chang, Y.-N., Deng, C.-H., Zhang, J., Liu, H.-Y., Huang, S.-Y., 2016. Magnetic chitosan−graphene oxide composite for anti-microbial and dye removal applications. Int. J. Biol. Macromol. 82, 702−710.

Le Cunff, J., Tomasic, V., Wittine, O., 2015. Photocatalytic degradation of the herbicide terbuthylazine : preparation, characterization and photoactivity of the immobilized thin layer of TiO_2/chitosan. J. Photochem. Photobiol. A 309, 22−29.

Leong, S., et al., 2014. TiO2 based photocatalytic membranes: a review. J. Membr. Sci. 472, 167−184.

Liu, Z., et al., 2012. Effects of chitosan molecular weight and degree of deacetylation on the properties of gelatine-based films. Food Hydrocolloids 26 (1), 311−317. Available at: https://doi.org/10.1016/j.foodhyd.2011.06.008.

Mahatmanti, F.W., Nuryono, Narsito, 2016. Adsorption of Ca (II), Mg (II), Zn (II), and Cd (II) on chitosan membrane blended with rice hull ash silica and polyethylene glycol. Indonesia J. Chem. 16 (1), 45−52.

Malini, M., et al., 2015. A versatile chitosan/ZnO nanocomposite with enhanced antimicrobial properties. Int. J. Biol. Macromol. 80, 121−129.

Massart, R., 1981. Preparation of aqueous magnetic liquids in alkaline and acidic media. IEEE Trans. Magn. 17 (2), 1247−1249.

Mourya, V.K., Inamdar, N.N., Tiwari, A., 2010. Carboxymethyl chitosan and its applications. Adv. Mater. Lett. 1 (1), 11−33.

Nagarpita, M.V., Roy, P., Shruthi, S.B., Sailaja, R.R.N., 2017. Synthesis and swelling characteristics of chitosan and CMC grafted sodium acrylate-co-acrylamide using modified nanoclay and examining its efficacy for removal of dyes. Int. J. Biol. Macromol. 102, 1226−1240.

Natarajan, S., et al., 2017. Antifouling activities of pristine and nanocomposite chitosan/TiO_2/Ag films against freshwater algae. RSC Adv 7, 27645−27655.

Nawi, M.A., Sabar, S., 2012. Photocatalytic decolourisation of Reactive Red 4 dye by an immobilised TiO_2/chitosan layer by layer system. J. Colloid Interface Sci. 372 (1), 80−87. Available at: https://doi.org/10.1016/j.jcis.2012.01.024.

Nowakowska, M., Moczek, L., Szczubialka, K., 2008. Photoactive modified chitosan. Biomacromolecules 9, 1631−1636.

Payne, G.F., Chaubal, M.V., Barbari, T.A., 1996. Enzyme-catalysed polymer modification: reaction of phenolic compounds with chitosan films. Polymer 37 (20), 4643−4648.

Pereira, F.A.R., Sousa, K.S., Cavalcanti, G.G.R.S., França, D.B., Queiroga, L.N.F., Santos, I.M.G., Fonseca, M.G., Jaber, M., 2017. Green biosorbents based on chitosan-montmorillonite beads for anionic dye removal. J. Environ. Chem. Eng. 5, 3309−3318.

Perez, J.V.D., et al., 2017. Response surface methodology as a powerful tool to optimize the synthesis of polymer-based graphene oxide nanocomposites for simultaneous removal of cationic and anionic heavy metal contaminants. RSC Adv. 7, 18480−18490. Available at: https://doi.org/10.1039/C7RA00750G.

Ravi Kumar, M.N., 2000. A review of chitin and chitosan applications. React. Funct. Polym. 46 (1), 1−27. Available at: http://linkinghub.elsevier.com/retrieve/pii/S1381514800000389.

Reijnders, L., 2008. Hazard reduction for the application of titania nanoparticles in environmental technology. J. Hazard. Mater. 152, 440−445.

Rinaudo, M., 2006. Chitin and chitosan: properties and applications. Progr. Polym. Sci. (Oxford) 31 (7), 603−632.

Rinaudo, M., Pavlov, G., Desbrie, J., Desbrières, J., 1999. Influence of acetic acid concentration on the solubilization of chitosan. Polymer 40, 7029−7032. Available at: http://www.cermav.cnrs.fr/monos/publi-pdf/P99-52.pdf.

Salah, T.A., Mohammad, A.M., Hassan, M.A., El-Anadouli, B.E., 2014. Development of nano-hydroxyapatite/chitosan composite for cadmium ions removal in wastewater treatment. J. Taiwan Inst. Chem. Eng. 45, 1571−1577.

Salehi, E., Daraei, P., Arabi, A., 2016. A review on chitosan-based adsorptive membranes. Carbohydr. Polym. 152, 419–432. Available at: https://doi.org/10.1016/j.carbpol.2016.07.033.

Shahzad, A., et al., 2017. Heavy metals removal by EDTA-functionalized chitosan graphene oxide nanocomposites. RSC Adv. 7, 9764–9771. Available at: https://doi.org/10.1039/C6RA28406J.

Shukla, S.K., Mishra, A.K., Arotiba, O.A., Mamba, B.B., 2013. Chitosan-based nanomaterials: a state-of-the-art review. Int. J. Biol. Macromol. 59, 46–58. Available at: https://doi.org/10.1016/j.ijbiomac.2013.04.043.

Soltani, R.D.C., Khataee, A.R., Safari, M., Joo, S.W., 2013. Preparation of bio-silica/chitosan nanocomposite for adsorption of a textile dye in aqueous solutions. Int. Biodeterior. Biodegrad. 85, 383–391. Available at: https://doi.org/10.1016/j.ibiod.2013.09.004.

Sureshkumar, V., Kiruba Daniel, S.C.G., Ruckmani, K., Sivakumar, M., 2016. Fabrication of chitosan–magnetite nanocomposite strip for chromium removal. Appl. Nanosci. 6, 277–285. Available from: https://doi.org/10.1007/s13204-015-0429-3.

Szabó, T., et al., 2013. Photocatalyst separation from aqueous dispersion using graphene oxide/TiO_2 nanocomposites. Colloids Surf. A 433, 230–239. Available at: https://doi.org/10.1016/j.colsurfa.2013.04.063.

Tanaka, H., Azuma, C., Sanui, K., Ogata, N., 1980. Synthesis of photosensitive polymers from chitosan. Polymer 12 (1), 63–66.

Tolba, A.A., et al., 2017. Synthesis and characterization of poly (carboxymethyl) -cellulose for enhanced La (III) sorption. Carbohydr. Polym. 157, 1809–1820. Available at: https://doi.org/10.1016/j.carbpol.2016.11.064.

Vakili, M., et al., 2017. Enhancing reactive blue 4 adsorption through chemical modification of chitosan with hexadecylamine and 3-aminopropyl triethoxysilane. J. Water Process Eng. 15, 49–54. Available at: https://doi.org/10.1016/j.jwpe.2016.06.005.

Wang, Y., Xia, G., Wu, C., Sun, J., Song, R., Huang, W., 2015. Porous chitosan doped with graphene oxide as highly effective adsorbent for methyl orange and amido black 10B. Carbohydr. Polym. 115, 686–693.

Wang, B., et al., 2017. Functionalized chitosan biosorbents with ultra-high performance, mechanical strength and tunable selectivity for heavy metals in wastewater treatment. Chem. Eng. J. 325, 350–359. Available at: https://doi.org/10.1016/j.cej.2017.05.065.

Wu, S., Zeng, F., Zhu, H., Tong, Z., 2005. Energy and electron transfers in photosensitive chitosan. J. Am. Chem. Soc. 127 (7), 2048–2049.

Xiao, B., et al., 2011. Preparation and characterization of antimicrobial chitosan- N -arginine with different degrees of substitution. Carbohydr. Polym. 83 (1), 144–150. Available at: https://doi.org/10.1016/j.carbpol.2010.07.032.

Zhang, W., et al., 2016. Membrane fouling in photocatalytic membrane reactors (PMRs) for water and wastewater treatment: a critical review. Chem. Eng. J. 302, 446–458. Available at: https://doi.org/10.1016/j.cej.2016.05.071.

Zinadini, S., et al., 2014. Novel high flux antifouling nano filtration membranes for dye removal containing carboxymethyl chitosan coated Fe_3O_4 nanoparticles. Desalination 349, 145–154. Available at: http://dx.doi.org/10.1016/j.desal.2014.07.007.

13

Application of Semiconductor Nanoparticles for Removal of Organic Pollutants or Dyes From Wastewater

Omme Kulsum Nayna[1,2], Shafi M. Tareq[1]

[1]DEPARTMENT OF ENVIRONMENTAL SCIENCES, JAHANGIRNAGAR UNIVERSITY, DHAKA, BANGLADESH [2]DEPARTMENT OF ENVIRONMENTAL SCIENCE AND ENGINEERING, EWHA WOMANS UNIVERSITY, SEOUL, REPUBLIC OF KOREA

13.1 Introduction

Increasing demand and shortage of pure water sources due to the rapid development of industrialization, population growth and long-term droughts have become an issue worldwide (Chong et al., 2010). Industrial growth produces a great variety of organic products and frequently these substances are complex to degrade. Industrial wastewater contains various chemicals especially synthetic dyes (Parsons, 2004). Example, Textile industry generate large amount of wastewater derived to different processes of color impregnation in textile fibers, which has a great amount of detergents, dyes, microfiber (cellulose, wool, and synthetic fibers), and inorganic salts. This residual water, with great load pollutants, generates the contaminations of natural waters bodies (Weber and Adams, 1995). Now a day, more than 50% of dyes used in the textile industry are monoazo, diazo, and triazo dyes, considering their chemical stability (recalcitrant) and negative influence on the ecological systems, the regulations of the removal color in the factory effluent is a current issue of discussion all over the world. The presence of even very low concentrations of dyes in effluent is highly visible and degradation products of these textile dyes are often carcinogenic. These effluent wastewaters have been recognized to have high color, high BOD and COD load, as well as high dissolve organic matter (DOM) concentration (Paul et al., 2012). Dyes are aromatic compounds which can absorb light in the visible wavelengths range (400−700 nm). The dye molecule is a combination of a chromophore which is a part of the molecule that can absorb light, that is, the color-absorbing coordination group and a conjugated system, and a structure with alternating double and single bonds. Chromophores are containing $C\!\!=\!\!C$ and $C\!\!=\!\!O$ (carbonyl), and azo group -N$=$N- or nitro group ($-NO_2$). A complex mixture containing various

Nanotechnology in Water and Wastewater Treatment. DOI: https://doi.org/10.1016/B978-0-12-813902-8.00013-7

organic materials, including carbohydrates or polysaccharides, amino acids or peptides or proteins, lipids, humic substances, and anthropogenic organic pollutants is known as Dissolved organic matter (DOM) (Leenheer and Croue 2003; Shon et al., 2006). It plays critical roles both in drinking water (DWTPs) and wastewater treatment plants (WWTPs) in determining the treatment performance and the distributed water quality and it is found in everywhere. The presence of DOM not only affects the current discharge standards, but also presents significant challenges in wastewater restoration (Guo et al., 2011).

Highly sensitive and reliable optical technique named fluorescence spectroscopy used for fast monitoring DOM in both natural and engineered systems (Henderson et al., 2009; Fellman et al., 2010). Excitation emission matrices (EEMs) of fluorescence comprise of fluorescence signals at several thousand pairs of excitation/emission wavelengths (Ex/Em) for each sample, which provide a lot of information about DOM (Coble 1996; Parlanti et al., 2000; Baker, 2001; Chen et al., 2003; Hudson et al., 2007). Including different functional groups DOM contains large amounts of unsaturated and aromatic structures that have fluorescence characteristics, which allows to use of fluorescence spectroscopy to extract information on the degradation of DOM at the time of Wastewater treatment. DOM contains organic molecules including chromophoric (light absorbing) and fluorophoric (light emitting) moieties (Wang et al., 2009). Therefore, three-dimensional excitation-emission matrix (3DEEM) fluorescence spectroscopy has been extensively used to detect detailed changes and transformations of organic matter in wastewater (Wang et al., 2009).

The main technologies available for the treatment of dyes involve the transfer of the pollutant from a liquid phase to another phase, that is, concentrating the dye on an adsorbent so that it can later be discarded in a landfill or incinerated (Rafatullah et al., 2010). This process known as conventional treatment technology and this phase exchange is not an ideal remedy. At present there is a continuously increasing worldwide concern for the development of wastewater treatment technologies. Selection of the best method and material for wastewater treatment is a highly complex task, which should consider many factors, such as the quality standards to be met and the efficiency as well as the cost (Huang et al., 2008; Oller et al., 2011). Therefore, the following four conditions must be considered in the decision on wastewater treatment technologies: (1) treatment flexibility and final efficiency, (2) reuse of treatment agents, (3) environmental security and friendliness, and (4) low cost (Zhang and Fang, 2010; Oller et al., 2011).

In recent years, photocatalysis, one of the advanced physico-chemical technology has attracted much attention in photodegradation of organic pollutants (Akhavan and Azimirad, 2009). The photocatalytic activity (PCA) depends on the ability of the catalyst to create electron−hole pairs, which generate highly reactive hydroxyl radicals, (OH•) for secondary reactions (Rahman et al., 2011). Hydroxyl radicals generated by photodegradation have emerged as a promising technology of water and wastewater treatment for the degradation or mineralization of a wide range of organic contaminants (Thiruvenkatachari et al., 2008). Commercial application of the process is called advanced oxidation process (AOP). The photoactivated reactions are characterized by the free radical mechanism initiated by the interaction of photons of a proper energy level with the catalyst (photocatalytic

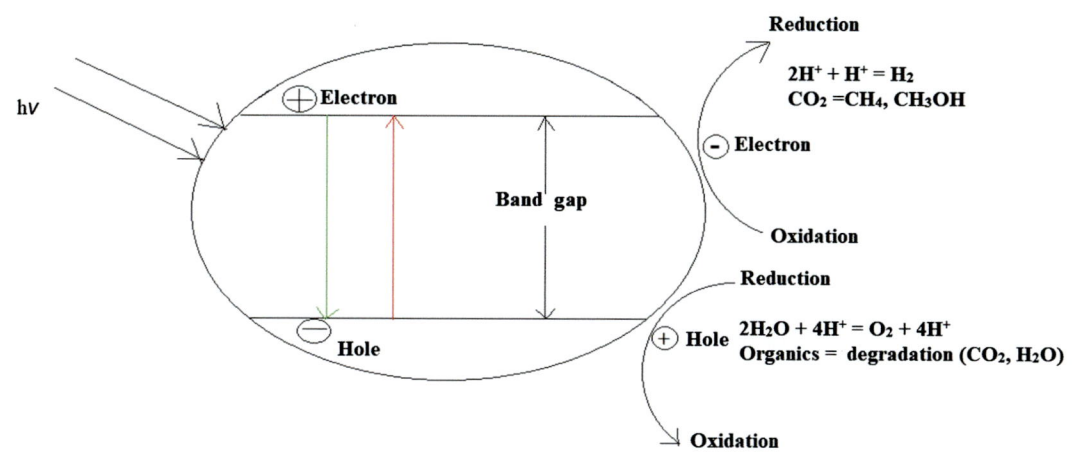

FIGURE 13-1 Photodegradation mechanism.

oxidation) (Fig. 13-1). Photocatalysis by TiO_2 considered as a green catalytic technology, which can degrade almost all organic pollutants without selection. More than 3000 kinds of difficult to degrade organic compounds can be degraded by TiO_2 photocatalytic technology investigated from different studies.

13.2 Semiconductor Nanoparticles

Semiconductors are act as sensitizers for light-induced redox-processes due to the electronic structure of the metal atoms in chemical combination, which is characterized by a filled valence band, and an empty conduction band (Hoffmann et al., 1995). Semiconductor-induced photocatalysis is a newly developed advanced oxidation process, which can be applied conveniently for the degradation of wastewater containing dyes. The semiconductors generate electron—hole pairs, to participate in reduction and oxidation processes. Among various oxide semiconductor photocatalysts (i.e., TiO_2, ZrO_2, ZnO, MoO_3, SnO_2, α-Fe_2O_3, etc.), anatase crystalline form of TiO_2 and Fe_2O_3 nanomaterials are the suitable photocatalyst with nontoxicity, strong oxidizing activity and long-term chemical stability (Park and choi, 2004; Nagaveni et al., 2004; Zou et al., 2001). Between these two nanomaterials TiO_2 has one limitations that it mainly absorbs UV light with wavelengths <380 nm which covers only 5% of the solar spectrum due to its wide band-gap of 3.2 eV (Table 13-1).

Fe_2O_3 nanoparticles is an *n*-type semiconducting material act as a suitable candidate of photocatalyst to absorb visible light with band-gap of 2.2 eV (Table 13-1). Akhavan and Azimirad 2009 investigated Fe-doped TiO_2 photocatalysts and Fe_2O_3—TiO_2 coupled semiconductor photocatalysts for photodegradation of toxic and organic pollutants under visible light. However, we are always looking for those techniques which are inexpensive and easy

Table 13-1 List of Some Common Semiconductor Photocatalysts Including Band Gap Energy

Photocatalyst	Bandgap Energy (eV)	Photocatalyst	Bandgap Energy (eV)
TiO_2 (rutile)	3.0	SnO_2	3.5
WO_3	2.7	Si	1.1
Fe_2O_3	2.2	TiO_2 (anatase)	3.2
ZnS	3.7	$SrTiO_3$	3.4
ZnO	3.2	Wse_2	1.2
CdS	2.4	α-Fe_2O_3	3.1

Included from Bhatkhande, D.S., Pangarkar, V.G., Beenackers, A. ACM, 2001. Photocatalytic degradation for environmental applications a review. J. Chem. Technol. Biotechnol. 77, 102.

to prepare. Hence, several laboratories have examined the efficiency of Fe_2O_3 nanoparticles (without doping) during the photocatalytic purification of dye mixed water (Bandara et al., 2007; Cunningham et al., 1988; Fernandez et al., 1998).

13.3 Synthesis of Semiconductor Nanoparticles

Semiconductor nanomaterials have interesting physical and chemical properties with useful functionalities, comparing with their conventional bulk counterparts and molecular materials. The most attractive properties of these materials are narrow and intensive emission spectra, continuous absorption bands, high chemical and photobleaching stability, processability, and surface functionality. The movement of electrons and holes in semiconductor nanomaterials is primarily governed by the well-known quantum confinement which is one of the unique properties of nanomaterials and the size and geometry of the materials largely affected the transport properties related to phonons and photons (Alivisatos, 1996a,b; Alivisatos, 1996a,b; Burda et al., 2005; Murray et al., 2000). The drastic increase of specific surface area and surface-to-volume ratio occurred while the size of the material decreases (Alivisatos, 1996a,b; Chen et al., 2005). Parameters including size, shape, and surface characteristics can be varied to control their properties for different applications of *interest* (Li and Jin, 2009). The uniform shape and size of the nanoparticles is a key issue for developing the design of novel advanced functional nanomaterials.

The nanoparticles can be prepared by different approaches such as chemical or physical including gaseous, liquid and solid media. By decreasing the size of the constituents of the bulk material is the general approach of the synthesis of nanostructures, known as physical methods, also called top-down approach. On the other hand, chemical methods tend to attempt to control the clustering of atoms or molecules at the nanoscale range, which known as bottom-up approach. The wet chemical methods are most popular because of their major advantages compared to other conventional methods. For example, they allow much more rigorous control of the shape and size of the nanoparticles with high reliability and cost effectiveness and the cluster of the resulting particles can be alleviated by functionalization with different capping ligands (Cushing et al., 2004). Different chemical methods have been successfully applied for developing various types of nanoparticles. They are: (1) Hydrothermal/Solvothermal, (2) Co-precipitation, and (3) Sol-Gel.

Among various available wet chemical methods, the sol-gel process is one of the simplest and the cheapest undoubtedly. This method is developed based on the phase transformation of a sol obtained from metallic alkoxides or organometallic precursors. The main advantages of the sol-gel method are versatility and probability to obtain high purity materials whose composition is perfectly controlled (Suresh, 2013). Solvothermal or hydrothermal reactions have been used to prepare micro or nanoparticles with different morphologies. The recent interest has been focused on the potentialities of solvothermal reactions in materials synthesis because of the necessity of developing new materials for basic research or industrial applications (Demazeau, 2008). Chemical and thermodynamical parameters are involved with high pressures in solvothermal reactions. In these cases, in an aqueous solution, hydrothermal technology has been used to synthesis novel materials (Demazeau, 1999; Feng and Xu, 2001; Demianets, 1991), extraction of minerals (Habashi, 2005), synthesis of geological materials (Hosaka, 1991), thin film deposition (Gogotsi and Yoshimura, 1994), elaboration of fine particles with well-defined size and morphology (Rajamathi and Seshadri, 2002). TiO_2 is an important semiconductor in applications of environmental remediations because of its photocatalytic properties which is synthesized by various processes (Table 13-2). Among these various synthesis processes of TiO_2, sol-gel process is considered as one of the important process recently because of its advantages like low processing temperature, versatility of processing, high homogeneity, and stability (Henrist et al., 2009; Hernandez−Martinez et al., 2013).

Table 13-2 Synthesis Processes of Widely Used TiO_2 Nanoparticles (Adopted from Eskandarloo and Badiei) and Green Synthesis Processes of Fe Nanoparticles

Semiconductor Nanoparticles	Synthesis Processes
TiO_2 nanoparticle	Sol−gel
	Reactive sputtering
	Solvothermal
	Liquid phase deposition
	Reverse micelle
	Hydrothermal treatment
	Electrochemical
Fe nanoparticles	Synthesis by leaf extract
	Hydrothermal synthesis using plant extract
	Synthesis using other plant materials (Alfalfa biomass, sorghums bran, plant peel extract)
	Synthesis by fruit extract
	Synthesis by seed extract

Adopted from Saif, S.,Tahir, A., Chen, Y., 2016. Review green synthesis of iron nanoparticles and their environmental applications and implications. Nanomaterials (Basel), 6(11), 209, in Environmental Remediations.

Among various semiconductor nanoparticles, iron oxides are available in various forms in nature like Magnetite (Fe_3O_4), maghemite (γ-Fe_2O_3), and hematite (α-Fe_2O_3) which are the most common forms (Cornel and Schwertmann, 1996; Chan and Ellis, 2004). In recent years, because of nano size, high surface area to volume ratios and superparamagnetism, the synthesis and utilization of iron oxide nanomaterials are widely spread (McHenry and Laughlin, 2000; Afkhami et al., 2010; Pan et al., 2010). Example, the low dimensional iron oxide nanoparticles have been synthesized by hydrothermal method by adding uni-molar concentration of ferric chloride and urea into hydrothermal cell (Teflon line autoclave).

13.3.1 Characterizations of Iron Oxide Nanoparticles

The optical property of iron oxide nanoparticles (Fe_2O_3) is one of the important characteristics for the evaluation of its optical and photocatalytic activity. The absorption spectrum of as-grown iron oxide nanoparticles solution was measured by UV-visible spectrophotometer in visible range between 200 to 800 nm wavelengths which displays an onset of absorption maxima at 444.0 nm. The lambda maxima of iron oxide nanoparticles reported by Cornell and Schwertmann, 2003 and Rahman et al., 2011 are very similar with as grown nanoparticles. This wavelength indicated the formation of reddish colored low dimensional Fe_2O_3 nanoparticles. Band gap energy is calculated based on the maximum absorption band (444 nm) of Fe_2O_3 nanoparticles. Band gap energy of nanoparticles obtained to be 2.792792793 eV, according to following equation:

$$Ebg = \frac{1240}{\lambda}(eV)$$

Where Ebg is the band-gap energy and λ_{max} is the wavelength (444.0 nm) of the nanoparticles.

FTIR, used for identifying types of chemical bonds in a molecule is a popular tool, was recorded for Fe_2O_3 nanoparticles (Fig. 13-2). It displays many bands at 447, 561, 935, 1400, 1637, 1751, and 3134 cm^{-1} and Fe_2O_3 nanoparticles give absorption bands at those wavelengths known from the literature (Rahman et al., 2011; Ma et al., 2007). The band observed at 447.49 and 561.29 represents Fe$-$O$-$Fe stretching vibration. The observed vibration bands at low frequencies regions suggest the formation of Fe_2O_3 nanoparticles. The vibration bands observed at 1400 and 1630 cm^{-1} assigned to O$=$C$=$O stretching and OH bending vibration. The band observed at 1751.36 cm^{-1} represents the C$=$O stretching of carbonyl. The strong absorption band at 3032 and at 3134.33 cm^{-1} represents the C-H stretching vibration of alkanes and alkenes. The absorption bands at these wavelengths normally comes from the carbon dioxide and water which generally nanomaterials absorbed from the environment due to their mesophorous structure.

To confirm the particle size and morphology of Fe_2O_3 nanoparticles scanning electron microscope (SEM) image used. From the SEM images it is clear that the synthesized products are nanoparticles, which grown in a very high-density and possessed almost uniform shape presented in Fig. 13-3 and most of the nanoparticles possessing spherical shapes. This

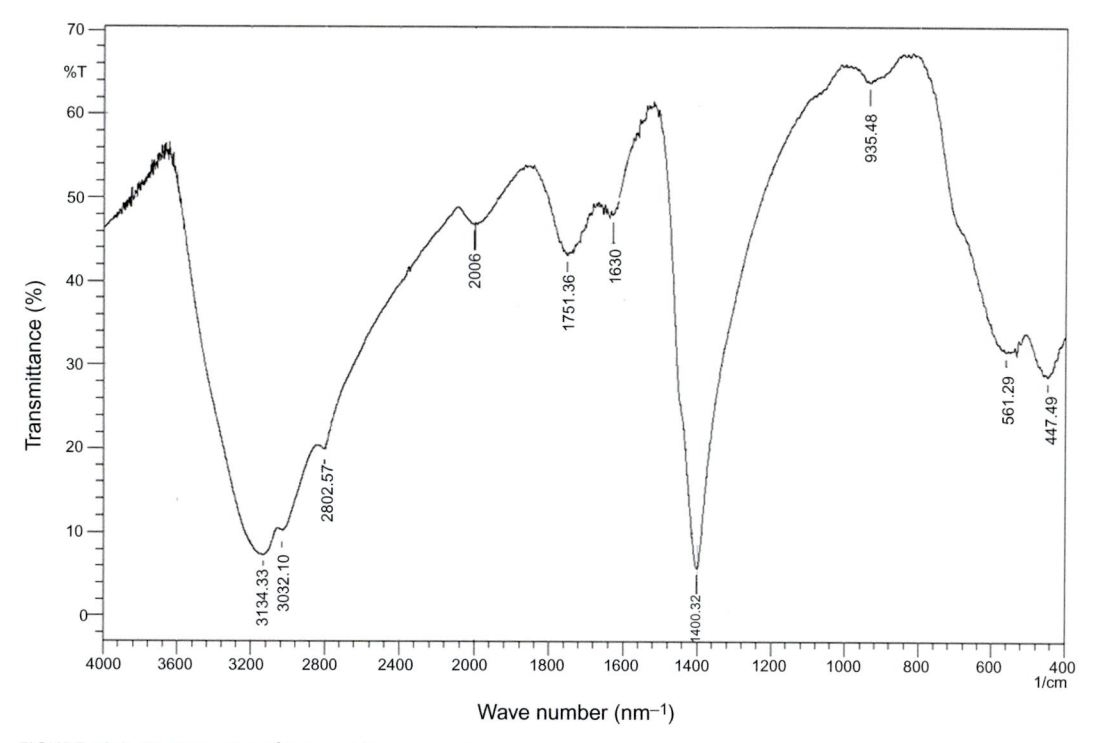

FIGURE 13-2 FT-IR spectra of iron oxide nanoparticles.

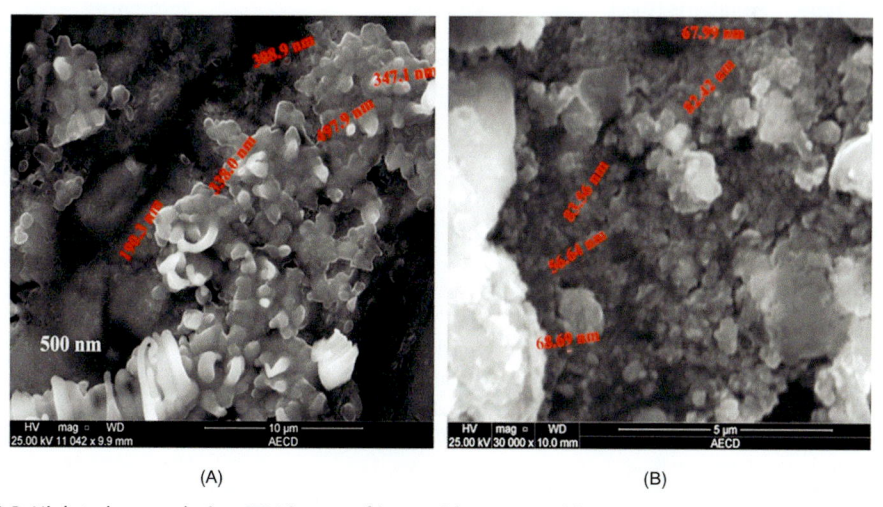

(A)   (B)

FIGURE 13-3 High to low resolution SEM image of iron oxide nanoparticles.

figure shows that the different diameter of nanoparticles grown by hydrothermal process. The particles are aggregated with small crystals with a broad size distribution up to 500 nm (Fig. 13-3A). The morphology of the synthesized iron oxide nanoparticles is identical. The diameter of iron nanoparticles is calculated in the range of 56–83 nm (Fig. 13-3B) where the average diameter of iron oxide nanoparticles is close to 71 nm (Rahman et al., 2011).

13.4 Application of Semiconductor Nanoparticles

Recently the interest in semiconductor photocatalysis has increased exponentially because of the potential and opportunities it offers in variety of fields including treatment of environmental pollution. The main advantages of semiconductor photocatalysis are that it has the capacity to use renewable and pollution free solar energy and it can offer a good substitute for the energy-intensive treatment methods. Unlike the conventional treatment methods, which transfer pollutants not only from one medium to another but also transform those to more refractory pollutants, semiconductor photocatalysis converts contaminants to innocuous products such as CO_2 and H_2O. Also, with mild reaction condition and modest reaction time and can be applied to aqueous, gaseous, and solid-phase treatments with the possibility of being both supplementary and complementary to the present technologies. Semiconductor photocatalysts not only minimize the cost but also generate the desired product efficiently and effectively. The applications of semiconductor nanoparticles induced photocatalytic degradation of organic pollutants in wastewater is shown in Table 13-3.

Table 13-3 List of Organic Pollutants (Dyes) Degraded by Semiconductor Nanoparticles Induced Photocatalytic Activation

Wastewater Type	Degraded Organic Pollutants (Dye)	Catalyst	References
Dye wastewater	Methyl orange	Y-TiO_2-HPW	Wang et al. (2011a,b)
	Alkaline red dye	TiO_2-Fenton	Khataee et al. (2011)
	Rhodamine6G	TiO_2	Abdullah et al. (2011)
	Anthraquinone dye	N-TiO_2	Veluru and Appukuttan (2011)
	Reactive Red 84	Fe_2O_3/H_2O_2	Cano-Guzman et al. (2014)
	Methyl orange (MO) Rhodamine 6G (R6G)	TiO_2, ZnO, CdS, ZnS, SnO_2	Kansal et al. (2007)
	Reactive yellow 81 and violet 1	TiO_2-P25 (Degussa)	Giwa et al. (2012)
	Rhodamine B	Fe–TiO_2 coated with side-glowing optical fibers	Lin et al. (2015)
	Methylene blue	AgBr/ZnO	Dai et al. (2014)
	Rhodamine B	TiO_2/trititanate	Chen et al. (2015)
	Acid Red 88 dye	CuO/ZnO	Sathishkumar et al. (2011)
	Methylene Blue	Al_2O_3/Fe_2O_3	Hassena (2016)

13.4.1 Photocatalytic Activity of TiO$_2$ Nanoparticles

Titania (TiO$_2$) nanoparticles found in three crystalline phases including anatase, rutile, and brookite. Anatase is the predominant commercial phase of TiO$_2$ which is widely used in photocatalytic process due to its high photocatalytic activity including crystal structure with a band gap of 3.23 eV (Nishanthi et al., 2013; Reyes−Coronado et al., 2008). The rate of photo-catalytic activity of TiO$_2$ catalyst is limited to the recombination rate of photo-produced elec-tron-hole pairs and the photocatalytic efficiency of TiO$_2$ catalyst decreases with quick recombination of photo-induced charge carriers (Xu et al., 2011; Zhang and Zhu 2012). Some techniques have been used to reduce the electron-hole recombination and improve the photocatalytic activity of TiO$_2$ including doping by various metallic species and coupling with other semiconductors (Song et al., 2008; Ambrus et al., 2008; Liqiang et al., 2006).

To prevent the recombination of charge carriers one option is doping of TiO$_2$ with various metallic species and on the surface of TiO$_2$, metal nanoparticles will act as an electron reser-voir to trap electrons which can greatly increase the efficiency of charge separation, resulting the improvement of TiO$_2$ photocatalytic activity. The addition of dopants can change the sur-face properties of TiO$_2$ catalyst including surface area and surface acidity (Pelaez et al., 2012). Castro et al. (2009) prepared doped anatase−structure TiO$_2$ nanoparticles (Sb/TiO$_2$, Nb/TiO$_2$, Fe/TiO$_2$, and Co/TiO$_2$) using hydrothermal methodology preceded by a precipita-tion doping step. The morphological characteristics of Fe/TiO$_2$ and Sb/TiO$_2$ nanoparticles, showed like that the un−doped pure TiO$_2$ sample, on the other hand Nb/TiO$_2$ and Co/TiO$_2$ nanoparticle, showed particles with an elongated morphology. The doped anatase−structure TiO$_2$ nanoparticles was used to degrade a model compound namely diquat dibromide monohydrate to examine the photocatalytic activity. The results revealed a significant enhancement in the photocatalytic activity of the Sb/TiO$_2$ and Nb/TiO$_2$ nanoparticles but for the Fe/TiO$_2$ and Co/TiO$_2$ nanoparticles there was no photocatalytic activity. This study stated that to determine the photocatalytic activity of doped TiO$_2$ nanoparticles both amount and valence of the doping ions are important factors.

However, there is no practical applications found on TiO$_2$ doped by a single species and it has been found that TiO$_2$ co−doping with different species leads to higher photocatalytic activity. Behnajady and Eskandarloo (2013a; 2013b) prepared TiO$_2$ nanoparticles using sol-gel method and commercially available TiO$_2$−P25 nanoparticles for direct and indirect dop-ing of silver and copper. Co-doped Ag-Cu/TiO$_2$ nanoparticles with a mean size of 10−35 nm and 20−30 nm is estimated for direct and indirect co-doped samples, respectively. The pres-ence of both silver and copper metals in chemical composition of the direct and indirect co-doped samples confirmed by the Energy−dispersive X-ray spectroscopy (EDX) analysis at the microscopic level. By the removal of C.I. Acid Orange 7 as a model pollutant from textile industry the photocatalytic activity of prepared nanoparticles was evaluated and co-doped Ag, Cu/TiO$_2$ nanoparticles showed the highest photocatalytic activity both direct and indirect doping methods compared to mono-doped Ag/TiO$_2$, Cu/TiO$_2$, and pure TiO$_2$ nanoparticles under both UV and visible light irradiation. To increase the lifetime of the photo-produced electron-hole pairs and increase the photocatalytic activity of TiO$_2$, coupling of TiO$_2$ with other metal oxides is another approach. Recently, many studies conducted with TiO_2 coupled with other metal oxides, including $ZnO-TiO_2$, ZrO_2-TiO_2, SnO_2-TiO_2, Cu_2O-TiO_2, WO_3-TiO_2, and CeO_2-TiO_2 (Kumar and Devi, 2011).

In a catalysis to determine the catalytic activity of catalyst supporting material is very important. Supporting materials of catalyst are porous with high surface areas. The rate of electron transfer increases and the rate of electron-hole combination decreases by the electronic interaction between the support and the catalyst which bring about slightly acidic conditions. Also, the rate of oxidation of organic pollutants increased due to the increase of adsorption ability and stability of the catalyst by supporting materials (Li et al., 2007). The catalyst supporting materials not only shift the band edge towards the visible region but also prevent the agglomeration of catalyst which help to improve the separation of catalyst from posttreatment wastes. These conditions are very important because they increase the photocatalytic activity. Most common types of supports used for catalysts including alumina supports, carbon supports, and carbon-covered, alumina (CCA) supports.

Though the alumina and carbon supports have high surface area and good mechanical properties, but they have some shortcomings. But the carbon covered alumina supports known as CCA can overcome these shortcomings. Reduction of the alumina acidity (Σ by 90%) due to the presence of carbon; increased electron-charge transfer and reduced metal-support interactions resulting in increased catalytic activity; and increased mechanical strength and increased surface area are the properties of carbon-covered alumina (CCA). Due to the combination of the properties of both carbon and alumina CCA supports are superior catalyst supports. Also, they have a high surface area and high adsorption affinity for both organic and inorganic compounds. CCA synthesized by Jana and Ganesan (2011) recently, in the form of foams and increased its surface area and also enhanced its adsorptive properties. Because of the high catalytic activity and stability, CCA supports have been used to support Ru catalysts in the synthesis of NH_3 (Masthan et al., 1991). CCA supported Ag nanoparticles have been used to remove bacteria in drinking water (Shashikala et al., 2007). Recently, CCA attached with Titania nanoparticles and used for the photocatalytic degradation of Rhodamine B under visible-light irradiation (Mahlambi et al., 2012). Metal-doped titania has also been supported on these CCA supports and Ag, Co, Ni, and Pd were used as the metal dopants (Mahlambi et al., 2013). The CCA supports were synthesized from glucose and an impregnation method was used to attach the nanoparticles on the supports and from the results it has been found that attaching the titania nanoparticles on the CCA supports greatly enhanced their photocatalytic activity. Both CCA supported TiO_2 and CCA supported metal doped TiO_2 nanoparticles had large surface area due to the porous nature of CCA supports and highly active under visible-light irradiation and exhibited less electron hole combination due to the presence of carbon. Giwa et al., 2012, investigated photocatalytic degradation of reactive yellow 81 and violet 1 dyes in aqueous solution using solar rays by TiO_2-P25 (Degussa) a semiconductor oxide act as a photocatalyst. To study the mineralization and degradation of the dyes chemical oxygen demand and UV-vis spectrophotometer were used respectively. The Degussa P25 the most effective catalyst which produced discoloration of 92% for reactive yellow 81 (50 mg/L) and 85% for reactive violet 1 (50 mg/L) within 20 minutes.

13.4.2 Photocatalytic Activity of Fe$_2$O$_3$ Nanoparticles

Photodegradation of organic matter using iron oxide nanoparticles as a semiconductor with solar energy is considering as a green technology. This process degrades at low or medium concentration of organic compounds containing water with negligible secondary pollution by utilizing solar energy. To enhance the degradation of organic matter by photodegradation it is most important to develop environmental friendly catalysts with high efficiency, less energy cost with low price. Iron is one of the abundant elements on earth and iron oxide nanoparticles is one of the photocatalysts with high efficiency and low price (Feng and Nansheng, 2000). The removal ability of contaminants by iron oxide nanomaterials has been demonstrated at both laboratory and field scale tests (White et al., 2009; Girginova et al., 2010). The applications of iron oxide nanomaterials in contaminated water treatment can be divided into two groups. Firstly, iron oxide nanomaterials used as a kind of nanosorbent or immobilization carrier for removal efficiency enhancement which is referred to here as adsorptive or immobilization technologies. Secondly, iron oxide nanomaterials used as a photocatalysts to break down or to convert contaminants into a less toxic form which is known as photocatalytic technologies and commercially known as advanced oxidation process. The photooxidation of halogenated acetic acids investigated on the surface of iron oxides by Pehkonen et al. (1995). The fastest rate of photoreduction of Fe(III) to Fe(II) was achieved with iron oxide as an electron acceptor and fluoroacetic acid as an electron donor. As a photocatalyst iron oxide nanomaterials is good in absorbing visible light compared with most widely used TiO$_2$.

Iron oxide nanoparticles as a representative semiconductor which can influenced the degradation of dissolved organic matter (DOM) in textile wastewater by photodegradation process and produce simple compounds such as CO$_2$, H$_2$O, NO$_3^-$, and NH$_4^+$, etc. The use of renewable energy (solar energy) make it more sustainable. The direct band-gap excitation of semiconductors generates electron−hole pairs, which participate in reduction and oxidation processes. The decolorization rate reflects the breakdown of dye molecules containing conjugated system by the attack of hydroxyl radicals (-OH) formed by the water hydrolysis with the action of solar radiation and consequently causes decolorization (Lucas and Peres, 2006). Illumination of iron oxide nanoparticles under visible light, electrons from valence band jump into conduction band, and generating electrons and positively charged holes leading to the formation of active oxidation species, which are responsible for enhanced dye degradation. In photodegradation process generally, the sites near the chromophore (for instance, C−N≡N− bond) is the attacked area and photocatalytic destruction of the C−N≡and −N≡N− bonds lead to fading of the dyes (Khataee and Kasiri, 2010).

Hoag et al. (2009) investigated the degradation of the organic contaminant (bromothymol blue) by green tea (GT) synthesized iron nanoparticles to catalyze hydrogen peroxide. The green synthesized nanoscale zero-valent iron was more efficient in catalysis process than that of Fe-EDTA and Fe-EDDS. During experiments, it was observed that by increasing concentrations of GT-nZVI, more hydrogen peroxide catalyzed, and finally increased the degradation of bromothymol blue. The rapid degradation of Bromothymol blue dye in presence of iron nanoparticles and H_2O_2, demonstrating that the iron nanoparticles catalyzes the reaction for production of free radicals from H_2O_2. The catalysis of H_2O_2 prompting the rate of reaction ultimately increases the rate of degradation of bromothymol blue (Njagi et al., 2011). Fe nanoparticles synthesized by green tea proved to be more effective as a Fenton-like catalyst both in terms of kinetics and percentage removal compared to iron nanoparticles produced by borohydride reduction (Shahwan et al., 2011).

Huang et al. (2013) employed oolong tea extract for synthesis of iron nanoparticles (OT-Fe NP) and used to degrade malachite green (MG). From the results it has been found firstly that polyphenols/caffeine in oolong tea extract acted as both reducing and capping agents in synthesis of Fe nanoparticles, leading to reduced aggregation and increased reactivity of OT-Fe NP. And secondly, OT-Fe NP proved to be efficient in the degradation of MG, resulting in 75.5% of MG (50 mg/L) being removed. Kuang et al. (2013) employed a study where used extracts of three different teas separately including green tea (GT), oolong tea (OT), and black tea (BT) for synthesis of iron nanoparticles. Iron nanoparticles synthesized from different green teas were used as a catalyst for Fenton-like oxidation of monochlorobenzene (MCB). Green tree synthesized Fe nanoparticles exhibited the highest degradation rate and was attributed to high polyphenol present in extract. From the results 69% degradation was observed for Green tea-Fe nanoparticles while 53% by Oolong tea-Fe nanoparticles and 39% by Black tea-Fe nanoparticles in 180 min.

Cano-Guzman et al. (2014), investigated kinetic study for reactive red 84 by photo degradation using iron (III) oxide nanoparticles. Iron oxide nanoparticles (Fe_2O_3) were synthesized by mixing iron (II) sulfate ($FeSO_4$) and hydrogen peroxide (H_2O_2) in aqueous solution at pH 3 and the presence of Fe_2O_3 nanoparticles was verified by HRTEM analyses. Chirita et al. in 2009 did experiment on Fe_2O_3-nanoparticles, physical properties and their photochemical and photoelectrochemical applications. This study described the use of photo electrochemistry such as solar energy conversion, water splitting, photocatalysts for the removal of organic and inorganic species from aqueous or gas phase. In 2016 Hassena investigated the photocatalytic degradation of methylene blue dye under visible radiation using Al_2O_3/Fe_2O_3 nano composite as photocatalyst and the degradation depends on the initial concentration of MB dye pH, amount of catalyst, and light intensity. Semiconductor induced photocatalysis of methylene blue dye is an effective method to remove organic pollutants from wastewater as a complete mineralization of dyes in to CO_2, H_2O and other oxides.

13.4.3 Photocatalytic Activity of ZnO Nanoparticles

ZnO has received much attention as a well-known photocatalyst to degrade and complete mineralization of environmental pollutants. Because of same band gap energy (3.2 eV) of ZnO and TiO_2, photocatalytic activity should be the same but due to the frequent photocorrosion of ZnO nanoparticles during illumination under UV light the photocatalytic activity of ZnO decreases in aqueous solutions. But some studies confirmed that ZnO exhibits better photocatalytic degradation of some dyes than TiO_2 and the results indicated that during the process of photocatalytic oxidation of gas phase chloridized hydrocarbon ZnO and TiO_2 exhibited higher activity than Fe_2O_3.

Recently Zhang et al., 2017 complied a study based on a facile synthetic method of ZnO nanoparticles and its role in photocatalytic degradation of refractory organic matters. They proposed optimized synthetic technology of the ZnO nanoparticles and use it for degradation of hazardous organic pollutants. In this study the as-grown ZnO nanoparticles exhibited excellent photocatalytic degradation performance for refractory dyes including crystal violet (CV) and Congo red (CR) in terms of rapid degradation rate, high degradation efficiency and broad pH range. This study is considered very useful for optimizing the synthetic conditions for ZnO nanoparticles and increasing the potential applications in photocatalytic degradation for refractory organic pollutants.

13.4.4 Photocatalytic Activity of Other Semiconductor Nanoparticles

Gopalappa et al. (2012), investigated the photocatalytic degradation of commercial azo dye acid orange 7 by synthesized $CaZnO_2$ nanoparticle as an effective catalyst using solar radiation. Calcium zincate nanoparticles were synthesized by solution combustion method and characterized by scanning electron microscopy (SEM) and X-ray diffraction (XRD). It was found from this experiment that 99 % degradation occurs with catalyst in presence of sunlight and it also found that calcium zincate appears to be a suitable alternative to TiO_2 for water treatment. Kansal et al. (2007) investigated the photocatalytic degradation of methyl orange (MO) and rhodamine 6 G (R6G), employing heterogeneous photocatalytic process in presence of various semiconductors such as titanium dioxide (TiO_2), zinc oxide (ZnO), stannic oxide (SnO_2), zinc sulfide (ZnS), and cadmium sulfide (CdS) under UV and solar radiations. The decolorization rate was estimated from residual concentration spectrophotometrically and maximum decolorization (more than 90%) occurred with ZnO catalyst. The performance of photocatalytic system employing ZnO under solar light was observed to be better than ZnO under UV system.

Hussen and Abass (2010) complied a study on solar photolysis and photocatalytic treatment of textile industrial wastewater and the decolorization of real and stimulated wastewater were carried out over a suspension of titanium dioxide or zinc oxide under solar irradiation. The progress of treatment stages was spectrophotometrically followed at different wavelength and under optimal conditions, the extent of decolorization was 100% after different periods of time ranging from 10 to 100 minutes and it is clearly indicated that titanium dioxide and zinc oxide could be used efficiently in photocatalytic treatments of textile industrial wastewater. The rate of decolorization of real and simulated textile industrial wastewater found both with and without light and catalyst. The decolorization rate is higher in presence of light and catalyst which demonstrate that decolorization of the dye depends on the presence of light and catalyst.

Molla et al. (2016), investigated a study on synthesis of tunable band gap semiconductor nickel sulfide nanoparticles, observed rapid and round the clock degradation of organic dyes. They synthesized and characterized the nanoparticles with controlled size and shape including tunable bandgap. The photocatalytic activity of the synthesized nanoparticles for round the clock (light and dark) decomposition of crystal violet (CV), rhodamine B (RhB), methylene blue (MB), nile blue (NB), and eriochrome black T (EBT) dyes performed both in presence and absence of light. The degradation rate was faster in presence of light than in absence. Bhakya et al. (2015) confirmed a study on catalytic degradation of organic dyes using synthesized silver nanoparticles, as a green approach. They synthesized silver nanoparticles (AgNPs) using the different parts of Vishanika or Indian screw tree, an ayurvedic medicinal tree and characterized by using UV−VIS spectroscopy, TEM, XRD, DLS, and FTIR. The catalytic activity of this size and extract dependent biosynthesized nanoparticles established in the degradation of organic dyes. The biosynthesized Ag nanoparticles exhibits a very high degradation activity under visible light and exhibited remarkable degradation properties in a reduction of organic dyes including methyl violet (MV), safranin (S), eosin methylene blue (EMB), and methyl orange (MO). This green synthesis technology is low cost and environment friendly which can help to get rid of hazards arising out of the use of chemical reducing agents and organic solvents.

13.5 Regeneration of Semiconductor Nanoparticles

Nano-sized photocatalyst particles are superior to microphotocatalyst particles due to their larger photocatalytic activity (Xu et al., 1999; Hanley et al., 2001; Jang et al., 2001). Though TiO_2 nanoparticles are toxic contrasting to larger TiO_2 particles which are used as food additives or abrasives in tooth paste. Nanoparticles not only enter cells but also mitochondria and nuclei and interact with cytoskeletal proteins which affecting a variety of cell functions (Gheshlagi et al., 2008). TiO_2 particles concentrations with about 100 nm diameter as low as $2 \, mg \, L^{-1}$ led to immobilization of Daphnia magna, while 200 nm TiO_2 particles were less toxic (Dabrunz et al., 2011). It has to be considered that municipal treatment plants cannot provide any safe barrier to control the spreading of nanoparticles into the aqueous environment (Limbach et al., 2008). In this situation, safe retention of photocatalyst nanoparticle in photocatalytic oxidation reactors is of eminent importance. Sedimentation, centrifugation and sand filtrations are not feasible technique to recovery photocatalyst nanoparticles because of photocatalyst nanoparticles can agglomerate depending on wastewater matrix (Armanious et al., 2011) However, it is extremely difficult to influence the wastewater constituents on photocatalyst in flocculation (Liriano-Jorge et al., 2014). Removal of TiO_2 nanoparticles possible by sand filtration process only flocculated conditions but not where no flocculation took place (Chowdhury et al., 2011). The suitable recovery of photocatalyst with particle size in a range of 1 to 100 μm diameter occurred by hydrocyclones (Bickley et al., 2005). Recently, sheet-like ZnO particles has been synthesized which consisting a particle size of 10 μm (Hong et al., 2013) and adding with silver particles in a mass ratio of 1%, which exhibited good photocatalytic activity and the particles were separated by centrifugation process. They were reused efficiently with less loss of activity than the reuse of TiO_2"P25" and also a porous titanium-niobium mixed oxide photocatalyst showed good settling properties

(Beydoun et al., 2001). The recovery of magnet photocatalyst occurred by coating on magnetic carrier particles (Saupe et al., 2005). Removal of nanoparticles by flotation using the coating of photocatalyst nanoparticles on particles with a lower density than water.

13.6 Environmental Impact of Semiconductor Nanoparticles

For the advancements in catalytic wastewater purification such as photocatalysis, electrocatalysis and fenton catalysis the use of nano-particles have a great value. The application of these materials in industries may cause exposure to UV light to the workers, thus causing serious risk to health including skin cancers and mutation in DNA (Lim et al., 2011) because most widely used TiO_2 and ZnO photocatalyst faced of limitation of requirements of UV light due to their wide band gap. TiO_2 is a potential carcinogenic material and may cause adenocarcinoma of the lung and pneumoconiosis in humans (NIOSH National Institute for Occupational Safety and Health, 2011). It is one of the requisite for industry to produce water of high quality, either for drinking or safe disposal. From this point of view, there is an obvious requirement to develop the stable materials and methods to overcome these challenges. Also, though iron nanoparticles have tremendous environmental application, but they also present a risk when the environment comes into direct contact with these nanomaterials. Presence of iron nanoparticles in environment creates many toxic impacts to microorganisms and soil fauna, which are directly and indirectly significant for environment. Vittori Antisari et al. (2013) employed a study to evaluate the impacts of engineered nanoparticles on soil microbial mass and observed the change in microbial mass of soil. Moreover, Fajardo et al., 2013 observed the transcriptional and proteomic stress responses to soil bacterium *Bacillus cereus* by nanosized zero-valent iron (nZVI) particles. Among the different species of iron nanoparticles, nano zero-valent iron (nZVI) is very reactive. Chen et al., 2013 observed the toxic effects of three different solutions containing carboxymethyl cellulose nZVI (CMC-nZVI), nFe3O4 and ferrous ion solution Fe(II)aq to evaluate the toxic effects by exposing to early life stages of medaka fish. From the results it has been found that nZVI caused a disturbance in the oxidative defence system for embryos and adults, as well as oxidative damage in embryos with some observed effects at concentrations as low as 0.5 mg/L. Adult fish also showed antioxidant balance disruption although they were able to recover afterwards.

13.7 Advantages and Limitations

The photocatalytic properties of semiconductors make its most important in applications for environmental remediation. Titanium dioxide is one of the most widely used semiconductor nanoparticles has shown tremendous ability not only as a sensor for chemical, biological, and various gases such as H_2, CO in low concentrations, but also in photocatalytic degradation and self-clean process for the contaminated environment (Chen and Mao, 2007). Moreover, the degradation of organic pollutants and reduction of metals to their zero oxidation states have been remarked as one of the peak applications of TiO_2 for the treatment of river water, groundwater, the drainage water from fish-feeding tanks, and industrial wastewater. Among numerous semiconductors photocatalysts (i.e., TiO_2, ZnO, ZrO_2, V_2O_5, WO_3, Fe_2O_3, SnO_2, CdSe, GaAs, GaP, and metal sulfides (CdS and ZnS)), titanium dioxide (TiO_2) is one of the most important and widely used photocatalysts, due to its suitable flat band potential, high chemical stability, nontoxicity, corrosion resistance, abundance, cheapness, and high photocatalytic activity. TiO_2 is not only used for the photocatalytic degradation of environmental pollutants but also applied to produce and storage of hydrogen gas. Except TiO_2 the use of other semiconductors such as ZnO, ZrO_2, V_2O_5, WO_3, Fe_2O_3, SnO_2, CdSe, GaAs, GaP, CdS, and ZnS have the limitation of instability and/or corrode easily, either dissolving or forming a thin film, which prevents the electrons from transferring across the semiconductor/liquid interface (Suresh, 2013).

Different process of limitations found from the use of nano-catalysts. One of the limitation of the use of photocatalyst in that hinders the photocatalytic activity is instability of various types of nano-catalysts i.e., AgBr, if it is dispersed in the solution then cannot be recycled for reuse (Dai et al., 2014). In this regard currently, the focus is on synthesizing new photocatalysts using metal oxide or composite with metals and semiconductor oxide to overcome the associated problems with conventional catalysts (Rashid et al., 2014; Mohaghegh et al., 2015).

Though TiO_2 suspension is used efficiently but due to its large surface area four major technical challenges present which are restrict its large-scale application and its use in water-treatment technologies. Firstly, the band gap of TiO_2 is $\sim$3.2 eV which falls in the UV range of the solar spectrum and unable to use visible light ruling part of sunlight to produce energy for photoactivation (Romero-Gomez et al., 2011; Wang et al., 2011a,b; Nawi et al., 2011; Hou et al., 2006; Zhao and Chen, 2011; Kao et al., 2010). Secondly, TiO_2 has low quantum efficiency because of the low rate of electron transfer to oxygen resulting the photogenerated electron-hole pairs with high recombination (Hou et al., 2006; Zhao and Chen, 2011; Ge, Xu and Fang, 2006). Thirdly, the use of TiO_2 in suspension form it aggregates rapidly due to its small size (i.e., 4−30 nm) which may cause scattering of the light beam resulting in loss of catalytic efficiency (Bhattacharyya et al., 2004; Nawi et al., 2011; Yu et al., 2002). And lastly, the application of TiO_2 catalysts in powder form requires posttreatment separation to recover the catalyst from water which is normally difficult, energy consuming, and not economically viable for use in water-treatment plants (Priya et al., 2009; Bhattacharyya et al., 2004; Lopez et al., 2010; Nawi et al., 2011; Ge et al., 2006).

13.8 Future Research

Currently, there is no doubt that the efficiency of utilization nanoparticles in wastewater treatment, but this technology has some serious drawbacks since nanoparticles might release into the environment during preparation and treatment processes; they can accumulate for long time and cause serious risks. In this regard to reduce the health risk there is need of future research to prepare such catalysts having least toxicity to the environment. Among various semiconductor nanoparticles TiO_2 is widely used and from different studies it has been found that this efficient photocatalyst has some major limitations and some harmful effects on environment and human health and which generating obstacle to develop sustainable degradation mechanism of environmental pollutants. In this regard, new research initiatives need to be explored to counter these challenges and so many research has been ongoing to overcome these obstacles. Some studies already developed new way to reduce the band gap to use solar energy efficiently (Zhao and Chen, 2011). Moreover, further work is required to develop the cost-effective methods for the synthesis of nanomaterials and for successful field application more testing need to find out the efficiency of the materials.

13.9 Conclusion

At present, there is a significant need to develop advance water treatment technologies to ensure the high quality of water, remove the chemical and biological pollutants from industrial wastewater and protect the environment from significant pollution and as well as intensify the industrial production process. From this point of view semiconductor nanoparticles induced advance wastewater treatment process is considered as a green technology or sustainable mechanism due to the use of renewable (solar) energy sources. Different semiconductor nanoparticles such as ZnO, TiO_2, CdS, ZnS, Fe_2O_3 successfully developed and investigated to remove organic pollutants as an industrial pollutant.

References

Abdullah, M.A., Muhammed, S.A., Tariq, A.A., 2011. Photodegradation of rhodamine 6g and phenol red by nanosized TiO_2 under solar irradiation. J. Saudi Chem. Soc. 15, 121.

Afkhami, A., Saber-Tehrani, M., Bagheri, H., 2010. Modified maghemite nanoparticles as an efficient adsorbent for removing some cationic dyes from aqueous solution. Desalination 263 (1−3), 240−248. s.

Akhavan, O., Azimirad, R., 2009. Photocatalytic property of Fe_2O_3 nanograin chains coated by TiO_2 nanolayer in visible light irradiation. Appl. Catal. A Gen. 369 (1−2), 77−82.

Alivisatos, A.P., 1996a. J. Phys. Chem. 100, 13226.

Alivisatos, A.P., 1996b. Science 271, 933.

Ambrus, Z., Balazs, N., Alapi, T., Wittmann, G., Sipos, P., Dombi, A., et al., 2008. Synthesis, structure and photocatalytic properties of Fe(III)−doped TiO_2 prepared from $TiCl_3$. Appl. Catal. B 81, 27−37.

Armanious, A., Ozkan, A., Sohmen, U., Gulyas, H., 2011. Inorganic greywater matrix impact on photocatalytic oxidation: does flocculation of TiO_2 nanoparticles impair process efficiency? Water Sci. Technol. 63, 2808−2813.

Baker, A., 2001. Fluorescence excitation-emission matrix characterization of some sewage-impacted rivers. Environ. Sci. Technol. 35, 948−953.

Bandara, J., Klehm, U., Kiwi, J., 2007. Raschig rings−Fe_2O_3 composite photocatalyst activate in the degradation of 4-chlorophenol and orange II under daylight irradiation. Appl. Catal. B Environ. 76 (1−2), 73−81.

Behnajady, M.A., Eskandarloo, H., 2013a. Silver and Copper Co−impregnated onto TiO_2−P25 nanoparticles and its photocatalytic activity. Chem. Eng. J. 228, 1207−1213.

Behnajady, M.A., Eskandarloo, H., 2013b. Characterization and photocatalytic activity of Ag−Cu/TiO_2 nanoparticles prepared by sol−gel method. J. Nanosci. Nanotechnol. 13 (1), 548−553.

Beydoun, D., Amal, R., Scott, J., Low, G., McEvoy, S., 2001. Studies on the mineralization and separation efficiencies of a magnetic photocatalyst. Chem. Eng. Technol. 24, 745−748.

Bhakya, S., Muthukrishnan, S., Sukumaran, M., Muthukumar, M., Senthil Kumar, T., Rao, M.V., 2015. Catalytic degradation of organic dyes using synthesized silver nanoparticles: a green approach. J. Bioremed. Biodeg. 6 (5). Available from: https://doi.org/10.4172/2155-6199.1000312.

Bhatkhande, D.S., Pangarkar, V.G., Beenackers, A., 2001. Photocatalytic degradation for environmental applications a reviewACM J. Chem. Technol. Biotechnol. 77, 102.

Bhattacharyya, A., Kawi, S., Ray, M.B., 2004. Photocatalytic degradation of orange II by TiO_2 catalysts supported on adsorbents. Catal. Today 98 (3), 431−439.

Bickley, R.I., Slater, M.J., Wang, W.J., 2005. Operating experience and performance evaluation of a photocatalytic reactor. Process Saf. Environ. Prot. 83, 217−223.

Burda, C., Chen, X., Narayanan, R., El-Sayed, M.A., 2005. Chem. Rev. 105, 1025.

Cano-Guzman, C.F., Perez-Orozco, J.P., Hernandez-Perez, I., Gonzalez- Reyes, L., Garibay-Febles, V., 2014. Kinetic study for reactive red 84 photo degradation using Iron (III) oxide nanoparticles in annular reactor. J. Textile Sci. Eng. 4, 155. Available from: https://doi.org/10.4172/2165-8064.1000155.

Castro, A.L., Nunes, M.R., Carvalho, M.D., Ferreira, L.P., Jumas, J.C., Costa, F.M., et al., 2009. Doped titanium dioxide nanocrystalline powders with high photocatalytic activity. J. Solid State Chem. 182 (7), 1838−1845.

Chan, H.B.S., Ellis, B.L., 2004. Carbon-encapsulated radioactive 99mTc nanoparticles. Adv. Mater. 16, 144−149.

Chen, C.C., Lin, C.L., Chen, L.C., 2015. Functionalized carbon nanomaterial supported palladium nanocatalysts for electrocatalytic glucose oxidation reaction. Electrochim. Acta 152, 408−416.

Chen, P.J., Wu, W.L., Wu, K.C., 2013. The zerovalent iron nanoparticle causes higher developmental toxicity than its oxidation products in early life stages of Medaka fish. Water Res. 47, 3899−3909.

Chen, W., Westerhoff, P., Leenheer, J.A., Booksh, K., 2003. Fluorescence excitation−emission matrix regional integration to quantify spectra for dissolved organic matter. Environ Sci Technol 37, 5701−5710.

Chen, X., Mao, S.S., 2007. Titanium dioxide nanomaterials: synthesis, properties, modifications and applications. Chem. Rev. 107 (7), 2891−2959.

Chen, X., Lou, Y., Dayal, S., Qiu, X., Krolicki, R., Burda, C., et al., 2005. Doped semiconductor nanomaterials. J. Nanosci. Nanotechnol. 5 (9), 1408−1420.

Chirita, M., Grozescu, I., 2009. Fe_2O_3 nanoparticles, physical properties and their photochemical and photoelectrochemical applications. Chem. Bull. Politehnica Univ. (Timisoara) 54 (68), 1.

Chong, M.N., Jin, B., Chow, C.W.K., Saint, C., 2010. Recent developments in photocatalytic water treatment technology: a review. Water Res. 44, 2997−3027.

Chowdhury, I., Hong, Y., Honda, R.J., Walker, S.L., 2011. Mechanism of TiO_2 nanoparticle transport in porous media: role of solution chemistry, nanoparticle concentration and flowrate. J. Colloid Interface Sci. 360, 548−555.

Coble, P.G., 1996. Characterization of marine and terrestrial DOM in seawater using excitation emission matrix spectroscopy. Mar. Chem. 51, 325−346.

Cornel, R.M., Schwertmann, U., 1996. The Iron Oxides: Structure, Properties, Reactions, Occurrences and Uses. New York, Weinheim.

Cornell, R.M., Schwertmann, U., 2003. The Iron Oxides: Structure, Properties, Reactions, Occurrences and Uses. Wiley–VCH GmbH & Co. KGaA, Darmstadt.

Cunningham, K., Goldberg, M.C., Weiner, E.R., 1988. Mechanisms for aqueous photolysis of adsorbed benzoate, oxalate, and succinate on iron oxyhydroxide (goethite) surfaces. Environ. Sci. Technol. 22, 1090−1097.

Cushing, B.L., Kolesnichenko, V.L., O' Connor, C.J., 2004. Recent advances in the liquid-phase syntheses of inorganic nanoparticles. Chem. Rev. 104, 3893−3946.

Dabrunz, A., Duester, L., Prasse, C., Seitz, F., Rosenfeldt, R., 2011. Biological surface coating and molting inhibition as mechanisms of TiO_2 nanoparticle toxicity in Daphnia magna. PLoS One.

Dai, K., Lv, J., Lu, L., Liu, Q., Zhu, G., Li, D., 2014. Synthesis of micro-nano heterostructure AgBr/ZnO composite for advanced visible light photocatalysis. Mater. Lett. 130, 5−8.

Demazeau, G., 1999. Solvothermal processes: a route to the stabilization of new materials. J. Mater. Chem. 9, 15−18.

Demazeau, G., 2008. Solvothermal reactions: an original route for the synthesis of novel materials. J. Mater. Sci. 43, 2104−2114.

Demianets, L.N., 1991. Hydrothermal synthesis of new compounds. Pro. Crystal Growth Charact. 21, 299−355.

Fajardo, C., Sacca, M.L., Martinez-Gomariz, M., Costa, G., Nande, M., Martin, M., 2013. Transcriptional and proteomic stress responses of a soil bacterium *Bacillus cereus* to nanosized zero-valent iron (nZVI) particles. Chemosphere 93, 1077−1083.

Fellman, J.B., Hood, E., Spencer, R.G.M., 2010. Fluorescence spectroscopy opens new windows into dissolved organic matter dynamics in freshwater ecosystems: a review. Limnol. Oceanogr. 55, 2452−2462.

Feng, S., Xu, R., 2001. New materials in hydrothermal synthesis. Acc. Chem. Res. 34, 239−247.

Feng, W., Nansheng, D., 2000. Photochemistry of hydrolytic iron (III) species and photoinduced degradation of organic compounds. a minireview. Chemosphere 41, 1137−1147.

Fernandez, J., Bandara, J., Lopez, A., Albers, P., Kiwi, J., 1998. Efficient photo-assisted Fenton catalysis mediated by Fe ions on Nafion membranes active in the abatement of non-biodegradable azo-dye. Chem. Commun. 14, 1493−1494.

Ge, L., Xu, M., Fang, H., 2006. Synthesis and characterization of the $Pd/InVO_4$-TiO_2 co-doped thin films with visible light photocatalytic activities. Appl. Surf. Sci. 253 (4), 2257−2263.

Gheshlagi, Z.N., Riazi, G.H., Ahmadian, S., Ghafari, M., Mahinpour, R., 2008. Toxicity and interaction of titanium dioxide nanoparticles with microtubule protein. ActaBiochimBiophys Sin. 40, 777−782.

Girginova, P.I., Daniel-da-Silva, A.L., Lopes, C.B., Figueira, P., Otero, M., Amaral, V.S., 2010. Silica coated magnetite particles for magnetic removal of Hg^{2+} from water. J. Colloid Interface Sci. 345 (2), 234−240.

Giwa, A., Nkeonye, P.O., Bello, K.A., Kolawole, E.G., 2012. Solar photocatalytic degradation of reactive yellow 81 and reactive violet 1 in aqueous solution containing semiconductors oxides. Int. J. Appl. Sci. Technol. 32, 2940−2946. 2 (4).

Gogotsi, Y.G., Yoshimura, M., 1994. Formation of carbon films on carbides under hydrothermal conditions. Nature 367, 628−630.

Gopalappa, H., Yogendra, K., Mahadevan, K.M., 2012. Solar photocatalytic degradation of commercial azo dye acid orange 7 by synthesized $CaZnO_2$ nanoparticle as an effective catalyst. Int. J. Res. Chem. Environ. 2, 39−43.

Guo, J., Peng, Y., Guo, J., Ma, J., Wang, W., Wang, B., 2011. Dissolved organic matter in biologically treated sewage effluent (BTSE): characteristics and comparison. Desalination 278 (1−3), 365−372.

Habashi, F., 2005. A short history of hydrometallurgy. Hydrometallurgy 79, 15−22.

Hanley, T., Krisnandi, Y., Eldewik, A., Luca, V., Howe, R., 2001. Nanosize effects in titania based photocatalyst materials. Ionics 7, 319−326.

Hassena, H., 2016. Photocatalytic degradation of methylene blue by using Al_2O_3/Fe_2O_3 nano composite under visible light. Mod. Chem. Appl. 4 (176). Available from: https://doi.org/10.4172/2329-6798.1000176.

Henderson, R.K., Baker, A., Murphy, K.R., Hambly, A., Stuetz, R.M., Khan, S.J., 2009. Fluorescence as a potential monitoring tool for recycled water systems: a review. Water Res. 43, 863−881.

Henrist, C., Dewalque, J., Mathis, F., Cloots, R., 2009. Control of the porosity of anatase thin films prepared by EISA: influence of thickness and heat treatment. Micropor. Mesopor. Mater. 117 (1), 292−296. 35.

Hernandez−Martinez, A.R., Estevez, M., Vargas, S., Rodriguez, R., 2013. New polyurethane−anatase titania porous hybrid composite for the degradation of azo−compounds wastes. Compos. Part B 44 (1), 686−691.

Hoag, G.E., Collins, J.B., Holcomb, J.L., Hoag, J.R., Nadagouda, M.N., Varma, R.S., 2009. Degradation of bromothymol blue by greener nano-scale zero-valent iron synthesized using tea polyphenols. J. Mater. Chem. 2009 (19), 8671−8677.

Hoffmann, M.R., Martin, S.T., Choi, W., Bahenemann, D.W., 1995. Chem. Rev 95, 69−96.

Hong, Y., Tian, C., Jiang, B., Wu, A., Zhang, Q., 2013. Facile synthesis of sheet-like ZnO assembly composed of small ZnO particles for highly efficient photocatalysis. J. Mater. Chem. A 1, 5700−5708.

Hosaka, M., 1991. Hydrothermal growth of gem stones and their characterization. Prog. Cryst. Growth Charact. Mater. 21, 71−96.

Hou, X., Wu, X., Liu, A., 2006. Studies on photocatalytic activity of Ag/TiO_2 films. Front. Chem. China 1 (4), 402−407.

Huang, D.L., Zeng, G.M., Feng, C.L., Hu, S., Jiang, X.Y., Tang, L., 2008. Degradation of leadcontaminated lignocellulosic waste by *Phanerochaete chrysosporium* and the reduction of lead toxicity. Environ. Sci. Technol. 42 (13), 4946−4951.

Huang, L., Weng, X., Chen, Z., Megharaj, M., Naidu, R., 2013. Synthesis of iron-based nanoparticles using Oolong tea extract for the degradation of malachite green. Spectrochim. Acta A 117, 801−804.

Hudson, N., Baker, A., Reynolds, D., 2007. Fluorescence analysis of dissolved organic matter in natural, waste and polluted waters—a review. River Res. Appl. 23, 631−649.

Hussen, F.H., Abass, T.A., 2010. Solar photolysis and photocatalytic treatment of textile industrial wastewater. Int. J. Chem. Sci. 8 (3), 1409−1420.

Jana, P., Ganesan, V., 2011. The production of a carbon-coated alumina foam. Carbon 49 (10), 3292−3298.

Jang, H.D., Kim, S.K., Kim, S.J., 2001. Effect of particle size and phase composition of titanium dioxide nanoparticles on the photocatalytic properties. J. Nanopart. Res. 3, 141−147.

Kansal, S.K., Singh, M., Sud, D., 2007. Studies on photodegradation of two commercial dyes in aqueous phase using different photocatalysts. J. Hazard. Mater. 141, 581−590.

Kao, M.C., Chen, H.Z., Young, S.L., Kung, C.Y., Lin, C.C., Hong, Z.Y., 2010. Microstructure and optical properties of tantalum modified TiO_2 thin films prepared by the sol-gel process. J. Superconduct. Novel Magn. 23 (5), 843−845.

Khataee, A.R., Kasiri, M.B., 2010. Photocatalytic degradation of organic dyes in the presence of nanostructured titanium dioxide: influence of the chemical structure of dyes. J. Mol. Catal. A 328 (1-2), 8−26.

Khataee, A.R., Zarei, M., Ordikhani, S.R., 2011. Heterogeneous photocatalysis of a dye solution using supported TiO_2 nanoparticles combined with homogeneous photoelectrochemical process: molecular degradation products. J. Mol. Catal. A 338, 84.

Kuang, Y., Wang, Q., Chen, Z., Megharaj, M., Naidu, R., 2013. Heterogeneous fenton-like oxidation of monochlorobenzene using green synthesis of iron nanoparticles. J. Colloid Interface Sci. 410, 67−73.

Kumar, S.G., Devi, L.G., 2011. Review on modified TiO_2 photocatalysis under UV/visible light: selected results and related mechanisms on interfacial charge carrier transfer dynamics. J. Phys. Chem. A 115 (46), 13211−13241. 49.

Leenheer, J.A., Croue, J.P., 2003. Characterizing aquatic dissolved organic matter. Environ. Sci. Technol. 37, 18A−26A.

Li, J., Jin, Z.Z., 2009. Coordination Chemistry Reviews Accepted proof.

Li, N., Descorme, C., Besson, M., 2007. Catalytic wet air oxidation of chlorophenols over supported ruthenium catalysts. J. Hazard. Mater. 146 (3), 602−609.

Lim, H.W., James, W.D., Rigel, D.S., Maloney, M.E., Spencer, J.M., Bhushan, R., 2011. Adverse effects of ultraviolet radiation from the use of indoor tanning equipment: time to ban the tan. J. Am. Acad. Dermatol. 64, 51−60.

Limbach, L.K., Bereiter, R., Muller, E., Krebs, R., Galli, R., 2008. Removal of oxide nanoparticles in a model wastewater treatment plant: influence of agglomerations and surfactants on clearing efficiency. Environ. Sci. Technol. 42, 5828−5833.

Lin, L., Wang, H., Luo, H., Xu, P., 2015. Enhanced photocatalysis using side-glowing optical fibers coated with Fe-doped TiO_2 nanocomposite thin films. J. Photochem. Photobiol. A. 307, 88−98.

Liqiang, J., Honggang, F., Baiqi, W., Dejun, W., Baifu, X., Shudan, L., et al., 2006. Effects of Sn dopant on the photoinduced charge property and photocatalytic activity of TiO_2 nanoparticles. Appl. Catal. B 62, 282−291.

Liriano-Jorge, C.F., Sohmen, U., Ozkan, A., Gulyas, H., Otterpohl, R., 2014. TiO_2 photocatalyst nanoparticle separation: flocculation in different matrices and use of powdered activated carbon as a precoat in low-cost fabric filtration. Advances in Materials Science and Engineering.

Lopez, A., Acosta, D., Martinez, A.I., Santiago, J., 2010. Nanostructured low crystallized titanium dioxide thin films with good photocatalytic activity. Powder Technol. 202 (1−3), 111−117.

Lucas, M.S., Peres, J.A., 2006. Decolorization of the azo dye reactive Black 5 Fenton and photo Fenton oxidation. J. Dyes. Pigm. 71, 236−244.

Ma, H., Qi, X., Maitani, Y., Nagai, T., 2007. Preparation and characterization of superparamagnetic iron oxide nanoparticles stabilized by alginate. Int. J. Pharm. 333 (1−2), 177−186.

Mahlambi, M.M., Mishra, A.K., Mishra, S.B., Krause, R.W., Mamba, B.B., Raichur, A.M., 2012. Synthesis and characterization of carbon-covered alumina (CCA) supported TiO_2 nanocatalysts with enhanced visible light photodegradation of Rhodamine B. J. Nanopart. Res. 14, 790.

Mahlambi, M.M., Mishra, A.K., Mishra, S.B., Krause, R.W., Mamba, B.B., Raichur, A.M., 2013. Effect of metal ions (Ag, Co, Ni, and Pd) on the visible light degradation of Rhodamine B by carbon-covered alumina-supported TiO_2 in aqueous solutions. Ind. Eng. Chem. Res. 52 (5), 1783−1794.

Masthan, S.K., Prasad, P.S.S., Rao, K.S.R., Rao, P.K., 1991. Hysteresis during ammonia synthesis over promoted ruthenium catalysts supported on carbon-covered alumina. J. Mol. Catal. 67 (2), L1−L5.

McHenry, M.E., Laughlin, D.E., 2000. Nano-scale materials development for future magnetic applications. Acta Mater. 48 (1), 223−238.

Mohaghegh, N., Tasviri, M., Rahimi, E., Gholami, M.R., 2015. Comparative studies on $Ag_3PO_4/BiPO_4$ metal-organic framework grapheme based nanocomposites for photocatalysis application. Appl. Surf. Sci. 351, 216−224.

Molla, A., Sahu, M., Hussain, S., 2016. Synthesis of tunable band gap semiconductor nickel sulphide nanoparticles: rapid and round the clock degradation of organic dyes. Sci. Rep. 6, 26034. Available from: https://doi.org/10.1038/srep26034.

Murray, C.B., Kagan, C.R., Bawendi, M.G., 2000. Annu. Rev. Mater. Sci. 30, 545.

NIOSH (National Institute for Occupational Safety and Health), 2011. Current intelligence bulletin 63. Occupational Exposure to Titanium Dioxide. DHHS (NIOSH) Publication No. 2011-160.

Nagaveni, K., Hegde, M.S., Ravishankar, N., Subbanna, G.N., Madrad, G., 2004. Synthesis and structure of nanocrystalline TiO_2 with lower band gap showing high photocatalytic activity. Langmuir 20 (7), 2900−20907.

Nawi, M.A., Jawad, A.H., Sabar, S., Ngah, W.S.W., 2011. Immobilized bilayer TiO_2/chitosan system for the removal of phenol under irradiation by a 45watt compact fluorescent lamp. Desalination 280 (1−3), 288−296.

Nishanthi, S.T., Henry Raja, D., Subramanian, E., Pathinettam Padiyan, D., 2013. Remarkable role of annealing time on anatase phase titania nanotubes and its photoelectrochemical response. Electrochim. Acta 89, 239–245. 27.

Njagi, E.C., Huang, H., Stafford, L., Genuino, H., Galindo, H.M., Collins, J.B., et al., 2011. Biosynthesis of iron and silver nanoparticles at room temperature using aqueous Sorghum bran extracts. Langmuir 27, 264–271.

Oller, I., Malato, S., Sanchez-Perez, J.A., 2011. Combination of advanced oxidation processes and biological treatments for wastewater decontamination: a review. Sci. Tot. Environ. 409 (20), 4141–4166.

Pan, B.J., Qiu, H., Pan, B.C., Nie, G.Z., Xiao, L.L., Lv, L., 2010. Highly efficient removal of heavy metals by polymer-supported nanosized hydrated Fe(III) oxides: behavior and XPS study. Water Res. 44 (3), 815–824.

Park, H., Choi, W., 2004. Effects of TiO_2 surface fluorination on photocatalytic reactions and photoelectrochemical behaviors. J. Phys. Chem. B 108 (13), 4086–4093.

Parlanti, E., Worz, K., Geoffroy, L., Lamotte, M., 2000. Dissolved organic matter fluorescence spectroscopy as a tool to estimate biological activity in a coastal zone submitted to anthropogenic inputs. Org. Geochem. 31, 1765–1781.

Parsons, S., 2004. Advanced Oxidation Processes for Water and Wastewater. IWA Publishing, London, UK.

Paul, S., Chavan, S.K., Khambe, S.D., 2012. Studies on characterization of textile industrial waste water in solapur city. Int. J. Chem. Sci. 10, 635–642.

Pehkonen, S.O., Siefert, R.L., Hofmann, M.R., 1995. Photoreduction of iron oxyhydroxides and the photooxidation of halogenated acetic acids. Environ. Sci. Technol. 29, 1215–1222.

Pelaez, M., Nolan, N.T., Pillai, S.C., Seery, M.K., Falaras, P., Kontos, A.G., et al., 2012. A Review on the visible light active titanium dioxide photocatalysts for environmental applications. Appl. Catal. B 125, 331–349.

Priya, D.N., Modak, J.M., Raichur, A.M., 2009. LbL fabricated poly (styrene sulfonate)/TiO_2 multilayer thin films for environmental applications. ACS Appl. Mater. Interfaces 1 (11), 2684–2693.

Rafatullah, M., Sulaiman, O., Hashima, R., Ahmad, A., 2010. Adsorption of methylene blue on low-cost adsorbents: a review. J. Hazard. Mat. 177, 70–80.

Rahman, M.M., Khan, S.B., Jamal, A., Faisal, M., Aisiri, A.M., 2011. In: Rahman, Mohammed (Ed.), Iron Oxide Nanoparticles, Nanomaterials. InTech978-953-307-913-4.

Rajamathi, M., Seshadri, R., 2002. Oxides and chalcogenides nanoparticles from hydrothermal/solvothermal reactions. Curr. Opin. Solid State Mater. Sci. 6, 337.

Rashid, J., Barakat, M.A., Salah, N., Habib, S.S., 2014. Ag/ZnO nanoparticles thin films as visible light photocatalyst. RSC Adv. 4, 56892–56899.

Reyes–Coronado, D., Rodriguez–Gattorno, G., Espinosa–Pesqueira, M.E., Cab, C., De Coss, R., Oskam, G., 2008. Phase–pure TiO_2 nanoparticles: anatase, brookite and rutile. Nanotechnology 19 (14), 145605–145614.

Romero-Gomez, P., Rico, V., Espinos, J.P., Gonzalez-Elipe, A.R., Palgrave, Egdell, R.G., 2011. Nitridation of nanocrystalline TiO_2 thin films by treatment with ammonia. Thin Solid Films 519 (11), 3587–3595.

Saif, S., Tahir, A., Chen, Y., 2016. Review green synthesis of iron nanoparticles and their environmental applications and implications. Nanomaterials (Basel) 6 (11), 209.

Sathishkumar, P., Sweena, R., Wu, J.J., Anandan, S., 2011. Synthesis of CuO-ZnO nanophotocatalyst for visible light assisted degradation of a textile dye in aqueous solution. Chem. Eng. J. 171, 136–140.

Saupe, G.B., Zhao, Y., Bang, J., Yesu, N.R., Carballo, G.A., 2005. Evaluation of a new porous titanium-niobium mixed oxide for photocatalytic water decontamination. Microchem. J. 81, 156–162.

Shahwan, T., Abu Sirriah, S., Nairat, M., Boyacı, E., Ero glu, A.E., Scott, T.B., et al., 2011. Green synthesis of iron nanoparticles and their application as a fenton-like catalyst for the degradation of aqueous cationic and anionic dyes. Chem. Eng. J. 172, 258−266.

Shashikala, V., Siva Kumar, V., Padmasri, A.H., 2007. Advantages of nano-silver-carbon covered alumina catalyst prepared by electro-chemical method for drinking water purification. J. Mol. Catal. A 268 (1-2), 95−100.

Shon, H.K., Vigneswaran, S., Snyder, S.A., 2006. Effluent organic matter (EfOM) in wastewater: constituents, effects, and treatment. Crit. Rev. Env. Sci. Tech. 36, 327−374.

Song, K., Zhou, J., Bao, J., Feng, Y., 2008. Photocatalytic activity of (copper, nitrogen)− codoped titanium dioxide nanoparticles. J. Am. Ceram. Soc. 91, 1369−1371.

Suresh, S., 2013. Semiconductor nanomaterials, methods and applications: a review. Nanosci. Nanotechnol. 3 (3), 62−74. Available from: https://doi.org/10.5923/j.nn.20130303.06.

Thiruvenkatachari, R., Vigneswaran, S., Il Shik Moon, I.S., 2008. A review on UV/TiO_2 photocatalytic oxidation process. Korean J. Chem. Eng. 25 (1), 64−72.

Veluru, J.B., Appukuttan, S.N., 2011. Synthesis and characterization of rice grains like nitrogen-doped TiO_2 nanostructures. Mater. Lett. 65, 3064.

Vittori Antisari, L., Carbone, S., Gatti, A., Vianello, G., Nannipieri, P., 2013. Toxicity of metal oxide (CeO_2, Fe_3O_4, SnO_2) engineered nanoparticles on soil microbial biomass and their distribution in soil. Soil Biol. Biochem. 60, 87−94.

Wang, P., Zhou, T., Wang, R., Lim, T.-T., 2011a. Carbon-sensitized and nitrogen-doped TiO_2 for photocatalytic degradation of sulfanilamide under visible-light irradiation. Water Res. 45 (16), 5015−5026.

Wang, Y., Lu, K., Feng, C., 2011b. Photocatalytic degradation of methyl orange by polyoxometalates supported on yttrium-doped TiO_2. J. Rare Earths 29, 866.

Wang, Z., Wu, Z., Tang, S., 2009. Characterization of dissolved organic matter in a submerged membrane bioreactor by using three-dimensional excitation and emission matrix fluorescence spectroscopy. Water Res. 43 (6), 1533−1540.

Weber, E.J., Adams, R.L., 1995. Chemical and sediment-mediated reduction of the azo dye disperse blue 79. Environ. Sci. Technol. 29, 1163−1170.

White, B.R., Stackhouse, B.T., Holcombe, J.A., 2009. Magnetic γ-Fe_2O_3 nanoparticles coated with poly-L-cysteine for chelation of As (III), Cu(II), Cd(II), Ni(II), Pb(II) and Zn(II). J. Hazard. Mater. 161 (2−3), 848−853.

Xu, N., Shi, Z., Fan, Y., Dong, J., Shi, J., 1999. Effects of particle size of TiO_2 on photocatalytic degradation of methylene blue in aqueous suspensions. Ind. Eng. Chem. Res. 38, 373−379.

Xu, T., Zhang, L., Cheng, H., Zhu, Y., 2011. Significantly enhanced photocatalytic performance of ZnO via graphene hybridization and the mechanism study. Appl. Catal. B 101 (3), 382−387. 41.

Yu, j, Yu, J.C., Cheng, B., Zhao, X., 2002. Photocatalytic activity and characterization of the sol-gel derived Pb-doped TiO_2 thin films. J. Sol-Gel Sci. Technol. 24 (1), 39−48.

Zhang, H., Songc, Z., Wanga, D., Tongc, Z., Qinb, Y., 2017. A facile synthetic method of ZnO nanoparticles and its role in photocatalytic degradation of refractory organic matters. Desalin. Water Treat. 90, 189−195. Available from: https://doi.org/10.5004/dwt.2017.21233.

Zhang, L., Zhu, Y., 2012. A review of controllable synthesis and enhancement of performances of bismuth tungstate visible−light−driven photocatalysts. Catal. Sci. Technol. 2 (4), 694−706.

Zhang, L.D., Fang, M., 2010. Nanomaterials in pollution trace detection and environmental improvement. Nano Today 5 (2), 128−142.

Zhao, B., Chen, Y.-W., 2011. Ag/TiO_2 sol prepared by a sol-gel method and its photocatalytic activity. J. Phys. Chem. Solids 72 (11), 1312−1318.

Zou, Z., Ye, J., Sayama, K., Arakawa, H., 2001. Direct splitting of water under visible light irradiation with an oxide semiconductor photocatalyst. Nature 414, 625−627.

Further Reading

Ana, L.G., Gustavo, A.P., Ricardo, A.T., 2010. Degradation of the antibiotic oxolinic acid by photocatalysis with TiO_2 in suspension. Water Res. 44, 5158.

Bai, A., Liang, W., Zheng, G., Xue, j, 2011. Preparation and enhanced daylight-induced photo-catalytic activity of transparent C-doped TiO_2 thin films. J. Wuhan Univ. Technol.-Mater. Sci. Ed. 25, 738−742.

Dionysios, D.D., Amid, P.K., Ann, M.K., 2000. Continuous-mode photocatalytic degradation of chlorinated phenols and pesticides in water using a bench-scale TiO_2 rotating disk reactor. Appl. Catal. B 24, 139.

Emad, S.E., Malay, C., 2010. Photocatalytic degradation of amoxicillin, ampicillin and cloxacillin antibiotics in aqueous solution using UV/TiO_2 and $UV/H_2O_2/TiO_2$ photocatalysis. Desalination 252, 46.

Mahmood, T., Wang, X., Chen, C., Ma, W., Zhao, J., 2007. Photocatalytic degradation of persistent and toxic organic pollutants. J. Catal. 28, 1117.

Venkateswarlu, S., Natesh Kumar, B., Prasad, C.H., Venkateswarlu, P., Jyothi, N.V.V., 2014. Bio-inspired green synthesis of Fe_3O_4 spherical magnetic nanoparticles using Syzygium cumini seed extract. Phys. B 449, 67−71.

Wang, W.P., Huang, Y.K., Yang, S.J., 2010. Photocatalytic degradation of nitrobenzene wastewater with H3PW12O40/TiO_2. IEEE Mech. Autom. Control Eng. 6, 1303.

Zheng, M., Shu, Y., Sun, J., Zhang, T., 2008. Carbon-covered alumina: a superior support of noble metal-like catalysts for hydrazine decomposition. Catal. Lett. 121 (1-2), 90−96.

14

Nanofiltration Membrane Technology Providing Quality Drinking Water

Amimul Ahsan[1,2], Monzur Imteaz[2]
[1]DEPARTMENT OF CIVIL ENGINEERING, UTTARA UNIVERSITY, DHAKA, BANGLADESH
[2]DEPARTMENT OF CIVIL AND CONSTRUCTION ENGINEERING, SWINBURNE UNIVERSITY OF TECHNOLOGY, MELBOURNE, VIC, AUSTRALIA

14.1 Introduction

Many studies and researches have been done to improve and find new and efficient water treatment technologies. Conventional water treatment method cost more and it needs big facilities. Several nanotechnologies are applied worldwide to fulfill fresh water demand such as microfiltration (MF), ultrafiltration (UF), nanofiltration (NF), and reverse osmosis (RO). NF and RO processes can supply potable water in huge quantity for industrial applications and the quality of water is better than conventional treatments. They can remove various organic and inorganic impurities from seawater, surface water, and groundwater. In surface water treatment, stringent regulations, and quality concerns are the major factors. The main drawback of spiral wound NF or RO system in treating surface water is NF/RO system foul rapidly (Reiss, 2005).

A hybrid method of flotation and membrane is designed using submerged MF modules in a flotation reactor for efficient separation. The removal of heavy metal ions involves bonding the metals firstly to an agent and then separating the loaded agents from any aqueous solutions. There are many of advantages of both flotation and membrane separation techniques (Blöcher et al., 2003).

Membranes are thin films of material to permit specific substances to pass through. Semipermeable membranes permit a few substances to pass through and are capable of separating very small particles, molecules, and ions in any aqueous solutions. A human body has various semi- permeable membranes that permit nutrients in and toxins out (ECW542, 2018).

The methods of removal through membranes are (Chapter 6, 2018):

- Microfiltration
- Ultrafiltration

- Nanofiltration
- Reverse osmosis filtration (ROF)
- Electrodialysis (ED)

In this study, a brief introduction on nanofiltration (NF) technique and the potential applications and concerns of NF are discussed.

14.2 Applications and Concerns of Nanofiltration

Another common name of NF is "loose" RO, which is able to reject/remove impurities $<0.002\ \mu m$. It is able to remove some dissolved elements from water and wastewater. It is mainly designed as an alternative to chemical softening, that's mean as a membrane softening process (Crites and Tchobangiglous, 1998; Membrane, 2018).

It can be used as a pretreatment of another treatment, for example, RO. The specific goals of NF pretreatment are to decrease the hardness ions (to avert scaling), particulate, turbidity, and microbial fouling of the RO membranes. It also decreases the RO operating pressure by dropping the TDS concentration of feed-water (Crites and Tchobangiglous, 1998; Membrane, 2018).

In separation processes, distinct features of membranes are using as an additional unit operation. Some benefits include (Membrane, 2018; Intratec. Membranes on Polyolefins Plants Vent Recovery, 2018):

- less power-intensive,
- no need adsorbents/solvents, which hard to manage or expensive, and
- simple equipment or system, where easy to add more efficient membranes.

In membrane filtration and in RO, membrane is used with pressure as the driving process. In dialysis and pervaporation, the driving force is the chemical potential along with concentration gradient. Also pertraction depends on the gradient of chemical potential. However, their awesome success in biological systems is not harmonized by their application. The major reasons are (Membrane, 2018; Chmiel, 2006):

- fouling,
- excessive cost,
- need of solvent resistant materials, and
- scaling problems.

The main application of NF is in the treatment of water and wastewater. Other applications of NF are quite common as well in the pulp and paper industry, in textile industry, in dairy and food industry, and in chemical processing. In water treatment, NF can be used to polish the product water at the end of conventional processes. It is an efficient method of water softening (Sutherland, 2018).

The NF is often used to soften hard water. It is completely dissimilar from UF. Its pore size is very tiny, so many substances that cannot pass include valence ions. It has a valence of two and has the same organic structure. In addition, it cannot allow any organic matter/salt to pass such as calcium and magnesium. It also cannot allow any viruses to go. It can be installed in each house as no chemical is used, so it has no adverse effect on the environment. The only concern is that it cannot remove any dissolved substances. So, the users need to have another facility to remove dissolved elements if that water contains those (Advantages and Disadvantages of Nanofiltration, 2018).

In NF membranes, the solute transport can be controlled by dielectric, Donnan, steric and transport effects and predictive modeling based on the extended Nernst—Planck equation has been developed. The model is successfully applied in many sectors. NF membranes are successful due to its varied selectivity in biotechnology, pharmaceutical, desalination, water and wastewater treatments, and food applications. But the major concern is to control fouling. In further study, better membranes can be developed to mitigate fouling problems (Mohammad et al., 2015).

Figure 14-1 shows innovative three-stage serial nanofiltration units under critical transmembrane pressure conditions (TMPc) to treat galacto-oligosaccharides. They found that TMPc are decreased with time to prevent fouling effects due to the increase of solute concentration. A relation between TMPc and the volume reduction factor (VRF) was also studied. They concluded that similar purities to that of commercial products can be obtained after three sequential batches (Córdova et al., 2017).

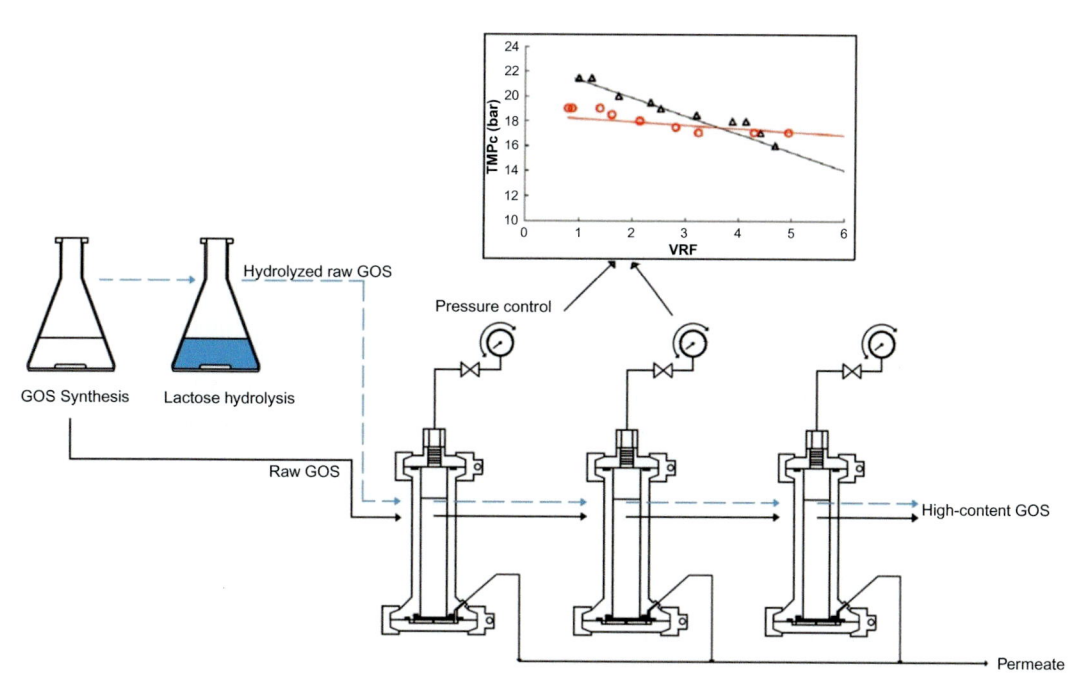

FIGURE 14-1 Three-stage serial nanofiltration units to treat galacto-oligosaccharides (Córdova et al., 2017).

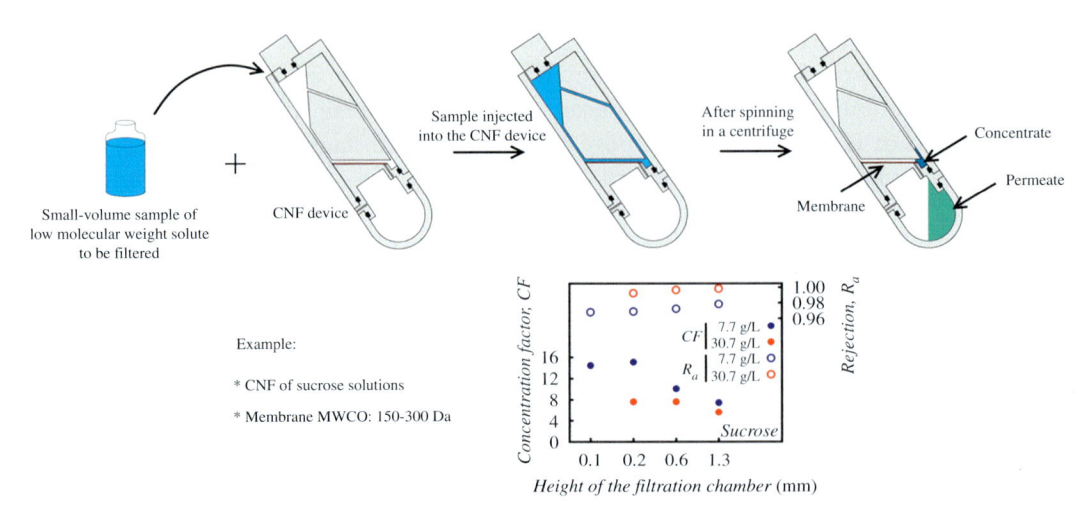

FIGURE 14-2 Centrifugal nanofiltration (CNF) for small-volume samples (Completo et al., 2017).

Figure 14-2 shows an innovative centrifugal nanofiltration (CNF) device for small-volume samples. They obtained concentration factors higher than 20 and reduced the sample volume from 3.2 mL to less than 100 μL of concentrate using three solute solutions such as potassium sulfate, sucrose, and polyethylene glycol. They found that the angle between centrifugal force and membrane surface affects the filtration performance, where the performance is better for negative angles resulting in the disruption of the concentration polarization layer (Completo et al., 2017).

14.3 Conclusions

The nanofiltration is a very effective method of particle separation. It has pros and cons as like any treatment techniques. It is efficient in the separation of impurities in water. It can remove various impurities such as viruses, organic elements, and ions. It has very tiny holes, so most moles cannot pass through it. The other advantages of NF include no chemicals are utilized, therefore it is a eco-friendly technique. It is an efficient technique of softening the hard water. However, soluble impurities in water cannot be separated and reduced by NF. Therefore, it can be incorporated with any other water treatment technique to remove soluble elements.

References

Advantages and Disadvantages of Nanofiltration. https://www.biotechwater.com/advantages-disadvantages-nanofiltration/ (accessed 02.18).

Blöcher, C., Dorda, J., Mavrov, V., Chmiel, H., Lazaridis, N.K., Matis, K.A., 2003. Hybrid flotation—membrane filtration process for the removal of heavy metal ions from wastewater. Water Res. 37 (16), 4018—4026.

Chapter 6: Reverse Osmosis and Nanofiltration - Presentation transcript. http://slideplayer.com/slide/10499074/ (accessed 02.18).

Chmiel, H., 2006. Bioprozesstechnik : Einführung in die Bioverfahrenstechnik (2nd ed.), Spektrum Akad. Verl. p. 279, ISBN 3827416078, München: Elsevier.

Crites, R., Tchobangiglous, G., 1998. Small and Decentralized Wastewater Management Systems. McGraw-Hill Book Company, New York.

Córdova, A., Astudillo, C., Santibañez, L., Cassano, A., Ruby-Figueroa, R., Illanes, A., 2017. Purification of galacto-oligosaccharides (GOS) by three-stage serial nanofiltration units under critical transmembrane pressure conditions. Chem. Eng. Res. Des. 117, 488−499.

Completo, C., Geraldes, V., Semião, V., Mateus, M., Rodrigues, M., 2017. Centrifugal nanofiltration for small-volume samples. J. Membr. Sci. 540, 411−421.

ECW542: Wastewater Engineering - Advanced Treatment, Lecture note, UiTM, Malaysia. https://issuu.com/starprime77/docs/ecw542-wk13-l38 (accessed 02.18).

Intratec. Membranes on Polyolefins Plants Vent Recovery. Improvement Economics Program of Intratec, ISBN 978-0615678917, 2012. https://books.google.com.bd/books?id = KAfPF-dYDaMC&printsec = frontcover&source = gbs_ge_summary_r&cad = 0 (accessed 02.18).

Membrane. https://en.wikipedia.org/wiki/Membrane (accessed 02.18).

Mohammad, A.W., Teow, Y.H., Ang, W.L., Chung, Y.T., Oatley-Radcliffe, D.L., Hilal, N., 2015. Nanofiltration membranes review: recent advances and future prospects. Desalination 356, 226−254.

Reiss, C.R., 2005. Mechanisms of Nanofilter Fouling and Treatment Alternatives for Surface Water Supplies (Ph.D. thesis), Electronic Theses and Dissertations, University of Central Florida, USA. http://stars.library.ucf.edu/etd/495 http://purl.fcla.edu/fcla/etd/CFE0000630 (accessed 02.18).

Sutherland, K. What is Nanofiltration? http://www.filtsep.com/water-and-wastewater/features/what-is-nanofiltration/ (accessed 02.18).

15

Application and Future Prospects of Reverse Osmosis Process

Amimul Ahsan[1,2], Monzur Imteaz[2]

[1]DEPARTMENT OF CIVIL ENGINEERING, UTTARA UNIVERSITY, DHAKA, BANGLADESH
[2]DEPARTMENT OF CIVIL AND CONSTRUCTION ENGINEERING,
SWINBURNE UNIVERSITY OF TECHNOLOGY, MELBOURNE, VIC, AUSTRALIA

15.1 Introduction

Many studies and researches have been done to improve and find new and efficient water treatment technologies. Conventional water treatment method cost more and it needs big facilities. Several nanotechnologies are applied worldwide to fulfill fresh water demand such as microfiltration (MF), ultrafiltration (UF), nanofiltration (NF), and reverse osmosis (RO). The NF and RO processes can supply potable water in huge quantity for industrial applications and the quality of water is better than conventional treatments. They can remove various organic and inorganic impurities from seawater, surface water, and groundwater. In surface water treatment, stringent regulations and quality concerns are the major factors. The main drawback is spiral wound NF or RO systems in treating surface waterfoul rapidly (Reiss, 2005).

In conventional desalination processes, the alternative energy sources such as wind energy, solar energy and nuclear energy can be used to save our environment. A hybrid process may be the suitable option as well to implement (Van der Bruggen and Vandecasteele, 2002).

In membrane technology, many advanced studies have been made in the last two decades and mass production has made wastewater reclamation feasible for large scale application. Two main membrane processes used in wastewater reclamation such as MF/UF and RO (Cheremisinoff, 2002; ECW542, 2018). In this study, a brief introduction on RO technique and the potential applications and concerns of RO are discussed (Table 15-1).

15.2 Reverse Osmosis

The RO is a process of forcing the water under pressure using a semipermeable membrane to produce potable water from polluted or saline water. Consequently, it allows the potable water molecules only, whereas it rejects salts and other dissolved elements (Reverse Osmosis, 2018). The RO membrane allows smaller components of the solution to pass freely

Nanotechnology in Water and Wastewater Treatment. DOI: https://doi.org/10.1016/B978-0-12-813902-8.00015-0

Table 15-1   Compares Membrane Structures (Cheremisinoff, 2002)

Technology	Structure	Driving Force	Mechanism
Microfiltration	Symmetric microporous (0.02–10 μm)	Pressure, 1–5 atm	Sieving
Ultrafiltration	Asymmetric microporous (1–20 nm)	Pressure, 2–10 atm	Sieving
Nanofiltration	Asymmetric microporous (0.01–5 nm)	Pressure, 5–50 atm	Sieving
Reverse osmosis	Asymmetric with homogeneous skin and microporous support	Pressure, 10–100 atm	Solution diffusion
Electrodialysis	Electrostatically charged membranes (cation and anion)	Electrical potential	Electrostatic diffusion

through the holes such as solvent molecules but not allows large molecules or ions (Warsinger et al., 2016; Reverse Osmosis, 2018). It is very popular technique for potable water production from saline water and seawater (Reverse Osmosis, 2018; Crittenden et al., 2005). It also can remove other common organics like Humic and Fulvic acids. Consequently, the RO process has many promising applications (Chapter 6, 2018).

15.3 Applications of Reverse Osmosis

Around the globe, the RO is a popular technique to produce drinking and cooking water (Reverse Osmosis, 2018).

15.3.1 Drinking Water

The RO process is used in worldwide, especially in Middle East, to produce pure water in huge quantity. For saline or seawater treatment, it may have the highest users compared to other techniques. However, in USA Marine Corps, the RO unit is replaced by both the Lightweight Water Purification System (LWPS) (470 L/h) and Tactical Water Purification Systems (TWPS) (4,500–5,700 L/h) (Fuentes, 2018).

15.3.2 Water and Wastewater

Singapore produces pure water (NEWater) from domestic wastewater using the RO process. In Los Angeles and other cities of United States, the water shortage problem is solved using rainwater collected from storm drains, which is purified by RO (Reverse Osmosis, 2018).

15.3.3 Wine Industry

About 60 RO machines were used in wine industry in Bordeaux, France, in 2002 (Reverse Osmosis, 2018; Reverse Osmosis in Wine Filtration, 2018).

15.3.4 Sirup Production

The RO process is used in maple sirup producing industry for removing water from sap, which is boiled to make sirup. It can remove about 75%—90% of the water from sap (Reverse Osmosis, 2018).

15.3.5 Hydrogen Production

The RO process is sometimes used in hydrogen production, to prevent formation of minerals in electrodes (Reverse Osmosis, 2018).

15.3.6 Reef Aquarium

The RO process is used in reef aquarium to mix seawater artificially (Reverse Osmosis, 2018).

15.3.7 Window Cleaning

The RO process is used in cleaning windows. As an alternative of washing the windows with detergent, they are scrubbed with the purified water (Reverse Osmosis, 2018).

15.3.8 Leachate Treatment

The use of RO process is limited in landfill leachate treatment (Reverse Osmosis, 2018).

15.3.9 Desalination

The RO process is used extensively in desalination due to its relatively low energy consumption. In 2011, the RO process was used in 66% of desalination pants, according to the International Desalination Association (IDA). In 2013, the largest RO plant was built in Sorek, Israel. It can produce 165 MGD and the water price is \$0.58 /m^3 (Reverse Osmosis, 2018; Next big future, 2018; Talbot, 2018).

15.3.10 Defluoridation

Defluoridation is a process used for the removal of fluoride from water/drinking water. Figure 15-1 shows that the RO integrated with adsorption for defluoridation. In India, this type of treatment was performed where two adsorbents, namely, a hybrid anion exchange resin embedded with zirconium oxide nanoparticles and activated alumina were used in column experiments. They found that fluoride can be successfully removed from the reject water of the RO unit using adsorption (Samrat et al., 2018).

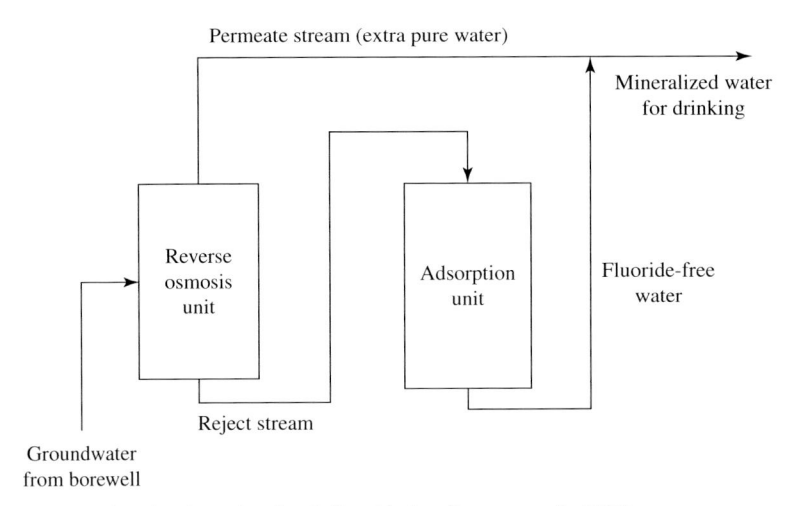

FIGURE 15-1 RO integrated with adsorption for defluoridation (Samrat et al., 2018).

15.4 Concerns in Reverse Osmosis Process

The fouling of RO systems is a major concern of many providers and users. The abundance of transparent exopolymer particles (TEP) in surface waters has been unnoticed for many years. Any newly developed technique may be installed to examine the role of TEP in the fouling of RO systems (Villacorte et al., 2009). In residential areas, RO units use a lot of water due to its low back pressure, only 5%–15% of the water used to treat, whereas the rest is released as wastewater (Reverse Osmosis, 2018; Singh, 2017). In commercial operations, the concentration of total dissolved solids in product water indicates that the effective contaminant removal rates are reduced (Reverse Osmosis, 2018; Health Risks From Drinking Demineralised Water, 2018).

The RO process is excellent in removing various impurities; however, it also removes beneficial minerals. Calcium and magnesium are very important minerals for our health, where the RO may remove over 90% minerals. It is great concern of World Health Organization (WHO). It may have adverse effects on human and animal health for drinking such demineralized water (Reverse Osmosis Water Exposed — What They Don't Tell You, 2018).

15.5 Conclusions

Nowadays, the RO is a common and extensively used technique in the world to purify salty water and wastewater. As like any other techniques, it has pros and cons. The RO not only removes dangerous impurities using its fine filtration process, but it also eliminates many necessary minerals such as calcium and magnesium. Hence, the health effects of drinking such demineralized water (although potable to drink) in long-term should be monitored to analyze. Therefore, the manufacturers of RO membrane modules can think to add some necessary minerals at the end of process or can find a suitable solution to reject the minerals during treatment. The fouling of RO systems is another major concern of many providers and users. Any newly developed technique may be installed to reduce/remove the fouling problem in RO systems.

References

Chapter 6: Reverse Osmosis and Nanofiltration - Presentation transcript. http://slideplayer.com/slide/10499074/ (accessed 02.18).

Cheremisinoff, N.P., 2002. Handbook of Water and Wastewater Treatment Technologies. Butterworth-Heinemann, Oxford, UK.

Crittenden, J., Trussell, R., Hand, D., Howe, K., Tchobanoglous, G., 2005. Water Treatment Principles and Design, second ed. John Wiley and Sons, New Jersey.

ECW542: Wastewater Engineering - Advanced Treatment, Lecture note, UiTM, Malaysia. https://issuu.com/starprime77/docs/ecw542-wk13-l38 (accessed 02.18).

Fuentes, G. Corps' plan for clean water downrange. Marine Corps Times. http://www.marinecorpstimes.com/news/2010/11/marine-afghanistan-water-purification-110410w/ (accessed 02.18).

Health risks from drinking demineralised water. http://www.who.int/water_sanitation_health/dwq/nutrientschap12.pdf (accessed 02.18).

Next big future: Israel scales up reverse osmosis desalination to slash costs with a fourth of the piping. http://nextbigfuture.com/2015/02/isreal-scales-up-reverse-osmosis.html (accessed 02.18).

Reiss, C.R., 2005. Mechanisms of Nanofilter Fouling and Treatment Alternatives for Surface Water Supplies (Ph.D. thesis), *Electronic Theses and Dissertations*, University of Central Florida, USA. http://stars.library.ucf.edu/etd/495 http://purl.fcla.edu/fcla/etd/CFE0000630 (accessed 02.18).

Reverse osmosis. https://en.wikipedia.org/wiki/Reverse_osmosis (accessed 02.18).

Reverse osmosis. Webster's New World College Dictionary, fifth ed., Houghton Mifflin Harcourt Publishing Company. https://www.collinsdictionary.com/dictionary/english/reverse-osmosis (accessed 02.18).

Reverse osmosis in wine filtration. http://www.thevintnervault.com/category/604/Reverse-Osomsis-Plants.html (accessed 02.18).

Reverse osmosis water exposed - What they don't tell you. https://www.aqualiv.com/reverse-osmosis-water-filter-health/ (accessed 02.18).

Samrat, M.V., Rao, K.K., SenGupta, A.K., Riotte, J., Mudakavi, J.R., 2018. Defluoridation of reject water from a reverse osmosis unit and synthetic water using adsorption. J. Water Process Eng. 23, 327–337.

Singh, G., 2017. Implication of household use of R.O. devices for Delhi's urban water scenario. J. Innov. Incl. Dev. 2 (1), 24–29.

Talbot, D. Megascale desalination. http://www.technologyreview.com/featuredstory/534996/megascale-desalination/ (accessed 02.18).

Van der Bruggen, B., Vandecasteele, C., 2002. Distillation vs. membrane filtration: overview of process evolutions in seawater desalination. Desalination 143 (3), 207–218.

Villacorte, L.O., Kennedy, M.D., Amy, G.L., Schippers, J.C., 2009. The fate of transparent exopolymer particles (TEP) in integrated membrane systems: removal through pre-treatment processes and deposition on reverse osmosis membranes. Water Res. 43 (20), 5039–5052.

Warsinger, D.M., Tow, E.W., Nayar, K.G., Maswadeh, L.A., Lienhard, V., John, H., 2016. Energy efficiency of batch and semi-batch (CCRO) reverse osmosis desalination. Water Res. 272–282.

16

Nanoengineered Materials for Water and Wastewater Treatments

Hasrinah Hasbullah, Nurul Shahira Mohd Sabri, Noresah Said,
Sarina Mat Rosid, Mohd Ikhram Roslan, Ahmad Fauzi Ismail,
Lau Woei Jye, Norhaniza Yusof

*ADVANCED MEMBRANE TECHNOLOGY RESEARCH CENTRE (AMTEC),
UNIVERSITI TEKNOLOGI MALAYSIA, JOHOR, MALAYSIA*

16.1 Introduction

Wastewater treatments are essential process to recover back the water that has being polluted. Since water resource had being consumed daily either from industries or household, the reliable and sustainable application is necessary to make sure the continuation supply of clean water. Currently, the requirements for clean water are very demanding for the purpose of drinking, sanitation and industrial usage. Water is considered as a very vital substance for all life because most of activities are depending on it. Nowadays, people are facing a major challenge to get clean water. As number of people increasing globally, a lot of wastewater is produced. This huge amount of wastewater is crucial to be treated for providing sufficient water quality to meet human and environmental needs. Furthermore, variety source that generate wastewater also produced different wastewater characteristic and composition. A wide range of contaminants are detected in wastewater including heavy metals, organic compounds and inorganic pollutants (Dubey et al., 2017). The releasing of these harmful contaminants into environment will effect on life of human being and other organism. Therefore, removal of unsafe contaminant from wastewater is priority.

Recent advance in nanoengineered material have shown as one of opportunities for the application in water purification. It offers potential advantages in removing the contaminants from wastewater in efficient and economically ways compared to conventional treatment. The good properties of this material had elevated the utilization of nanotechnology. Nanoengineered material certainly being use in small size material that less than 100 nm. At such small nano-sized, the materials show certain extraordinary and unique properties which is not possess from material at micron size scale (Kyzas and Matis, 2015). This nanoengineered material has been successfully applied in water and wastewater treatment either been using in membrane filtration or as a nanosorbent. This achievement is due to the availability of super high surface area over volume and high reactivity of the nanoengineered material. A great performance of wastewater treatment reported to achieve when

Nanotechnology in Water and Wastewater Treatment. DOI: https://doi.org/10.1016/B978-0-12-813902-8.00016-2

using this technology. Beside the promising enhancement, the nanoengineered material show some limitation and drawbacks that need to be concern.

16.2 Nanoengineered Materials: Merits and Limitations

Nanoengineered materials are one of the emerging sources that have attracted many attentions since it can be used as alternative method for water and wastewater treatment. It is a material that manufactured in the size range from 1 to 100 nm. Due to having a small size and provide high surface area, this material is widely being applied in various applications. In this section, the explanation on the use of this material as nanosorbent, nanometals/nanometal oxides, nanocarbon and membrane/membrane process are detailed out. Then, the merits and limitations of nanoengineered materials in each application are briefly described.

16.2.1 Nanosorbents

Nanosorbent is a material in nanometer sized range that applied in adsorption process. As a nanosorbent material, they should be in chemically active and can perform a great sorption capacity. Use of nanoengineered materials in adsorption application has been widely practice considering to its reasonable cost, ease to operate and effective method (Thekkudan et al., 2016). Meanwhile, compared to other process, adsorption is one of the feasible techniques for removing contaminants from wastewater by accumulate pollutant at the interface of nanosorbents. This process occurs on the solid surface which is called adsorbent while the materials that adhere at the interface of adsorbent are called adsorbate (Bhatnagar et al., 2011). The adherence of adsorbate involves of differ mechanism such as surface adsorption, ion-exchange and electrostatic interaction as illustrates in Fig. 16-1. In adsorption process, there are two modes depending on the nature of force between adsorbate and adsorbent, which is in chemical sorption and physical sorption (Yagub et al., 2014). Table 16-1 summarizes the difference force involves between this two modes.

Usually, adsorption process is study through isotherm graph to describe adsorption equilibrium. This adsorption equilibrium is important in order to determine the most correlation for gaining an ideal adsorption system (Kyzas and Matis, 2015). Formation of monolayer or multilayer adsorption can determine through this isotherm. Different isotherm occurs for each adsorbent and evaluated by using equation. Typical isotherm with equation involves in adsorption are listed as the following:

- Langmuir isotherm

$$Q_e = \frac{Q_m K_L C_e}{1 \pm K_L C_e} \tag{16.1}$$

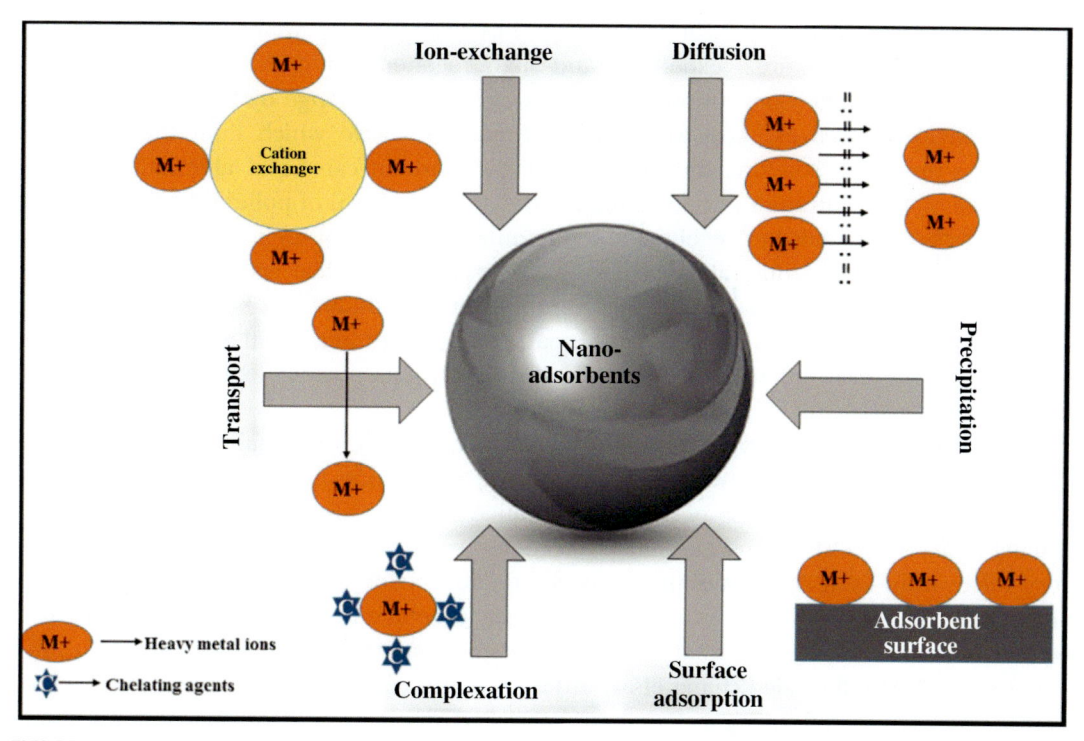

FIGURE 16-1 Various adsorption mechanism of metal ion adsorbate on nanosorbent. *Reprinted with permission. Copyright 2017 Elsevier.*

Table 16-1   Typical Modes in Adsorption

Physical Sorption	Chemical Sorption
Involves weak van der Waals forces	Involves chemical bonds (ionic or covalent bond)
Process reversible	Process irreversible
Forms monolayer or multilayer	Form monolayer
Not very specific	Highly specific

- Temkin isotherm

$$Q_e = \left(\frac{RT}{b_T}\right)\ln(A_T C_e) \tag{16.2}$$

- Frendlich isotherm

$$Q_e = K_F(C_e)^{\frac{1}{n}} \tag{16.3}$$

Conventionally, activated carbon adsorbent has been usually applied in industries but due to limited surface area, expensive and high regeneration cost, their application in adsorption is restricted. Conventional adsorbent also faces challenge by having low capacity and selectivity as well as short adsorption-regeneration cycle which limit their using in adsorption process. Therefore, the existence of nanosorbent can be as alternative to solve this problem. The utilization of nanosorbents in dealing removal of pollutant from wastewater such as heavy metals has received much attention because of their unique physical and chemical properties. At this range of size, it shows high chemical activity and adsorption capacity on the surface of nanomaterials from properties exhibited in bulk size (Kyzas and Matis, 2015). The distinctive active surface area and abundant functional group on the nanosorbents surface also enhances the reactivity and sorption process to treat contaminants from wastewater (Hua et al., 2012).

In adsorption there two type of nanosorbent which is based on metal or metal oxide and carbon based sorbent. Nanometal and metal oxide are among the most used materials in adsorption application because of high performance and low cost for contaminants removal (Zhang et al., 2016a,b). Nanoscale metal and metal oxide sorbent mainly include nanosized zero-valent iron, ferric oxide, aluminum oxide, titanium oxide, magnesium oxide, zinc oxide and copper oxide. The unique properties of this nanometal and metal oxide enable them to be used widely in wastewater treatment that contains heavy metal (Ghilou et al., 2016), dye (Saini et al., 2018), fluoride (Zhang et al., 2014a,b) and organic materials (Naghizadeh et al., 2016). Extensive study among researchers suggested that this type of nanosorbent shows excellent sorption toward heavy metal contaminant such as chromium (Ghosh et al., 2015), arsenic (Alswata et al., 2017), cadmium (Ghilou et al., 2016), lead (Verma et al., 2017) and copper (Rafiq et al., 2014). Benefits of having large surface area, high reactivity, fast kinetics and specific affinity to various contaminants made the metal and metal oxide nanosorbents as potentially excellent materials as adsorbent to gives outstanding performance in adsorption.

Generally, adsorption capability is depend on the adsorption coefficient for each of material. Availability of hydroxyl group and electrostatic charge by metal and metal oxide nanosorbent influence the adsorption of pollutant. Another factor is having high surface area and free adsorption sites. This prevalence of vacant adsorption sites functional groups present on the surface of nanosorbent provides a greater number of active sites for adsorption. There are numerous studies on synthesizing nanosorbent with tunable surface in different shape. Instead of using nanoparticle, nanosorbent in form of nanoblades and nanoflower have been explored to get high adsorption efficiency (Bhanjana et al., 2017). With various shape, it can be utilized as an efficient adsorbent material for wastewater treatment because of plentiful adsorption sites and high adsorption capacity.

Discoveries of carbon based nanosorbent in adsorption shows as a potential material for removal contaminant in wastewater. Carbon nanosorbent, especially carbon nanotube (CNTs), multi-walled carbon nanotube (MWCNTs) and graphene have been widely used due to high specific surface areas and large pore volumes. In addition, it is noncorrosive property, nontoxicity, tunable surface chemistry and presence of surface oxygen-containing functional groups on carbon materials which contributed to be better adsorbent (Xu et al., 2018). CNTs material exhibits highly porous and hollow structure. It is reported that CNTs capable to adsorb heavy metals (Sigdel et al., 2016) and dye (Robati et al., 2016). The adsorption mechanism by CNTs with chemical functional group of dyes involved hydrophobic effect, hydrogen bond, covalent and electrostatic interactions. However, the adsorption capacities is very low (Lata and Samadder, 2016). Surface functionalized with O, N and P containing groups are required to overcome low adsorption capacities of pristine CNTs (Lu and Astruc, 2018). Graphene and graphene oxide (GO) is another carbon-based material that been prepared and developed for the removal of pollutant from water. This material shows efficient adsorption through electrostatic attraction by abundant functional group on the GO surface.

Various improvement and development has been addressed to improve decontamination of water by using nanosorbents. Improvement of nanosorbent by using as bimetallic material and surface functionalized lead to achieve high adsorption performance. Bimetallic nanosorbent involved the combination of two or more different material consist metal and metal oxide nanosorbent that capable to scavenging contaminant from water more efficient. It is found to be excellent adsorption for removal lead ion when utilization of GO and magnesium oxide (Mohan et al., 2017). This is because well dispersed magnesium oxide on GO surface help maintaining the high surface area. Moreover, the functionalization of CNTs with thiol and ferric oxide provide efficient recycling of nanosorbent by using magnet and the thiol group forming as donor center with strong affinity to heavy metal pollutant (Zhang et al., 2012).

Instead of having positive site by advance nanosorbent, this material shows some shortcoming. The main limitations of using nanosorbent in wastewater treatment are listed as following aspect:

- Regeneration of nanosorbent
- Separation of nanosorbent from aqueous solution
- Adsorption process condition
- Toxicity
- Function to treat combined pollutant

Commonly in adsorption process, regeneration of adsorbent is an important part to improve their sustainability. The capability of nanosorbent to be reused after desorption need to be maintained. Some nanosorbent shows difficult regeneration due to strong interaction between adsorbent and adsorbate. Besides that, the active site of adsorbent were blocked throughout wash and dry during the regeneration process (Li et al., 2017). The selection of desorption reagent also need to be considered depends on adsorbate structure as well as removal mechanism. Other aspect is on the process during separation of nanosorbent from aqueous solution or water. A suitable technique is essential to replace the tedious wash and overnight dry process.

Nanosorbent adsorption efficiency strongly depends on the condition process which is pH, temperature, amount of adsorbent and contact time (Yagub et al., 2014). Different materials have different pH to achieve maximum adsorption. As well as contact time, when increases time it provides longer time for the adsorption of unwanted particle from aqueous solution. This is taking to account that initially high adsorption occurred because of high vacant active site on nanosorbent. The adsorption capacity will decreases as time increased. Besides that, with high amount of adsorbent in the solution also improve sorption of material to nanosorbent due to abundant active site is provided. Therefore, nanosorbent that able to function without making large pH alteration and temperature of water is needed to create energy saving process.

Further considerations are taken regarding on safety parameter of the nanosorbent toxicity. Surface modification of nanosorbent provides opportunity to prevent this disadvantages but it might reduce the activity of nanosorbent (Mohammed et al., 2017). Advance development needed to explore in modification technique to sustain both properties. Treating of combined pollutant instead single type molecule using same adsorbent is still a challenge. Feasible techniques such as functionalized and using bimetallic material nanosorbent are introduced to overcome this limitation. In contrast with all the limitation, this nanosorbent have reported to show an outstanding performance in wastewater treatment. For these reasons, the use of nanosorbent is beneficial in adsorption process.

16.2.2 Nanometals and Nanometal Oxides

There are various kinds of metal-based nanoparticles that have been produced by different methods and utilized in water and wastewater treatments. Metal and metal oxide nanoparticles include the famous nano-zero-valent iron or nano-zero-valent zinc, titanium oxides nanoparticles (TiO_2), zinc oxide (ZnO) nanoparticles, copper oxide (CuO) nanoparticles, iron oxide nanoparticles (IONPs) and noble metal nanoparticles. The studies on these nanoparticles have increased in recent years due to their remarkable impacts on the water and waste water treatments.

Zero valent iron (ZVI) or zero valent zinc (ZVZn) for example is an efficient medium for water treatments and its reactivity has been mostly upgraded by the advance of nanoscale zero valent iron (NZVI) (Zhang et al., 2016a,b). NZVI has many positive properties such as a high reactivity to a wide range of pollutants due to their large total specific areas, excellent adsorption properties, precipitation and oxidation, and low cost for treatment of contaminated areas (Tosco et al., 2014). As a result of the large specific surface area and small size, NZVI possesses good adsorption capacity and strong reducing ability. These characteristics contributed most on the excellent performance in the removal of pollutants. NZVI usually presents as a core-shell structure with the outer layer of iron oxides and the inner layer of Fe^0. The Fe^0 core could be oxidized to form iron oxides upon reactions with water, and transforms to a series of corrosion products, including goethite (α-FeOOH), akaganeite (β-FeOOH), lepidocrocite (γ-FeOOH), magnetite (Fe_3O_4), maghemite (γ-Fe_2O_3), green rusts, and siderite ($FeCO_3$). The corrosion products also offered influential adsorption capacity towards large quantities of pollutants (Liu et al., 2014). Despite many advantages, NZVI also has its own disadvantages, for instance aggregation, oxidation and separation trouble from the degraded system. To resolve these problems, several modification methods have been set up to improve the performance of NZVI wastewater treatment.

Although most reported studies on pollutants degradation in water and wastewater treatment have been focused more on NZVI, NZVZn has also been counted as competitors to the NZVI. With more negative reduction potential, Zn (-0.762 E^0/V) is a stronger reductant compared to iron (-0.440 E^0/V) (Bratsch, 1989). Therefore, the pollutant degradation rate of NZVZn may be faster than that of NZVI. Although several studies have demonstrated that contamination reduction by NZVZn could be success, the application of NZVZn is mainly restricted in the degradation of halogenated compounds, especially CCl_4 (Tratnyek et al., 2010).

Apart from becoming reducing agents, metal oxide nanoparticles are also used for photocatalytic degradation of harmful compounds. TiO_2 nanoparticles are among photocatalysts that has been extensively investigated in recent years due to its high photocatalytic activity, low cost, photostability, chemical and biological inertness, mechanical robustness, and flexible in its surface functions (Rawal et al., 2013; Imamura et al., 2013). They become the most excellent photocatalyst to date that allows outstanding photocatalytic activity on many pollutants. The large band gap energy (3.2 eV) of TiO_2 requires ultraviolet (UV) excitation to activate charge separation within the particles. Upon the UV irradiation, TiO_2 will produce reactive oxygen species (ROS) which can fully degrade impurities in short reaction time. Furthermore, TiO_2 nanoparticles display low selectivity and thus are appropriate for the degradation of all types of pollutants, such as phenols in wastewater, organic pollutants such as pesticides, arsenic, chlorinated organic compounds and heavy metal (Ahmed et al., 2011).

However, TiO_2 nanoparticles also have some disadvantages. As discussed above, TiO_2 owned high band gap energy that makes it need the excitation of UV. Thus, the photocatalytic activities by TiO_2 nanoparticles under visible light are almost impossible and insignificant. Hereafter, studies have been conducted to improve the photocatalytic properties of TiO_2 nanoparticles under visible light and UV. For example, metal doping to improve the visible light absorbance of TiO_2 nanoparticles (Anpo et al., 2001) and increase their photocatalytic activity under UV irradiation (Mu et al., 1989). Above all, the production of TiO_2 nanoparticles is rather tedious and complex. It is also problematic to recover TiO_2 nanoparticles from the treated wastewater, especially when it is used as suspension. In recent years, more efforts have been dedicated to solve this problem.

On the other hand, ZnO nanoparticles have been established as alternative applicant in water and wastewater treatment because of their unique properties, wide band gap in the near-UV spectral region, antibacterial, good photocatalytic feature, and cheaper (Janotti and Van de Walle, 2009; Reynolds et al., 1999). ZnO nanoparticles are environmental friendly thus they are compatible with organisms (Schmidt and MacManus, 2007), which makes them relevant for the treatment of water and wastewater. Besides, Daneshvar et al. (2004) revealed that the photocatalytic capability of ZnO nanoparticles is similar to that of TiO_2 nanoparticles because their band gap energies are almost the same. Moreover, ZnO nanoparticles can adsorb a broader range of solar spectra and lighter than some metal oxides (Behnajady et al., 2006). Nonetheless, similar to that of TiO_2 nanoparticles, the light absorption of ZnO nanoparticles is also restricted in the ultraviolet light section because of their large band gap energies. In addition, the utilization of ZnO nanoparticles is hindered by photo corrosion, which resulted in fast recombination of photogenerated charges and thus causes low photocatalytic efficiency (Gomez et al., 2015).

Copper oxide nanoparticle is another of potential nanoparticles to be used in wastewater treatment. This nanoparticle shows excellent properties due to their unique biological, chemical and physical characteristic (Dizaj et al., 2014). Besides that, this nanoparticle has a crystal structure which offered high surface area for wastewater treatment application. Compared to silver, CuO is cheaper and easily to access (Ren et al., 2009). Incorporation of hydrophobic polyethersulfone (PES) polymeric membrane with Cu NPs reported to alter the water contact angle of the membrane. Addition of Cu NPs in PES membrane resulted in decrease of contact angle (Akar et al., 2013). Baghbanzadeh et al. (2015) discovered same result with the lowered down contact angle value and improved in pure water flux of polyvinylidene fluoride (PVDF) membrane when utilized with CuO NPs. In the current study, when incorporating CuO NPs in polysulfone (PSf) membrane, the result for bovine serum albumin (BSA) removal achieved around 90% rejection and water permeation was significantly increases as shown in Fig. 16-2. This could be due to the development of hydrophilic CuO NPs in PSf that has been reported in detailed in our previous study that shown high improvement of pure water flux across the membrane (Sabri et al., 2017). Since BSA possesses hydrophobic characteristic in nature, the hydrophilic membrane is repelling it very well. Thus, a high protein removal can be achieved by the incorporation of hydrophilic CuO NPs in the developed membranes.

In recent years, there is a rising interest in the use of IONPs due to their availability and excellent properties for the removal of heavy metal. Magnetic magnetite (Fe_3O_4) and magnetic maghemite (Fe_2O_4) and nonmagnetic hematite (α-Fe_2O_3) and hydrous ferric oxide (HFO) are often used as nano-adsorbents. Normally, due to the large total surface area of nano-adsorbent materials, their separation and recovery from polluted water are bigger

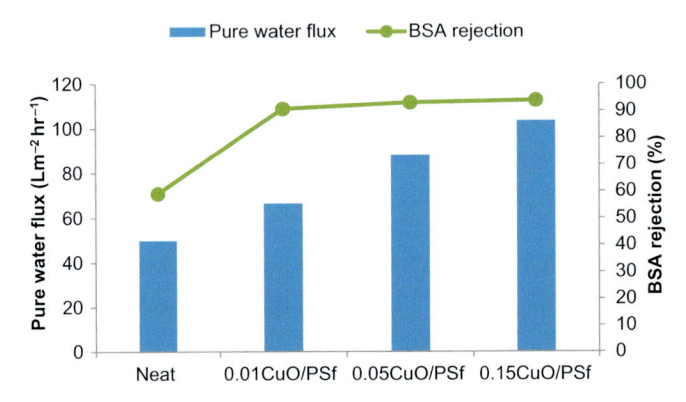

FIGURE 16-2 Pure water flux measurement and BSA rejection at different loading of CuO NP incorporated in PSf membrane.

challenges for water treatment. However, magnetic magnetite (Fe_3O_4) and magnetic maghemite (Υ-Fe_2O_4) can be easily separated and recovered from the system with the assistance of an external magnetic field. Consequently, they have been successfully used as sorbent materials in the removal of various heavy metals from water systems (Lei et al., 2014; Tan et al., 2014; Ngomsik et al., 2012). In order to increase adsorption efficiency and to avoid interference from other metals ions, IONPs have been functionalized to alter their adsorption properties by adding various ligands (Warner et al., 2010; Ge et al., 2012). A flexible ligand shell has been reported to enable the incorporation of a wide range of functional groups into the shell and ensured the goods of Fe_3O_4 nanoparticles are undamaged (Khaydarov et al., 2010).

Noble metals are normally referred to transition metals like Gold (Au), Silver (Ag), Platinum (Pt) and Palladium (Pd). They usually have high ionization energy due to their small atom size and low oxidation potential (Pradeep and Anshup, 2009). At nanoscale size, their ionization energy and oxidation allowing numerous novel reactions with noble metal viable (Schmid, 1992). The synthesis of noble metal nanoparticles is commonly through reduction of the metal salts, controlled nanocrystal nucleation and development with a stabilizing agent. Au and Ag nanoparticles have been widely employed for the discovery of trace levels of organic contaminants because of their exclusive optical features (Sajanlal and Pradeep, 2008). By modifying gold nanoparticle surface with indoxyl groups, the signature of the functional groups under the presence of pesticides was altered, with the detection limit even reaching part per trillion (ppt) levels. Till now, several Ag/Pt, Au/Pt or Ag/Au bimetallic nanoparticles-based electrodes were studied for trace contaminants sensing, monitoring and photocatalysis (Tan et al., 2015; Zhang et al., 2016a,b). Other than contaminant monitoring, noble metal nanoparticles could also be useful to adsorb the pollutants and to deactivate bacterium. The biocidal activity of silver nanoparticles was exploited for water disinfection, whereby water borne micro-organisms such as *E. coli* could be inactivated when in contact with Ag nanoparticles (Xiu et al., 2011). The cell wall will change its permeability resulting in the cell lysis and death subsequently once the negatively charged silver ions adhere to the cell (Lv et al., 2009). Ag nanoparticles are now widely used as disinfectant for surgical masks, textile fibers and even mouthwash. Noble metal nanoparticles are also being used for photocatalytic degradation of many water pollutants for instance pesticides, dyes and halogenated organics (Zhang et al., 2016a,b). However, Ag nanoparticles also have some disadvantages which are highly reactive. The reactivity of silver may cause the leach out of Ag especially if it is incorporated with polymeric materials in drinking water filtration system. Additionally, in vitro study revealed that the Ag nanoparticles triggered DNA impairment in mouse embryonic stem cells and fibroblasts (Peng et al., 2012).

16.2.3 Nanocarbon

Carbon-based nanoparticles emerge as very useful materials that can treat water and wastewater in many ways. Among the renowned allotropes of carbon are carbon nanotubes (CNTs) and graphene. The CNTs can be classified into three kinds, single-walled CNTs (SWCNTs), multi-walled CNTs (MWCNTs), and CNTs composites which is including different types of functionalized CNTs (Bahgat et al., 2011, Goh et al., 2014). The CNTs exist in tubular shape, comprised of six-member carbon rings, illustrated similar with roll graphene sheet and perform outstanding physical features such as mechanical and thermally. Other than their nanoscale dimension and large shape anisotropy, CNTs and graphene possessed high modulus and strength (Zhang et al., 2014a,b). Besides, these two carbon nanoparticles have the ability to cause oxidative damage to bacteria or, in other words, they possess the antimicrobial properties (Perreault et al., 2015). CNTs could be conveniently functionalized and modified to enable favorable accessibility and attachment of various organic and inorganic pollutants (Allen et al., 2009; Georgakilas et al., 2012).

CNTs were generally used as a combination of suspended solids and adsorption, in which CNTs display high capacities for different types of organic mixtures including phenolic compound (Yang et al., 2006), endocrine disruptive compound (Pan et al., 2008) and antibiotics (Wang et al., 2010b). CNT also has been commonly discovered for its huge potential in water and waste water treatments. The modification with oxygen-rich species (-COOH and -OH) allows the surface of CNTs to be introduced with desired hydrophilic groups. In recent years, both MWCNTs and SWCNT have been utilized for removal of pollutant.

The high fluid/water transport capacity are contributed by the atomic-scale smoothness of the CNTs walls, the curvature of the surface, the pore's shape and the molecular ordering phenomena such as carbon-water interaction inside the nanopores (Noy et al., 2007). Additionally, the helicity of CNTs has also made important impacts on the dynamic behaviors and molecular motion of water confined in different nanotubes which in turn determine the transport properties in CNTs (Liu et al., 2005). The smooth and frictionless CNTs surfaces bring about in weak carbon-water contact and thus enabled the very high flow velocity. Other vital properties discovered is the outcome of tube diameter, in which the magnitude of the flow rate is decrease with the reducing CNTs tube diameter lead to the failing of water-CNT contact. Hummer et al. (2001) have found that the water fills the empty cavity of the CNTs within a few tens of picoseconds and the filled stated continuous over the entire simulation time (66 ns). Further analysis of the simulations results of Hummer et al. (2001) shows that water molecules inside and outside a nanotube are in thermodynamics equilibrium. In spite of the exceptional properties of CNTs, the development and application of CNTs are mainly restricted by their low volume of manufacture and expensive. However, CNTs cannot be used alone without any supporting medium or matrix to form structural components (Chatterjee and Deopura, 2002).

Graphene naturally prevents water, but when narrow pores are made in it, quick water permeation is induced and sparked the idea of incorporating graphene for water filtration. Over the past decade, study on graphene has grown due to its superior features wastewater treatment (Sophia et al., 2016). The interest in graphene also because of its remarkable physiochemical features such as large surface area, unique structure, chemically and thermally stable, and mechanical strength. It can be used to reduce pollutant load by adsorption or via incorporating in the host. Graphene is able to adsorb various pollutants because of its large surface area, delocalized π-electron system, which can form strong interactions with other contaminants. Kyzas et al. (2014) used graphite oxide for the removal of dorzolamide (dorzo) in pharmaceutical waste water. The study also discovered that the adsorption was due to the complex combination of force, bond formation, or electrostatic interactions.

The most common derivative of graphene is graphene oxides (GO). The surface of GO, attained by the chemical oxidation technique is dispersed with many polar oxygen-containing functional groups, hydroxyl group, and carboxyl at the edges (Perreault et al., 2015). Due to the presence of these oxygen-containing groups, it has a good hydrophilicity. GO also possess outstanding antimicrobial properties where several studies have revealed the strong antimicrobial features of GO versus various types of microorganisms, phytopathogens, and bacteria (Chen et al., 2014; Mejías Carpio et al., 2012; Akhavan and Ghaderi, 2010). The bacteria cell membrane has been damaged by the GO in this process. Cell membrane ruptured may be due to the atomically sharp edges of graphene, which could cross the cell membrane thus physically disrupt its integrity. The attachment of bacterial cells on the GO-coated surfaces has exposed to induce an interruption of the membrane cell integrity and damage of cell viability, thus controlling the bacterial growth (Mejías Carpio et al., 2012; Akhavan and Ghaderi, 2010). Graphene based materials have been revealed to have potentials to be one of the most consistent and multipurpose materials in upcoming water and wastewater treatment. Nevertheless, graphene materials adsorptive magnitude is depending on the experimental condition and there are many difficulties and problems that must be overcome before they can be used in different application fields especially water and waste water treatment (Sophia et al., 2016).

16.2.4 Membrane and Membrane Processes

The incorporation of nanoparticles in membrane matrix for surface modification and performance enhancement of membranes is an emerging trend in membrane technology. The enhancement in nanotechnology has provided great chances to improve synergistic effects on water and wastewater treatment using membrane and thus, getting a higher chance to develop a new generation of membranes. The new generation of membranes combined the versatility of polymer or ceramic with the unique properties of nanoparticles. The produced membranes offer great advantages such as their ability to work in a harsh condition in water and wastewater processes. Furthermore, the improved membrane characteristics in terms of hydrophilicity, mechanical strength, chemical stability, bio-durability offer a great potential wastewater treatment (Kim and Bruggen, 2010; Ng et al., 2013; Goh et al., 2014).

Generally, the most highlighted problem or critical problem of the current membranes is fouling caused by organic, inorganic, chemical, colloids, and microbes which resulted in flux deterioration during the process (Qu et al., 2013; Goh et al., 2014). To enhance the properties of these membranes against flux deterioration and fouling without weakening the membrane separation, for examples in terms of permeability and rejection, many attempts have been made to modify the membrane surface by chemical or physical modification. Recently, many varieties of inorganic nanoparticles have been used as nanofillers in preparing the membranes from metal, metal oxide and carbon-based nanoparticles. The small nanoparticles size affecting the surface area of the nanoparticles which is relatively superior than in their bulk size which provides a higher reactivity to create alteration on the membrane morphology (Ng et al. 2013).

In recent years, polymeric membrane is not centering on the polymer alone. The exceptional properties of nanoparticles have attracted membranologists to adopt the nanoparticles as membrane fillers (Favvas et al., 2014) to solve the common issues faced by polymeric membranes. Polymeric membranes blended with inorganic nanoparticles are normally called nanocomposite membrane or mixed matrix membranes and sometimes thin film nanocomposite membrane if the nanoparticles are embedded in a separate thin layer on top of the membrane. Most polymeric membranes are hydrophobic in nature and it is known that membrane fouling is directly proportional to hydrophobicity. Thus, in some cases, hydrophilic nanoparticles are placed specifically in the membrane pores, where they have a promising effect on the flux improvement and fouling mitigation.

For example, finely dispersed metal and metal oxide nanoparticles have been incorporated in polymeric membranes for a broad range of membrane applications from gas separation, nanofiltration, pervaporation, to ultrafiltration (Bottino et al., 2001) for desalination, water and wastewater treatment (Ng et al., 2013). Most studies have shown a good impact on the membrane in terms of mechanical strength and water permeability. Previous study has also revealed the remarkable performance of modified membranes over the pristine polymeric membrane in treating water pollutants such as organic, inorganic, and biological (Das et al., 2014).

Ag nanoparticles for instance have been studied for a variety of membrane water treatment. Other than improving the membrane hydrophilicity, the membrane also has very good and long lasting antibacterial properties (Zodrow et al., 2009; Basri et al., 2010). In the previous research conducted by Mansourpanah et al. (2009), polyethersulfone (PES)/TiO_2 nanoparticles ultrafiltration membrane was fabricated by surface coating of TiO_2 nanoparticles on the PES membrane surface. The method was found to be more favorable to encounter membrane fouling since the coating layer modified the surface characteristics of the PES membrane and offered more active sites, thus inhibited the developing foulant layer on the surface of membrane. Besides, most membranes used for photocatalytic degradation process combined with separation process such as polyvinylidene fluoride (PVDF) membranes utilize TiO_2 nanoparticles as filler due to their photocatalytic behavior and superhydrophilicity.

IONPs in the form of Fe_2O_3 or Fe_3O_4 could also be potentially used as nanofiller in the fabrication of membrane to improve the performance of the membrane due to their large surface area, porous structures, and abundance of binding sites which act as adsorbent of contaminant in water (Gohari et al., 2013; Abdullah et al., 2016). Additionally, ZnO nanoparticles has been used as substitute nanofiller in the fabrication of membranes to improve the hydrophilicity of the membranes, perform photocatalytic activity and antibacterial property through the release of zinc ions, causing the bacteria's cell membrane damage (Liang. et al. (2012). Apart from that, PVDF membranes incorporated with silica, SiO_2 nanoparticles produced higher water permeability and good filterability to treat activated sludge, microalgae solution, secondary effluent, and raw sewage (Fernandes et al., 2017).

Apart from metal and metal oxide nanoparticles, interest has also been shown towards the use of carbon-based nanoparticles. The direct incorporation of modified CNTs has been reported to induce tremendous increments in water flux, antifouling and mechanical strength properties of the polymeric membranes (Abidin et al., 2017; Irfan et al., 2014, Nie et al., 2015). The modification with oxygen-rich species (-COOH and -OH) allows the surface of CNTs to be introduced with desired hydrophilic groups. In other case, Zhao et al. (2013) revealed that with the addition of 2 wt% of GO in PVDF ultrafiltration membrane, the water permeability and protein rejection were increased by 79% and 99%, respectively due to the increased membrane surface hydrophilicity.

Poor distribution, agglomeration and leaching of nanoparticles have been classified as the critical challenges for the operation of the nanocomposite membranes in water and wastewater treatment. Possible consequences for human and environmentally well-being due to the nanoparticles contamination become among the raised issues nowadays. On the other hands, the poor dispersion of nanoparticles became one of the limiting factors for the incorporation of nanoparticles in polymer matrix. Therefore, in the preparation of the new functional membrane incorporated with nanoparticles, the aggregation should be monitored especially for the nanoparticles with less than 100 nm in diameter because of the very high surface energy and interactions. Hence, the discussed challenges must be taken into attention for practical executions. The nanocomposite membrane could become a favorable option wastewater treatment by means of addressing the existing problems in a manner that offers significant enhancements in membrane separation performance.

16.3 Overcoming Inherent Limitations of Nanomaterials

Application of nanoengineered materials result in both positive and negative effect. The nano-scale size materials had shown a great alternative for water treatment by providing high surface area and reactivity. At nanometer range, the material also can enter into contamination area while the micro size cannot (Alswata et al. 2017). However, at such a small size the materials favor to be unstable and tend to agglomerate (Thekkudan et al. 2016). This is happening due to increase of surface energy when the size decreases to nano-scale level. Another case is leaching of nanoengineered material into environment due to incomplete incorporation of the chemical binding between the nanoengineered material and the membrane. It is not a new issue, during filtration process some of it will eventually come out and washed away. This arising issue always hinders the progress of membrane production and optimization. There are several possible solutions that can be done in overcoming the limitation of this material, in this section method by chemical modification and physical modifications are discussed.

16.3.1 Chemical Modifications

Limitation of nanoengineered material can be overcome through chemical modifications. This method involve the modification through chemical reaction either by addition or removal certain element from polymer and nanoengineered materials. The introducing of various functional groups occurs at the surface of membrane and nanoengineered material, when it is treated in this technique. A deep discussion on the chemical method by using encapsulation, grafting and cross-link are reported in this part.

16.3.1.1 Encapsulation

Encapsulation technique is a step to layer or cover a thin film of material on the surface of nanoengineered material. According by Yang et al. (2017), it occurred when a defined vesicular shell structure formed incorporated to the synthesis of nanoengineered material with specific size during self-assembly approach. The main aim of this process is to chemically entrap the nanoengineered material within formation of outer layer or shell. Encapsulation process involved by using various method including self-assembly, chemical and physical chemistry (Ladj et al. 2013). Recently, this method has being progressively introduced since it can create stability, prevent occurrence of oxidation process and improve dispersion.

Usually, formation of demixing solution occurred during encapsulation with capping material due to incompatibility (Ladj et al. 2013). A surface modification needed prior to encapsulation process to ensure homogeneity between nanoengineered material and capping agent solution. Encapsulation involves the using of different type material as a capping agent such as polymer, hydrocarbon, amino acid and fatty acid. Table 16-2 shows the materials that have been used previously to encapsulate nanoengineered material.

Encapsulation technique has been proven to confront the drawback of nanoengineered material in water remediation. According by Angamuthu et al. (2017), encapsulation of Fe_2O_3 nanoparticles by carbon nanodisk improves the regeneration nanoengineered material for reduction of methylene blue in water. The particle undergoes ultrasonic treatment during regeneration and show unchanged of structure that can be reuse back for water treatment.

Table 16-2 List of Material Used to Encapsulate Nanoengineered Material

Nanoengineered Material	Encapsulation Material	Method	References
Fe_3O_4	Carbon nanodisk	Catalytic carbonization and air treatment	Angamuthu et al. (2017)
Fe_3O_4	L-Serine	Coprecipitation	Belachew et al. (2017)
Nano zero valent iron	Ethylenediamine and diethylenetriamine	Reflux	Mahmoud et al. (2017)
ZnO	Silica	Sol-gel	El-naggar et al. (2017)
ZnO	Cysteine	Chemical vapor	Sandmann et al. (2015)
Meghamite	Silica	Hydrolysis and condensation	Kralj et al. (2011)

This is due to the formation of stable and high mechanical strength mesopores carbon nano-disk particles. Besides that, it is also reported that leaching of iron to water are detected in small amount through this modification.

Aggregation and agglomeration of nanoengineered material is another limitation that hinders their effectiveness for wastewater treatment. Through surface capping, it can arrest this problematic phenomenon. Recently, the uses of amino acid have been showing fascinating interest because of their biocompatibility and absence of toxic solvent during encapsulation process. For an example, Belachew et al. (2017) had done their study encapsulating the Fe_2O_3 nanoparticles with L-Serine for adsorption of dye from wastewater. From the characterization result, the modification shows effective capping through interaction of carbonyl group from L-Serine to nanoparticle. Another study by Sandmann et al. (2015), using other type of amino acid such as L-Cysteine have been studied as capping agent for zinc oxide nanoparticle. A good dispersion and homogeneous particle size is obtained. This is because the thiol groups from cysteine act as a stabilizer through formation of shell surrounded on nanoparticle. Therefore, from all the research that has done, encapsulation is one of the good approaches that can be applied in nanoengineered material for wastewater treatment to overcome the challenges and limitation.

16.3.1.2 Grafting

Grafting method on nanoparticles are comprises of two method which is "grafting from" (increasing a polymeric chain on the surface of the nanoparticles) and "grafting to" (a reaction of nanoparticles and functionalized polymeric chains). For the grafting of polymeric chains onto the nanoparticles, a surface initiated polymerization technique is needed such as atom transfer radical polymerization (ATRP), reversible addition-fragmentation chain transfer (RAFT), nitroxide mediated polymerization (NMP), surface-initiated polymerization (SIP), cationic, anionic, free radical, frontal and ring opening polymerization (Francis et al., 2014).

"Grafting from" methodology offer the polymeric brush layer that of high grafting density to arrange for polymerization of the monomer from the initiator immobilized on membrane surface, the disadvantage of "grafting from" is that it is an extended polymerization time and usage of a toxic monomer that might disrupt the membrane. Grafting to method another alternative method to make a brush of polymer on the surface of the membrane, the performed polymer is grafted on the surface polymer of choice. The polymer can be prepared in a number of ways before the modification and the surface modification can incorporate any antimicrobial, hydrophilic, hydrophobic, and charged group agents (Burtovyy et al., 2009).

Wang et al. (2007) reported that the preparation of synthesizing poly(methyl methacrylate)-grafted TiO_2 nanoparticles with photocatalytic polymerization process on PMMA chain was grafted directly on the surface of TiO_2 nanoparticles on sunlight under water. Whereas Tang et al. (2006) have modified the surface of ZnO nanoparticles and grafting the polymethacrytic acid chains on the surface have created a better dispersion in aqueous systems, the $-OH$ of the nanoparticles surface interacts with carbonyl (COO-) in PMMA to form the poly(zinc methacrylate complex).

From Paris et al. (2015) on the research of developing a new ultrasound-responsive system on the mesoporous silica nanoparticles by grafting a copolymer that acts as a gate keeper for the pores in drug delivery. As the ultrasound enables the sensitive copolymer to change its hydrophobicity which a conformation to the coil-like and open gate and releasing the content. From Foster et al. (2014) with the research of mechanism in a polymers that grafted to inorganic nanoparticles, the straightforward but extremely versatile atom "grafting through" technique used that bond the high organic fractions of poly(oligo(ethylene oxide) monomethyl ether methacrylate) on iron oxide without catalysts. The grafting contribute to the high local concentration of grafted polymer to the dodecane/water interface that manufacture low interfacial tension of solely 0.003 w/v% (of polymer and particle core), whereas this influence the interfacial activity and rheology of polymers in grafting which will be applicable in several stabilization of emulsions and foam fields.

From Zeng et al. (2016) on the examine the duration of which the surface of PVDF membrane become hydroxylated via helium plasma evoked polyethylene glycol (PEG) grafting process then dealt with three-aminopropyl-trimethoxysilane (APTMS) which study the graphene oxide quantum dots (GOQDs) covalent bonding onto polyvinylidene fluoride (PVDF) membrane which is amino modified in order to introduce the formation of covalent bonds among the carboxylic businesses on GOQDs and the amine group on the PVDF surface by the amine group and form covalent linkage of GOQDs through it. The very last GOQDs functionalized membranes flux has been improved as the 10 hours permeation experimental with bacterium-containing feed water have been has be conducted with the increasing the concentration of GOQDs. It is reported that usage of antibacterial and resistance of biofouling towards GOQDs-PVDF membrane had been elevated with the overall increasing of GOQDs from amount of 0 to 1 mg/mL.

Whereas from Huang et al. (2016) with the purpose to enhance antifouling and antiadhesion of particles, to brought on antifouling and antiadhesion residences by using GO that is a hydrophilic, one atom-thick two-dimensional structure that possesses antimicrobial properties and grafting graphene oxide (GO) onto commercial polyamide RO membranes. Azide functionalized grapheme oxide (AGO) was applied to alter industrial conventional membranes because of the one-of-a-kind chemistry of chemical compound. AGO produces an extraordinarily reactive singlet nitrene intermediate that reacts with the plentiful aromatic ring discovered a few of the polymer membrane energetic layer when photo activated. Graphene oxide is mentioned to inactivate bacteria via bodily disruption, formation of reactive oxygen species, and extraction of lipids from cell membranes. Consequently, by way of anchoring GO molecules on the polymeric amide RO membrane floor, the take a look at have with fulfillment altered the surface houses of a commercial polymeric amide RO membrane, creating it quite a few hydrophilic, easy, and antibacterial with massive resistance to protein fouling and biofouling.

From Nie et al. (2015) study regarding manufacturing of composite membrane incorporates heparin-mimicking polymer brush functionalized carbon nanotubes (F-CNTs) and polyethersufone (PES) that is surface activated by atom switch polymerization (SI-ATRP) to synthesize heparin-mimicking chemical compound brush (contained the sodium styrene sulfonate (SS) and methyl ether methacrylate (EGMA) devices) grafted CNT. The addition of f-CNT, the composite membranes showed remarkably stronger elimination performance of uremic poisonous substance exhibited antifouling capability in ultrafiltration, improved blood and cellular compatibility and cost effective toxic molecules elimination ratio, consequently being extremely good capability for several applications, like hemodialysis and bio-artificial liver guide.

Vantapour et al. (2017) research on the modification of conventional seawater reverse osmosis membranes, the membrane hydrophilicity was increased due to the grafting process of hydrophilic acidic monomer and thermal initiator. Carboxylated multiwalled carbon nanotubes (MWCNTs) was the nanomaterial that were distributed within the grafting solution and placed on membrane surface to lowered fouling by making polymeric brushes and fluid mechanics resistance. The nanotubes with different weight percentages were spread in the acrylic acid monomer solution as the optimum grafting condition was chosen. The COOH-MWCNTs showed the best fouling resistance membrane was in the concentration of 0.25 wt%.

16.3.1.3 Cross-link

Cross-linking is a process of improving the mechanical and structural characteristic of the membrane. Cross-link is a bonding that links another polymer chain to one another, either by covalent or ionic bond. For example in Tsai and Wang (2007) stated that chitosan nanoparticle has low tensile strength that limit the usage, but CS has active amino group and hydroxyl group that form hydrogen bonding with water which enable CS to be chemically modified. Cross-linking is where the polymer chain are joined chemically in places by covalent bonding, by having these joined attachments the polymer molecule cannot slide over each other thus formed and tougher and less flexible polymer. A crossed-linked polymer is mechanically stronger and has high resistance to heat, wear and the attack of solvents

Li et al. (2016) research that the preparation of chitosan nanoparticles incorporated into a quaternized poly(vinyl alcohol) matrix for the direct methanol alkaline fuel cells, it showed that the cross-linking process improved the methanol barrier properties as the concentration of chitosan of 10%. Rajaeian et al. (2015) studied on a thin film nanocomposite membrane fabricated by surface-modified porous poly (vinylidene fouride) (PVDF) that was supported poly (vinyl alcohol) (PVA) dope in solution of TiO_2 nanoparticles. The TiO_2 was crosslinking with chloroacetic acid hence a carboxylation that reduces the particle agglomeration.

In Lin et al. (2003) study was the fabrication on the functional nanostructured materials for sensing, encapsulation and delivery, which uses Cadmium selenide (Cdse) nanoparticles that was cross-link with functionalized ligands that help to stabilize nanoparticle assembly on the interface droplets. Interfacial cross-linking at droplet surfaces enables the encapsulation of water-soluble or oil-soluble materials inside the resulting nanocontainers.

A post-treatment by cross-linking chitosan hollow fiber membrane with Glutaraldehye (GA) solution was administered to improve the strong mechanical properties of the membranes. GA concentration within the vary 50−1000 mg/L was used and elongation at break point reduced with increasing concentration of GA (Tasselli et al., 2013). The reason for the changes of mechanical properties is that a cross-linked network could also be established whereby the cross-linking agent establishes bridges in between the chitosan molecules (Knaul et al., 1999).

From Ertas and Uyar (2017) the study on polybenzoxazine that is based primarily on cross-linked the membrane of cellulose acetate nanofibrous that exhibits increased improved adsorption capability and also the enhance thermal/mechanical properties. Thermal curing was performed by step-wise to acquired cross-linked of composite nanofibrous membranes. CA10/PolyBA-a5/CTR1 a pristine nanofibrous membrane has been shown improvement in their mechanical properties having tensile strength and young's modulus of 8.64 ± 0.63 MPa and 213.87 ± 30.79 MPa. While for pristine CA and CA10/PolyBA-a5/CTR1 nanofibrous membranes adsorption capability was studied by a model polycyclic aromatic hydrocarbon (PAH) compound (i.e., phenanthrene) in aqueous solution, which exhibit the cross-linked nanofibrous membrane shown with higher removal capability (98.5%) and capacity of adsorption which is (592 μg/g).

Morelos-Gomez et al. (2017) did a research on graphene based membrane that was sprayed a coating of graphene oxide/ few superimposed graphene/deoxycholate dispersion of it. The polyvinyl alcohol (PVA) have altered the membranes were deposited onto polysulfone support membranes, followed by Ca^{2+} cross-linking with thermal treatment. The experimental test of strong cross-flow shear for a prolonged amount (120 hours) showed that the membranes were strong to resist the flow and at the same time the anionic dye was NaCl rejected close to 85% and 96% in a maintain. The chlorine resistance shows that the hybrid-layered membranes showcase elevated chlorine resistance compared to graphene oxide membranes of its pure.

Soyekwo et al. (2017) study about the polydopamine grafted CNTs (PDA-CNTs) are rendered dispersible through chemical crosslinking via bioinspired borate chemistry. Nanolayer coatings of polydopamine are deposited on the side walls of CNTs forming aggregates that lead to highly dense and agglomered PDA-CNTs sediments in aqueous media. Borate crosslinking facilitates the obtainment of a homogeneous dispersion of borate cross-linked PDA-CNTs (B-PDA-CNTs) by forming strong hydrogen bonds in solution and thus permitting the facile fabrication of B-PDA-CNTs membranes with an interconnected nanoporous structure on a macroporous support. The obtained membranes with a controllable thickness are stable and display good ultrafiltration performances of ferritin molecules ($>90\%$) and high water permeability of more than 2.8×103 $Lm^{-2}h^{-1}bar^{-1}$. The borate chemical cross-linking highlights a versatile approach for the preparation of stable polydopamine coated nanomaterials for wide applications with bio-macromolecules and in nanocomposite materials.

From Lv et al. (2018) study on the nanocomposite membrane of a novel NCMs have been fabricated via codeposition of polydopamine (PDA), polyethyleneimine (PEI) and electropositive gold nanoparticles (GNPs) followed by cross-linking. The GNPs distribute in the formed selective layer uniformly without obvious aggregation due to their good dispersion and compatibility with the positively charged PDA/PEI matrix. The membrane have high retention ratio ($>90\%$) for bivalent cations, such as Mg^{2+}, Ca^{2+}, and various heavy metal ions, the permeate flux of the NCMs doubles compared with the PDA/PEI codeposited nanofiltration membranes (NFMs) attributed to the hydrophilicity of the embedded GNPs and the loosened selective layer structures.

Based on the study done by Guo et al. (2017) used on cancer therapy which is doxorubicin (DOX) that is prepared topical implantable delivery device for controlled drug release and site-specific treatment. The core region consisted of poly (lactic co-glycolic acid) and poly-caprolactone, as the shell region was composed of cross-linked gelatin that was enclosed DOX in the core region of a core-shell nanofiber obtained by electrospinning. This implantable delivery device was implanted on the top of the melanoma in a mouse model, which had shown a DOX controlled release profile with sustained and sufficient local concentration against melanoma growth in mice with negligible side effects. The implantable device allows precisely localized treatment and therefore can reduce the dose, decrease the injection frequency, and ensure antitumor efficacy associated with lower side effects to normal tissues.

Liu et al. (2017) had study on nanohybrid membranes for oil/water separation and antifouling by using carbon nanotubes (CNTs). The PEI is used as nanocoating on CNTs surface to improve their hydrophilicity through PEI and CNT multiple interactions in between. Then a technique utilized to fabricate the free-standing membranes of filtrating CNTs dispersion by using a vacuum power-assisted self-assembly technique. Using the alteration of the surface charge through the reaction between acyl chloride groups and amino groups as well as the hydrolysis of acyl chloride groups into carboxyl groups, trimesoyl chloride (TMC) is used to cross-link the membranes, which it improves the mechanical strength and hydrophilicity. The evaluation of the data by filtration experiments of numerous oil-in-water emulsions indicates that the membrane have a good high hydrophilicity and negatively charged antifouling properties that help counters antifouling problems.

16.3.2 Physical Modifications

Physical modifications are another method to overcome the limitation of nanoengineered materials. It is usually carried out by incorporating hydrophilic materials either in substrates or membrane surfaces to increase the membrane performance. This method does not involve any chemical reaction between the materials. Detailed process in physical modification that focus on method of coating and composite membrane preparation is discusses.

16.3.2.1 Coating

Coating is a type of covering that is applied to the surface of an object called substrate. The coatings have a few purposes such as decorative, functional or both. The functional coating has been applied to change the surface properties of the substrate such as adhesion, wettability, corrosion resistance, or abrasion and/or wear resistance (Gordon, 2016). Coating can be applied in various form of substrate such as liquids, gases, or solids and this make coating can be used in several of industries and application. A coating can be formed through bulk material or substrate, interface (interaction between substrate and covering), and a modified surface layer (Vargas-Bernal, 2017).

Coating refers as one of the methods in physical modification to overcome the limitation of nanoengineered material in water and wastewater treatment. There are different kinds of coatings have been applied to decrease the agglomeration and dissolution of nanomaterial and at the same time modify their own biological activity (Yang et al., 2012; Chappell et al., 2011). According to Suresh et al. (2012) also, different type of coatings may give different result of surface charges. The surface charge may affect the nanoparticles interactions between living systems (Choi and Hu, 2008) and materials of dispersant used for coating (Chappell et al., 2011). Moreover, nanocoating can also improve the biocompatibility and stability by using various surfactants as coating agents like poly (ethylene glycol), oleic acid, poly (acrylic acid), gluconic acid, liposome, and fatty acid.

To date, coating or encapsulation of nanoparticles or nanomaterials have been caught interest for researcher in water reclamation application especially to treat the wastewater and remove the heavy metal in water (Liu et al., 2008; Limbach et al., 2008; Xu et al., 2012; Sutisna et al., 2017). This is because by coating, nanomaterials can enhance more adsorption capacity, prevent agglomeration, increase hydrophilicity, and improve antibacterial or biocompatibility properties (Zhang et al., 2011; Liu et al., 2011). On other hand, its core or shell structure itself has a few of attractive properties such as high adsorption capacity and chemical and thermal stability (Tran et al., 2013). Besides that, nanocoatings also have been introduced for improvement of performance or replacing conventional coatings, since these materials possess unique physicochemical properties and also to satisfy the environmental fulfillment (Makhlouf, 2014).

Polymer-based coatings typically consist of fragment of pigments and additives that will disperse in a continuous polymer matrix. In addition, the polymer ensures the ability of the coating film is outstanding in terms of chemical resistance, elasticity, durability in the presence of environmental stresses, gloss properties, mechanical and thermal properties depending on the type of polymer used (Shukla et al., 2015). Polymer-based surface coating has been applied in different kind of applications. For an example, a study by Phong et al. (2009) has reported, silver nanoparticles have been coated with flexible polyurethane foams to produce an antibacterial water filter. From Raman, FESEM/EDS and ICP-AAS data, it was presented that silver nanoparticles were stable on the polyurethane foam and no leaching occur in water. From microbiological tests also, the bacteria was killed completely thus proven 100% of antibacterial efficiency.

Nevertheless, other materials beside polymer also can be used as coating agents to improve the performance of nanoparticles. This is proven by Lizzy et al. (2012), that used five different types of substrates which are zeolite, sand, fibreglass, anion, and cation resin to coat silver nanoparticles in increasing the removal rate of microorganisms in groundwater. From results obtained, Ag/cation resin showed the highest removal of bacteria with 100% completely remove of targeted bacteria compared to Ag/zeolite with only 8% to 67% of removal rate.

Table 16-3   List of Material Used to Coating Nanoengineered Material

Type of Nanoengineered Materials Used	Type of Coating Materials Used	Application	References
Titanium dioxide (TiO_2)	Plastic sheets	Degraded dye in Batik industries	Sutisna et al., 2017
Magnetite (Fe_3O_4)	Humic acid	Removal of heavy metals toxic Hg(II), Pb(II), Cd(II), and Cu(II) in water	Liu et al., 2008
Silver nitrate ($AgNO_3$)	Polyurethane (PU) foams	Antibacterial water filter from contaminated drinking water	Phong et al., 2009
Zero-valent iron (ZVI)	Copper	Hexavalent chromium removal from natural groundwater	Hu et al., 2010
Zero-valent iron (ZVI)	Chitosan	Removal of Cr(VI) from water	Geng et al., 2009a,b
Silver (Ag)	Polyvinyl pyrrolidone (PVP)	Simulated wastewater treatment and effect in microbial community	Doolette et al., 2013
Titanium dioxide (TiO_2)	Coated fibers	Decomposing of organic pollutants in water	Isnaeni et al., 2013

Table 16-3 show the type of materials that have been used as coating agents to improve the performance of nanomaterials in water and wastewater treatment.

16.3.2.2 Composite Membrane Preparation

In water and wastewater treatment, the demand for membrane separation processes has been increased rapidly. Additionally, membranes will act as a physical barrier for substances based on their pore size and molecule size. This can be achieved by introduced nanoparticles or nanomaterials into membrane producing polymeric nanocomposite membrane. Nanocomposite membrane has become a new trend in worldwide as a group of filtration materials consist of mixed matrix membranes and surface functionalized membrane.

Mostly, mixed matrix membranes utilize the inorganic nanofillers as its main substrate and incorporated in a polymeric or inorganic oxide matrix. This is due to its characteristic have large surface area lead to higher surface to mass ratio (Wegmann et al., 2008; Feng et al., 2013). The properties of composites membrane depend on the type of nanoparticles used, their size and shape, their concentration and interaction between nanoparticles itself with polymer matrix (Kango et al., 2013). The metal oxide nanoparticles can increase the mechanical and thermal stability as well as permeate flux of polymeric membranes. Meanwhile, zeolites can improve the hydrophilicity; and photocatalytic nanomaterials like bimetallic nanoparticles as well as antimicrobial nanoparticles like CNTs can increase the resistance towards membrane fouling. However, the main problem of polymer nanocomposites is the prevention of the aggregation of particles. To address this problem, some modification of the surface of inorganic particles should be done. Automatically, this will improve the interactions between polymer matrix and inorganic particles.

Nowadays, polymer-supported metal or metal oxide nanoparticles have been new trend that enhance the activity of polymer as well as increase the properties of nanoparticles itself. Nanoparticles have been immobilized following these two steps (Sarkar et al., 2012):

- The sorption of metal ion or metal-precursor in ion exchange resin or functionalized membrane; and
- The reduction of metal ion or metal-precursor by a suitable reducing agent

On other hand, there are two route involved to synthesis polymer-supported nanoparticles which are ex situ and in situ. Ex situ route involved synthesizing the inorganic nanoparticles first before dispersing them in a polymer matrix. However, this method has a drawback as homogeneity and well dispersed between blending polymers and nanoparticles hard to get. To overcome this problem, in situ route has been proposed. In in situ method, polymer matrix itself will provide confined medium for synthesizing nanoparticles of metal and metal oxides. Additionally, polymers also will act as stabilizer and prevent the synthesized nanoparticles from aggregation. Moreover, in in situ method, dispersion of nanoparticles can be regenerated many times within the micropores of the host material (Sarkar et al., 2011a,b).

According to Kim and Deng (2011), ordered mesoporous carbons also have been used as nanofiller in thin film polymeric matrices applied in reverse osmosis process. By applying atmospheric pressure, ordered mesoporous carbons have increase hydrophilicity thus raise pure water permeability. Mostly, nanoparticles or nanomaterials are designed to attract water and highly porous while rejected dissolved salts and others impurities. Besides that, it can also prevent the clogging of organic compounds and bacteria in conventional membrane.

Furthermore, surface functionalization also another way used by researcher to prevent membrane clogging. In this method, it uses chemical substances that have ability to oxidize organic contaminants thus prevent membrane fouling forming layers. Gehrke et al. (2012) has inserted photocatalytic TiO_2 nanoparticles (P25, Evonik) on metallic filter material using dip coating method. This is proven can degrade the water impurities on the surface before a dense cake layer is formed. However, this method is limited only to chemically robust materials exclude the polymeric materials as this material itself will degrade by oxidation process. To overcome this problem, a novel approach has been made to use bionanocomposite membranes with highly selective protein immobilized on their surfaces (Gehrke, 2014).

Ben-Sasson et al. (2016) has been conducted an approach method to create in-situ attachment of biocidal copper nanoparticles (Cu-NPs) onto the surface of a thin-film composite reverse osmosis (TFC RO) membrane. The process and product of modified membrane are shown in Fig. 16-3. From the image of scanning electron microscopy (SEM) and X-ray photoelectron spectroscopy (XPS) analysis in this study, Cu nanoparticles have been successfully deposited on the membrane either in form of metallic copper or copper oxide. The in situ Cu-NP modification on membrane surface showed slightly raising of water and salt permeability as well as strong antibacterial activity reached to 90% reduction in the number of attached live *Escherichia coli* bacteria on the surface membrane compared to pristine reverse osmosis membrane.

On other hand, Prince et al. (2013) has proposed triple layer composite membranes consist of polyvinylidene fluoride (PVDF) thin hydrophobic nanofiber layer, conventional casted microporous layer which is translucent, white and flat membrane and polyethylene

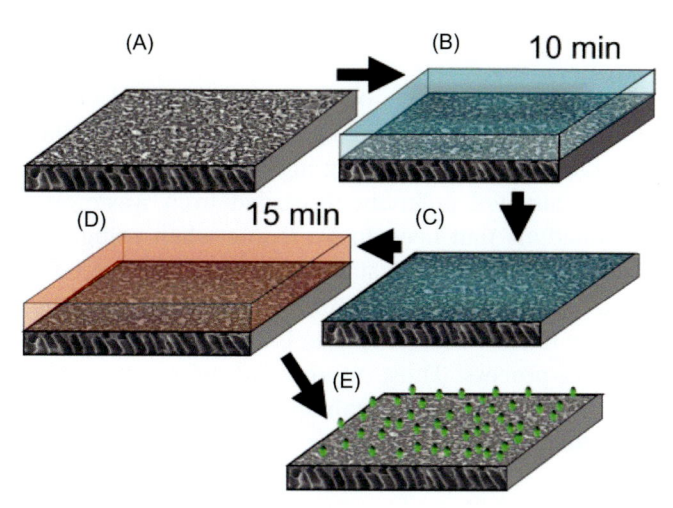

FIGURE 16-3 A schematic diagram and process for in situ formation of Cu-NPs on a TFC RO membrane: (A) pristine RO membrane (B) covered with $CuSO_4$ solution (pale blue) for 10 min (C) discarded $CuSO_4$ solution leaving a thin layer of the $CuSO_4$ solution on the RO membrane surface (D) the membrane recoated with $NaBH_4$ solution for 15 min (E) TFC RO membrane surface cover with Cu-NPs. *Reprinted with permission. Copyright 2016 Elsevier.*

terephthalate (PET) as support layer. Each layer is bound by solvent binding and heat pressing. Electrospun PVDF nanofibers which are highly hydrophobic have been chosen as top selective layer to prevent water molecules from passing through the membranes. Meanwhile, second layer will increase the liquid entry pressure of water (LEPw) by preventing long term pore wetting. Lastly, third hydrophilic layer will help to draw more water vapor from second layer through absorption process (Qtaishat et al., 2009; Khayet et al., 2005; Khayet and Matsuura, 2003; Khayet et al., 2003).

Besides that, thin film nanocomposite (TFN) also has been proposed as nanofiltration membrane. The TFN have been prepared via in situ interfacial consist of copolyamide that has been fabricated on a polyimide support incorporated with TiO_2 nanoparticles (NPs). To achieve compatibility and homogeneous mixture, both amine and chloride compounds have been used to functionalize TiO_2 NPs. From the results, higher methanol flux and dye rejection as well as low in swelling degree have been accomplished through TFN membranes (Peyravi et al., 2014).

Polymeric nanoadsorbents like dendrimers often used to remove organics and heavy metals in water. Theoretically, organic compounds will be adsorbed by hydrophobic shells, while heavy metals will be adsorbed by tailored exterior branches (Hajeh et al., 2013). This is similar with a study by Diallo et al. (2005) that used dendrimers integrated into ultrafiltration system to remove copper from water. On other hand, this adsorbent also can be regenerated through a pH shift as dendrimers itself are difficult to synthesize. Moreover, bioadsorbent consist of chitosan-dendrimer nanostructure also can remove anionic compounds like dye from textile wastewater efficiently up to 99% (Sadeghi-Kiakhani et al., 2013). Nevertheless, bio adsorbent properties usually show biodegradable, biocompatible, and nontoxic.

16.4 Summary

Nanoengineered materials are extensively being explored as one of potential alternative in water and wastewater treatment technology. In general, this material shows valuable benefits in water purification application such as having high reactivity to specific contaminants. This will allowed the harmful targeted pollutant discarded from wastewater effluent. However there are still many limitations that need to overcome when using this material. The problems arise due to low dispersion and stability of nanoengineered material which had caused agglomeration phenomena. Another problem, occurred when this material is leached when utilize in membrane filtration. Besides that, issues of regeneration by this material also happen during adsorption process. Thus, appropriate approaches are required to address this multiple problems for ensuring the sustainability of nanoengineered material in wastewater treatment application.

References

Abdullah, N., Gohari, R.J., Yusof, N., Ismail, A.F., Jaafar, J., Lau, W.J., et al., 2016. Polysulfone/hydrous ferric oxide ultrafiltration mixed matrix membrane: preparation, characterization and its adsorptive removal of lead (II) from aqueous solution. Chem. Eng. J. 289, 28−37.

Abidin, M.N.Z., Goh, P.S., Ismail, A.F., Othman, M.H.D., Hasbullah, H., Said, N., et al., 2017. Development of biocompatible and safe polyethersulfone hemodialysis membrane incorporated with functionalized multi-walled carbon nanotubes. Mater. Sci. Eng. C 77, 572−582.

Ahmed, S., Rasul, M., Brown, R., Hashib, M., 2011. Influence of parameters on the heterogeneous photocatalytic degradation of pesticides and phenolic contaminants in wastewater: a short review. J. Environ. Manage. 92 (3), 311−330.

Akar, N., Asar, B., Dizge, N., Koyuncu, I., 2013. Investigation of characterization and biofouling properties of PES membrane containing selenium and copper nanoparticles. J. Memb. Sci. 437, 216−226.

Akhavan, O., Ghaderi, E., 2010. Toxicity of graphene and graphene oxide nanowalls against bacteria. ACS Nano 4, 5731−5736.

Allen, M.J., Tung, V.C., Kaner, R.B., 2009. Honeycomb carbon: a review of graphene. Chem. Rev. 110 (1), 132−145.

Alswata, A.A., Ahmad, M., Al-hada, N.M., Kamari, H.M., Hussein, M.Z., Ibrahim, N.A., 2017. Preparation of zeolite/zinc oxide nanocomposites for toxic metals removal from water. Results Phys. 7, 723−731.

Angamuthu, M., Satishkumar, G., Landau, M.V., 2017. Precisely controlled encapsulation of Fe_3O_4 nanoparticles in mesoporous carbon nanodisk using iron based MOF precursor for effective dye removal. Microporous Mesoporous Mater. 251, 58−68.

Anpo, M., Kishiguchi, S., Ichihashi, Y., Takeuchi, M., Yamashita, H., Ikeue, K., et al., 2001. The design and development of second-generation titanium oxide photocatalysts able to operate under visible light irradiation by applying ametal ion-implantation method. Res. Chem. Intermed. 27 (4-5), 459−467.

Baghbanzadeh, M., Rana, D., Matsuura, T., Lan, Q.C., 2015. Effects of hydrophilic CuO nanoparticles on properties and performance of PVDF VMD membranes. Desalination 369, 75−84.

Bahgat, M., Farghali, A., El Rouby, W., Khedr, M., 2011. Synthesis and modification of multi-walled carbon nanotubes (MWCNTs) for water treatment applications. J. Anal. Appl. Pyrolysis 92 (2), 307−313.

Basri, H., Ismail, A.F., Aziz, M., Nagai, K., Matsuura, T., Abdullah, M.S., et al., 2010. Silver-filled polyethersulfone membranes for antibacterial applications-effect of pvp and tap addition on silver dispersion. Desalination 261 (3), 264–271.

Behnajady, M.A., Modirshahla, N., Hamzavi, R., 2006. Kinetic study on photocatalytic degradation of C.I. Acid Yellow 23 by ZnO photocatalyst. J. Hazard. Mater. 133 (1–3), 226–232.

Belachew, N., Devi, D.R., Basavaiah, K., 2017. Green synthesis and characterisation of L-serine capped magnetite nanoparticles for removal of Rhodamine B from contaminated water. J. Exp. Nanosci. 12 (1), 114–128.

Ben-Sasson, M., Lu, X., Nejati, S., Jaramillo, H., Elimelech, M., 2016. In situ surface functionalization of reverse osmosis membranes with biocidal copper nanoparticles. Desalination 388, 1–8.

Bhanjana, G., Dilbaghi, N., Singhal, N.K., Kim, K., Kumar, K., 2017. Copper oxide nanoblades as novel adsorbent material for cadmium removal. Ceram. Int. 43 (8), 6075–6081.

Bhatnagar, A., Kumar, E., Sillanpää, M., 2011. Fluoride removal from water by adsorption—a review. Chem. Eng. J. 171, 811–840.

Bottino, A., Cappennli, G., D'Asti, V., Piaggio, P., 2001. Preparation and properties of novel organic-inorganic porous membranes. Sep. Purif. Technol. 22-23, 269–275.

Burtovyy, O., Klep, V., Turel, T., Gowayed, Y. and Luzinov, I. (2009). Polymeric membranes: surface modification by "grafting to" method and fabrication of multilayered assemblies. In: Nanoscience and Nanotechnology for Chemical and Biological Defense, In: ACS Symposium Series, vol. 1016, ACS Publications, pp. 289–305.

Chappell, M.A., Miller, L.F., George, A.J., Pettway, B.A., Price, C.L., Porter, B.E., et al., 2011. Simultaneous dispersion–dissolution behavior of concentrated silver nanoparticle suspensions in the presence of model organic solutes. Chemosphere 84, 1108–1116.

Chatterjee, A., Deopura, B.L., 2002. Carbon nanotubes and nanofibre: an overview. Fibers Polym. 3 (4), 134–139.

Chen, J., Peng, H., Wang, X., Shao, F., Yuan, Z., Han, H., 2014. Graphene oxide exhibits broad-spectrum antimicrobial activity against bacterial phytopathogens and fungal conidia by intertwining and membrane perturbation. Nanoscale 6, 1879–1889.

Choi, O., Hu, Z., 2008. Size dependent and reactive oxygen species related nanosilver toxicity to nitrifying bacteria. Environ. Sci. Technol. 42, 4583–4588.

Daneshvar, N., Salari, D., Khataee, A.R., 2004. Photocatalytic degradation of azo dye acid red 14 in water on ZnO as an alternative catalyst to TiO_2. J. Photochem. Photobiol. A 162 (2-3), 317–322.

Das, R., Hamid, S.B.A., Ali, M.E., Ismail, A.F., Anuar, M.S.M., Ramakrishna, S., 2014. Multifuntional carbon nanotubes in water treatment: the present, past and future. Desalination 354, 160–179.

Diallo, M.S., Christie, S., Swaminathan, P., Johnson, J.H., Goddard, W.A., 2005. Dendrimer enhanced ultrafiltration. 1. Recovery of Cu(II) from aqueous solutions using PAMAM dendrimers with ethylenediamine core and terminal NH_2 groups. Environ. Sci. Technol. 39, 1366–1377.

Dizaj, S.M., Lotfipour, F., Barzegar-Jalali, M., Zarrintan, M.H., Adibkia, K., 2014. Antimicrobial activity of the metals and metal oxide nanoparticles. Mater. Sci. Eng. C 44, 278–284.

Doolette, C.L., McLaughlin, M.J., Kirby, J.K., Batstone, D.J., Harris, H.H., Ge, H., et al., 2013. Transformation of PVP coated silver nanoparticles in a simulated wastewater treatment process and the effect on microbial communities. Chem. Cent. J. 7 (46), 1–18.

Dubey, S., Banerjee, S., Upadhyay, S.N., Sharma, Y.C., 2017. Application of common nano-materials for removal of selected metallic species fromwater and wastewaters: a critical review. J. Mol. Liq. 240, 656–677.

El-naggar, M.E., Hassabo, A.G., Mohamed, A.L., Shaheen, T.I., 2017. Surface modification of SiO_2 coated ZnO nanoparticles for multifunctional cotton fabrics. J. Colloid. Interface. Sci. 498, 413–422.

Ertas, Y., Uyar, T., 2017. Fabrication of cellulose acetate/polybenzoxazine cross-linked electrospun nanofibrous membrane for water treatment. Carbohydr. Polym. 177, 378−387.

Favvas, E.P., Nitodas, S.F., Stefopoulos, A.A., Papageogiou, S.K., Stefanopoulos, K.L., Mitropoulos, A.C., 2014. High purity multi-walled carbon nanotubes: preparation, characterization and performance as filler materials in co-polyimide hollow fiber membranes. Sep. Purif. Technol. 122, 262−269.

Feng, C., Khulbe, K.C., Matsuura, T., Tabe, S., Ismail, A.F., 2013. Preparation and characterization of electrospun nanofiber membranes and their possible applications in water treatment. Sep. Purif. Technol. 102, 118−135.

Fernandes, C.S., Bilad, M.R., Nordin, N.A.H.M., 2017. Silica incorporated membrane for wastewater-based filtration. AIP Conf. Proc. 1891, 020041.

Foster, L.M., Worthen, A.J., Foster, E.J., Dong, J., Roach, C.M., Metaxas, A.E., Hardy, C.D., Larsen, E.S., Bollinger, J.A., Truskett, T.M., Bielawski, C.W., Johnston, K.P., 2014. High interfacial activity of polymers "grafted through" functionalized iron oxide nanoparticle clusters. Langmuir 30 (34), 10188−10196.

Francis, R., Joy, N., Aparna, E.P., Vijayan, R., 2014. Polymer grafted inorganic nanoparticles, preparation, properties, and applications: a review. Polym. Rev. 54 (2), 268−347.

Ge, F., Li, M.-M., Ye, H., Zhao, B.-X., 2012. Effective removal of heavy metal ions Cd^{2+}, Zn^{2+}, Pb^{2+}, Cu^{2+} from aqueous solution by polymer-modified magnetic nanoparticles. J. Hazard. Mater. 211-212, 366−372.

Gehrke, I., 2014. Environmental friendly recycling of strategic metals. Fraunhofer UMSICHT Annual Report. 2013 (accessed 24.07.14.).

Gehrke, I., Keuter, V., Groß, F., 2012. Development of nanocomposite membranes with photocatalytic surfaces. J. Nanosci. Nanotechnol. 12, 9163−9168.

Geng, B., Jin, Z., Li, T., Qi, X., 2009a. Preparation of chitosan-stabilized Fe^{0} nanoparticles for removal of hexavalent chromium in water. Sci. Total Environ. 407 (18), 4994−5000.

Geng, B., Jin, Z., Li, T., Qi, X., 2009b. Kinetics of hexavalent chromium removal from water by chitosan-Fe^{0} nanoaprticles. Chemosphere 75 (6), 825−830.

Georgakilas, V., Otyepka, M., Bourlinos, A.B., Chandra, V., Kim, N., Kemp, K.C., et al., 2012. Functionalization of graphene: covalent and non-covalent approaches, derivatives and applications. Chem. Rev. 112 (11), 6156−6214.

Ghilou, I., Ghoul, J.E., Modwi, A., Mir, L.E., 2016. Ga-Doped ZnO for adsorption of heavy metals from aqueous solution. Mater. Sci. Semiconductor Process. 42, 102−106.

Ghosh, A., Pal, M., Biswas, K., Ghosh, U.C., Manna, B., 2015. Manganese oxide incorporated ferric oxide nanocomposites (MIFN): a novel adsorbent for effective removal of Cr (VI) from contaminated water. J. Water Process Eng. 7, 176−186.

Goh, P.S., Ng, B.C., Lau, W.J., Ismail, A.F., 2014. Inorganic nanomaterials in polymeric ultrafiltration membranes for water treatment. Sep. Purif. Rev. 44 (3), 216−249.

Gohari, R.J., Lau, W.J., Matssura, T., Halakoo, E., Ismail, A.F., 2013. Adsorptive removal of Pb(II) from aqueous solution by novel PES/HMO ultrafiltration mixed matrix membrane. Sep. Purif. Technol. 120, 59−68.

Gomez-Solıs, C., Ballesteros, J.C., Torres-Martınez, L.M., Juarez-Ramirez, I., Torres, L.A.D., ZarazuapMorin, M.E., et al., 2015. Rapid synthesis of ZnO nano-corncobs from Nital solution and its application in the photodegradation of methyl orange. J. Photochem. Photobiol. A 298, 49−54.

Gordon, P.B., 2016. Surface Coating. *Encyclopedia Britannica*, State of Washington, pp. 1−12.

Guo, M., Zhou, G., Liu, Z., Liu, J., Tang, J., Xiao, Y., et al., 2017. Direct site-specific treatment of skin cancer using doxorubicin-loaded nanofibrous membranes. Sci. Bull. 63 (2), 92−100.

Hajeh, M., Laurent, S., Dastafkan, K., 2013. Nanoadsorbents: classification, preparation, and applications (with emphasis on aqueous media). Chem. Rev. 113, 7728−7768.

Hu, C.Y., Lo, S.L., Liou, Y.H., Hsu, Y.W., Shih, K., Lin, C.J., 2010. Hexavalent chromium removal from near natural water by copper-iron bimetallic particles. Water Res. 44, 3101−3108.

Hua, M., Zhang, S., Pan, B., Zhang, W., Lv, L., Zhang, Q., 2012. Heavy metal removal from water/wastewater by nanosized metal oxides: a review. J. Hazard. Mater. 212−212, 317−331.

Huang, X., Marsh, K., McVerry, B., Hoek, E., Kaner, R., 2016. Low-fouling antibacterial reverse osmosis membranes via surface grafting of graphene oxide. ACS Appl. Mater. Interfaces 8 (23), 14334−14338.

Hummer, G., Rasaiah, J.C., Noworyta, J.P., 2001. Water conduction through the hydrophobic channel of a carbon nanotube. Nature 414, 188−190.

Imamura, K., Yoshikawa, T., Hashimoto, K., Kominami, H., 2013. Stoichiometric production of aminobenzenes and ketones by photocatalytic reduction of nitrobenzenes in secondary alcoholic suspension of titanium(IV) oxide under metal-free conditions. Appl. Catal. B 134-135, 193−197.

Irfan, M., Idris, A., Yusof, N.M., Mohd Khairuddin, N.F., Akhmal, H., 2014. Surface modification and performance enhancement of nano-hybrid F-MWCNT/PVP90/PES hemodialysis membranes. J. Memb. Sci. 467, 73−84.

Isnaeni, V.A., Arutanti, O., Sustini, E., Aliah, H., Khairurrijal, Abdul, M., 2013. A novel system for producing photocatalytic titanium dioxide-coated fibers for decomposing organic pollutants in water. Am. Inst. Chem. Eng. Environ. Prog. 32, 42−51.

Janotti, A., Van de Walle, C.G., 2009. Fundamentals of zinc oxide as a semiconductor. IOP Sci. 72 (12).

Kango, S., Kalia, S., Celli, A., Njugunad, J., Habibi, Y., Kumar, R., 2013. Surface modification of inorganic nanoparticles for development of organic−inorganic nanocomposites—A review. Prog. Polym. Sci. 38, 1232−1261.

Khaydarov, R.A., Khaydarov, R.R., Gapurova, O., 2010. Water purification from metal ions using carbon nanoparticle-conjugated polymer nanocomposites. Water Res. 44 (6), 1927−1933.

Khayet, M., Matsuura, T., 2003. Application of surface modifying macromolecules for the preparation of membranes for membrane distillation. Desalination 158, 51−56.

Khayet, M., Suk, D.E., Narbaitz, R.M., Santerre, J.P., Matsuura, T., 2003. Study on surface modification by surface-modifying macromolecules and its applications in membrane-separation processes. J. Appl. Polym.Sci. 89, 2902−2916.

Khayet, M., Mengual, J.I., Matsuura, T., 2005. Porous hydrophobic/hydrophilic composite membranes: application in desalination using direct contact membrane distillation. J. Membr. Sci. 252, 101−113.

Kim, E.S., Deng, B., 2011. Fabrication of polyamide thin-film nano-composite (PA-TFN) membrane with hydrophilized ordered mesoporous carbon (H-OMC) for water purifications. J. Memb. Sci. 375, 46−54.

Kim, J., Bruggen, B.V.D., 2010. The use of nanoparticles in polymeric and ceramic membrane structures: review of manufacturing procedures and performance improvement for water treatment. Environ. Pollut. 158 (7), 2335−2349.

Knaul, J.Z., Hudson, S.M., Creber, K.A.M., 1999. Improved mechanical properties of chitosan fibers. J. Appl. Polym. Sci. 72 (13), 1721−1732.

Kralj, S., Miha, D., Makovec, D., 2011. Controlled surface functionalization of silica-coated magnetic nanoparticles with terminal amino and carboxyl groups. J. Nanopart. Res. 13, 2829−2841.

Kyzas, G.Z., Matis, K.A., 2015. Nanoadsorbents for pollutants removal: a review. J. Mol. Liq. 203, 159−168.

Kyzas, G.Z., Bikiaris, D.N., Seredych, M., Bandosz, T.J., Deliyanni, T.J., 2014. Removal of dorzolamide from biomedical wastewaters with adsorption onto graphite oxide/poly(acrylic acid) grafted chitosan nanocomposite. Bioresour. Technol. 152, 399−406.

Ladj, R., Bitar, A., Eissa, M.M., Fessi, H., Mugnier, Y., Dantec, R.L., et al., 2013. Polymer encapsulation of inorganic nanoparticles for biomedical applications. Int. J. Pharm. 458, 230−241.

Lata, S., Samadder, S.R., 2016. Removal of arsenic from water using nano adsorbents and challenges : a review. J. Environ. Manage. 166, 387−406.

Lei, Y., Chen, F., Luo, Y., Zhang, L., 2014. Three-dimensional magnetic graphene oxide foam/Fe_3O_4 nano-composite as an efficient absorbent for Cr(VI) removal. J. Mater. Sci. 49 (12), 4236−4245.

Li, P., Liao, G., Kumar, S.R., Shih, C., Yang, C., Wang, D., Lue, S.J., 2016. Fabrication and characterization of chitosan nanoparticle-incorporated quaternized poly(vinyl alcohol) composite membranes as solid electrolytes for direct methanol alkaline fuel cells. Electrochim. Acta 187, 616−628.

Li, Z., Qi, M., Tu, C., Wang, W., Chen, J., Wang, A., 2017. Highly efficient removal of chlorotetracycline from aqueous solution using graphene oxide/TiO_2 composite : properties and mechanism. Appl. Surf. Sci. 425, 765−775.

Liang., S., Xiao, K., Mo, Y., Huang, X., 2012. A novel ZnO nanoparticle blended polyvinylidene flouride membrane for anti-irreversible fouling. J. Memb. Sci. 394-395, 184−192.

Limbach, L.K., Bereiter, R., Muller, E., Krebs, R., Galli, R., Stark, W.J., 2008. Removal of oxide nanoparticles in a model wastewater treatment plant: influence of agglomeration and surfactants on clearing efficiency. Environ. Sci. Technol. 42 (15), 5828−5833.

Lin, Y., Skaff, H., Boker, A., Dinsmore, A.D., Emrick, T., Russell, T.P., 2003. Ultrathin cross-linked nanoparticle membranes. J. Am. Chem. Soc. 125 (42), 12690−12691.

Liu, A., Liu, J., Pan, B., Zhang, W.-X., 2014. Formation of lepidocrocite (γ-FeOOH) from oxidation of nano-scale zero-valent iron (nZVI) in oxygenated water. RSC Adv. 4 (101), 57377−57382.

Liu, J., Zhao, Z., Jiang, G., 2008. Coating Fe_3O_4 magnetic nanoparticles with humic acid for high efficiency removal of heavy metals in water. Environ. Sci. Technol. 42 (18), 6949−6954.

Liu, S., Zeng, T.H., Hofmann, M., 2011. Anti bacterial activity of graphite, graphite oxide, graphene oxide, and reduced graphene oxide: membrane and oxidative stress. ACS Nano 5 (9), 6971−6980.

Liu, T.-Y., Lin, W.-C., Huang, L.-Y., Chen, S.-Y., Yang, M.-C., 2005. Hemocompatibility and anaphylatoxin formation of protein-immobilizing polyacrylonitrile hemodialysis membrane. Biomaterials 26 (12), 1437−1444.

Liu, Y., Wang, Q., Zhang, L., Wu, T., 2005. Dynamics and density profile of water in nanotubes as one-dimensional fluid. Langmuir 21 (25), 12025−12030.

Liu, Y., Su, Y., Cao, J., Guan, J., Xu, L., Zhang, R., et al., 2017. Synergy of the mechanical, antifouling and permeation properties of a carbon nanotube nanohybrid membrane for efficient oil/water separation. Nanoscale 9 (22), 7508−7518.

Lizzy, M., Mthombeni, N.H., Onyango, M.S., Nomba, M.N.B., 2012. Cost-effective filter materials coated with silver nanoparticles for the removal of pathogenic bacteria in groundwater. Int. J. Environ. Res. Public Health 9, 244−271.

Lu, F., Astruc, D., 2018. Nanomaterials for removal of toxic elements from water precipitaion adsorption heavy metals removal ion exchange filtration coagulation bio-sorption. Coord. Chem. Rev. 356, 147−164.

Lv, Y., Liu, H., Wang, Z., Liu, S., Hao, L., Sang, Y., et al., 2009. Silver nanoparticle-decorated porous ceramic composite for water treatment. J. Memb. Sci. 331, 50−56.

Lv, Y., Du, Y., Chen, Z., Qiu, W., Xu, Z., 2018. Nanocomposite membranes of polydopamine/electropositive nanoparticles/polyethyleneimine for nanofiltration. J. Memb. Sci. 545, 99−106.

Mahmoud, M.E., Saad, E.A., Soliman, M.A., Abdelwahab, M.S., 2017. Encapsulation of nano zerovalent iron with ethylenediamine and diethylenetriamine for removing cobalt and zinc and their radionuclides from water. J. Environ. Chem. Eng. 5, 5157−5168.

Makhlouf, A.S.H., 2014. Handbook of Smart Coatings for Materials Protection, first ed. Woodhead Publishing, USA.

Mansourpanah, Y., Madaeni, S.S., Rahimpour, A., Farhadian, A., Taheri, A.H., 2009. Formation of appropriate sites on nanofiltration membrane surface for binding tio$_2$ photo-catalyst: performance, characterization, and fouling-resistant capability. J. Memb. Sci. 330, 297–306.

Mejías Carpio, I.E., Santos, C.M., Wei, X., Rodrigues, D.F., 2012. Toxicity of a polymer-graphene oxide composite against bacterial planktonic cells, biofilms, and mammalian cells. Nanoscale 4, 4746–4756.

Mohammed, L., Gomaa, H.G., Ragab, D., Zhu, J., 2017. Particuology magnetic nanoparticles for environmental and biomedical applications : a review. Particuology 30, 1–14.

Mohan, S., Kumar, V., Singh, D.K., Hasan, S.H., 2017. Effective removal of lead ions using graphene oxide-MgO nanohybrid from aqueous solution : isotherm, kinetic and thermodynamic modeling of adsorption. J. Environ. Chem. Eng. 5, 2259–2273.

Morelos-Gomez, A., Cruz-Silva, R., Muramatsu, H., Ortiz-Medina, J., Araki, T., Fukuyo, T., et al., 2017. Effective NaCl and dye rejection of hybrid graphene oxide/graphene layered membranes. Nat. Nanotechnol. 12 (11), 1083–1088.

Mu, W., Herrmann, J.-M., Pichat, P., 1989. Room temperature photocatalytic oxidation of liquid cyclohexane into cyclohexanone over neat and modified TiO$_2$. Catal. Lett. 3 (1), 73–84.

Naghizadeh, A., Shahabi, H., Ghasemi, F., Zarei, A., 2016. Synthesis of walnut shell modified with titanium dioxide and zinc oxide nanoparticles for efficient removal of humic acid from aqueous solutions. J. Water Health 14 (6), 989–997.

Ng, L.Y., Mohammad, A.W., Leo, C.P., Hilal, N., 2013. Polymeric membranes incorporated with metal/metal oxide nanoparticles: a comprehensive review. Desalination 308, 15–33.

Ngomsik, A.-F., Bee, A., Talbot, D., Cote, G., 2012. Magnetic solid-liquid extraction of Eu(III), La(III), Ni(II) and Co(II) with maghemite nanoparticles. Sep. Purif. Technol. 86, 1–8.

Nie, C., Ma, L., Xia, Y., He, C., Deng, J., Wang, L., et al., 2015. Novel heparin-mimicking polymer brush grafted carbon nanotube/pes composite membranes for safe and efficient blood purification. J. Memb. Sci. 475, 455–468.

Noy, A., Park, H.G., Fornaseiro, F., Holt, J.K., Grigoropolos, C.P., Bakajin, O., 2007. Review: Nanofluidics in Carbon Nanotubes. Nano Today 2, 22–29.

Pan, B., Lin, D., Mashayekhi, H., Xing, B., 2008. Adsorption and hysteresis of Bisphenol A and 17α-ethinyl estradiol on carbon nanomaterials. Environ. Sci. Technol. 42 (15), 5480–5485.

Paris, J.L., Cabañas, M.V., Manzano, M., Vallet-Regí, M., 2015. Polymer-grafted mesoporous silica nanoparticles as ultrasound-responsive drug carriers. ACS Nano. 9 (11), 11023–11033.

Peng, H., Zhang, X., Wei, Y., Liu, W., Li, S., Yu, G., Fu, X., Cao, T., Deng, X., 2012. Cytotoxicity of silver nanoparticles in human embryonic stem cell-derived fibroblasts and an L-929 cell line. J. Nanomater. 2012, 1–9.

Pereira, F.A.R., Sousa, K.S., Cavalcanti, G.G.R.S., França, D.B., Queiroga, L.N.F., Santos, I.M.G., Fonseca, M.G., Jaber, M., 2017. Green biosorbents based on chitosan-montmorillonite beads for anionic dye removal. Journal of Environmental Chemical Engineering 5, 3309–3318.

Perreault, F., De Faria, A.F., Nejati, S., Elimelech, M., 2015. Antimicrobial properties of graphene oxide nanosheets: why size matters. ACS Nano 9 (7), 7226–7236.

Peyravi, M., Jahanshahi, M., Rahimpour, A., Javadi, A., Hajavi, S., 2014. Novel thin film nanocomposite membranes incorporated with functionalized TiO$_2$ nanoparticles for organic solvent nanofiltration. Chem. Eng. J. 241, 155–166.

Phong, N.T.P., Thanh, N.V.K., Phuong, P.H., 2009. Fabrication of antibacterial water filter by coating silver nanoparticles on flexible polyurethane foams. J. Phys. 187, 1–8.

Pradeep, T., Anshup, 2009. Noble metal nanoparticles for water purification: a critical review. Thin. Solid. Films. 517 (24), 6441–6478.

Prince, J.A., Anbharasi, V., Shanmugasundaram, T.S., Singh, G., 2013. Preparation and characterization of novel triple layer hydrophilic−hydrophobic composite membrane for desalination using air gap membrane distillation. Sep. Purif. Technol. 118, 598−603.

Qtaishat, M., Rana, D., Khayet, M., Matsuura, T., 2009. Effect of surface modifying macromolecules stoichiometric ratio on composite hydrophobic/hydrophilic membranes characteristics and performance in direct contact membrane distillation. AIChE J. 55, 3145−3151.

Qu, X., Alvarez, P.J.J., Li, Q., 2013. Applications of nanotechnology in water and wastewater treatment. Water Res. 47 (12), 3931−3946.

Rafiq, Z., Nazir, R., Shahwar, D., Shah, M.R., Ali, S., 2014. Utilization of magnesium and zinc oxide nano-adsorbents as potential materials for treatment of copper electroplating industry wastewater. J. Environ. Chem. Eng. 2, 642−651.

Rajaeian, B., Heitza, A., Tade, M.O., Liu, S., 2015. Improved separation and antifouling performance of PVA thin film nanocomposite membranes incorporated with carboxylated TiO_2 nanoparticles. J. Memb. Sci. 485, 48−59.

Rawal, S.B., Bera, S., Lee, D., Jang, D.-J., Lee, W.I., 2013. Design of visible-light photocatalysts by coupling of narrow bandgap semiconductors and TiO_2: effect of their relative energy band positions on the photocatalytic efficiency. Catal. Sci. Technol. 3 (7), 1822−1830.

Ren, G., Hu, D., Cheng, E.W.C., Vargas-Reus, M.A., Reip, P., Allaker, R.P., 2009. Characterisation of copper oxide nanoparticles for antimicrobial applications. Int. J. Antimicrob. Agents 33, 587−590.

Reynolds, D.C., Look, D.C., Jogai, B., Litton, C.W., Cantwell, G., Harsch, W.C., 1999. Valence-band ordering in ZnO. Phys. Rev. B 60 (4), 2340−2344.

Robati, D., Mirza, B., Ghazisaeidi, R., Rajabi, M., Moradi, O., Tyagi, I., et al., 2016. Adsorption behavior of methylene blue dye on nanocomposite multi-walled carbon nanotube functionalized thiol (MWCNT-SH) as new adsorbent. J. Mol. Liq. 216, 830−835.

Sabri, N.S.M., Azlan, M.Q.A.Z.Z., Hasbullah, H., Said, N., Lau, W.J., Ibrahim, N., et al., 2017. Effect of copper oxide nanoparticles loading on polysulfone ultrafiltration membrane. J. Eng. Appl. Sci. 12 (18), 4745−4751.

Sadeghi-Kiakhani, M., Mokhtar, A.M., Gharanjig, K., 2013. Dye removal from colored-textile wastewater using chitosan-PPI dendrimer hybrid as a biopolymer: optimization, kinetic, and isotherm studies. J. Appl. Polym. Sci. 127, 2607−2619.

Saini, J., Garg, V.K., Gupta, R.K., 2018. Removal of methylene blue from aqueous solution by Fe_3O_4@Ag/SiO_2 nanospheres: synthesis, characterization and adsorption performance. J. Mol. Liq. 250, 413−422.

Sajanlal, P.R., Pradeep, T., 2008. Electric-field-assisted growth of highly uniform and oriented gold nanotriangles on conducting glass substrates. Adv. Mater. 20 (5), 980−991.

Salah, T.A., Mohammad, A.M., Hassan, M.A., El-Anadouli, B.E., 2014. Development of nano-hydroxyapatite/chitosan composite for cadmium ions removal in wastewater treatment. Journal of the Taiwan Institute of Chemical Engineers 45, 1571−1577.

Sandmann, A., Kompch, A., Mackert, V., Liebscher, C.H., Winterer, M., 2015. Interaction of L-cysteine with ZnO: structure, surface chemistry and optical properties. Langmuir 31 (21), 5701−5711.

Sarkar, S., Chatterjee, P.K., Cumbal, L.H., SenGupta, A.K., 2011a. Hybrid ion exchanger supported nanocomposites: sorption and sensing for environmental applications. Chem. Eng. J. 166 (3), 923−931.

Sarkar, S., Prakash, P., SenGupta, A.K., 2011b. Polymeric/inorganic hybrid ion exchanger: preparation, characterization and environmental application. In: SenGupta, A.K. (Ed.), Ion Exchange and Solvent Extraction: A Series of Advances, 2. CRC Press, pp. 292−342.

Sarkar, S., Gulbal, E., Quignard, F., 2012. Polymer-supported metals and metal oxide nanoparticles: synthesis, characterization, and applications. J. Nanopart. Res. 14 (715), 1−24.

Schmid, G., 1992. Large clusters and colloids-metals in the embryonic state. Chem. Rev. 92 (8), 1709−1727.

Schmidt, M.L., MacManus, D.J.L., 2007. ZnO-nanostructures, defects, and devices. Mater. Today 10, 40−48.

Shukla, S.K., Srivastava, K., Srivastava, D., 2015. Studies on the thermal, mechanical and chemical resistance properties of natural resource derived polymers. Mater. Res. 18 (6), 1217−1223.

Sigdel, A., Park, J., Kwak, H., Park, P., 2016. Arsenic removal from aqueous solutions by adsorption onto hydrous iron oxide-impregnated alginate beads. J. Ind. Eng. Chem. 35, 277−286.

Sophia, A.C., Lima, E.C., Allaudeen, N., Rajan, S., 2016. Application of graphene based materials for adsorption of pharmaceutical traces from water and wastewater- a review. Desalin. Water Treat. 57, 27573−27586.

Soyekwo, F., Zhang, Q., Zhen, L., Ning, L., Zhu, A., Liu, Q., 2017. Borate crosslinking of polydopamine grafted carbon nanotubes membranes for protein separation. Chem. Eng. J. 337, 110−121.

Suresh, A.K., Pelletier, D., Wang, W., Morrell-Falvey, J.L., Gu, B., Doktycz, M.J., 2012. Cytotoxicity induced by engineered silver nanocrystallites is dependent on surface coatings and cell types. Langmuir 28, 2727−2735.

Sureshkumar, V., Kiruba Daniel, S.C.G., Ruckmani, K., Sivakumar, M., 2016. Fabrication of chitosan−magnetite nanocomposite strip for chromium removal. Appl. Nanosci 6, 277−285. Available from: https://doi.org/10.1007/s13204-015-0429-3.

Sutisna, Wibowo, E., Rokhmat, M., Rahman, O.Y., Murniati, R., Khairurrijal, Abdullah, M., 2017. Batik wastewater treatment using TiO_2 nanoparticles coated on the surface of plastic sheet. Procedia Eng. 170, 78−83.

Tan, L., Xu, J., Xue, X., Lou, Z., Zhu, J., Baig, S.A., et al., 2014. Multifunctional nanocomposite $Fe_3O_4@SiO_2$-mPD/SP for selective removal of Pb(II) and Cr(VI) from aqueous solutions. RSC Adv. 4 (86), 45920−45929.

Tan, L.Y., Li, L.D., Peng, Y., Guo, L., 2015. Synthesis of Au@Pt bimetallic nanoparticles with concave Au nanocuboids as seeds and their enhanced electrocatalytic properties in the ethanol oxidation reaction. Nanotechnology 26 (50).

Tang, E., Cheng, G., Ma, X., Pang, X., Zhao, Q., 2006. Surface modification of zinc oxide nanoparticle by PMAA and its dispersion in aqueous system. Appl. Surf. Sci. 252 (14), 5227−5232.

Tasselli, F., Mirmohseni, A., Seyed Dorraji, M.S., Figoli, A., 2013. Mechanical, swelling and adsorptive properties of dry−wet spun chitosan hollow fibers crosslinked with glutaraldehyde. React. Funct. Polym. 73 (1), 218−223.

Thekkudan, V.N., Vaidyanathan, V.K., Ponnusamy, S.K., Charles, C., Sundar, S., Vishnu, D., et al., 2016. Review on nanoadsorbents : a solution for heavy metal removal from wastewater. J. IET Nanobiotechnol. 1−12.

Tosco, T., Petrangeli-papini, M., Cruz Viggi, C., Sethi, R., 2014. Nanoscale zerovalent iron particles for groundwater remediation: a review. J. Cleaner Prod. 77, 10−21.

Tran, Q.H., Nguyen, V.Q., Le, A.T., 2013. Silver nanoparticles: synthesis, properties, toxicology, applications and perspectives. Adv. Nat. Sci. 4 (3).

Tratnyek, P.G., Salter, A.J., Nurmi, J.T., Sarathy, V., 2010. Environmental applications of zerovalent metals: iron vs. zinc,". Nanoscale Mater. Chem. 1045, 165−178.

Tsai, H., Wang, Y., 2007. Properties of hydrophilic chitosan network membranes by introducing binary cross-link agents. Polym. Bull. 60 (1), 103−113.

Vargas-Bernal, R., 2017. Advances in functional nanocoatings applied in the aerospece industry. In: Hersey, P.A. (Ed.), Materials Science and Engineering: Concepts, Methodologies, Tools and Applications. IGI Global, USA.

Vatanpour, V., Zoqi, N., 2017. Surface modification of commercial seawater reverse osmosis membranes by grafting of hydrophilic monomer blended with carboxylated multiwalled carbon nanotubes. Appl. Surf. Sci. 396, 1478−1489.

Verma, M., Tyagi, I., Chandra, R., Gupta, V.K., 2017. Adsorptive removal of Pb (II) ions from aqueous solution using CuO nanoparticles synthesized by sputtering method. J. Mol. Liq. 225, 936−944.

Wang, X., Song, X., Lin, M., Wang, H., Zhao, Y., Zhong, Y., Du, Q., 2007. Surface initiated graft polymerization from carbon-doped TiO2 nanoparticles under sunlight illumination. Polymer 48 (20), 5834−5838.

Wang, Y., Xia, G., Wu, C., Sun, J., Song, R., Huang, W., 2015. Porous chitosan doped with graphene oxide as highly effective adsorbent for methyl orange and amido black 10B. Carbohydrate Polymers 115, 686−693.

Wang, Z., Yu, X., Pan, B., Xing, B., 2010b. Norfloxacin sorption and its thermodynamics on surface-modified carbon nanotubes. Environ. Sci. Technol. 44 (3), 978−984.

Warner, C.L., Addleman, R.S., Cinson, A.D., Timothy, C., Droubay, T.C., Engelhard, M.H., et al., 2010. High-performance, superparamagnetic, nanoparticle-based heavy metal sorbents for removal of contaminants from natural waters. Chem. Sustain. Energy Mater. 3 (6), 749−757.

Wegmann, M., Michen, B., Graule, T., 2008. Nanostructured surface modification of microporous ceramics for efficient virus filtration. J. Eur. Ceram. Soc. 28, 1603−1612.

Xiu, Z.M., Ma, J., Alvarez, P.J.J., 2011. Differential effect of common ligands and molecular oxygen on antimicrobial activity of silver nanoparticles versus silver ions. Environ. Sci. Technol. 45 (20), 9003−9008.

Xu, J., Cao, Z., Zhang, Y., Yuan, Z., Lou, Z., Xu, X., et al., 2018. A review of functionalized carbon nanotubes and graphene for heavy metal adsorption from water: preparation, application, and mechanism. Chemosphere 195, 351−364.

Xu, P., Zeng, G.M., Huang, D.L., Feng, C.L., Hu, S., Zhao, M.H., et al., 2012. Use of iron oxide nanomaterials in wastewater treatment: a review. Sci. Total Environ. 424, 1−10.

Yagub, M.T., Sen, T.K., Afroze, A., Ang, H.M., 2014. Dye and its removal from aqueous solution by adsorption: a review. Adv. Colloid. Interface. Sci. 209, 172−184.

Yang, K., Zhu, L., Xing, B., 2006. Adsorption of polycyclic aromatic hydrocarbons by carbon nanomaterials. Environ. Sci. Technol. 40 (6), 1855−1861.

Yang, J., Hu, Y., Wang, R., Xie, D., 2017. Nanoparticle encapsulation in vesicles formed by amphiphilic diblock copolymers. Soft Matter 13 (43), 7840−7847.

Yang, X., Gondikas, A.P., Marinakos, S.M., Auffan, M., Liu, J., Hsu-Kim, H., et al., 2012. Mechanism of silver nanoparticle toxicity is dependent on dissolved silver and surface coating in caenorhabditis elegans. Environ. Sci. Technol. 46, 1119−1127.

Zeng, Z., Yu, D., He, Z., Liu, J., Xiao, F., Zhang, Y., Wang, R., Bhattacharyya, D., Tan, T.T.Y., 2016. Graphene oxide quantum dots covalently functionalized PVDF membrane with significantly-enhanced bactericidal and antibiofouling performances. Sci. Rep. 6, 20142.

Zhang, C., Sui, J., Li, J., Tang, Y., Cai, W., 2012. Efficient removal of heavy metal ions by thiol-functionalized superparamagnetic carbon nanotubes. Chem. Eng. J. 210, 45−52.

Zhang, C., Chen, L., Wang, T., Su, C., Jin, Y., 2014a. Synthesis and properties of a magnetic core-shell composite nano-adsorbent for fluoride removal from drinking water. Appl. Surf. Sci. 317, 552−559.

Zhang, C., Zhang, Y., Du, X., Chen, Y., Dong, W., Han, B., et al., 2016b. Facile fabrication of Pt-Ag bimetallic nanoparticles decorated reduced graphene oxide for highly sensitive non-enzymatic hydrogen peroxide sensing. Talanta 159, 280−286.

Zhang, X., Niu, H., Yan, J., Cai, Y., 2011. Immobilizing silver nanoparticles onto the surface of magnetic silica composite to prepare magnetic disinfectant with enhanced stability and antibacterial activity. Colloids Surf. A 375 (1-3), 186−192.

Zhang, Y., Wu, B., Xu, H., Liu, H., Wang, M., He, Y., et al., 2016a. Nanomaterials-enabled water and wastewater treatment. Nano Impact 3−4, 22−39.

Zhang, Y.J., Chi, H.J., Zhang, W.H., Sun, Y.Y., Liang, Q., Gu, Y., et al., 2014b. Highly efficient adsorption of copper ions by a PVP-reduced graphene oxide based on a new adsorptions mechanism. Nano-Micro Lett. 6 (1), 80−87.

Zhao, Y., Xu, Z., Shan, M., Min, C., Zhou, B., Li, Y., et al., 2013. Effect of graphite oxide and multiwalled carbon nanotubes on the microstructure and performance of PVDF membranes. Sep. Purif. Technol. 103, 78−83.

Zodrow, K., Brunet, L., Mahendra, S., Li, D., Zhang, A., Li, Q., et al., 2009. Polysulfone ultrafiltration membranes impregnated with silver nanoparticles show improved biofouling resistance and virus removal. Water Res. 43, 715−723.

Further Reading

Abbasi, M., 2017. Synthesis and characterization of magnetic nanocomposite of chitosan/SiO2/carbon nanotubes and its application for dyes removal. Journal of Cleaner Production 145, 105−113.

Akhavan, O., Ghaderi, E., Esfandiar, A., 2011. Wrapping bacteria by graphene nanosheets for isolation from environment, reactivation by sonication, and inactivation by near- infrared irradiation. J. Phys. Chem. B 115, 6279−6288.

Aliabadi, M.M., Irani, M., Ismaeili, J., Piri, H., Parnian, M.J., 2013. Electrospun nanofiber membrane of PEO/Chitosan for the adsorption of nickel, cadmium, lead and copper ions from aqueous solution. Chemical Engineering Journal 220, 237−243.

Amit, B., Kumar, E., Sillanpää, M., 2011. Fluoride removal from water by adsorption - a review. Chem. Eng. J. 171, 811−840.

Cao, X.C., Shi, J., Ma, X.H., Ren, Z.J., 2006. Effect of TiO$_2$ nanoparticle size on the performance of PVDF membrane. Appl. Surf. Sci. 253 (4), 2003−2010.

Chen, A., Zeng, G., Chen, G., Hu, X., Yan, M., Guan, S., Shang, C., Lu, L., Zou, Z., Xie, G., 2012. Novel thiourea-modified magnetic ion-imprinted chitosan/TiO2 composite for simultaneous removal of cadmium and 2,4-dichlorophenol. Chemical Engineering Journal 191, 85−94.

Jiang, Y., Gong, J.-L., Zeng, G.-M., Ou, X.-M., Chang, Y.-N., Deng, C.-H., Zhang, J., Liu, H.-Y., Huang, S.-Y., 2016. Magnetic chitosan−graphene oxide composite for anti-microbial and dye removal applications. International Journal of Biological Macromolecules 82, 702−710.

Souza, V.C., Quadri, M.G.N., 2013. Organic-inorganic hybrid membranes in separation processes: a 10-year review. Braz. J. Chem. Eng. 30 (4), 683−700.

17

Nanotechnology Based Solutions for Wastewater Treatment

Arabinda Baruah[1], Vandna Chaudhary[2], Ritu Malik[3], Vijay K. Tomer[4]

[1]INDIAN INSTITUTE OF SCIENCE EDUCATION AND RESEARCH, MOHALI, PUNJAB, INDIA [2]CENTER OF EXCELLENCE FOR ENERGY AND ENVIRONMENT STUDIES, D.C.R. UNIVERSITY OF SCIENCE & TECHNOLOGY, SONEPAT, HARYANA, INDIA [3]SYNTHESIS AND REAL STRUCTURE GROUP, TECHNICAL FACULTY, INSTITUTE OF MATERIALS SCIENCE, UNIVERSITY OF KIEL, KIEL, GERMANY [4]BERKELEY SENSOR & ACTUATOR CENTER (BSAC), UNIVERSITY OF CALIFORNIA, BERKELEY, UNITED STATES

17.1 Introduction

Clean water is essential for the survival and growth of mankind. It is, in fact, a basic necessity for all living beings. But, due to extensive water pollution, availability of safe drinking water has become an issue of global concern in the present times. The United Nations has stated that contaminated water is killing more people annually, than the total number of people dying of all sorts of violence combined, including war (United Nations News Centre, 2010). In its World Water Development Report (2016), the United Nations says that, currently more than 8% of the total world population lives in water-scarce areas; but, by 2025, the world population is expected to reach nearly 8.1 billion and as high as 38% of it will suffer from fresh water scarcity (Jennifer, 2005). Regarding this impending global water crisis, *Nature* has highlighted in its web focus as: "more than one billion people in the world lack access to clean water, and things are getting worse. Over the next two decades, the average supply of water per person will drop by a third, possibly condemning millions of people to an avoidable premature death" (Nature publishing group, web focus, 2018).

Purified water is essential not only for human consumption, but also for a number of different purposes. It is a critical feedstock for medical, pharmacological, food processing and chemical industries. Commonly used methods for water purification can be classified into two broad categories; physical methods and chemical methods. Physical methods include, boiling, filtration, sedimentation, distillation, desalination, reverse osmosis and irradiation with ultraviolet light. Among chemical methods, coagulation, flocculation, and chlorination are most commonly used. Photocatalytic degradation of dissolved water pollutants under irradiation is a kind of physico-chemical process, which has also been widely investigated over the past decade for wastewater treatment. The traditional approaches for purifying

Nanotechnology in Water and Wastewater Treatment. DOI: https://doi.org/10.1016/B978-0-12-813902-8.00017-4

water such as, sand filtration, sedimentation, flocculation, coagulation, chlorination and adsorption on activated carbon etc., are not much efficient in removing dissolved organic compounds and toxic metal ions. Irradiation with ultraviolet light, ozone treatment and incineration are various alternative methods for the removal of pollutants present in water, though they are not cost-effective for elimination of trace contaminants.

Revolutionary progress in the area of nanoscience and nanotechnology over the last couple of decades has motivated the scientific community to explore this highly potential area of research, where, the unique and advantageous properties of novel nanostructured materials could be exploited to deliver more efficient and sustainable solutions to the prevailing water related problems (Lu et al., 2016). Nanomaterials have typical dimensions between 1 and 100 nm. Because of their smaller size, they contain quite lesser number of atoms, which in turn, give rise to very different properties in them as compared to the bulk materials. Owing to their small size, nanomaterials possess very high surface area to volume ratio, which leads to more surface dependent properties. This has been found that, due to their advantageous physicochemical properties, nanomaterials are excellent adsorbents for a variety of water pollutants (Brumfiel, 2003; Gupta et al., 2015; Das et al., 2014; Urban et al., 2010; Theron et al., 2008).

Several types of nanomaterials that could be used in water remediation are reported in literature, namely, metal nanoparticles, polymer nanoparticles, zeolites, carbon based nanomaterials, self-assembled monolayer on mesoporous supports (SAMMS), biopolymers, iron nanoparticles and nanoscale semiconductor photocatalysts (Baruah et al., 2013; Baruah et al., 2015; Baruah et al., 2017; Sharma et al., 2014; Ganguli et al., 2014; Kumar et al., 2014; Kumar et al., 2013; Kumar et al., 2013; Kumar et al., 2014; Kumar et al., 2014) etc. Several different nanotechnology based pathways for wastewater treatment have been developed so far. Among them, most commonly used techniques can be classified into following broad categories:

I. Adsorption based technology
II. Membrane based technology
III. Anti-microbial nanomaterials based technology
IV. Photocatalytic water treatment technology
V. Sensing and monitoring technology

In the following sections of the chapter, we have illustrated the various aspects of these methods of water treatment. We have also provided a brief discussion on the current challenges in the path of their practical usage and potential hazards of using these nanomaterials based technologies.

17.2 Adsorption Based Techniques for Wastewater Treatment

Adsorption is the process of binding ions, atoms or molecules of a liquid, gas or a dissolved solid to the surface of a solid material. It is a surface phenomenon. Like surface tension, it is also an outcome of surface energy. Inside a bulk material, the atoms are always surrounded by other atoms without leaving any more bonding possibility. But the atoms present on the surface of a material are not entirely surrounded by other atoms and therefore they possess potential sites for binding other ions, atoms or molecules. During adsorption, the adsorbate gets adsorbed on the surface of the adsorbent. However, the type of bonding between the adsorbent and the adsorbate may vary depending on the kind of species involved. Adsorption is divided into the two sub-categories, namely, physical adsorption (physisorption) and chemical adsorption (chemisorption).

Physisorption is possible to all adsorbate-adsorbent systems if the conditions of pressure and temperature are appropriate, while, chemisorption may only happen if the system is capable of making a chemical bond. Physisorption is frequently observed in case of high surface area porous materials like activated carbon, zeolites and porous silica nanoparticles, whereas, chemisorption is predominant in case of surface functionalized nanoparticles. In order to develop suitable adsorbents for the removal of various pollutants from water, one must consider the following points: (1) higher surface area and porous structure, (2) appropriate surface functionalization for binding toxic metal ions and organic pollutants from water, (3) stability and recyclability of the material.

Nanomaterials provide abundant scopes for surface modification. Functionalization with organic ligands enables the nanomaterials to specifically bind metal ions or other contaminants from water. A large number of nanostructured materials with surface modification have been reported in the literature for application in wastewater purification via adsorption of pollutants. Surface modified carbon nanotubes, graphene oxide, graphene-Fe_3O_4 nanocomposites, Fe_3O_4/C core−shell nanoparticles, amino and hydroxyl functionalized silica gel etc., have been tested for the removal of metal ions and dye molecules from water (Zhang et al., 2013; Zhang et al., 2004; Wang et al., 2013; Hu et al., 2010; Zhao et al., 2011; Zhao et al., 2011). A large number of silica based adsorbents, such as, amino-functionalized SBA-15, magnetic mesoporous silica, multiamine-grafted mesoporous silica, amino-functionalized silica hollow spheres and silica gel, amino-functionalized mesoporous silica nanoparticles and silica gel functionalized with EDTA and/or DTPA have been reported for efficient adsorption of heavy metal ions from water (Tomer et al., 2016). Magnetic nanoparticles and their composite with high surface area materials are also excellent adsorbents for organic and inorganic water pollutants. They can be easily separated after adsorption using a simple bar magnet, which makes them advantageous over other adsorbents (Gupta et al., 2017; Ambashta and Sillanpaa, 2010).

The most widely applied nanomaterials can be classified into four major groups:

a. Carbon based nano-adsorbents
b. Metal based nano-adsorbents
c. Polymer based nanoadsorbents
d. Zeolites

In the following section, we have provided brief discussions on each of these classes of nano-adsorbents.

17.2.1 Carbon Based Nanoadsorbents

Owing to their low cost, nontoxicity and advantageous physico-chemical properties, this class of nanoadsorbents holds high potential for applications in wastewater treatment (Mauter and Elimelech, 2008). Depending upon the coordination number of carbon and its atomic arrangement in the lattice structure, carbon allotropes can be divided into four different dimensional categories. Carbon dots and fullerenes fall in the group of zero dimensional material, whereas, carbon nanotubes (CNTs) are included in one dimensional materials. Two dimensional allotrope of carbon is graphene, which has a planer structure with all carbon atoms having sp2 hybridization. Such variation in the dimensionalities leads to unique surface chemistry in the carbon based nanomaterials favourable for adsorption of various entities. Apart from these nanostructured materials, three dimensional amorphous carbon or porous carbon is also crucial from the application point of view. Due to its higher surface area, porous structure and surface chemistry, activated charcoal exhibits excellent adsorption capacity towards fine dust, toxic organic compounds and other inorganic pollutants present in water (Tomer et al., 2014).

In scientific literature, there are numerous reports on application of fullerenes, graphene, carbon nanotubes (CNTs) and surface functionalized activated carbon materials for the efficient removal of organic pollutants and various heavy metal ions, like, iron, copper, nickel, chromium, lead, and zinc etc. (Stafiej and Pyrzynska, 2007). In the case of CNTs, high adsorption efficiency has been attributed to the hollow and multilayered structure of these high surface area nanotubes. However, acid or alkali treatment has been found to be very effective in further enhancing the adsorption capacity of the CNTs. In fact, treatment with strong oxidizing agents like, nitric acid, sulphuric acid, and potassium permanganate leads to functionalization of the CNT surface with $-COOH$, -OH and $-C=O$ groups. This enables the nano-adsorbents to readily bind heavy metal ions through strong electrostatic bonding (Fu and Wang, 2011). C-60, which is an icosahedral fullerene, has also been extensively used for the adsorption of hydrophobic organic compounds. Graphene and its composites with a variety of metals and metal oxides are reported to be excellent agents for photocatalytic dye degradation and metal ion adsorption. Abundance of oxygen containing functionalities present on the graphene surface leads to rapid adsorption of the toxic cations from water. Porous activated carbon sheets and fibres, derived from biomass like, plant sources or agricultural wastes, have been found to be highly effective in the treatment of wastewater containing organic dyes, chlorinated toxic compounds as well as hazardous metal ions like, lead, cadmium and mercury etc. They are advantageous over other nano-adsorbents owing to the low cost, outstanding adsorption capacity and easy recyclability (Clark et al., 2000).

17.2.2 Metal Based Nanoadsorbents

Application of metal and metal oxide-based nanoadsorbents for wastewater treatment has attracted much attention in the recent years (Hua et al., 2012). For example, metallic iron and its various oxides, titanium dioxide, magnesium oxide, alumina, and zinc oxide have been widely explored for the purification of wastewater. Owing to their intrinsic affinity towards hazardous metal ions, magnetic behavior and tunable size or morphology, these metal oxides exhibit several advantageous features, like, faster adsorption kinetics, high efficiency and facile recovery after adsorption (Xu et al., 2012).

Pure metallic iron has been demonstrated to detoxify wastewater containing pesticides, polychlorinated biphenyls and other organic compounds (Wang and Zhang, 1997). Oxides, hydroxides and oxy-hydroxides of iron have been reported to be highly effective in the adsorption of different arsenic species from water (Lafferty and Loeppert, 2005; Dixit and Hering, 2003; Andjelkovic et al., 2015). Higher surface area of the nanoscale iron oxide particles and their inherent ability to bind As(III) and As(V) species through electrostatic interaction, make these magnetically separable nano-adsorbents very popular among the researchers working in the area of arsenic removal. In addition to that, they have also been used for the removal of cobalt, chromium, copper, lead, and nickel from water (Kumari et al., 2015; Lee et al., 2017). Surface functionalized iron oxides have been investigated in the literature for the specific binding of chromium from water (Burks et al., 2014). Nanotubes and nanorods of γ-Fe_2O_3 are found to be highly effective in binding iron, nickel, lead, copper, zinc, and cadmium. Hematite nanoparticles deposited on graphene oxide have been successfully used for the specific binding of hazardous Hg(II) ions from water (Gautam et al., 2015; Diagboya et al., 2015).

Apart from all these iron-based materials, nanostructured ZnO, TiO_2, and MgO are also well studied and reported to be highly efficient in removing various toxic elements from water (Hu et al., 2015; Srivastava et al., 2015; Ray and Shipley, 2015). Hydrated amorphous titania is found to be highly efficient in the removal of arsenic, copper, lead, cadmium, manganese, and iron (Hu et al., 2015). Nano-flowers of magnesium oxide having high surface area are reported to be efficient adsorbents for arsenic, lead and cadmium (Srivastava et al., 2015). Zinc oxide and alumina are other low cost nano-adsorbents which have been frequently used for the adsorption and separation of mercury, cadmium and chromium ions from water (Ray and Shipley, 2015; Sankararamakrishnan et al., 2014).

17.2.3 Polymer Based Nanoadsorbents

Over the past decade, polymeric nanoadsorbents have emerged as a new class of material for water treatment (Yin and Deng, 2015). Various organic as well as organic-inorganic hybrid polymers have been developed and used for the adsorptive removal of different types of water pollutants. Metal doped polymeric materials having porous structure, high specific surface area and appropriate functional groups on the surface are reported to be capable of efficiently binding organic dyes as well as heavy metal ions like, lead, cadmium, and zinc from water (Lofrano et al., 2016). A large number of adsorbents for the removal of metal ions and dyes from water have been developed; however, the issues like lower adsorption capacity, lack of specificity and poor recyclability hinder one from commercializing majority of them. These hurdles can be overcome by preparing smart and multifunctional materials like, organic-inorganic hybrid polymers with stronger adsorption capacity, greater thermal stability and higher recyclability (Lofrano et al., 2016).

Various different techniques, like sol-gel, surfactant assisted self assembly formation followed by co-precipitation, forming hierarchical structures and interpenetrating networks have been used for designing such polymeric materials. For example, high surface area nano-adsorbents like, mesoporous silica having organic functional groups on the surface have been developed by sol-gel method and used for efficient removal of different toxic metal ions from water (Min et al., 2014). In a very interesting report by Fei et al. Fei et al. (2012), Fe_3O_4 magnetic nanoparticles modified with 3-aminopropyltriethoxysilane and copolymers of acrylic acid and crotonic acid were used for removing heavy metal ions like, Cd^{2+}, Zn^{2+}, Pb^{2+} and Cu^{2+} from aqueous solution at different pH. They have also studied the metal ion uptake capacity as a function of contact time and metal ion concentration. These magnetic polymer nano-composites exhibited highly efficient adsorption of the metal ions even under acidic pH.

17.2.4 Zeolites

Zeolites are among the most popular natural adsorbents, mainly because of their porous structure which enable them to accommodate a large variety of metal ions. Both, naturally occurring as well as synthetic zeolites have been widely used for the adsorption based wastewater treatment. Naturally occurring zeolites, like, analcime, chabazite, clinoptilolite, heulandite, natrolite, phillipsite, and stilbite, are basically three dimensional microporous alumina silicate structures wherein, small cations such as sodium, potassium, calcium, and magnesium etc., are loosely bound. This facilitates rapid exchange of these cations with other pollutant cations present in water (Dąbrowski et al., 2004).

Synthetic zeolites used for metal ion adsorption are generally functionalized with different organosilane moieties such as, aminopropyl, mercaptopropyl etc. (Dąbrowski et al., 2004). Zeolites functionalized with cationic surfactants have been extensively used for the simultaneous removal of carcinogenic organic pollutants and harmful anions from water (Li and Bowman, 1998). In a few reports, zeolites loaded with silver and lead were used for the removal of pathogenic microbes as well as hazardous anions like, arsenite, arsenate, chromate and cyanide from water via formation of insoluble complexes (Kwakye et al., 2008; Eva et al., 2003). Synthetic zeolites are found to be advantageous over natural zeolites in terms of specificity, thermal stability, structural rigidity, and adsorption efficiency because of their engineered channel networks, controlled pore shape and size as well as calculatingly modified surfaces.

17.3 Membrane Based Techniques for Wastewater Treatment

A membrane is a very thin semipermeable sheet of material that allows selective permeation of entities through its pores based on their size and shape. Now-a-days, in wastewater treatment, membrane based technologies have been finding wide applications. Basically, if a membrane is used for water filtration, it blocks the contaminants (having size greater than its pores) from passing through, thereby giving out cleaner water. Starting from desalination process to antimicrobial filtration, membrane technology has grown very rapidly in the modern times. Highly efficient membranes have been developed for the separation of microorganisms, particulate matter, micropollutants, and other organic materials that impart color, tastes, and odors to the water.

Based on their porosity, composition and mode of application, membranes can be divided into several types, such as nanofiltration membranes, nanocomposite membranes, nanofiber membranes, self-assembling membranes, thin-film composite membranes, aquaporin based membranes, biological membranes, and membranes for forward and reverse osmosis. The following section deals with the preparation of different types of membranes and their application in wastewater treatment.

Nanofiltration is a pressure driven process, where impure water is flown through a thin polymer membrane containing pores of $1-10$ nm so that the contaminants having size bigger than the size of the pores is separated from the water. This is useful for the removal of dissolved solids, polyvalent cations and natural organic matter from water. The most popular use of nano-filtration membrane is for water softening as it blocks the calcium and magnesium ions, while passing only hydrated monovalent ions along with water. In pharmaceutical and fine chemical industries, nanofiltration membranes are being used for solvent recovery. They are also used in petrochemical industries to purify gas condensates. In life sciences, nanofiltration membranes are utilized for the extraction of amino acids and lipids from blood and other cell cultures (Mohammad et al., 2015).

A number of processes for the fabrication and surface modification of filtration membranes have been developed. For example, Yang and co-workers (Xi et al., 2017) have reported the preparation of a thin film composite nanofiltration membrane by depositing an interlayer of polydopamine-polyethylenimine on an ultrafiltration substrate (Fig. 17-1). These membranes show 97% removal of Na_2SO_4 and salt rejection order of $Na_2SO_4 \approx MgSO_4 > MgCl_2 > NaCl$.

For the removal of small ions like sodium, potaasium, and chloride reverse osmosis (RO) membranes are very useful. Therefore, they are finding extensive applications in desalination of sea water. Over the past few decades, a considerable amount of effort has been directed towards the development of cost effective and energy efficient desalination membrane filtration technologies. Majority of the existing commercial RO membranes utilizes composite thin film structure, where, a layer of polyamide is coated on the top of a microporous support membrane (Jiang et al., 2017). The permeation selectivity is efficiently controlled by the polyamide films. However, their performance declines with time due to membrane fouling caused by the accumulation of organic matter and microbes in the pores. Membranes are often deteriorated by chlorine assisted oxidative degradation, too. To overcome these problems, the most common practice is to do the surface modification of the membranes by grafting or coating with appropriate functional groups (Jiang et al., 2017).

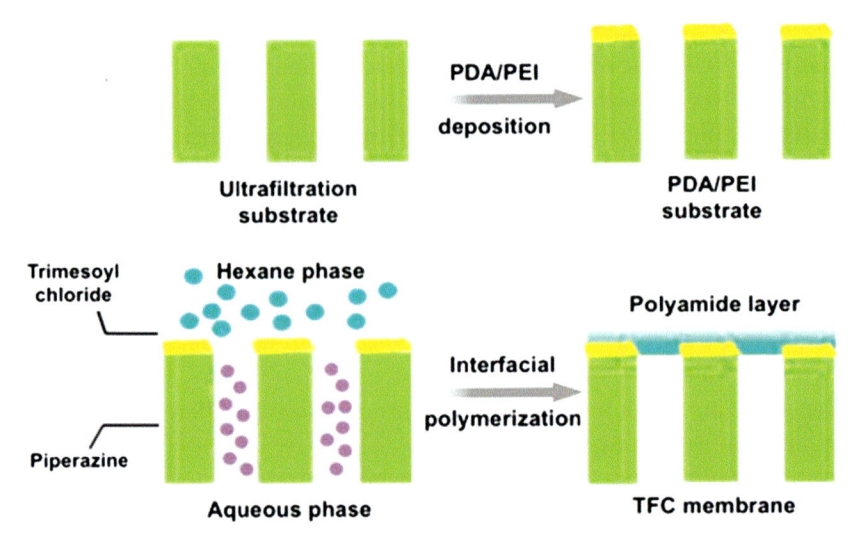

FIGURE 17-1 Schematic illustration of the interfacial polymerization mediated by PDA/PEI mussel-inspired interlayer and the resulting thin film composite nanofiltration membrane polyamide layer. *Adapted with permission from Xi, Y, Du, Y, Zhang, X, He, A, Xu, Z., 2017. Nanofiltration membrane with a mussel-inspired interlayer for improved permeation performance. Langmuir 33,:2318—2324. Copyright 2017 The American Chemical Society.*

People have also reported the incorporation of nanomaterials in the fabrication of novel membrane materials. Choi et al. (2013) used graphene oxide (GO) and showed that its coating on the surface of the membrane served as a dual-functional protective layer to enhance both membrane antifouling and chlorine resistance, while maintaining its separation performance (Fig. 17-2). Use of functionalized carbon nanotubes in composite RO membranes has gained much attention in the recent past. In a very interesting study by Chan et al. (2013) it has been shown that the water flow and the salt rejection ratio can be significantly enhanced by incorporating zwitterion functionalized carbon nanotubes in the polyamide membrane. Kim et al. (2014) have used functionalized carbon nanotubes on polyamide RO membranes to obtain higher water flux. Additionally, the membranes loaded with carbon nanotubes demonstrated higher robustness and greater chemical resistance against sodium chlorides than bare membranes. These advantageous features of the composite membranes are credited to the hydrophobic nano-channels created by the carbon nanotubes and homogeneity in the membrane originating from the uniform interactions between CNTs and polyamide in the active layers (Ma et al., 2017).

Most of the commercially available RO membranes are of mixed-matrix type, wherein, super-hydrophilic nanoparticles are embedded on a polyamide film to repel microbes, dissolved salts and impurities. Nanosilver and TiO_2 coating on composite RO membranes are usually applied to prevent fouling through oxidative degradation of organic contaminants (Mohammad et al., 2017). Recently, membranes modified with biomolecules like proteins have been developed to increase selectivity (Miller et al., 2017). Self-assembling membranes

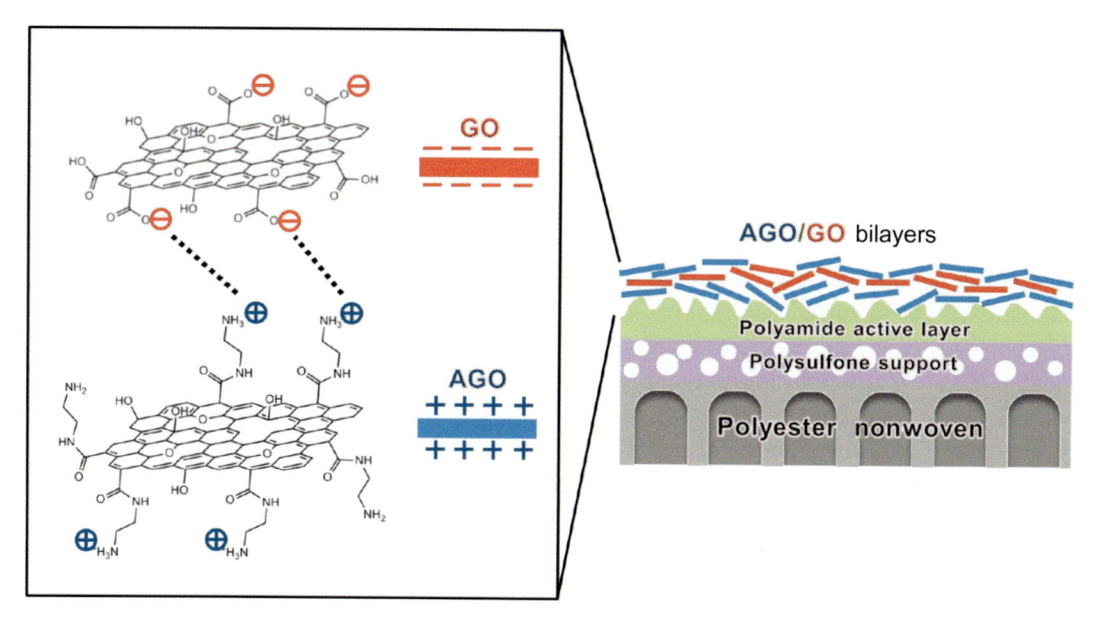

FIGURE 17-2 Schematic illustration of a multilayered graphene oxide (GO) coating on a polyamide thin-film composite membrane surface via layer-by-layer (LbL) deposition of oppositely charged GO and aminated-GO (AGO) nanosheets. *Adapted with permission from Choi, W, Choi, J, Bang, J, Lee, J.H., 2013. Layer-by-layer assembly of graphene oxide nanosheets on polyamide membranes for durable reverse-osmosis applications. ACS Appl. Mater. Interfaces 5, 12510–12519. Copyright 2013 The American Chemical Society.*

from block copolymers have also been investigated for ultra-filtration and found to have higher selectivity and permeation efficiency (Miller et al., 2017).

Another important class of membranes is fabricated using polymeric nanofibers and ceramics obtained via electro-spinning. Interconnected open pore structures over an optimized thickness of polymer membrane impart high water flux in polymer nanofiber based membranes (Gopakumar et al., 2017). Polyurethane, polylactic acid and polyethylene oxide based bio-nanofiber membranes are also developed to obtain higher antifouling property (Rana and Matsuura, 2010). Bio-mimetic membranes are known to have better selectivity and higher water flux. An example of such a membrane is the vesicle incorporated polymer supported aquaporin based nanofiltration membrane, which has been commercialized by Aquaporin Inside (Aquaporin A/S, Copenhagen, Denmark) and it is highly efficient in the desalination of sea water (Tang et al., 2013).

17.4 Nanomaterials for Disinfection and Microbial Control

Water infected with pathogenic microbes is lethal to human health. As per World Health Organization (WHO), annually about 3.4 million people die as a result of various water borne diseases around the world (Ashbolt, 2004). Prolonged use of antibiotics has allowed the microbes to undergo genetic mutation and develop resistance against them. Such evolution of antibiotic resistant microbes is a serious threat to all the living beings. Therefore, as an alternative to the traditional techniques for disinfection, recently, nanotechnology-based solutions have started emerging, utilizing the bio-toxicity of certain nanomaterials (Li et al., 2008). Use of metal nanoparticles for microbial control is advantageous over antibiotics, because, metal nanoparticles can simultaneously proceed through different mechanisms of attack, making it very difficult for the microbes to develop resistance by undergoing several different genetic transformations simultaneously (Rai et al., 2009). So far, a large number of nanostructured materials having anti-microbial property have been synthesized and used for microbial control. Among them silver, TiO_2, ZnO, and carbon based nanomaterials are the most extensively studied nanostructured anti-microbial agents (Malik et al., 2018). Here, we will briefly discuss about the recent nanotechnology based solutions against microbial water pollution and their mechanism of action.

Since long time, silver is known for its antibacterial and sensing property (Prabhu and Poulose, 2012). But, when used in nano form rather than bulk, its activity gets enhanced manifold. In literature, metallic silver nanoparticles, silver oxide, silver phosphate and many silver doped nanomaterials have been explored for their antimicrobial activity (Morones et al., 2005). Silver nanoparticles exhibit strong and broad-spectrum antimicrobial activity. It kills both gram positive and gram negative bacteria. Most importantly, it has very less toxicity towards human cell and it is easy to use. This makes nano-silver the most popular material for wastewater disinfection, pharmaceutical applications, and drinking water treatment (Morones et al., 2005). It has been used in many commercial water purifiers like, Tata Swach, Blue Star, and Aquapure systems, etc.

Carbon nanotubes (CNTs) are another class of nanomaterials possessing very strong activity against microbes and thus, suitable for many anti-microbial applications. Their fibrous structure enables removal of microbes by size exclusion and depth filtration. Electrical conductivity of CNTs allows inactivation of adsorbed microbes in few seconds by applying even a very small voltage of $2-3$ V (Upadhyayula et al., 2009). TiO_2 nanoparticles with different sizes have also been used to kill pathogenic microbes, such as, *Escherichia Coli, Staphylococcus Aureus*, and *Klebsiella Pneumoniae* (Dizaj et al., 2014). Upon illumination with a light source of suitable energy, titanium dioxide nanoparticles can generate reactive oxygen species in cells, thereby, efficiently killing microbes. Their activity is found to be dependent on the size; smaller ones being more effective (Dizaj et al., 2014). Silver doped titania nano-fibers and sol-gel thin films have also been reported to have high anti-bacterial activity against E. Coli (Kudhier et al., 2017; Epifani et al., 2000). Disinfection of water using nanostructured zinc oxide and its composites with silver and TiO_2 have also been investigated by several research groups (Epifani et al., 2000; Padmavathy and Vijayaraghavan, 2008) and found to have prominent antimicrobial activity over a broad spectrum of bacterial species. Since zinc oxide nanoparticles are reported to be non-toxic to human cells, they are finding applications in antibacterial medicines and commercial water purifiers. Several other metal oxide nanoparticles like copper oxide, magnesium oxide, and iron oxide and their composites with silver have also been applied as antibacterial agents (Azam et al., 2012).

Nano-sized particles invariably exhibit greater cytotoxicity than bigger ones, because of their stronger interaction with the cell membrane. They alter the structure and permeability of the cell membrane by forming strong bonds, thereby causing membrane rupture (Karlsson et al., 2013).

17.4.1 Mechanism of Antimicrobial Action of Nanoparticles

It is difficult to propose a general mechanism of antimicrobial action for different types of nanoparticles as it varies with their nature and chemistry. Some of the widely accepted pathways are oxidative stress induction, metal ion release, and nonoxidative mechanisms. Silver nanoparticles are known for their strong antibacterial activity. Their mechanism of action is well investigated in the literature (Karlsson et al., 2013; Marambio and Hoek, 2010). It starts with release of the silver ions which readily bind to the thiol groups of some crucial proteins and cause enzyme damage. Silver also blocks DNA replication process and cleaves the cell membrane. Fig. 17-3 shows the schematic drawing showing the various mechanisms of antibacterial activities exerted by silver nanoparticles. Nanostructured silver, ZnO, TiO$_2$ and C$_{60}$ etc. generate reactive oxygen species (ROS) and H$_2$O$_2$ inside the cell which help in killing the microbe via membrane cleavage and protein damage. Carbon based materials like carbon nanotubes, graphene etc. are quite effective in disinfection and microbial control as they can induce oxidative stress in the cells and kill the microbes by oxidizing essential cellular components (Perreault et al., 2015; Kang et al., 2008). The activity is of CNTs is a function of their physicochemical properties and it has been reported that nanotubes of smaller dimensions are more effective against microbes (Kang et al., 2008).

17.4.2 Potential Applications in Wastewater Treatment

Nanostructured materials having strong antimicrobial activity are highly promising candidates for the treatment of water infected with pathogenic bacteria and viruses. Nano silver, TiO$_2$, and ZnO are being used in water filters and RO membranes to resist bacterial

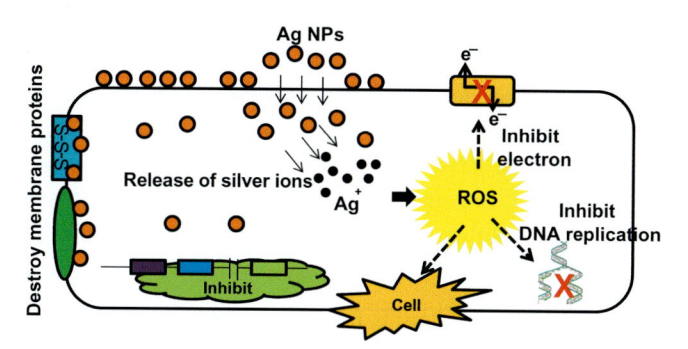

FIGURE 17-3 Schematic drawing showing the various mechanisms of antibacterial activities exerted by silver nanoparticles.

contamination. In addition to that, they have already been used in food packaging, surface coating of medical devices, textiles, and dental implants. Application of nanomaterials for microbial control is advantageous over other chemical treatment processes, because, they are relatively inert in water and produce minimal quantity of toxic degradation by products.

17.5 Heterogeneous Photocatalysis for Wastewater Treatment

Dye manufacturing and textile industries are the chief contributors to the problem of water pollution. It is reported that, textile pigmentation and treatment processes contribute 17%—20% to the global industrial water pollution. Annually $\approx 7 \times 10^5$ tons of dyestuffs are produced worldwide and $>10\%-15\%$ of which used is released into the environment during their synthesis and dyeing (Kant, 2012). The release of such untreated wastes into the water bodies is resulting in the destruction of aquatic life and disturbance in the ecological balance. It also leads to various diseases and health problems in human beings. Degradation of such water pollutants using light absorbing materials under solar radiation is an economic and highly efficient way of wastewater treatment. Through photocatalysis, one can easily get rid of the harmful organic dyes present in wastewater utilizing the solar radiation. In the following section we will discuss about the basics of photocatalysis, properties and types of photocatalyst materials and the mechanism of photocatalytic dye degradation.

17.5.1 What is Photocatalysis?

Photocatalysis is the process of accelerating a light induced reaction in the presence of suitable heterogeneous catalyst. Semiconductor nanoparticles having appropriate band gap are commonly used in photocatalysis as the catalyst material. This technique has been widely used for carrying out two major purposes, photocatalytic water splitting and photocatalytic degradation of organic molecules in water. During photocatalysis, the photocatalyst (present in the reaction medium) is illuminated with UV or visible radiation having energy equivalent or greater than its band gap, which excites the electrons present in the valence band and the excited electrons eventually move to the conduction band, leaving the holes in the valence band. These photo-generated holes and electrons initiate oxidation and reduction reactions respectively.

Photocatalytic degradation of dissolved water pollutants under irradiation is a kind of physico-chemical process which has also been widely investigated for wastewater treatment. The technology used for photocatalytic destruction of toxic components present in water can be briefly described as below:

1. Dispersion of the photocatalyst materials in the contaminated water
2. Irradiation with light source having appropriate wavelength
3. Monitoring of the water samples
4. Separation of the catalyst nanoparticles from the treated water

Photocatalytic dye degradation occurs in presence of visible as well as ultra-violet radiation depending upon the band gap of the catalyst material. For visible light driven photocatalysis people generally use xenon arc lamp as the light source and for UV light active photocatalysts, high pressure mercury lamp is usually used. In order to cut down the IR radiation and to avoid any thermal effects a water filter is used. Illumination is carried out over an external pyrex container. For the evaluation of the photocatalytic efficiency of a catalyst material, generally highly fluorescent organic dyes like, Rhodamine B, Methyl Orange, Congo Red, Methylene Blue, and Melachite Green etc are being used as model water pollutants. In a typical photocatalytic process, the dispersed catalyst nanoparticles absorb the photons and produce charge carriers (photogenerated electron hole pairs). These charge carriers may undergo recombination or they may react with the dye molecules.

17.5.2 Properties of Photocatalyst Materials

While choosing materials for photocatalytic application, one needs to take several points into consideration. Most important of them is the band gap of the material. The radiation source must have energy greater than or equivalent to the energy gap between the conduction and valence band of the semiconductor. The structural and electronic properties of the catalyst also affect the generation of electron-hole pairs and their recombination possibilities. High degree of crystallinity is desirable for a photocatalyst, as the defects present on the material surface act as recombination centres for the photo-generated electron-hole pairs, thereby, reducing the rate of catalysis. Nanoparticles usually possess higher surface area. Additionally, due to their smaller dimensions, the diffusion length for the photo-generated electron-hole pair also reduces which in turn enhances the catalytic efficiency. Therefore, semiconductor nanoparticles have been widely used as efficient photocatalysts for water splitting as well as dye degradation reactions (Malik et al., 2015).

Photocatalytic reactions are initiated by the photo-generated electron-hole pairs. Thus, their mobility and life-time are crucial factors that determine the efficiency of a photocatalytic process. For performing the reactions, excitons need to reach the catalyst surface. If their mobility is high, it becomes easy for them to reach the surface and take part in the photocatalytic degradation process. Lifetime of electron-hole pairs depend on the crystallinity of the photocatalyst material. In case of highly crystalline materials, lifetime and mobility of the photo-generated electron-hole pairs are high due to lesser the number of surface defects (recombination sites for charge carriers) and thus, the photocatalytic activity of such materials is also high.

Separation of photo-generated electrons and holes is very much important during the photocatalytic process, so as to prevent them from recombination. If the charge separation is higher, the photocatalytic activity is also enhanced. Surface properties of the photocatalyst materials greatly influence their photocatalytic degradation efficiency. A material having higher surface area can offer greater number of active sites for adsorption and catalysis. Addition of co-catalysts (like, NiO, Pt, RuO$_2$, etc.) further increases the number of active sites on the surface and therefore increases the photocatalytic activities (Mills et al., 1993; Serpone and Emeline, 2012; Mills and Hunte, 1997).

17.5.3 Mechanism of Photocatalysis

The mechanisms of the photocatalytic dye degradation in presence of semiconductor photo-catalysts have been well investigated in the literature (Malik et al., 2015). In general, it involves following reactions (Eqs. (17.1−17.4)):

$$H^+ + O_2 + 2e^- \rightarrow HO_2^- \tag{17.1}$$

$$\text{Catalyst} \xrightarrow{h\nu} h^+ + e^- \tag{17.2}$$

$$2e^- + O_2 \rightarrow O_2^{\bullet -} \tag{17.3}$$

$$h^+ + OH^- \rightarrow HO^\bullet \tag{17.4}$$

When irradiated with an ultraviolet or visible light source (having energy greater than or equal to the band gap of the photocatalyst), the electrons present in the valence band of the semiconductor get excited and move to the conduction band creating holes in the valence band. These photo-induced electrons then react with dissolved oxygen to produce super oxide radicals ($O_2^{\bullet -}$). These radicals ultimately react with water to produce $^\bullet OH$. On the other hand, photo-generated holes interact with hydroxide ions of water and generate $^\bullet OH$ radicals. Hydroxide radical produced in these two processes eventually attack organic molecules present in water and decompose them. A schematic representation of the photocatalytic dye degradation mechanism is shown in Fig. 17-4. As the formation of $^\bullet OH$ is the most crucial

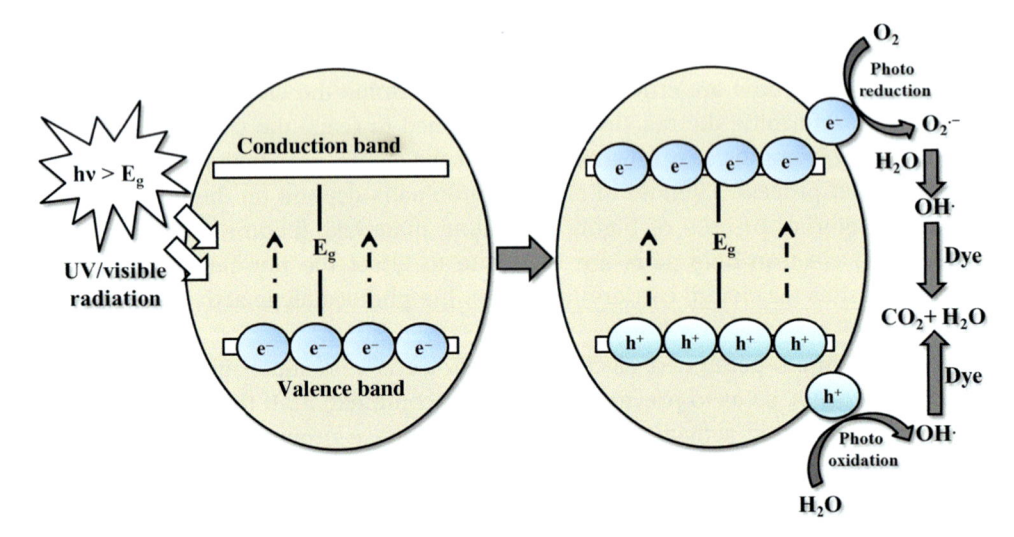

FIGURE 17-4 Schematic representation of the photocatalytic dye degradation process.

step of the photocatalytic degradation reactions, the basic requirement for a good semiconductor photocatalyst is to have the band gap coupled with the redox potential of the reaction of $H_2O/{}^{\bullet}OH$ ($OH^- = {}^{\bullet}OH + e^-$; $E^{\circ} = -2.8$ V).

17.5.4 Effect of Experimental Parameters on Photocatalysis

Experimental conditions, like, intensity of radiation, reaction temperature, humidity, availability of oxygen and the amount of catalyst have considerable effects on the rate of photocatalytic decomposition reactions.

1. **Irradiation intensity:** it has been observed that, in general, the rate of photocatalytic dye degradation reaction increases with the increase in the intensity of irradiation.
2. **Photocatalyst concentration:** the rate of photocatalytic dye degradation is usually greater when the concentration of the catalyst is increased.
3. **Humidity:** amount of water vapour present in the surrounding atmosphere has a critical role to play in determining the rate of photocatalytic dye degradation reaction. The water molecules adsorbed on the surface of the photocatalyst get transformed into hydroxyl radicals (${}^{\bullet}OH$) when react with photo-generated holes (h^+) or superoxide radical anions ($O_2^{-\bullet}$) and eventually take part in the dye degradation process. Thus, as the amount of humidity increases, the rate of the reaction also increases. However, at much higher concentration of water vapor, due to competitive adsorption between water vapour and dye molecules on the active sites, degradation rate may decrease.
4. **Reaction temperature:** the rate of reaction and the adsorption of dye on catalyst surface are strongly influenced by reaction temperature. Increasing temperature generally accelerates the rate of degradation. However, photocatalytic reactions are preferably carried out at a constant temperature in order to minimize the effect of thermal activation.
5. **Oxygen:** as oxygen is an effective electron acceptor, oxygen molecules pre-adsorbed onto the photocatalyst surface can instantaneously trap the photo-generated electrons and eventually suppressing the rate of electron-hole recombination process. This helps to increase the rate of dye degradation. Therefore, with increasing oxygen concentration, the rate of photocatalysis increases.

17.5.5 Different Types of Photocatalysts

Recently, photocatalysis-based technologies have emerged as some of the most promising tools for addressing the global energy and environmental issues (Malik et al., 2017). Based on the phase of the photocatalyst and the medium, photocatalysis can be divided into two broad categories, viz, homogeneous photocatalysis, and heterogeneous photocatalysis.

If the photocatalyst and the reaction medium are in the same phase, it is called as homogeneous photocatalysis. For instance, the photo-assisted degradation of aqueous organic dye using water soluble carbon dots is homogeneous photocatalysis. On the other hand, if the photocatalyst and the reaction medium are not in the same phase, then it is called as heterogeneous photocatalysis. It is the most frequently used technique for the photocatalytic water treatment, as one can easily separate the catalyst material after application. In case of homogeneous photocatalysis, separation of the used catalyst is a difficult task. An example of heterogeneous photocatalysis is the visible light driven degradation of Rhodamine B dye using silver phosphate nanoparticles. Semiconductor based photocatalytic reactions usually come under heterogeneous catalysis, that is, the photocatalyst and targeted molecules remain in different phases. Most commonly, semiconductor materials form the solid phase over which the liquid phase reactants get adsorbed and the degradation reaction takes place.

Several semiconductor nanoparticles including TiO_2, Ta_2O_5, ZnO, Fe_2O_3, CdS, Ag_3PO_4, ZnS, and their composites with other substrates have been reported as sensitizers for light-induced redox reactions (Malik et al., 2016). However, it is evident that the efficiency of a photocatalyst is limited by the rate of recombination of its charge carriers and the amount of light being absorbed. Therefore, in order to circumvent these issues of speedy neutralization of the photo-generated electron hole pairs and insufficient photon absorption, several approaches have been innovated and implemented in the past few decades; and the most extensively used one is to develop suitable heterojunctions of different photo-active materials (Adhyapak et al., 2013). For example, Kumar et al. have prepared a series of semiconductor heterojunctions based on graphitic carbon nitride (g-C_3N_4) to achieve superior photocatalytic dye degradation efficiency (Kumar et al., 2014; Kumar et al., 2013; Kumar et al., 2013; Kumar et al., 2014; Kumar et al., 2014).

It has been reported that N-doped ZnO/g-C_3N_4 hybrid core-shell nano-plates show greatly enhanced visible-light photocatalysis for the degradation of Rhodamine B compared to that of the pure N-doped ZnO and g-C_3N_4. This has been attributed to the possibility that, direct contact of the N-doped ZnO surface and g-C_3N_4 shell without any adhesive interlayer introduced a new carbon energy level in the N-doped ZnO band gap and thereby effectively lowered the band gap energy (Fig. 17-5; Kumar et al., 2014).

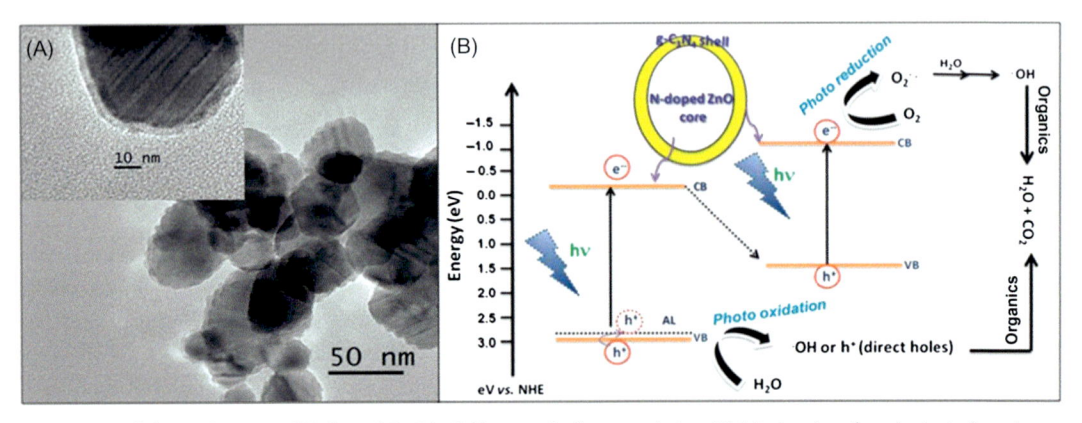

FIGURE 17-5 (A) Tem images of N-doped ZnO/g-C_3N_4 core-shell nano-plates, (B) Mechanism for photo-induced charge carrier transfer in the photocatalytic system (Kumar et al., 2014).

Based on their constituents, the hetero-structured materials can be classified into four broad types, namely, semiconductor-semiconductor, semiconductor-metal, semiconductor-carbon and the multicomponent systems. Here, various aspects of these four junction systems have been discussed with respect to their energy and environmental applications, such as photocatalytic degradation of pollutants, water splitting and photocatalytic extermination of pathogenic microbes.

1. Semiconductor-semiconductor type: this class has been further divided into two sub-groups based on the band alignment of p-type and n-type semiconductors; one is the p-n heterojunction photocatalyst, where p type and n type semiconductors are in direct contact with each other, forming a p-n junction with a space-charge region at the interfaces due to the diffusion of electrons and holes, and thus creating a built-in electrical potential that can direct the electrons and holes to travel in the opposite direction. Consequently, the recombination process is hindered and the efficiency of the photocatalyst is higher. The another type is non p-n heterojunctions, where, two semiconductors with suitable band potentials are tightly bound in such a way that, under irradiation, excited electrons from the conduction band of one semiconductor goes to that of the another, which is having its conduction band at a lower position. Again, the position of the valence bands in them should be aligned in such a way that the photogenerated holes move in a direction opposite to that of the electrons, resulting in efficient charge separation.

Few crucial examples of semiconductor-semiconductor hetero-structures with improved photocatalytic efficiency are: Bi_2WO_6-TiO_2, SnO_2-TiO_2, CdS-TiO_2, Bi_2O_3-Bi_2WO_6, WO_3-$BiVO_4$, C_3N_4-sulfur-modified-C_3N_4, C_3N_4-MoS_2, g-C_3N_4-Ag_3PO_4, g-C_3N_4/N-doped $SrTiO_3$, g-C_3N_4-Fe_3O_4, and N-doped ZnO/g-C_3N_4 (Kumar et al., 2014; Kumar et al., 2013; Kumar et al., 2013; Kumar et al., 2014; Kumar et al., 2014; Xiang et al., 2012; Malik et al., 2016). In addition to efficient charge separation, by creating semiconductor-semiconductor heterojunctions with suitable band gaps, one can also boost the light absorption and promote the surface reaction kinetics; subsequently, enhancing the photocatalytic efficiency. For example, AgX/Ag_3PO_4 (X = Cl, Br, I) core-shell hetero-structures shows enhanced solar absorption as compared to pure Ag_3PO_4. Further, they exhibited much higher photocatalytic activities, structural stabilities and photoelectric conversion performances than the pure Ag_3PO_4 catalyst (Bi et al., 2011).

2. Semiconductor-metal type: this category includes heterostructures designed by combining a semiconductor with a metal. It is well-known that electrons flow from one material to the other (from the higher to the lower Fermi level) at the interface of the two materials to align the Fermi energy levels. For an ideal semiconductor-metal heterostructure, the work function of the metal should be higher than that of the n-type semiconductor (such as TiO_2), so that electrons can flow from the semiconductor into the metal to adjust the Fermi energy levels. The formation of the Schottky barrier results in the fact that the metal has excess negative charges and the semiconductor has excess positive charges. In addition, the Schottky barrier can serve as an efficient electron trap preventing electron-hole recombination in photocatalysis, which often results in an enhanced photocatalytic performance. In the recent years, many such heterojunction photocatalysts have been successfully prepared by different research groups. Some highly interesting examples are Au/TiO$_2$, Ag-AgCl, Ag-ZnO, Ag-Ag$_3$PO$_4$, N-doped carbon-metal, and C$_3$N$_4$-metal heterojunctions (Malik et al., 2016; Malik et al., 2016).

3. Semiconductor-carbon type: here, for the construction of a hetero-junction, different types of carbon materials like activated carbon, CNTs and graphene etc are used along with semiconductors (Qu and Duan, 2013). Usually, the increase of the surface area leads to the improvement of the photocatalytic activity, which can be concluded from the Langmuir-Hinshelwood mechanism. Therefore, it can be expected that combining semiconductors with the activated carbon yields an increase in adsorbed amounts of pollutants and thus enhances their photocatalytic activity. CNTs also provide a larger specific surface area and thus the coupling of semiconductors on it can enhance the photocatalytic degradation efficiency. Furthermore, similar to the metals above, CNTs also exhibit metallic conductivity as one of the possible electronic structures and just like the semiconductor-metal heterojunction system. Semiconductor-CNTs can form a Schottky barrier junction which is an effective method of increasing recombination time. Furthermore, CNTs have a large electron-storage capacity and therefore can accept photon-excited electrons in nanocomposites with semiconductors, thus slowing down the recombination process (Woan et al., 2009).

In literature, plenty of reports are there, where, people have made heterojunctions of graphene with photocatalysts to enhance their photocatalytic performance (Malik et al., 2016). The graphene in the composites can promote charge separation, restrain the hole-electron recombination as well as provide a large surface for heterogeneous reactions at the interface, which results in an enhanced photocatalytic activity. It is worth mentioning that in some cases, the electronic interactions and charge equilibration between graphene and the semiconductor can lead to a shift in the Fermi level and decrease the conduction band potential of the semiconductor. Thus, the negative shift in the Fermi level of semiconductor-graphene and the elevated migration efficiency of photo-induced electrons can restrain the charge recombination efficiently, resulting in the enhanced photocatalytic activity (Malik et al., 2016).

4. Multicomponent systems: in order to address the issue of insufficient light absorption by the photocatalysts, multicomponent heterojunction systems with components having suitable band alignments are fabricated. In such systems, visible-light sensitive materials are spatially integrated with an electron-transfer system. An appropriate example is the preparation of Au-WO3/SBA-15 three component junctions by Malik et al. Malik et al. (2016) via a facile hydrothermal method. This heterojunction demonstrated better photocatalytic performance than individual components of the system. This can be attributed to efficient separation of photo-generated charge carriers across the heterojunction interfaces and improved light absorption due to favorable band alignment of its components.

Another highly fascinating and newly emerged method to achieve enhanced photon absorption and higher rate of photocatalysis is the application of transparent microfluidic flow reactors for the degradation of organic pollutants. Recently, Baruah et al. (2017) have reported a facile surfactant-free microfluidic route for the preparation of ZnO nanostructures with varying morphology (spindles, sheets, and spheres) using polydimethylsiloxane based

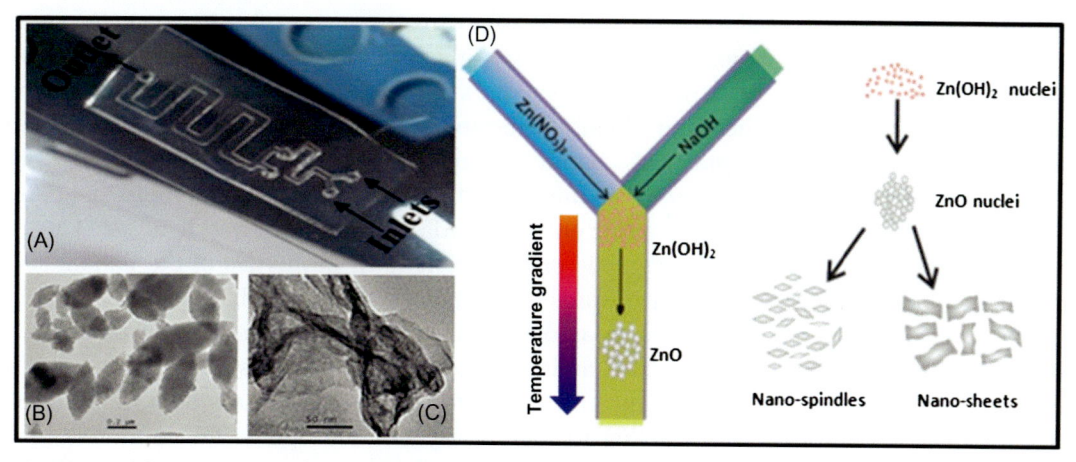

FIGURE 17-6 (A) Digital image of the microreactor used for synthesis and catalysis, TEM images of (B) nano-spindles and (C) nano-sheets, (D) Schematic diagram showing the formation of ZnO nanostructures inside the microchannels (Baruah et al., 2017).

microreactors having different channel geometry (Fig. 17-6). ZnO nano-sheets and spindles, immobilized on the bed of the channels of the microreactor were found to have very high rate of photocatalysis with an overall degradation efficiency of 82% within 30 minutes of irradiation.

17.6 Sensing and Monitoring of Water Pollutants

Nanotechnology based sensors have been widely applied for detecting a variety of different chemical and biological species present in the environment (Tomer et al., 2017; Tomer et al., 2017; Tomer et al., 2016; Tomer and Duhan, 2016; Tomer et al., 2016; Tomer and Duhan, 2016; Tomer et al., 2016; Tomer and Duhan, 2015; Tomer and Duhan, 2015; Tomer et al., 2015; Tomer et al., 2015; Tomer et al., 2015; Tomer et al., 2014; Malik et al., 2018; Malik et al., 2017; Malik et al., 2017; Singh et al., 2017; Chaudhary et al., 2016; Tomer and Duhan, 2015; Tomer et al., 2015; Tomer et al., 2015; Duhan and Tomer, 2014). In wastewater treatment, nanosensors are being frequently used to detect trace pollutants as well as to monitoring the level of common microbial contaminants. Sensing and monitoring of these toxic entities present in water is extremely crucial from the safety point of view. It is a highly challenging and exciting area of research, as it requires highly sensitive technologies to detect the minuscule quantities of inorganic, organic, and biological contaminants present in water.

17.6.1 Pathogen Detection

To ensure proper public health, it is utmost important to detect microbial contaminants present in water bodies. Also, the detection of the contaminating species is a pre-requisite in order to undertake appropriate control measures. Conventional microbial growth indicators are not very efficient and fail to detect some deadly virus, bacteria and protozoan (Gu et al., 2006). Therefore, novel nanomaterial based sensing and monitoring technologies have started evolving over the past decade. Researchers around the world have been putting enormous effort in synthesizing highly efficient pathogen sensing nanostructured materials. In general, nanomaterials based pathogen sensors are functionalized with suitable biomolecules like, antibodies, aptamers, carbohydrates, and antimicrobial peptides, so as to selectively recognize the targeting entities (Doria et al., 2012; Tiwari et al., 2011).

Exploiting the unique optical, magnetic or electrochemical properties of the nanomaterials, people have achieved sensitivity towards microbial cells and other biomolecules present in water. For sensing pathogens, nanomaterials such as quantum dots, carbon nanotubes, magnetic nanoparticles, noble metals and dye-doped nanoparticles have been widely explored (Saha et al., 2012; Vikesland and Wigginton, 2010). Quantum dots hold strong potential for sensing application owing to their easily tunable optical and photo-electrochemical properties. Monitoring the change in the fluorescence spectra or the photo-electrochemical behaviour of the quantum dots upon interaction with pathogens and biomolecules, one can easily detect the level of microbial contaminants present in water. Quantum dots like, tin oxide, cadmium sulphide, cadmium selenide, zinc sulphide, and TiO_2 etc. are commonly used for the detection of various pathogens in wastewater (Hahn et al., 2005; Slowing et al., 2007). In order to enhance sensitivity and selectivity of quantum dot based sensors, proteins, antibodies and other biomolecules have been used to modify their surface (Sapsford et al., 2006).

Gold nanoparticles are being extensively used for sensing applications mainly due to their size and shape dependent surface plasmon resonance phenomenon. For example, cystein functionalized gold nanoparticles are reported to have high sensitivity for *E. Coli* bacteria (Raj et al., 2015). Tb^{3+} and Eu^{3+} chelated Au nanoparticles have been reported for the sensing of dipicolinic acid which is a unique biomarker of bacterial spores (Cable et al., 2007). In another report, Jin and co-workers have employed aptamer functionalized gold nanoparticles in combination with lanthanide-doped up-conversion nanoparticles functionalized with complementary DNA (Fig. 17.5) for the fluorescence resonance energy transfer (FRET) based sensing of E. coli in actual food and water samples (Birui et al., 2017).

Fig. 17-7 shows the schematic illustration of up-conversion nanoparticles based FRET aptasensor for rapid detection of bacteria ((A) The amino-modified complementary DNA of the aptamer is attached to the carboxyl-functionalized UCNPs by condensation reaction. (B) Conjugating thiol-modified aptamers to the AuNPs through Au-S chemistry. (C) The FRET system is established between a donor-acceptor pair: UCNPs-cDNA hybridized with AuNPs-aptamers. (D) By introducing target bacteria into the FRET system, aptamers preferentially bind to target bacteria resulting in the dissociation of cDNA, thereby aptamers-DNA pairs are destroyed and the green fluorescence recovers).

Fluorescent dye functionalized silica nanoparticles have also been used for sensing applications (Slowing et al., 2007). For electrochemical sensing of pathogens, carbon nanotubes provide ample scopes owing to their high conductivity, easy surface modification and excellent adsorption capacity (Guillén et al., 2009). CNT based biosensors are also very popular. The basic structure of any CNT-based biosensors comprises of a biologically sensitive

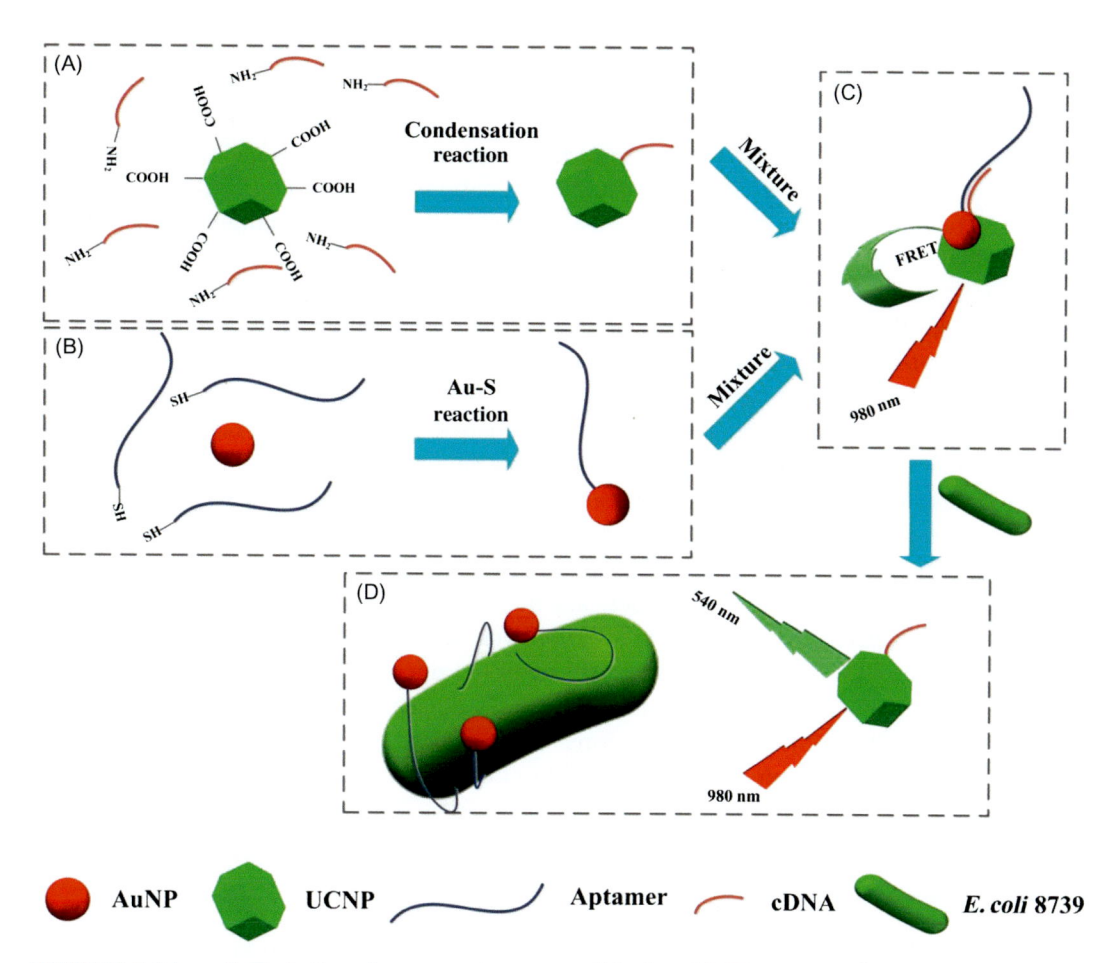

FIGURE 17-7 Schematic illustration of up conversion nanoparticles-based FRET aptasensor for rapid and ultrasensitive bacteria detection. *Adapted with permission from Birui, J, Wang, S, Lin, M, Jin, Y, Zhang, S, Cui, X, et al., 2017. Upconversion nanoparticles based FRET aptasensor for rapid and ultrasenstive bacteria detection. Biosens. Bioelectron. 90, 525−533. Copyright 2017 Elsevier.*

element (such as, proteins, cell receptors, enzymes, antibodies, oligo or polynucleotides, microorganisms, or even whole biological tissues) and a transducer that can convert the analyte concentration to some other measurable physical signals, such as currents, absorbance, mass, or acoustic variables (García et al., 2010). Nanosensors are advantageous due to their photo-stability and high sensitivity, however, specificity is an issue in most of the existing nano-based sensors.

17.6.2 Trace Contamination Detection

For the detection of trace organic or inorganic pollutants like, metal ions, toxic anions, and carcinogenic organic molecules, several nanosensors based on carbon nanotubes, noble metals and quantum dots have been reported in the literature. Preconcentration of the pollutants is the initial step of sensing trace contaminants. This requires faster adsorption kinetics. Carbon based materials like CNTs and graphene are found to be suitable candidates for these applications predominantly because of their higher adsorption capacity and electrical conductivity. Based on the variation in their electrical conductance after binding the trace pollutants, people have used them for sensing a wide variety of contaminants (He et al., 2012; Kauffman and Star, 2008).

Arsenic is one of the most hazardous trace water pollutants. Recently, Vaishanav et al. (2017) have developed an L-cysteine capped CdTe Quantum dot based optical sensor for the fluorometric detection of arsenic (III) in real water sample. The method is based on the fluorescence quenching of QDs with the addition of arsenic solution that caused the reduction in fluorescence intensity due to strong interaction between As^{3+} and L-cysteine to form As $(Cys)_3$. Titania nanotubes functionalized with CdTe quantum dots have been reported by Yang et al. (2010) for the detection of extremely low concentrations of carcinogenic polycyclic aromatic hydrocarbons (PAHs) utilizing the fluorescence resonance energy transfer process (Yang et al., 2010). Gold nanoparticles and clusters have also been widely used to detect various trace contaminants like pesticides, arsenic, mercury, lead, and cadmium with very high sensitivity and selectivity (Serena et al., 2006; Yen and Tseng, 2010). In a very interesting report, Nath et al. (2014) have fabricated a paper-based microfluidic device for arsenic sensing using gold functionalized with biomolecules. This device develops a visible bluish-black precipitate upon interaction with As^{3+} ions on a paper substrate. It is reported to be extremely selective for arsenic with a detection limit of 1.0 ppb.

17.7 Challenges and Future Outlook

Currently, several nanotechnology based products for water purification and pollutant sensing are commercially available. However, it is suspected that they may unveil some adverse effects on human health in the long run, as with the growing number of water purification systems containing nanomaterials, we are increasingly exposing ourselves to the risk of toxicity caused by the leached out nanoparticles. These nanoparticles are finding applications based on their sized, shape, reactivity and antimicrobial activity. But, they may become toxic to human body and harmful for the other living beings due to these very same properties. Nanoparticles can easily enter our body through skin, inhalation and ingestion etc and reach blood stream. Nanoparticles adsorbed onto the human skin are capable of affecting the activity of the enzymes and the skin proteins. Through blood they can move into all the body parts, including brain, kidney, liver, and heart, where they are able to exhibit their high reactivity disturbing the normal functioning of the organs. Nanomaterials such as carbon nanotubes and metal oxides show high reactivity in biological systems by producing various reactive oxygen species (ROS). These ROS are capable of causing oxidative stress and inflammation in the cells (Baun et al., 2008).

Though the fate of these nanoparticles and their response to the human health is still unclear, processes like the treatment of freshwater bodies with antimicrobial nanomaterials or the application of nanoadsorbents to remove organic dyes and metal ions from wastewater are suspected to have undesirable effects on the aquatic environment. Once entered into the food cycle, nanoparticles gradually undergo bio-concentration. Their accumulation in the body is the major cause of their toxicity. Grieger et al. have analysed and discussed the environmental risk of various nanoparticles (Grieger and Baun, 2011). However, the researchers are still not fully aware of the potential health hazards and exposure risk of nanomaterials as we lack sufficient reports on the same. Therefore, prior to building up a large scale nanotechnology based water filtration project, the real challenge before the scientists is to carryout proper assessment of the risk and explore potential hazard management plans. Safe disposal of the wastes of the water treatment process is another challenging task, which has not been systematically addressed in majority of the scientific literature. Most of the nanotechnology based wastewater treatment or drinking water purification processes are expensive and not energy efficient. Commonly used membrane filtration systems for water purification are also not free from these drawbacks. In addition to that, nearly 75% of the water gets wasted during the process of reverse osmosis based filtration. Because of these major challenges, large-scale application of nanotechnology for water purification is getting hindered.

17.7.1 Retention and Regeneration of Used Nanomaterials

From the perspective of cost and public health, it is crucial to design the nanotechnology based devices in such a way that the active nanomaterials should not escape out and must be regenerated through cost and energy efficient methods after a certain period of usage. There are several different techniques available for the regeneration of the used nanomaterials. Application of pH-dependent solvents to extract adsorbed components is one of them. To obtain higher retention values, commonly used practices are to either immobilize the nanomaterials on a solid matrix or to pack them between two porous supports. However, the current immobilization techniques are not very efficient. It is necessary to explore some low-cost yet efficient methods for the immobilization of nanoparticles keeping the efficiency of the nanoadsorbents intact.

From the retention and regeneration point of view, membrane filtration is superior to other techniques of water treatment due to the ease of separation and reactivation. Use of magnetic nanoparticles enables most facile separation of the used adsorbents from the solution phase. After adsorption, using a simple bar magnet it is possible to recover the nanoparticles. In case of metal ion adsorption, washing of the used nanoadsorbents with dilute acids and hot water is sufficient to recover the adsorption sites. For the recovery of nanomaterials after adsorption of organic compounds, people have used various simple organic solvents like acetone and ethyl acetate to extract adsorbed species. With the number of adsorption/desorption cycles, the adsorption capacity of the nanomaterials are found to gradually diminish. This has been attributed to the blocking of some active adsorption sites by the pollutants which did not come off while washing. In some other cases, partial removal of the active binding groups from the surface of the nanoparticles during the regeneration process results in lowering of the efficiency of the materials (Jonathon et al., 2011).

17.8 Summary and Conclusion

In summary, this chapter highlights the various aspects of nanotechnology based wastewater treatment techniques. Though adsorption-based purification of water is an age old traditional technique, but now, the incorporation of high surface area and chemically modified nanoadsorbents has increased the efficiency of this method by manifold. Membrane based filtration of water is an extremely efficient process and currently, it is the most popular technology for domestic as well as industrial applications. However, it is an expensive and energy intensive process. Photocatalytic degradation of organic contaminants present in water using nanoscale catalysts is another very effective technique for wastewater treatment. But it is also not free from several drawbacks that hinder its large scale application. Application of antimicrobial nanomaterials for killing pathogens in water is a highly effective method of disinfection and this has been finding wide applications in the real world.

Nanotechnology not only provides methods for purifying contaminated water, but also offers ample scopes for sensing and monitoring water pollutants. Commercial nano-based sensor kits have been widely used to check the levels of inorganic, organic and microbial contaminants present in water. Although nanomaterials hold immense potential for solving the global water problems, they are equally likely to cause severe environmental toxicity. However, to assess the effect of nanomaterials on health and the environment, methods and tools are not so well developed till date. Therefore, it is utmost important that before bringing into public use, we must analyze the long term effects of these nanomaterials on human as well as environment. Anyhow, it is assured that novel nanomaterials are going to play vital roles in providing safe drinking water in the near future, so that the ever-increasing demand for potable water can be met.

References

Adhyapak, P.V., Meshram, S.P., Tomer, V.K., Amalnerkar, D., Mulla, I.S., 2013. Effect of preparation parameters on the morphologically induced photocatalytic activities of hierarchical zinc oxide nanostructures. Ceram. Int 39, 7367−7378.

Ambashta, R.D., Sillanpaa, M., 2010. Water purification using magnetic assistance: a review. J. Hazard. Mater. 180, 38−49.

Andjelkovic, I., Tran, D.N., Kabiri, S., Azari, S., Markovic, M., Losic, D., 2015. Graphene aerogels decorated with α-FeOOH nanoparticles for efficient adsorption of arsenic from contaminated waters. ACS Appl. Mater. Interfaces 7, 9758−9766.

Ashbolt, N.J., 2004. Microbial contamination of drinking water and disease outcomes in developing regions. Toxicology 198, 229−238.

Azam, A., Ahmed, A.S., Oves, M., Mohammad, S.K., Habib, S.S., Memic, A., 2012. Antimicrobial activity of metal oxide nanoparticles against Gram-positive and Gram-negative bacteria: a comparative study. Int. J. Nanomed. 7, 6003–6009.

Baruah, A., Kumar, S., Vaidya, S., Ganguli, A.K., 2013. Efficient entrapment of dye in hollow silica nanoparticles: direct evidence using fluorescence spectroscopy. J. Fluoresc. 23, 1287–1292.

Baruah, A., Jha, M., Kumar, S., Ganguli, A.K., 2015. Enhancement of photocatalytic efficiency using heterostructured SiO_2-Ta_2O_5 thin films. Mater. Res. Exp. 2, 056404.

Baruah, A., Jindal, A., Acharya, C., Prakash, B., Basu, S., Ganguli, A.K., 2017. Microfluidic reactors for the morphology controlled synthesis and photocatalytic study of ZnO nanostructures. J. Micromech. Microeng. 23, 035013.

Baun, A., Hartmann, N.B., Grieger, K., Ole, K.K., 2008. Ecotoxicity of engineered nanoparticles to aquatic invertebrates: a brief review and recommendations for future toxicity testing. Ecotoxicology 17, 387–395.

Bi, Y., Ouyang, S., Cao, J., Ye, J., 2011. Facile synthesis of rhombic dodecahedral AgX/Ag_3PO_4 (X = Cl, Br, I) heterocrystals with enhanced photocatalytic properties and stabilities. Phys. Chem. Chem. Phys. 21, 10071–10075.

Birui, J., Wang, S., Lin, M., Jin, Y., Zhang, S., Cui, X., et al., 2017. Upconversion nanoparticles based FRET aptasensor for rapid and ultrasenstive bacteria detection. Biosens. Bioelectron. 90, 525–533.

Brumfiel, G., 2003. Nanotechnology: a little knowledge. Nature. 424, 246–248.

Burks, T., Avila, M., Akhtar, F., Göthelid, M., Lansåker, P.C., Toprak, M.S., et al., 2014. Studies on the adsorption of chromium (VI) onto 3-mercaptopropionic acid coated superparamagnetic iron oxide nanoparticles. J. Colloid Interface Sci. 425, 36–43.

Cable, M.L., Kirby, J.P., Sorasaenee, K., Gray, H.B., Ponce, A., 2007. Bacterial spore detection by [Tb^{3+} (macrocycle)(dipicolinate)] luminescence. J. Am. Chem. Soc. 129, 1474–1475.

Chan, W.F., Chen, H., Surapathi, A., Taylor, M.G., Shao, X., Marand, E., et al., 2013. Zwitterion functionalized carbon nanotube/polyamide nanocomposite membranes for water desalination. ACS Nano 7, 5308–5319.

Chaudhary, V., Malik, R., Tomer, V.K., Nehra, S.P., Duhan, S., 2016. Enhanced relative humidity sensing performance using TiO_2 loaded SiO_2 nanocomposite. Energy Environ. Focus 5, 234–239.

Choi, W., Choi, J., Bang, J., Lee, J.H., 2013. Layer-by-layer assembly of graphene oxide nanosheets on polyamide membranes for durable reverse-osmosis applications. ACS Appl. Mater. Interfaces. 5, 12510–12519.

Clark, S., Brown, P., Pitt, R., 2000. Wastewater treatment using low-cost adsorbents and waste materials. Proc. Water Environ. Fed 5, 11–34.

Das, R., Ali, M.E., Hamida, S.B., Ramakrishna, S., Chowdhury, Z.Z., 2014. Carbon nanotube membranes for water purification: a bright future in water desalination. Desalination 336, 97–109.

Dąbrowski, A., Hubicki, Z., Podkościelny, P., Robens, E., 2004. Selective removal of the heavy metal ions from waters and industrial wastewaters by ion-exchange method. Chemosphere 56, 91–106.

Diagboya, P.N., Olu-Owolabi, B.I., Adebowale, K.O., 2015. Synthesis of covalently bonded graphene oxide–iron magnetic nanoparticles and the kinetics of mercury removal. R. Sci. Adv. 5, 2536–2542.

Dixit, S., Hering, J.G., 2003. Comparison of arsenic (V) and arsenic (III) sorption onto iron oxide minerals: implications for arsenic mobility. Environ. Sci. Technol. 37, 4182–4189.

Dizaj, S.M., Lotfipour, F., Barzegar-Jalali, M., Mohammad, H.Z., Adibkia, K., 2014. Antimicrobial activity of the metals and metal oxide nanoparticles. Mater. Sci. Eng. 44, 278–284.

Doria, G., Conde, J., Veigas, B., Giestas, L., Almeida, C., Assunção, M., et al., 2012. Noble metal nanoparticles for biosensing applications. Sensors. 12, 1657–1687.

Duhan, S., Tomer, V.K., 2014. *Mesoporous Silica: Making "Sense" of Sensors*, in Advanced Sensor and Detection Materials. Wiley-Scrivener, USA, pp. 149–192.

Epifani, M., Giannini, C., Tapfer, L., Vasanelli, L., 2000. Sol−gel synthesis and characterization of Ag and Au nanoparticles in SiO_2, TiO_2, and ZrO_2 thin films. J. Am. Ceram. Soc. 83, 2385−2393.

Eva, C., Karol, J., Gáplovská, K., 2003. Arsenate and chromate removal with cationic surfactant-loaded and cation-exchanged clinoptilolite-rich tuff vs montmorillonite. Collection Czechoslovak Chem. Commun. 68, 823−836.

Fei, G., Li, M., Ye, H., Zhao, B., 2012. Effective removal of heavy metal ions Cd^{2+}, Zn^{2+}, Pb^{2+}, Cu^{2+} from aqueous solution by polymer-modified magnetic nanoparticles. J. Hazard. Mater. 211, 366−372.

Fu, F., Wang, Q., 2011. Removal of heavy metal ions from wastewaters: a review. J. Environ. Manage. 92, 407−418.

Ganguli, A.K., Kumar, S., Baruah, A., Vaidya, S., 2014. Nanocrystalline silica from termite mounds. Curr. Sci. 106, 83−88.

García, A.C., Cella, L.N., Shirale, D.J., Park, M., Muñoz, F.J., Yates, M.V., et al., 2010. Carbon nanotubes-based chemiresistive biosensors for detection of microorganisms. Biosens. Bioelectron. 26, 1437−1441.

Gautam, R.K., Soni, S., Chattopadhyaya, M.C., 2015. Functionalized magnetic nanoparticles for environmental remediation. Handbook of research on diverse applications of nanotechnology in biomedicine, chemistry, and engineering. IGI Global 518−551.

Gopakumar, D.A., Pasquini, D., Henrique, M.A., Morais, L.C., Grohens, Y., Thomas, S., 2017. Meldrum's acid modified cellulose nanofiber-based polyvinylidene fluoride microfiltration membrane for dye water treatment and nanoparticle removal. ACS Sustain. Chem. Eng. 5, 2026−2033.

Grieger KD, Baun A., 2011. Understanding and assessing potential environmental risks of nanomaterials: Emerging tools for emerging risks. Technical University of Denmark.

Gu, H., Xu, K., Xu, C., Xu, B., 2006. Biofunctional magnetic nanoparticles for protein separation and pathogen detection. Chem. Commun. 9, 941−949.

Guillén, Z., Gustavo, A., Riu, J., Düzgün, A., Xavier, F.R., 2009. Immediate detection of living bacteria at ultra-low concentrations using a carbon nanotube based potentiometric aptasensor. Angew. Chem. Int. Ed. 48, 7334−7337.

Gupta, N., Pant, P., Gupta, C., Goel, P., Jain, A., Anand, S., et al., 2017. Engineered magnetic nanoparticles as efficient sorbents for wastewater treatment: a review. Mater. Res. Innov. 1−17.

Gupta, V.K., Tyagi, I., Sadegh, H., Shahryari-Ghoshekand, R., Makhlouf, A.S.H., Maazinejad, B., 2015. Nanoparticles as adsorbent: a positive approach for removal of noxious metal ions: a review. Sci. Technol. Dev. 34, 195−214.

Hahn, M.A., Tabb, J.S., Krauss, T.D., 2005. Detection of single bacterial pathogens with semiconductor quantum dots. Anal. Chem. 77, 4861−4869.

Hu, H., Wang, Z., Pan, L., 2010. Synthesis of monodisperse $Fe_3O_4@$ silica core-shell microspheres and their application for removal of heavy metal ions from water. J. Alloys. Comp. 492, 656−661.

Hu, S., Shi, Q., Jing, C., 2015. Groundwater arsenic adsorption on granular TiO_2: integrating atomic structure, filtration, and health impact. Environ. Sci. Technol. 49, 9707-1.

Hua, M., Zhang, S., Pan, B., Zhang, W., Lv, L., Zhang, Q., 2012. Heavy metal removal from water/wastewater by nanosized metal oxides: a review. J. Hazard. Mater. 211, 317−331.

Jennifer, Norman M. (Ed.), 2005. The New Atlas of Planet Management. University of California Press: Nature.

Jiang, S., Li, Y., Bradley, P.L., 2017. A review of reverse osmosis membrane fouling and control strategies. Sci. Total Environ. 595, 567−583.

Jonathon, B., Li, Q., Alvarez, P.J.J., 2011. Nanotechnology-enabled water treatment and reuse: emerging opportunities and challenges for developing countries. Trends Food Sci. Technol. 22, 618−624.

Kang, S., Herzberg, M., Rodrigues, D.F., Elimelech, M., 2008. Antibacterial effects of carbon nanotubes: size does matter. Langmuir. 24, 6409−6413.

Kant, R., 2012. Textile dyeing industry an environmental hazard. Nat. Sci. 4, 22−26.

Karlsson, H.L., Cronholm, P., Hedberg, Y., Tornberg, M., Battice, L.D., Svedhem, S., et al., 2013. Cell membrane damage and protein interaction induced by copper containing nanoparticles-Importance of the metal release process. Toxicology 313, 59−69.

Kauffman, D.R., Star, A., 2008. Carbon nanotube gas and vapor sensors. Angew. Chem. Int. Ed 47, 6550−6570.

Kim, H.J., Choi, K., Baek, Y., Kim, D.G., Shim, J., Yoon, J., et al., 2014. High-performance reverse osmosis CNT/polyamide nanocomposite membrane by controlled interfacial interactions. ACS Appl. Mater. Interfaces. 6, 2819−2829.

Kudhier, M.A., Sabry, R.S., Al-Haidarie, Y.K., Al-Marjani, M.F., 2017. Significantly enhanced antibacterial activity of Ag-doped TiO_2 nanofibers synthesized by electrospinning. Mater. Technol. 1−7.

Kumar, S., Kumar, B., Baruah, A., Shanker, V., 2013. Synthesis of magnetically separable and recyclable g-C_3N_4−Fe_3O_4 hybrid nanocomposites with enhanced photocatalytic performance under visible-light irradiation. J. Phys. Chem. C. 49, 26135−26143.

Kumar, S., Surendar, T., Baruah, A., Shanker, V., 2013. Synthesis of a novel and stable gC_3N_4−Ag_3PO_4 hybrid nanocomposite photocatalyst and study of the photocatalytic activity under visible light irradiation. J. Mater. Chem. A. 17, 5333−5340.

Kumar, S., Surendar, T., Baruah, A., Kumar, B., Shanker, V., 2014. Synthesis of novel and stable g-C_3N_4/N-doped $SrTiO_3$ hybrid nanocomposites with improved photocurrent and photocatalytic activity under visible light irradiation. Dalton Transac. 43, 16105−16114.

Kumar, S., Surendar, T., Kumar, B., Baruah, A., Shanker, V., 2014. Synthesis of highly efficient and recyclable visible-light responsive mesoporous g-C_3N_4 photocatalyst via facile template-free sonochemical route. RSC Adv. 4, 8132−8137.

Kumar, S., Baruah, A., Tonda, S., Kumar, B., Shanker, V., Sreedhar, B., 2014. Cost-effective and eco-friendly synthesis of novel and stable N-doped ZnO/gC_3N_4 core−shell nanoplates with excellent visible-light responsive photocatalysis. Nanoscale 6, 4830−4842.

Kumari, M., Pittman, C.U., Mohan, D., 2015. Heavy metals [chromium (VI) and lead (II)] removal from water using mesoporous magnetite (Fe_3O_4) nanospheres. J. Colloid Interface Sci. 442, 120−132.

Kwakye, A., Bright, C.W., Kenward, M.A., Radecka, I., 2008. Antimicrobial action and efficiency of silver-loaded zeolite X. J. Appl. Microbiol. 104, 1516−1524.

Lafferty, B.J., Loeppert, R.H., 2005. Methyl arsenic adsorption and desorption behavior on iron oxides. Environ. Sci. Technol. 39, 2120−2127.

Lee, C.G., Lee, S., Park, J.A., Park, C., Lee, S.J., Kim, S.B., et al., 2017. Removal of copper, nickel and chromium mixtures from metal plating wastewater by adsorption with modified carbon foam. Chemosphere 166, 203−211.

Li, Q., Mahendra, S., Lyon, D.Y., Brunet, L., Liga, M.V., Li, D., et al., 2008. Antimicrobial nanomaterials for water disinfection and microbial control: potential applications and implications. Water Res. 42, 4591−4602.

Li, Z., Bowman, R.S., 1998. Sorption of chromate and PCE by surfactant-modified clay minerals. Environ. Eng. Sci. 15, 237−245.

Lofrano, G., Maurizio, C., Giovanni, L., Rute, F.D., Arjen, M., Luciana, D., et al., 2016. Polymer functionalized nanocomposites for metals removal from water and wastewater: an overview. Water Res. 92, 22−37.

Lu, H., Wang, J., Stoller, M., Wang, T., Bao, Y., Hao, H., 2016. An overview of nanomaterials for water and wastewater treatment. Adv. Mater. Sci. Eng. 2016, 1−10.

Ma, L., Dong, X., Chen, M., Zhu, L., Wang, C., Yang, F., et al., 2017. Fabrication and water treatment application of carbon nanotubes (CNTs)-based composite membranes: a review. Membranes 7, 16.

Malik, R., Tomer, V.K., Rana, P.S., Nehra, S.P., Duhan, S., 2015. Effect of annealing temperature on the photocatalytic performance of SnO_2 nano-flowers towards degradation of Rhodamine B. Adv. Sci. Eng. Med. 7, 448−456.

Malik, R., Tomer, V.K., Rana, P.S., Nehra, S.P., Duhan, S., 2015. One-pot hydrothermal synthesis of porous SnO_2 nanostructures for photocatalytic degradation or organic pollutants. Energy Environ. Focus 4, 340−345.

Malik, R., Chaudhary, V., Rana, P.S., Tomer, V.K., Nehra, S.P., Duhan, S., 2016. Lanthanide ions doped-SnO_2: a stable and efficient photocatalyst for dye decontamination. Energy Environ. Focus 5, 35−42.

Malik, R., Tomer, V.K., Chaudhary, V., Dahiya, M.S., Rana, P.S., Nehra, S.P., et al., 2016. Facile synthesis of hybridized mesoporous Au@TiO_2/SnO_2 as efficient photocatalyst and selective VOC sensor. Chem. Select 1, 3247−3258.

Malik, R., Tomer, V.K., Rana, P.S., Nehra, S.P., Duhan, S., 2016. Facile preparation of TiO_2/SnO_2 catalysts using TiO_2 as an auxiliary for gas sensing and advanced oxidation processes. MRS Adv. 46, 3157−3162.

Malik, R., Rana, P.S., Tomer, V.K., Chaudhary, V., Nehra, S.P., Duhan, S., 2016. Visible light-driven mesoporous Au-TiO_2/SiO_2 photocatalysts for advanced oxidation process. Ceram. Int. 42, 10892−10901.

Malik, R., Rana, P.S., Tomer, V.K., Chaudhary, V., Nehra, S.P., Duhan, S., 2016. Nano gold supported on ordered mesoporous WO_3/SBA-15 hybrid nanocomposite for oxidative decolorization of Azo dye. Micropor. Mesopor. Mater. 225, 245−254.

Malik, R., Chaudhary, V., Rana, P.S., Tomer, V.K., Nehra, S.P., Duhan, S., 2016. Nanostructured WO_3/SnO_2 and TiO_2/SnO_2 heterojunction with enhanced photocatalytic activity. Energy Enviro. Focus 5, 108−115.

Malik, R., Tomer, V.K., Chaudhary, V., Dahiya, M.S., Sharma, A., Nehra, S.P., et al., 2017. Excellent humidity sensor based on in-SnO_2 loaded mesoporous graphitic carbon nitride. J. Mater. Chem. A 5, 14134−14143.

Malik, R., Tomer, V.K., Chaudhary, V., Dahiya, M.S., Nehra, S.P., Rana, P.S., et al., 2017. Ordered mesoporous In-(TiO_2/WO_3) nanohybrid: an ultrasensitive n-butanol sensor. Sens. Actuat. B 239, 364−373.

Malik, R., Chaudhary, V., Tomer, V.K., Nehra, S.P., 2017. Nanocasted synthesis of Ag/WO_3 nanocomposite with enhanced sensing and photocatalysis applications. Energy Environ. Focus 6, 43−48.

Malik, R., Tomer, V.K., Chaudhary, V., Dahiya, M.S., Nehra, S.P., Duhan, S., et al., 2018. A low temperature, highly sensitive and fast response toluene gas sensor based on In(III)-SnO_2 loaded cubic mesoporous graphitic carbon nitride. Sens. Actuat. B 255, 3564−3575.

Malik, R., Tomer, V.K., Kienle, L., Chaudhary, V., Nehra, S.P., Duhan, S., 2018. Ordered mesoporous Ag-ZnO@g-CN nanohybrid as highly efficient bifunctional sensing material. Adv. Mater. Interfaces 5, 1701357.

Marambio, J.C., Hoek, E.M.V., 2010. A review of the antibacterial effects of silver nanomaterials and potential implications for human health and the environment. J. Nanopart. Res. 12, 1531−1551.

Mauter, M.S., Elimelech, M., 2008. Environmental applications of carbon-based nanomaterials. Environ. Sci. Technol. 42, 5843−5859.

Miller, D.J., Dreyer, D.R., Christopher, W.B., Paul, D.R., Freeman, B.D., 2017. Surface modification of water purification membranes. Angew. Chem. Int. Ed. 56, 4662−4711.

Mills, A., Hunte, S.L., 1997. An overview of semiconductor photocatalysis. J. Photochem. Photobiol. A 108, 1−35.

Mills, A., Davies, R.A., Worsley, D., 1993. Water purification by semiconductor photocatalysis. Chem. Soc. Rev. 22, 417−425.

Min, L., Li, M., Feng, C., Zhang, Q., 2014. Preparation and characterization of multi-carboxyl-functionalized silica gel for removal of Cu (II), Cd (II), Ni (II) and Zn (II) from aqueous solution. Appl. Surf. Sci. 314, 1063–1069.

Mohammad, A.W., Teow, Y.H., Ang, W.L., Chung, Y.T., Oatley-Radcliffe, D.L., Hilal, N., 2015. Nanofiltration membranes review: recent advances and future prospects. Desalination 356, 226–254.

Mohammad, K., Peyravi, M., Jahanshahi, M., 2017. The potential of nanoparticles for upgrading thin film nanocomposite membranes-a review. J. Memb. Sci. Res. 3, 2–12.

Morones, J.R., Elechiguerra, J.L., Camacho, A., Holt, K., Kouri, J.B., Ramírez, J.T., et al., 2005. The bactericidal effect of silver nanoparticles. Nanotechnology 16, 2346.

Nath, P., Ravi, K.A., Chanda, N., 2014. A paper based microfluidic device for the detection of arsenic using a gold nanosensor. RSC Adv. 4, 59558–59561.

Nature publishing group, web focus, 2018. Available from: URL: http://www.nature.com/nature/focus/water/index.html.

Padmavathy, N., Vijayaraghavan, R., 2008. Enhanced bioactivity of ZnO nanoparticles-an antimicrobial study. Sci. Technol. Adv. Mater. 9, 035004.

Perreault, F., Faria, A.F.D., Nejati, S., Elimelech, M., 2015. Antimicrobial properties of graphene oxide nanosheets: why size matters. ACS Nano 9, 7226–7236.

Prabhu, S., Poulose, E.K., 2012. Silver nanoparticles: mechanism of antimicrobial action, synthesis, medical applications, and toxicity effects. Int. Nano Lett. 2, 32.

He, Q., Wu, S., Yin, Z., Zhang, H., 2012. Graphene-based electronic sensors. Chem. Sci. 3, 1764–1772.

Qu, Y., Duan, X., 2013. Progress, challenge and perspective of heterogeneous photocatalysts. Chem. Soc. Rev. 42, 2568–2580.

Rai, M., Yadav, A., Gade, A., 2009. Silver nanoparticles as a new generation of antimicrobials. Biotechnol. Adv. 27, 76–83.

Raj, V., Vijayan, A.N., Joseph, K., 2015. Cysteine capped gold nanoparticles for naked eye detection of *E. coli* bacteria in UTI patients. Sens. Bio-Sens. Res. 5, 33–36.

Rana, D., Matsuura, T., 2010. Surface modifications for antifouling membranes. Chem. Rev. 110, 2448–2471.

Ray, P.Z., Shipley, H.J., 2015. Inorganic nano-adsorbents for the removal of heavy metals and arsenic: a review. RSC Adv. 5, 29885–29907.

Saha, K., Agasti, S.S., Kim, C., Li, X., Vincent, M.R., 2012. Gold nanoparticles in chemical and biological sensing. Chem. Rev. 112, 2739–2779.

Sankararamakrishnan, N., Jaiswal, M., Verma, N., 2014. Composite nanofloral clusters of carbon nanotubes and activated alumina: an efficient sorbent for heavy metal removal. Chem. Eng. J. 235, 1–9.

Sapsford, K.E., Pons, T., Medintz, I.L., Mattoussi, H., 2006. Biosensing with luminescent semiconductor quantum dots. Sensors 6, 925–953.

Serena, L., Palchetti, I., Mascini, M., 2006. Gold-based screen-printed sensor for detection of trace lead. Sens. Actuat. B. 114, 460–465.

Serpone, N., Emeline, A.V., 2012. Semiconductor photocatalysis: past, present, and future outlook. J. Phys. Chem. Lett. 3, 673–677.

Sharma, M., Das, D., Baruah, A., Jain, A., Ganguli, A.K., 2014. Design of porous silica supported tantalum oxide hollow spheres showing enhanced photocatalytic activity. Langmuir 30, 3199–3208.

Singh, H., Tomer, V.K., Jena, N., Bala, I., Sharma, N., Nepak, D., et al., 2017. Truxene based porous, crystalline covalent organic frameworks and it's applications in humidity sensing. J. Mater. Chem. A 5, 21820–21827.

Slowing, I.I., Trewyn, B.G., Giri, S., VSY, Lin, 2007. Mesoporous silica nanoparticles for drug delivery and bio-sensing applications. Adv. Funct. Mater. 17, 1225−1236.

Srivastava, V., Sharma, Y.C., Sillanpää, M., 2015. Green synthesis of magnesium oxide nanoflower and its application for the removal of divalent metallic species from synthetic wastewater. Ceram. Int. 41, 6702−6709.

Stafiej, A., Pyrzynska, K., 2007. Adsorption of heavy metal ions with carbon nanotubes. Sep. Purif. Technol. 58, 49−52.

Tang, C.Y., Zhao, Y., Wang, R., Hélix-Nielsen, C., Fane, A.G., 2013. Desalination by biomimetic aquaporin membranes: review of status and prospects. Desalination 308, 34−40.

Theron, J., Walker, J.A., Cloete, T.E., 2008. Nanotechnology and water treatment: applications and emerging opportunities. Crit. Rev. Microbiol. 34, 43−69.

Tiwari, P.M., Vig, K., Vida, A.D., Singh, S.R., 2011. Functionalized gold nanoparticles and their biomedical applications. Nanomaterials 1, 31−63.

Tomer, V.K., Duhan, S., 2015. In-situ synthesis of SnO_2/SBA-15 hybrid nanocomposite as highly efficient humidity sensor. Sens. Actuat. B 212, 517−525.

Tomer, V.K., Duhan, S., 2016. Ordered mesoporous Ag-doped TiO_2/SnO_2 nanocomposite based highly sensitive and selective VOC sensors. J. Mater. Chem. A 4, 1033−1043.

Tomer, V.K., Duhan, S., 2015. Nano titania loaded mesoporous silica: preparation and application as high performance humidity sensor. Sens. Actuat. B 220, 192−200.

Tomer, V.K., Duhan, S., 2015. Highly sensitive and stable relative humidity sensors based on WO_3 modified mesoporous silica. Appl. Phys. Lett. 106, 063105.

Tomer, V.K., Duhan, S., 2016. A facile nanocasting synthesis of mesoporous Ag-doped SnO_2 nanostructures with enhanced humidity sensing performance. Sens. Actuat. B 223, 750−760.

Tomer, V.K., Adhyapak, P.V., Duhan, S., Mulla, I.S., 2014. Humidity sensing properties of Ag-loaded mesoporous silica SBA-15 nano composites prepared via hydrothermal process. Micropor. Mesopor. Mater. 197, 140−147.

Tomer, V.K., Ritu, M., Jangra, Nehra, S., Duhan, S.P., One Pot, S., 2014. Direct synthesis of mesoporous SnO_2/SBA-15 nano-composite by the hydrothermal Method. Mater. Lett. 132, 228−230.

Tomer, V.K., Duhan, S., Adhyapak, P.V., Mulla, I.S., 2015. Mn loaded mesoporous silica nanocomposite: a highly efficient humidity sensor. J. Am. Ceram. Soc. 98, 741−747.

Tomer, V.K., Duhan, S., Sharma, A.K., Malik, R., Jangra, S., Nehra, S.P., et al., 2015. Humidity sensing properties of Ag nano particles supported WO_3-SiO_2 with super rapid response and excellent stability. Eur. J. Inorg. Chem. 31, 5232−5240.

Tomer, V.K., Jangra, S., Malik, R., Duhan, S., 2015. Effect of in-situ loading of nano Titania particles on structural ordering of mesoporous SBA-15 framework. Colloids Surf. A 466, 160−165.

Tomer, V.K., Duhan, S., Malik, R., Nehra, S.P., Devi, S., 2015. A novel highly sensitive humidity sensor based on ZnO/SBA-15 hybrid nanocomposite. J Am. Ceram. Soc. 98, 3719−3725.

Tomer, V.K., Duhan, S., Sharma, A.K., Malik, R., Nehra, S.P., Devi, S., 2015. One pot synthesis of mesoporous ZnO-SiO_2 nanocomposite for room temperature relative humidity sensor. Colloids Surf. A 483, 121−128.

Tomer, V.K., Devi, S., Malik, R., Nehra, S.P., Duhan, S., 2016. Fast response with high performance humidity sensing of Ag-$SnO2$/SBA-15 nanohybrid sensors. Micropor. Mesopor. Mater. 219, 240−248.

Tomer, V.K., Devi, S., Malik, R., Duhan, S., 2016. Mesoporous Materials & Their Nanocomposites, in Nanomaterials and Nanocomposites. Wiley-VCH Verlag, Germany, pp. 223−254.

Tomer, V.K., Devi, S., Malik, R., Nehra, S.P., Duhan, S., 2016. Highly sensitive and selective sensors to volatile organic amines (VOAs) using mesoporous WO_3-SnO_2 nanohybrids. Sens. Actuat. -B. 229, 321−330.

Tomer, V.K., Thangaraj, N., Gahlot, S., Kailasam, K., 2016. Cubic mesoporous Ag@CN: a high performance humidity sensor. Nanoscale 8, 19794−19803.

Tomer, V.K., Singh, K., Kaur, H., Shorie, M., Sabherwal, P., 2017. Rapid acetone detection using indium loaded WO_3/SnO_2 nanohybrid sensor. Sens. Actuat. B 253, 703−713.

Tomer, V.K., Malik, R., Kailasam, K., 2017. Near room temperature ethanol detection using Ag-loaded mesoporous carbon nitrides. ACS Omega 2, 3658−3668.

United Nations News Centre, Web update; March, 2010. Available from: http://www.un.org/apps/news/story. asp?NewsID = 34150#.WlSW3FSWbIU.

Upadhyayula, V.K.K., Deng, S., Mitchell, M.C., Smith, G.B., 2009. Application of carbon nanotube technology for removal of contaminants in drinking water: a review. Sci. Total Environ. 408, 1−13.

Urban, I., Ratcliffe, N.M., Duffield, J.R., Elderb, G.R., Patton, D., 2010. Functionalized paramagnetic nanoparticles for waste water treatment. Chem. Commun. 46, 4583−4585.

Vaishanav, S.K., Korram, J., Pradhan, P., Chandraker, K., Nagwanshi, R., Ghosh, K.K., et al., 2017. Green luminescent CdTe quantum dot based fluorescence nano-sensor for sensitive detection of arsenic (III). J. Fluoresc. 27, 781−789.

Vikesland, P.J., Wigginton, K.R., 2010. Nanomaterial enabled biosensors for pathogen monitoring-a review. Environ. Sci. Technol. 44, 3656−3669.

Wang, C.B., Zhang, W.X., 1997. Synthesizing nanoscale iron particles for rapid and complete dechlorination of TCE and PCBs. Environ. Sci. Technol. 31, 2154−2156.

Wang, S., Sun, H., Ang, H.M., Tade, M.O., 2013. Adsorptive remediation of environmental pollutants using novel graphene-based nanomaterials. Chem. Eng. J. 226, 336−347.

Woan, K., Pyrgiotakis, G., Sigmund, W., 2009. Photocatalytic carbon nanotube-TiO_2 composites. Adv. Mater. 21, 2233−2239.

Xi, Y., Du, Y., Zhang, X., He, A., Xu, Z., 2017. Nanofiltration membrane with a mussel-inspired interlayer for improved permeation performance. Langmuir 33, 2318−2324.

Xiang, Q., Yu, J., Jaroniec, M., 2012. Graphene-based semiconductor photocatalysts. Chem. Soc. Rev. 41, 782−796.

Xu, P., Zeng, G.M., Huang, D.L., Feng, C.L., Hu, S., Zhao, M.H., et al., 2012. Use of iron oxide nanomaterials in wastewater treatment: a review. Sci. Total Environ. 424, 1−10.

Yang, L., Chen, B., Luo, S., Li, J., Liu, R., Cai, Q., 2010. Sensitive detection of polycyclic aromatic hydrocarbons using CdTe quantum dot-modified TiO_2 nanotube array through fluorescence resonance energy transfer. Environ. Sci. Technol. 44, 7884−7889.

Yen, H.L., Tseng, W.L., 2010. Ultrasensitive sensing of Hg^{2+} and CH_3Hg^+ based on the fluorescence quenching of lysozyme type VI-stabilized gold nanoclusters. Analyt. Chem. 82, 9194−9200.

Yin, J., Deng, B., 2015. Polymer-matrix nanocomposite membranes for water treatment. J. Membrane Sci. 479, 256−275.

Zhang, J., Zhai, S., Li, S., Xiao, Z., Song, Y., An, Q., et al., 2013. Pb(II) removal of Fe_3O_4@SiO_2-NH_2 core-shell nanomaterials prepared via a controllable sol-gel process. Chem. Eng. J. 215, 461−471.

Zhang, M., Xie, X., Tang, M., Criddle, C.S., Cui, Y., Wang, S.X., 2004. Magnetically ultraresponsive nanoscavengers for next-generation water purification systems. Nat. Commun. 4, 1−4.

Zhao, G., Li, J., Ren, X., Chen, C., Wang, X., 2011. Few-layered graphene oxide nanosheets as superior sorbents for heavy metal ion pollution management. Environ. Sci. Technol. 45, 10454−10462.

Zhao, G., Ren, X., Gao, X., Tan, X., Li, J., 2011. Removal of Pb(II) ions from aqueous solutions on few-layered graphene oxide nanosheets. Dalton Trans. 40, 10945−10952.

Further Reading

Malik, R., Tomer, V.K., Rana, P.S., Nehra, S.P., Duhan, S., 2015. Surfactant assisted Hydrothermal Synthesis of porous 3-D hierarchical SnO_2 Nanoflowers for photocatalytic degradation of Rose Bengal. Mater. Lett. 154, 124–127.

Nassar, N.N., 2010. Rapid removal and recovery of Pb (II) from wastewater by magnetic nanoadsorbents. J. Hazard. Mater. 184, 538–546.

Index

Note: Page numbers followed by "*f*" and "*t*" refer to figures and tables, respectively.

Printed in the United States
By Bookmasters